Birkhäuser

Operator Theory: Advances and Applications
Volume 267

Founded in 1979 by Israel Gohberg

More information about this series at http://www.springer.com/series/4850

Carlos André • M. Amélia Bastos
Alexei Yu. Karlovich • Bernd Silbermann
Ion Zaballa
Editors

Operator Theory, Operator Algebras, and Matrix Theory

Birkhäuser

Editors
Carlos André
Departamento de Matemática
Faculdade de Ciências
Universidade de Lisboa
Lisboa Portugal

M. Amélia Bastos
Departamento de Matemática
Instituto Superior Técnico
Universidade de Lisboa
Lisboa Portugal

Alexei Yu. Karlovich
Departamento de Matemática
Faculdade de Ciências e Tecnologia
Universidade Nova de Lisboa
Lisboa, Portugal

Bernd Silbermann
Fakultät für Mathematik
Technische Universität Chemnitz
Chemnitz, Germany

Ion Zaballa
Departamento de Matemática Aplicada y EIO
Universidad del País Vasco (UPV/EHU)
Leioa, Spain

ISSN 0255-0156 ISSN 2296-4878 (electronic)
Operator Theory: Advances and Applications
ISBN 978-3-030-10202-9 ISBN 978-3-319-72449-2 (eBook)
https://doi.org/10.1007/978-3-319-72449-2

Mathematics Subject Classification (2010): 47-xx, 15-xx, 46-xx

This book is published under the imprint Birkhäuser, www.birkhauser-science.com by the registered company Springer Nature Switzerland AG part of Springer Nature.
The registered company address is: Gewerbestrasse 11, 6330 Cham, Switzerland

Contents

Operator Theory:
Advances and Applications, Vol. 267, vii–viii

Editorial Introduction

This book is dedicated to the International Workshop on Operator Theory and Operator Algebras – WOAT 2016, which took place in Lisbon at Instituto Superior Técnico, from July 5 to July 8 of 2016. This workshop continued a series of conferences organized in Lisboa IST since 2006 and aimed to promote research exchanges among Operator Theory, Operator Algebras and Matrix Theory areas.

The book consists of 17 chapters that cover research fields in Operator Theory and Operator Algebras as well as in Matrix Theory and Representation Theory.

The research field in Operator Theory and Operator Algebras is mainly represented in chapters that cover the following different topics:

- Fredholm theory for non-local C^*-algebras of singular integral operators with piecewise quasicontinuous coefficients and the local trajectory method. Explicit Fredholm conditions for classes of pseudodifferential operators on singular and non compact spaces and the new concept of Fredholm groupoid.
- A Korovkin type approximation theorem. Weighted statistical relative approximation by positive linear operators.
- Study of general, possibly non-commuting, tuples using geometric ideas based on the newly emerged concept of projective joint spectrum.
- Factorization of singular integral operators with a Carleman backward shift of linear fractional type, on L_p^n. Sufficient conditions for semiFredholmness on $L^p(\mathbb{R}_+)$ of weighted singular integral operators with shifts and slowly oscillating data with discontinuities at 0 and ∞.
- The structure of every separable C^*-algebra of approximation sequences and the C^*-algebra of the finite sections discretization for Toeplitz operators with continuous generating functions. Banach algebras of Toeplitz like operators defined in an axiomatic way and the classical Toeplitz and Hankel operators. Overview of the historical development of more than 40 years of algebraic techniques in Numerical Analysis.
- A formula for the distance from an orthogonal projection, on some Hilbert space, to a set of orthogonal projections.

The research field in Matrix Theory is covered in several chapters devoted to the following topics:

- Spectral analysis of non-Hermitian operators appearing in quantum physics. The diagonalization of such operators and their adjoints is proved.

- Simultaneous feedback and output injection on descriptor linear system described by a quadruple of matrices. Description of the possible Kronecker invariants of the resulting pencil. A constructive and explicit solution is given over algebraically closed fields.
- Spectral representations of the dilation and translation operators used to construct wavelets acting on the Hilbert space of square integrable functions on the real line. The spectral analysis concerns some particular orthonormal bases: the Haar basis, the Walsh–Paley basis and the trigonometric basis.
- A survey on the nonnegative inverse eigenvalue problem (NIEP), and its several variants, with an emphasis on recent results.
- Linearization of nonlinear control systems. Necessary and sufficient conditions are found in a particular case (local A-linearizability) of the more general case (local feedback linearization).

Finally, in the research field of Representation Theory, the notion of an indecomposable (extreme) supercharacter is defined for infinite algebra groups and a description of these is given for the infinite unitriangular group in terms of supercharacters of the finite unitriangular groups.

The editors of the volume are grateful for the support of the Portuguese Foundation for Science and Technology and the Center for Function Analysis, Linear Structures and Applications. They also deeply thank the Birhäuser's editorial team, Dorothy Mazlum and Sabrina Höcklin, for their availability during the preparation of the volume.

The editors December 2017

Operator Theory:
Advances and Applications, Vol. 267, 1–24

Indecomposable Supercharacters of the Infinite Unitriangular Group

Carlos A.M. André, Filipe Gomes and Jocelyn Lochon

Abstract. Let $U_\infty(\Bbbk)$ be the locally finite unitriangular group defined over a finite field $\Bbbk$ with q elements. We define the notion of an indecomposable supercharacter and describe these indecomposable supercharacters in terms of the supercharacters of the finite unitriangular groups $U_n(\Bbbk)$.

Mathematics Subject Classification (2010). Primary 20C15, 20G40; Secondary 05E10, 43A35.

Keywords. Unitriangular group; indecomposable supercharacter; branching graph; multiplicative graph; multiplicative linear function.

1. Introduction

Let G be a group. A complex-valued function $\phi\colon G \to \mathbb{C}$ is said to be *positive definite* if

1. $\phi(g^{-1}) = \overline{\phi(g)}$ for all $g \in G$, and
2. For any finite collection $g_1, \dots, g_m$ of elements of G, the Hermitian matrix $\left[\phi(g_i g_j^{-1})\right]_{1 \leq i,j \leq m}$ is nonnegative.

A function $\phi\colon G \to \mathbb{C}$ is said to be *central* (or a *class function*) if it is constant on conjugacy classes, that is, if $\phi(ghg^{-1}) = \phi(h)$ for all $g, h \in G$; and it is said to be *normalized* if $\phi(1) = 1$. We will denote by $\mathrm{Ch}(G)$ the set consisting of all central, positive definite, normalized functions on G (if G is a topological group, we additionally require the functions to be continuous). If $\phi, \psi \in \mathrm{Ch}(G)$, then

This research was made within the activities of the Group for Linear, Algebraic and Combinatorial Structures of the Center for Functional Analysis, Linear Structures and Applications (University of Lisbon, Portugal), and was partially supported by the Portuguese Science Foundation (FCT) through the Strategic Project UID/MAT/04721/2013.
The second and third authors were partially supported by the Lisbon Mathematics PhD program (funded by the Portuguese Science Foundation). The first half of the work is part of the third author Ph.D. thesis, whereas the second half is part of the second author Ph.D. thesis.

for every real number $0 \leq t \leq 1$ the function $(1-t)\phi + t\psi$ is also an element of $\mathrm{Ch}(G)$, which means that $\mathrm{Ch}(G)$ is a convex set. A point of a convex set is called an *extreme point* if it is not contained in the interior of any interval entirely contained in the set; we denote by $\mathrm{ExCh}(G)$ the subset of $\mathrm{Ch}(G)$ consisting of extreme points of $\mathrm{Ch}(G)$. We refer to an element of $\mathrm{Ch}(G)$ as a *character* of G and to an element of $\mathrm{ExCh}(G)$ as an *indecomposable character* of G. (Indeed, $\mathrm{Ch}(G)$ is a Choquet simplex which means that every character of G is uniquely representable as a convex mixture of indecomposable characters; see, for example, [18] or [31].)

If G is a compact group (in particular, a finite group), then the extreme points of $\mathrm{Ch}(G)$ are exactly the *normalized irreducible characters* of G; the character of an irreducible (complex) representation π of the group G is the function $\chi\colon G \to \mathbb{C}$ given by the trace $\chi(g) = \mathrm{tr}(\pi(g))$ for all $g \in G$, and we define the normalized irreducible character $\widehat{\chi}$ by

$$\widehat{\chi}(g) = \frac{\chi(g)}{\chi(1)}$$

for all $g \in G$. This definition makes sense because every irreducible representation of a compact group is finite-dimensional, but it does not make sense in general for non-compact groups. A classical construction associates indecomposable characters with factor representations of finite von Neumann type, and the representation theory is then mainly focused on the classification of indecomposable characters and on the decomposition of a generic character as a convex sum of indecomposable characters (the latter being as a counterpart of the decomposition of a representation into the irreducible ones). In fact, the classification of finite factor representations of the group G reduces, up to quasi-equivalence, to the classification of $\mathrm{ExCh}(G)$ (see, for example, [19]).

In [34], Thoma obtained an explicit description of the indecomposable characters of the infinite symmetric group S_∞, and his description of $\mathrm{ExCh}(S_\infty)$ was later explained in terms of infinite paths on Young diagrams by Kerov–Vershik [28]; these works opened up a new field of research in asymptotic representation theory (see [26]). There are a few more examples of groups whose indecomposable characters have been explicitly classified (see, for example, [9], [15], [16], [20], [21], [22], [32]); all these examples are restricted to inductive limits of compact groups (and their completion in some topology), and they depend on the knowledge of the indecomposable characters of those compact groups. On the other hand, in many examples the indecomposable characters may be complicated functions, and the set of indecomposable characters may be too large; it is therefore of interest to study smaller and more tractable families of reducible characters (for instance, those which have some kind of symmetry or depend on some simple statistic on the group). In this paper, we consider the standard supercharacter theory for algebra groups (defined over finite fields), and describe all the indecomposable supercharacters of the locally finite unitriangular group $U_\infty(\Bbbk)$ defined over a finite field $\Bbbk$. Indeed, the group $U_\infty(\Bbbk)$ can be naturally realized as the inductive limit of the inductive family $\{U_n(\Bbbk)\}_{n=1}^{\infty}$ of all finite unitriangular groups over $\Bbbk$, and thus the Kerov–Vershik's ergodic method (as introduced in [28]) allows for the description of the indecomposable characters of $U_\infty(\Bbbk)$ in terms of the irreducible characters of the finite groups $U_n(\Bbbk)$ for all $n \in \mathbb{N}$. However, it is well known that the classification of the irreducible characters of the finite unitriangular group is a wild problem (and probably impossible to accomplish), and thus it is worthy to approximate irreducible characters (and conjugacy classes) by the more tractable family of supercharacters (and superclasses). The notion of a supercharacter theory will be defined in Section 2 below where we also describe the (standard) superclasses and supercharacters of the finite unitriangular groups and introduce the notion of a supercharacter and of an indecomposable supercharacter of $U_\infty(\Bbbk)$. (We observe that this notion may be naturally extended to the larger family of locally finite algebra groups; however, for the purposes of this paper, we prefer to skip this more general notion.)

2. Superclasses and supercharacters of unitriangular groups

The notion of a supercharacter theory of a finite group was introduced by Diaconis–Isaacs [14] to generalise an approach used by André (*e.g.*, [2, 4, 5]) and Yan [35] to study the irreducible characters of the finite unitriangular groups. The basic idea is to coarsen the usual irreducible character theory of a group by replacing irreducible characters with integer linear combinations of irreducible characters that are constant on a set of clumped conjugacy classes.

By a *supercharacter theory* of a finite group G we mean a pair $(\mathcal{X}, \mathcal{Y})$ where $\mathcal{X}$ is set partition of G, and $\mathcal{Y}$ an orthogonal set of characters of G (not necessarily irreducible), such that:

1. $|\mathcal{X}| = |\mathcal{Y}|$,
2. every character $\chi \in \mathcal{Y}$ takes a constant value on each member $\mathcal{K} \in \mathcal{X}$, and
3. each irreducible character is a constituent of one of the characters $\chi \in \mathcal{Y}$.

We refer to the members $\mathcal{K} \in \mathcal{X}$ as *superclasses* and to the characters $\chi \in \mathcal{Y}$ as *supercharacters* of G. We note that the superclasses of G are always unions of conjugacy classes; moreover, $\{1\}$ forms a superclass and the principal character 1_G is always a supercharacter of G.

Let p be an arbitrary (but fixed) prime number, let $q = p^e$ ($e \geq 1$) be a power of p, and let $\Bbbk$ be a finite field with q elements. Let $\mathcal{A}$ be a nilpotent associative $\Bbbk$-algebra (which is not assumed to be finite-dimensional), and let $G = 1 + \mathcal{A}$ be the set of all formal expressions of the form $1 + a$ for $a \in \mathcal{A}$. Then, G is a group under the operation $(1+a)(1+b) = 1 + (a+b+ab)$ for all $a, b \in \mathcal{A}$; following [23], we say that $G = 1 + \mathcal{A}$ is an *algebra group over* $\Bbbk$ (or simply an $\Bbbk$-*algebra group*). A prototype example is the unitriangular group $U_n(\Bbbk)$ consisting of all unipotent uppertriangular $n \times n$ matrices over $\Bbbk$; in this case, we have $U_n(\Bbbk) = 1 + \mathfrak{u}_n(\Bbbk)$ where $\mathfrak{u}_n(\Bbbk)$ is the nilpotent $\Bbbk$-algebra consisting of all nilpotent uppertriangular $n \times n$ matrices over $\Bbbk$.

Throughout the paper, we set $U_n = U_n(\Bbbk)$ and $\mathfrak{u}_n = \mathfrak{u}_n(\Bbbk)$. Moreover, we consider the group $B_n = B_n(\Bbbk)$ consisting of all invertible uppertriangular $n \times n$ matrices over $\Bbbk$; then, U_n is a normal subgroup of B_n, and B_n is indeed the semidirect product $B_n = T_n U_n$ where $T_n = T_n(\Bbbk)$ is the subgroup of B_n consisting of all diagonal matrices. Left multiplication naturally defines an action of B_n on the left of the $\Bbbk$-algebra $\mathfrak{u}_n$, whereas right multiplication defines an action of B_n on the right of $\mathfrak{u}_n$. Since the left action of B_n on $\mathfrak{u}_n$ obviously commutes with the right action, we can define an action of the group $B_n \times B_n$ on the left of $\mathfrak{u}_n$ by the rule $(g,h) \cdot a = gah^{-1}$ for all $g, h \in B_n$ and all $a \in \mathfrak{u}_n$. Thus, the $\Bbbk$-algebra $\mathfrak{u}_n$ is partitioned into "two-sided" orbits $B_n a B_n$ for $a \in \mathfrak{u}_n$, and this determines a partition of the unitriangular group U_n into subsets $1 + B_n a B_n$ for $a \in \mathfrak{u}_n$; these are precisely what we define as the *superclasses* of U_n. Notice that every element of $\mathfrak{u}_n$ may be transformed (or normalized) under the action of $B_n \times B_n$ to such an element $a \in \mathfrak{u}_n$ that:

1. for every $1 \leq i < j \leq n$, the (i,j)th coefficient $a_{i,j}$ is either 0 or 1;
2. if $a_{i,j} = 1$ for $1 \leq i < j \leq n$, then $a_{i,l} = 0$ for all $i < l \leq n$, $l \neq j$, and $a_{k,j} = 0$ for all $1 \leq k < j$, $k \neq i$.

The superclasses of the group U_n are parametrised by *set partitions* of $[n] = \{1, 2, \ldots, n\}$; we denote by $\mathcal{SP}(n)$ the set consisting of all set partitions of $[n]$. For simplicity of writing, we also define $[[n]] = \{(i,j) \colon 1 \leq i < j \leq n\}$. If $\pi \in \mathcal{SP}(n)$, then a pair $(i,j) \in [[n]]$ is said to be an *arc of* π if i and j occur in the same block B of π and there is no $k \in B$ with $i < k < j$; we denote by $\mathcal{D}(\pi)$ the set consisting of all arcs of π. Every set partition of $[n]$ is uniquely determined by its set of arcs, and hence by a subset of $[[n]]$. We say that $\mathcal{D} \subseteq [[n]]$ is a *basic subset* if $\mathcal{D} = \mathcal{D}(\pi)$ for some $\pi \in \mathcal{SP}(n)$; notice that the empty subset $\emptyset \subseteq [[n]]$ is basic, and that it corresponds to the set partition $\pi = 1/2/\ldots/n$ of $[n]$. (As usual, we write $\pi = B_1/B_2/\ldots/B_t$ where $B_1, B_2, \ldots, B_t$ are the blocks of a set partition $\pi \in \mathcal{SP}(n)$ which we agree to be ordered by increasing value of the smallest element in the block.)

With every set partition $\pi \in \mathcal{SP}(n)$ we associate the matrix

$$e_\pi = \sum_{(i,j) \in \mathcal{D}(\pi)} e_{i,j} \in \mathfrak{u}_n$$

where $\{e_{i,j} \colon (i,j) \in [[n]]\}$ is the standard $\Bbbk$-basis of $\mathfrak{u}_n$; as usual, for every $(i,j) \in [[n]]$, we denote by $e_{i,j}$ the matrix in $\mathfrak{u}_n$ with 1 in the (i,j)th entry and zeroes elsewhere. It is obvious that every two-sided orbit of B_n on $\mathfrak{u}_n$ contains a unique matrix of the form e_π for some $\pi \in \mathcal{SP}(n)$, and thus the superclasses of U_n are parametrised by set partitions of $[n]$. For every $\pi \in \mathcal{SP}(n)$, we write $\mathcal{K}_\pi = 1 + B_n\, e_\pi\, B_n$; in the case where π is the unique set partition of $[n]$ such that $\mathcal{D}(\pi) = \{(i,j)\}$ for $1 \leq i < j \leq n$, we simplify the notation and write $\mathcal{K}_{i,j}$ instead of $\mathcal{K}_\pi$. It can be proved that

$$\mathcal{K}_\pi = \prod_{(i,j) \in \mathcal{D}(\pi)} \mathcal{K}_{i,j}$$

for all $\pi \in \mathcal{SP}(n)$; this follows easily from [5, Proposition 2.4] (see also [7, Theorem 3.1]) because

$$\mathcal{K}_\pi = \bigcup_{t \in T_n} \mathcal{K}_\pi(t)$$

where $\mathcal{K}_\pi(t) = 1 + U_n(t^{-1}e_\pi t)U_n$ for all $t \in T_n$. We observe that the double U_n-orbits $\mathcal{K}_\pi(t)$, for $\pi \in \mathcal{SP}(n)$ and $t \in T_n$, are precisely the superclasses in the standard supercharacter theory of U_n as defined in [14]. Although our results are valid for this standard supercharacter theory, we prefer to use the above coarsening (which is natural and avoids considering $\Bbbk^\times$-coloured set partitions); as we will see below, this supercharacter theory has integer-valued supercharacters, and hence its use might be combinatorially more natural and advantageous. (This slightly coarser supercharacter theory has been somewhat explored and used in several papers; see for example [8], [10], [33], or [13]; it is also worth mentioning that by a result of J. Keller [24], for each finite group G, there exists a unique finest supercharacter theory with integer values.)

We next define the set of supercharacters of U_n; our construction is based on the results of [14], and the details can be found in [7]. Let $\mathfrak{u}_n^\circ$ be the dual group of the additive group $\mathfrak{u}_n^+$ of $\mathfrak{u}_n$; hence, $\mathfrak{u}_n^\circ$ consists of linear (additive) characters of $\mathfrak{u}_n$. For each of the natural actions of B_n on $\mathfrak{u}_n$, there is a corresponding action of B_n on $\mathfrak{u}_n^\circ$: given $\vartheta \in \mathfrak{u}_n^\circ$ and $g \in B_n$, we define $g\vartheta, \vartheta g \in \mathfrak{u}_n^\circ$ by the formulas $(g\vartheta)(a) = \vartheta(g^{-1}a)$ and $(\vartheta g)(a) = \vartheta(ag^{-1})$ for all $a \in \mathfrak{u}_n$. It is routine to check that these actions commute, and thus we have left orbits $B_n\vartheta$, right orbits ϑB_n, and "two-sided" orbits $B_n\vartheta B_n$ for any $\vartheta \in \mathfrak{u}_n^\circ$.

For notational simplicity, we introduce the (bijective) map $\nu\colon U_n \to \mathfrak{u}_n$ given by $\nu(g) = g - 1$ for all $g \in U_n$, and observe that $\nu(gh) = gh - 1 = gh - g + g - 1 = g\nu(h) + \nu(g)$ for all $g, h \in U_n$. Then, U_n acts on left of $\mathfrak{u}_n$ by $g \cdot a = ga + \nu(g)$ for all $g \in U_n$ and all $a \in \mathfrak{u}_n$; we note that $g \cdot \nu(h) = \nu(gh)$ for all $g, h \in U_n$. This action can be extended by linearity to the group algebra $\mathbb{C}\mathfrak{u}_n$ of $\mathfrak{u}_n$, and so $\mathbb{C}\mathfrak{u}_n$ becomes a left $\mathbb{C}U_n$-module. Moreover, it is clear that the map $\nu\colon U_n \to \mathfrak{u}_n$ extends linearly to an isomorphism of $\mathbb{C}U_n$-modules $\nu\colon \mathbb{C}U_n \to \mathbb{C}\mathfrak{u}_n$, and thus the $\mathbb{C}U_n$-module $\mathbb{C}\mathfrak{u}_n$ affords the regular character ρ_{U_n} of U_n, that is, the character afforded by the left (regular) $\mathbb{C}U_n$-module $\mathbb{C}U_n$ (where U_n acts naturally by left multiplication); recall that the degree $\rho_{U_n}(1) = \dim_{\mathbb{C}} \mathbb{C}U_n$ equals the order $q^{n(n-1)/2}$ of U_n, whereas $\rho_{U_n}(g) = 0$ for all $g \in U_n$, $g \neq 1$.

Besides its natural $\mathbb{C}$-basis, the group-algebra $\mathbb{C}\mathfrak{u}_n$ also has a $\mathbb{C}$-basis $\{\varepsilon_\vartheta \colon \vartheta \in \mathfrak{u}_n^\circ\}$ where

$$\varepsilon_\vartheta = \frac{1}{|\mathfrak{u}_n|} \sum_{a \in \mathfrak{u}_n} \overline{\vartheta(a)}\, a$$

is the central primitive idempotent of $\mathbb{C}\mathfrak{u}_n$ which corresponds to the linear character $\vartheta \in \mathfrak{u}_n^\circ$. For every $g \in U_n$ and every $\vartheta \in \mathfrak{u}_n^\circ$, we have $g \cdot \varepsilon_\vartheta = \overline{\vartheta(\nu(g^{-1}))}\, \varepsilon_{g\vartheta}$, and thus

$$\mathfrak{L}_\vartheta = \sum_{\tau \in B_n\vartheta} \mathbb{C}\varepsilon_\tau$$

is a $\mathbb{C}U_n$-submodule of $\mathbb{C}\mathfrak{u}_n$; in fact, we have

$$\mathfrak{L}_\vartheta = \sum_{t\in T_n} \mathfrak{L}_\vartheta(t)$$

where $\mathfrak{L}_\vartheta(t) = \mathbb{C}U_n \cdot \varepsilon_{t\vartheta}$ for all $t \in T_n$. Then, we define the *supercharacter* ξ_ϑ of U_n associated with $\vartheta \in \mathfrak{u}_n^\circ$ to be the character of U_n afforded by $\mathfrak{L}_\vartheta$; in particular, ξ_ϑ has degree $\xi_\vartheta(1) = |B_n\vartheta|$. We observe that, if χ_ϑ is the character afforded by the left $\mathbb{C}U_n$-module $\mathfrak{L}_\vartheta(1) = \mathbb{C}U_n\cdot\varepsilon_\vartheta$, then for every $t \in T_n$ the left $\mathbb{C}U_n$-module $\mathfrak{L}_\vartheta(t)$ affords the T_n-conjugate character $(\chi_\vartheta)^t$; moreover, we have $(\chi_\vartheta)^t = \chi_{t\vartheta}$ for all $t \in T_n$. It is not difficult to justify that the characters $(\chi_\vartheta)^t$, for $\vartheta \in \mathfrak{u}_n^\circ$ and $t \in T_n$, are precisely the standard supercharacters of U_n (as defined by Diaconis–Isaacs in [14]); in particular, it follows from [14, Theorem 5.6] that

$$\xi_\vartheta(g) = \frac{|B_n\vartheta|}{|B_n\vartheta B_n|} \sum_{\tau\in B_n\vartheta B_n} \tau(g-1)$$

for all $\vartheta \in \mathfrak{u}_n^\circ$ and all $g \in U_n$. As a consequence, we see that, for all $\vartheta, \vartheta' \in \mathfrak{u}_n^\circ$, the supercharacters ξ_ϑ and $\xi_{\vartheta'}$ have a common irreducible constituent if and only if $B_n\vartheta B_n = B_n\vartheta' B_n$, in which case we must have $\xi_\vartheta = \xi_{\vartheta'}$.

If $\mathcal{X}(n) = \{\mathcal{K}_a : a \in \mathfrak{u}_n\}$ and $\mathcal{Y}(n) = \{\xi_\vartheta : \vartheta \in \mathfrak{u}_n^\circ\}$, then the pair $(\mathcal{X}(n), \mathcal{Y}(n))$ is a supercharacter theory of U_n. We know that

$$\mathcal{X}(n) = \{\mathcal{K}_\pi : \pi \in \mathcal{SP}(n)\}$$

(in fact, the mapping $\pi \mapsto \mathcal{K}_\pi$ defines a bijection between $\mathcal{SP}(n)$ and $\mathcal{X}(n)$), and thus $|\mathcal{Y}(n)| = |\mathcal{X}(n)| = |\mathcal{SP}(n)|$; indeed, there is also a natural bijection between $\mathcal{SP}(n)$ and $\mathcal{Y}(n)$. To describe this, it is convenient to consider the $\Bbbk$-bilinear form on $\mathfrak{u}_n$ defined by $a \cdot b = \operatorname{tr}(a^\mathsf{T} b)$ for all $a, b \in \mathfrak{u}_n$. On the other hand, we fix a non-trivial linear character $\vartheta : \Bbbk \to \mathbb{C}^\times$ of the additive group of $\Bbbk$, and for every $a \in \mathfrak{u}_n$ we define the map $\vartheta_a : \mathfrak{u}_n \to \mathbb{C}^\times$ by $\vartheta_a(b) = \vartheta(a \cdot b)$ for all $b \in \mathfrak{u}_n$. It is routine to check that $\vartheta_a \in \mathfrak{u}_n^\circ$; in fact, $\mathfrak{u}_n^\circ = \{\vartheta_a : a \in \mathfrak{u}_n\}$. For every set partition $\pi \in \mathcal{SP}(n)$, we set $\vartheta_\pi = \vartheta_{e_\pi}$ and denote by ξ_π the supercharacter of U_n which corresponds to ϑ_π. Since every two-sided B_n-orbit contains ϑ_π for a unique $\pi \in \mathcal{SP}(n)$, we conclude that

$$\mathcal{Y}(n) = \{\xi_\pi : \pi \in \mathcal{SP}(n)\};$$

in fact, the mapping $\pi \mapsto \xi_\pi$ defines a bijection between $\mathcal{SP}(n)$ and $\mathcal{Y}(n)$.

Finally, we mention the formula below for the supercharacter values; a proof can be found in [7, Theorem 3.9] (in fact, it is a consequence of the corresponding formula for the standard supercharacter values; see [5, Theorem 5.1] and [6, Theorem 3]). For every set partition $\pi \in \mathcal{SP}(n)$, we say that $(k, l) \in [[n]]$ is a π-*singular arc* if $i \le k < l \le j$ for some arc $(i, j) \in \mathcal{D}(\pi)$ and if these two arcs share one and only one common vertex (that is, either $k = i$, or $l = j$, but not both); otherwise, we say that (k, l) is a π-*regular arc*. We denote by $\operatorname{Sing}(\pi)$ the subset of $[[n]]$ consisting of all π-singular arcs, and by $\operatorname{Reg}(\pi)$ the subset of $[[n]]$ consisting of all π-regular arcs; in particular, we have $\mathcal{D}(\pi) \subseteq \mathrm{Reg}(\pi)$. Now, if $\pi, \sigma \in \mathcal{SP}(n)$ and $g \in \mathcal{K}_\sigma$, then $\xi_\pi(g) = 0$ unless $\mathcal{D}(\sigma) \subseteq \mathrm{Reg}(\pi)$, in which case we have

$$\xi_\pi(g) = \frac{(-1)^{|\mathcal{D}(\pi)\cap\mathcal{D}(\sigma)|} q^{\dim(\pi)} (q-1)^{|\mathcal{D}(\pi)|}}{q^{\mathrm{nst}_\pi(\sigma)} q^{|\mathcal{D}(\pi)|} (q-1)^{|\mathcal{D}(\pi)\cap\mathcal{D}(\sigma)|}} \tag{2.1}$$

where

$$\dim(\pi) = \sum_{(i,j)\in\mathcal{D}(\pi)} (j-i)$$

and

$$\mathrm{nst}_\pi(\sigma) = \sum_{(i,j)\in\mathcal{D}(\pi)} \big|\{(k,l) \in \mathcal{D}(\sigma) \colon i < k < l < j\}\big|.$$

(There is at least one different way of defining the dimension $\dim(\pi)$. Our definition is slightly less natural in light of the papers [11, 12] on set partitions statistics where the sum is considered over $j-i-1$ instead of $j-i$. However, this has no essential effect other than simplifying some of the formulas. We note that our definition is also used in other papers; see for example [8] or [13].)

In the case where $\pi \in \mathcal{SP}(n)$ is such that $\mathcal{D}(\pi) = \{(i,j)\}$, we simplify the notation and write $\xi_{i,j}$ instead of ξ_π. It follows easily from the above formula that

$$\xi_\pi = \prod_{(i,j)\in\mathcal{D}(\pi)} \xi_{i,j}, \qquad \pi \in \mathcal{SP}(n); \tag{2.2}$$

indeed, this is an easy consequence of [7, Theorem 3.6]. We note that, if $(i,j) \in [[n]]$, $\sigma \in \mathcal{SP}(n)$ and $g \in \mathcal{K}_\sigma$, then

$$\xi_{i,j}(g) = \begin{cases} -\dfrac{q^{j-i-1}}{q^{\mathrm{nst}_{i,j}(\sigma)}}, & \text{if } (i,j) \in \mathcal{D}(\sigma), \\ \dfrac{q^{j-i-1}(q-1)}{q^{\mathrm{nst}_{i,j}(\sigma)}}, & \text{if } \mathcal{D}(\sigma) \cap (\{(i,j)\} \cup \mathrm{Sing}(i,j)) = \emptyset, \\ 0, & \text{if } \mathcal{D}(\sigma) \cap \mathrm{Sing}(i,j) \neq \emptyset. \end{cases} \tag{2.3}$$

where we define $\mathrm{nst}_{i,j}(\sigma) = \mathrm{nst}_\pi(\sigma)$ and $\mathrm{Sing}(i,j) = \mathrm{Sing}(\pi)$ whenever $\pi \in \mathcal{SP}(n)$ is such that $\mathcal{D}(\pi) = \{(i,j)\}$.

Next, we introduce the notions of superclasses and supercharacters of the locally finite unitriangular group $U_\infty(\Bbbk)$; for simplicity, we shall write $U_\infty = U_\infty(\Bbbk)$. For every $n \in \mathbb{N}$, we naturally identify B_n with the subgroup

$$\widetilde{B}_n = \left\{ \begin{bmatrix} g & 0 \\ 0 & 1 \end{bmatrix} : g \in B_n \right\}$$

of B_{n+1}; hence, we consider U_n as a subgroup of U_{n+1} and T_n as a subgroup of T_{n+1}. Then,

$$U_\infty = \bigcup_{n\in\mathbb{N}} U_n;$$

moreover, if we set

$$B_\infty = \bigcup_{n\in\mathbb{N}} B_n \qquad \text{and} \qquad T_\infty = \bigcup_{n\in\mathbb{N}} T_n,$$

then U_∞ is a normal subgroup of B_∞, and B_∞ is the semidirect product $B_\infty = T_\infty U_\infty$. Moreover, we consider the locally finite nilpotent $\Bbbk$-algebra $\mathfrak{u}_\infty = \bigcup_{n\in\mathbb{N}} \mathfrak{u}_n$ (where we also agree that $\mathfrak{u}_n \subseteq \mathfrak{u}_{n+1}$ for all $n \in \mathbb{N}$); hence, U_∞ is the $\Bbbk$-algebra group $U_\infty = 1 + \mathfrak{u}_\infty$.

As in the finite case, for every $a \in \mathfrak{u}_\infty$, we may consider the two-sided B_∞-orbit $B_\infty a B_\infty \subseteq \mathfrak{u}_\infty$, and define $\mathcal{K}_a = 1 + B_\infty a B_\infty$ to be a *superclass* of U_∞. We note that, since $a \in U_n$ for some $n \in \mathbb{N}$, there exists a unique set partition $\pi \in \mathcal{SP}(n)$ such that $e_\pi \in B_n a B_n$, and thus we must have $B_\infty a B_\infty = B_\infty e_\pi B_\infty$. Therefore, the superclasses of U_∞ are parametrised by set partitions $\pi \in \bigcup_{n\in\mathbb{N}} \mathcal{SP}(n)$; throughout the paper, we set

$$\mathcal{SP} = \bigcup_{n\in\mathbb{N}} \mathcal{SP}(n).$$

For every $\pi \in \mathcal{SP}$, we write $\mathcal{K}_\pi(\infty) = 1 + B_\infty e_\pi B_\infty$; thus, $\mathcal{K}_\pi(\infty)$ is the superclass of U_∞ which contains the element $g_\pi = 1 + e_\pi \in U_\infty$.

A function $\phi\colon U_\infty \to \mathbb{C}$ is said to be *supercentral* (or a *superclass function*) if it is constant on superclasses, that is, if $\phi(1+gah) = \phi(1+a)$ for all $g, h \in B_\infty$ and all $a \in \mathfrak{u}_\infty$. We will denote by $\mathrm{SCh}(U_\infty)$ the set consisting of all supercentral, positive definite and normalized functions on U_∞; it is clear that $\mathrm{SCh}(U_\infty) \subseteq \mathrm{Ch}(U_\infty)$. As in the case of characters, $\mathrm{SCh}(U_\infty)$ is a convex set; we denote by $\mathrm{ExSCh}(U_\infty)$ the subset of $\mathrm{SCh}(U_\infty)$ consisting of all extreme points of $\mathrm{SCh}(U_\infty)$. We refer to an element of $\mathrm{SCh}(U_\infty)$ as a *supercharacter* of U_∞, and to an element of $\mathrm{ExSCh}(\infty)$ as an *indecomposable supercharacter* of U_∞. (As in the case of characters, $\mathrm{SCh}(U_\infty)$ is a Choquet simplex, and thus every supercharacter of U_∞ is uniquely representable as a convex mixture of indecomposable characters.)

Similarly, we may also define supercharacters and indecomposable supercharacters of the finite unitriangular group. For every $n \in \mathbb{N}$, the indecomposable supercharacters are precisely the normalized supercharacters $\widehat{\xi_\pi} = \xi_\pi(1)^{-1}\xi_\pi$ for $\pi \in \mathcal{SP}(n)$, whereas a supercharacter is a positive integer linear combination of the indecomposable supercharacters. We should note that we use the word "indecomposable" in the sense that an indecomposable supercharacter cannot be expressed as a sum of two other supercharacters (hence, an indecomposable supercharacter is not necessarily an indecomposable character). On the other hand, if $m \in \mathbb{N}$ is such that $n \leq m$, then a set partition $\pi \in \mathcal{SP}(n)$ can be naturally identified with the (unique) set partition $\pi' \in \mathcal{SP}(m)$ such that $\mathcal{D}(\pi') = \mathcal{D}(\pi)$, and it is easy to see that the supercharacter $\xi_\pi \in \mathcal{Y}(n)$ can be naturally extended to the supercharacter $\xi_{\pi'} \in \mathcal{Y}(m)$; hence, without loss of generality, we may assume that $\mathcal{Y}(n) \subseteq \mathcal{Y}(m)$. More generally, every normalized supercharacter $\widehat{\xi_\pi}$, for $\pi \in \mathcal{SP}(n)$, can be naturally extended to an indecomposable supercharacter of U_∞, and hence we may assume that $\widehat{\xi_\pi} \in \mathrm{ExSCh}(U_\infty)$ for all $n \in \mathbb{N}$ and all $\pi \in \mathcal{SP}(n)$.

The set $\mathrm{ExSCh}(U_\infty)$ is larger than the union $\bigcup_{n\in\mathbb{N}} \{\widehat{\xi_\pi} : \pi \in \mathcal{SP}(n)\}$. As a first example, let $m \in \mathbb{N}$ be fixed, and define the supercentral function $\xi_m \colon U_\infty \to \mathbb{C}$

on an element $g_\pi = 1 + e_\pi \in U_\infty$, for $\pi \in \mathcal{SP}$, by the rule

$$\xi_m(g_\pi) = \begin{cases} q^{-\mathrm{nst}_m(\pi)}, & \text{if } \mathcal{D}(\pi) \cap \{(m,n) \colon n \in \mathbb{N},\ m < n\} = \emptyset, \\ 0, & \text{otherwise,} \end{cases} \tag{2.4}$$

where $\mathrm{nst}_m(\pi) = \big|\{(i,j) \in \mathcal{D}(\pi) \colon m < i\}\big|$. It is clear that ξ_m is a supercharacter of U_∞; in fact, we have the following.

Proposition 2.1. *In the notation as above, $\xi_m \in \mathrm{ExSCh}(U_\infty)$ for all $m \in \mathbb{N}$.*

Proof. Let $m \in \mathbb{N}$ be arbitrary and suppose that $\xi_m \notin \mathrm{ExSCh}(U_\infty)$. Then, $\xi_m = t\zeta + (1-t)\zeta'$ for some $t \in \mathbb{R}$, $0 < t < 1$, and some $\zeta, \zeta' \in \mathrm{SCh}(U_\infty)$. Let $n \in \mathbb{N}$ be such that $m < n$. Then,

$$\mathrm{Res}^{U_\infty}_{U_n}(\xi_m) = \frac{1}{q^{n-m}}\big(1 + \xi_{m,m+1} + \cdots + \xi_{m,n}\big),$$

in fact, we have

$$\begin{aligned}\mathrm{Res}^{U_\infty}_{U_n}(\xi_m) &= \xi_{m,n+1}(1)^{-1}\mathrm{Res}^{U_{n+1}}_{U_n}(\xi_{m,n+1}) \\ &= \frac{1}{(q-1)q^{n-m}}\,\mathrm{Res}^{U_{n+1}}_{U_n}(\xi_{m,n+1}),\end{aligned}$$

whereas

$$\mathrm{Res}^{U_{n+1}}_{U_n}(\xi_{m,n+1}) = (q-1)\big(1 + \xi_{m,m+1} + \cdots + \xi_{m,n}\big).$$

Therefore, the restriction $\mathrm{Res}^{U_\infty}_{U_n}(\zeta)$ decomposes as a nonnegative linear combination (with rational coeficients) of some of the supercharacters 1 and $\xi_{m,k}$ for $m < k \leq n$; moreover, if $\xi_{m,n+1}$ is a constituent of $\mathrm{Res}^{U_\infty}_{U_{n+1}}(\zeta)$, then

$$\mathrm{Res}^{U_\infty}_{U_n}(\zeta) = \mathrm{Res}^{U_{n+1}}_{U_n}(\xi_{m,n+1}) = \mathrm{Res}^{U_\infty}_{U_n}(\xi_m).$$

It follows that, if $\mathrm{Res}^{U_\infty}_{U_n}(\zeta) \neq \mathrm{Res}^{U_\infty}_{U_n}(\xi_m)$, then $\xi_{m,n+1} \in \mathrm{SCh}(U_{n+1})$ must be a constituent of the restriction $\mathrm{Res}^{U_\infty}_{U_{n+1}}(\zeta')$, and thus

$$\mathrm{Res}^{U_\infty}_{U_n}(\zeta') = \mathrm{Res}^{U_\infty}_{U_n}(\xi_m)$$

(because $\mathrm{Res}^{U_{n+1}}_{U_n}(\xi_{m,n+1})$ must be a constituent of $\mathrm{Res}^{U_\infty}_{U_n}(\xi_m)$). Therefore, we conclude that

$$\begin{aligned}\mathrm{Res}^{U_\infty}_{U_n}(\xi_m) &= t\mathrm{Res}^{U_\infty}_{U_n}(\zeta) + (1-t)\mathrm{Res}^{U_\infty}_{U_n}(\zeta') \\ &= t\mathrm{Res}^{U_\infty}_{U_n}(\zeta) + (1-t)\mathrm{Res}^{U_\infty}_{U_n}(\xi_m),\end{aligned}$$

which implies that $\mathrm{Res}^{U_\infty}_{U_n}(\zeta) = \mathrm{Res}^{U_\infty}_{U_n}(\xi_m)$, a contradiction. It follows that $\xi_m \in \mathrm{ExSCh}(U_\infty)$, as required. □

We observe that equation (2.3) implies that, for every $m \in \mathbb{N}$ and every $g \in U_\infty$, the sequence $\big\{\widehat{\xi}_{m,n}(g)\big\}_{n=m+1}^{\infty}$ is convergent and that

$$\xi_m(g) = \lim_{n\to\infty} \widehat{\xi}_{m,n}(g) = \lim_{n\to\infty} \frac{\xi_{m,n}(g)}{\xi_{m,n}(1)}.$$

3. The superbranching graph

A *Bratteli diagram* is an infinite directed graph Γ such that the vertex set $\mathcal{V}(\Gamma)$ and the edge set $\mathcal{E}(\Gamma)$ can be partitioned into finite sets

$$\mathcal{V}(\Gamma) = \mathcal{V}_0(\Gamma) \sqcup \mathcal{V}_1(\Gamma) \sqcup \mathcal{V}_2(\Gamma) \sqcup \cdots \quad \text{and} \quad \mathcal{E}(\Gamma) = \mathcal{E}_1(\Gamma) \sqcup \mathcal{E}_2(\Gamma) \sqcup \cdots$$

satisfying the following properties:

1. $\mathcal{V}_0(\Gamma)$ consists of a single vertex which we denote by $\varnothing$.
2. If $t\colon \mathcal{E}(\Gamma) \to \mathcal{V}(\Gamma)$ is the associated target map and $s\colon \mathcal{E}(\Gamma) \to \mathcal{V}(\Gamma)$ is the associated source map, then
   (a) $t(\mathcal{E}_n(\Gamma)) \subseteq \mathcal{V}_n(\Gamma)$ and $s(\mathcal{E}_n(\Gamma)) \subseteq \mathcal{V}_{n-1}(\Gamma)$ for all $n \in \mathbb{N}$;
   (b) the set $s^{-1}(\lambda)$ is nonempty for all $\lambda \in \mathcal{V}(\Gamma)$, and the set $t^{-1}(\lambda)$ is nonempty for all $\lambda \in \mathcal{V}(\Gamma) \setminus \mathcal{V}_0(\Gamma)$.

To simplify the notation, we identify the graph Γ with its vertex set $\mathcal{V}(\Gamma)$, and write $\Gamma_n = \mathcal{V}_n(\Gamma)$ for all $n \in \mathbb{N}_0$[1]. The *level* of a vertex $\lambda \in \Gamma$ is the unique $n \in \mathbb{N}_0$ such that $\lambda \in \Gamma_n$; we write $n = |\lambda|$. If $e \in \mathcal{E}(\Gamma)$ is an edge with $\lambda = s(e)$ and $\mu = t(e)$, then we represent e as $\lambda \nearrow \mu$; notice that we must have $|\mu| = |\lambda| + 1$. Notice that we are implicitly assuming that a Bratelli diagram Γ is oriented so that any vertex of level n is always "below" some vertex of level $n+1$, and any vertex of level $n+1$ is "above" some vertex of level n. We also assume that there is an *edge multiplicity function* $\varkappa\colon \mathcal{E}(\Gamma) \to \mathbb{R}^+$ which assigns to an edge $\lambda \nearrow \mu$, for $\lambda, \mu \in \Gamma$, a strictly positive real number $\varkappa(\lambda,\mu)$; we refer to $\varkappa(\lambda,\mu)$ as the multiplicity of the edge $\lambda \nearrow \mu$, and we emphasise that $\varkappa(\lambda,\mu)$ is not necessarily a positive integer. Following [25], we refer to the pair $(\Gamma, \varkappa)$ as a *branching graph.*

Our main example occurs when we consider the locally finite group $U_\infty = \bigcup_{n\in\mathbb{N}} U_n$. We define the *superbranching graph* of U_∞ to be the Bratelli diagram $\Gamma = \mathcal{SP}$ with set of vertices

$$\mathcal{SP} = \mathcal{SP}(0) \sqcup \mathcal{SP}(1) \sqcup \mathcal{SP}(2) \sqcup \cdots \sqcup \mathcal{SP}(n) \sqcup \cdots$$

where $\mathcal{SP}(0)$ consists of a single vertex $\varnothing$ (the unique partition of the empty set $\emptyset$). For simplicity, we set $U_0 = \{1\}$ (the trivial group); notice that U_0 is the trivial $\Bbbk$-algebra group $U_0 = 1 + \mathfrak{u}_0$ where we agree that $\mathfrak{u}_0 = \{0\}$, and the supercharacter $\xi_\varnothing$ is the trivial character of U_0. In order to define the edges of Γ, we first prove the following result (see [14, Theorem 6.4], or [7, Theorem 2.21], for the case of the standard supercharacter theory of finite algebra groups).

Proposition 3.1. *For each $n \in \mathbb{N}$, the restriction of any supercharacter of U_n is a nonnegative integer linear combination of supercharacters of U_{n-1}.*

Proof. It follows easily from the results of [2] (namely, Lemma 10 and Lemma 7, respectively) that the following formulae holds for every $1 \leq i < j < k \leq n$:

- $$\operatorname{Res}^{U_n}_{U_{n-1}}(\xi_{i,j}) = \begin{cases} \xi_{i,j}, & \text{if } j < n, \\ (q-1)\Big(1_{U_{n-1}} + \sum_{i<j'<n} \xi_{i,j'}\Big), & \text{if } j = n. \end{cases}$$

[1]Throughout the paper, we set $\mathbb{N}_0 = \mathbb{N} \sqcup \{0\}$.

- $\xi_{i,j}\xi_{k,j} = (q-1)\Big(\xi_{i,j} + \sum_{k<j'<j} \xi_{i,j}\xi_{k,j'}\Big)$.

The result now follows using the factorisation equation (2.2) and applying successively these formulae; notice that, if $\pi \in \mathcal{SP}(n)$ and $\sigma \in \mathcal{SP}(n-1)$ are such that the supercharacter ξ_σ of U_{n-1} occurs in the restriction of the supercharacter ξ_π of U_n, then $\mathcal{D}(\sigma) \subseteq \big\{(i,j) \in [[n]] \colon (i,k) \in \mathcal{D}(\pi) \text{ for some } i < k \leq n\big\}$. □

It follows from the previous proposition that, for every $n \in \mathbb{N}_0$ and every $\pi' \in \mathcal{SP}(n+1)$ there exist nonnegative integers $\varkappa(\pi,\pi')$, for $\pi \in \mathcal{SP}(n)$, such that

$$\mathrm{Res}^{U_{n+1}}_{U_n}(\xi_{\pi'}) = \sum_{\pi \in \mathcal{SP}(n)} \varkappa(\pi,\pi')\xi_\pi.$$

Then, two vertices $\pi, \pi' \in \mathcal{SP}$ are joined by a directed edge $\pi \nearrow \pi'$ if and only if there exists $n \in \mathbb{N}_0$ such that $\pi \in \mathcal{SP}(n)$, $\pi' \in \mathcal{SP}(n+1)$ and $\varkappa(\pi,\pi') \neq 0$; moreover, we obtain the edge multiplicity function $\varkappa \colon \mathcal{E}(\mathcal{SP}) \to \mathbb{N}$ which assigns to each edge $\pi \nearrow \pi'$ the (positive) multiplicity $\varkappa(\pi,\pi')$ of ξ_π in the decomposition of $\mathrm{Res}^{U_{n+1}}_{U_n}(\xi_{\pi'})$. By the way of example, consider the set partition $\pi = 16/24/35$ of $[6]$ and the supercharacter $\xi_\pi = \xi_{1,6}\xi_{2,4}\xi_{3,5}$ of U_6. Using the formulae above, we deduce that

$$\begin{aligned}\mathrm{Res}^{U_6}_{U_5}(\xi_\pi) &= (q-1)\xi_{2,4}\xi_{3,5} + (q-1)\xi_{1,2}\xi_{2,4}\xi_{3,5} + (q-1)^2\xi_{1,4}\xi_{3,5} \\ &\quad + (q-1)^2\xi_{1,4}\xi_{2,3}\xi_{3,5} + q(q-1)^2\xi_{1,5}\xi_{2,4},\end{aligned}$$

and thus we have the following connections in the superbranching graph

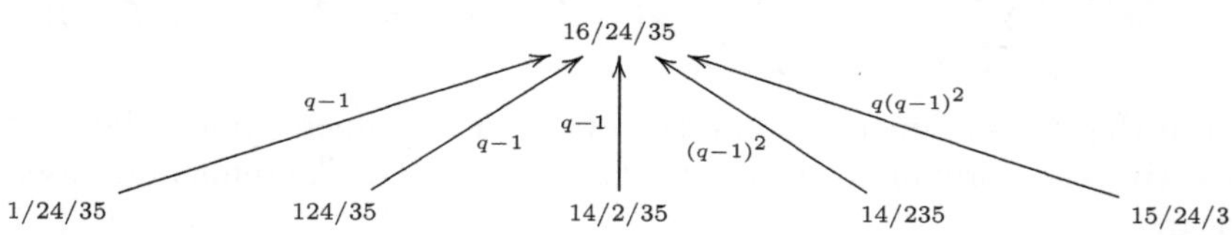

where we label the edges according to their multiplicities.

A complex-valued function $\varphi \colon \Gamma \to \mathbb{C}$ defined on the vertices of a branching graph $(\Gamma, \varkappa)$ is said to be *harmonic* if it satisfies the condition

$$\varphi(\lambda) = \sum_{\mu \in \Gamma,\, \lambda \nearrow \mu} \varkappa(\lambda,\mu)\varphi(\mu) \tag{3.1}$$

for all $\lambda \in \Gamma$. We denote by $\mathcal{H}(\Gamma)$ the set of all harmonic functions defined on Γ, and consider $\mathcal{H}(\Gamma)$ as a topological space with respect to the topology of pointwise convergence. We denote by $\mathcal{H}^+(\Gamma)$ the subset of $\mathcal{H}(\Gamma)$ consisting of nonnegative harmonic functions (that is, harmonic functions with nonnegative real values), and by $\mathcal{H}^+_1(\Gamma)$ the subset consisting of all functions $\varphi \in \mathcal{H}^+(\Gamma)$ which satisfy the normalising condition $\varphi(\varnothing) = 1$. It is clear that $\mathcal{H}^+_1(\Gamma)$ is a convex subset of $\mathcal{H}(\Gamma)$, and it is also a compact measurable space; we refer to [31] for the general theorems about convex compact measurable sets. Let $\mathrm{Ex}(\Gamma)$ denote the set of extreme points of $\mathcal{H}^+_1(\Gamma)$; this is a set of type G_δ, and hence it is a Borel measurable set.

In our running example, every supercharacter ξ of U_∞ determines uniquely a harmonic function $\varphi \in \mathcal{H}_1^+(\mathcal{SP})$ on its superbranching graph $\mathcal{SP}$ by means of the formula

$$\operatorname{Res}_{U_n}^{U_\infty}(\xi) = \sum_{\pi \in \mathcal{SP}(n)} \varphi(\pi)\xi_\pi;$$

in fact, it is clear that $\operatorname{Res}_{U_n}^{U_\infty}(\xi)$ is a superclass function of U_n (hence, it is a $\mathbb{C}$-linear combination of indecomposable supercharacters of U_n). Conversely, every harmonic function $\varphi \in \mathcal{H}_1^+(\mathcal{SP})$ determines uniquely a supercharacter of U_∞. Indeed, the following result is true.

Theorem 3.2. *Let $(\mathcal{SP}, \varkappa)$ be the superbranching graph of U_∞. Then, for every harmonic function $\varphi \in \mathcal{H}_1^+(\mathcal{SP})$, there exists a unique supercharacter $\xi_\varphi \in \operatorname{SCh}(U_\infty)$ such that*

$$\xi_\varphi(g) = \sum_{\pi \in \mathcal{SP}(n)} \varphi(\pi)\xi_\pi(g)$$

for all $g \in U_n$ and all $n \in \mathbb{N}$. Moreover, the mapping $\varphi \mapsto \xi_\varphi$ establishes an affine homeomorphism between $\mathcal{H}_1^+(\mathcal{SP})$ and $\operatorname{SCh}(U_\infty)$.

Proof. Let $\varphi \in \mathcal{H}_1^+(\mathcal{SP})$. Then, the harmonicity of φ shows that ξ_φ is well defined: indeed, for every $g \in U_n$ for $n \in \mathbb{N}$, we have

$$\begin{aligned}\xi_\varphi(g) &= \sum_{\pi \in \mathcal{SP}(n)} \varphi(\pi)\xi_\pi(g) = \sum_{\pi \in \mathcal{SP}(n)} \Bigg(\sum_{\pi' \in \mathcal{SP}(n+1)} \varkappa(\pi, \pi')\varphi(\pi') \Bigg) \xi_\pi(g) \\ &= \sum_{\pi' \in \mathcal{SP}(n)} \varphi(\pi') \Bigg(\sum_{\pi \in \mathcal{SP}(n)} \varkappa(\pi, \pi')\xi_\pi(g) \Bigg) = \sum_{\pi' \in \mathcal{SP}(n)} \varphi(\pi')\xi_{\pi'}(g).\end{aligned}$$

It is clear that ξ_φ is a superclass function of U_∞; it is a supercharacter because φ is nonnegative and normalized (hence, $\xi_\varphi(1) = \varphi(\varnothing) = 1$). The remaining assertions are clear. □

Let $(\Gamma, \varkappa)$ be a branching graph. If $\tau = (\lambda_1, \ldots, \lambda_k) \in \Gamma^k$ is a path in Γ (hence, $\lambda_i \nearrow \lambda_{i+1}$ is an edge of Γ for all $1 \leq i < k$), then we define the *weight* of τ to be the product

$$w(\tau) = \varkappa(\lambda_1, \lambda_2) \cdots \varkappa(\lambda_{k-1}, \lambda_k).$$

On the other hand, if $\lambda, \mu \in \Gamma$ are such that $|\lambda| < |\mu|$, then we define the *dimension of the interval* $[\lambda, \mu] \subseteq \Gamma$ to be the sum

$$\dim(\lambda, \mu) = \sum_{\tau \in \mathcal{P}(\lambda, \mu)} w(\tau)$$

where $\mathcal{P}(\lambda, \mu)$ denotes the set of all paths joining λ and μ; we extend this definition to all vertices $\lambda, \mu \in \Gamma$ by agreeing that $\dim(\lambda, \lambda) = 1$ and that $\dim(\lambda, \mu) = 0$ if $\lambda \neq \mu$ and $\mathcal{P}(\lambda, \mu)$ is empty. Finally, we define the *dimension of a vertex* $\lambda \in \Gamma$ to be the dimension of the interval $[\varnothing, \lambda] \subseteq \Gamma$, that is,

$$\dim(\lambda) = \dim(\varnothing, \lambda), \qquad \lambda \in \Gamma.$$

We note that, for every $\lambda, \mu \in \Gamma$ with $|\lambda| \leq |\mu|$, the dimension $\dim(\lambda, \mu)$ may be computed recursively by the formulae

$$\dim(\lambda, \mu) = \sum_{\nu \in \Gamma,\ \nu \nearrow \mu} \dim(\lambda, \nu) \varkappa(\nu, \mu) = \sum_{\nu \in \Gamma,\ \lambda \nearrow \nu} \varkappa(\lambda, \nu) \dim(\nu, \mu).$$

If $(\mathcal{SP}, \varkappa)$ is the superbranching graph of U_∞ and if $m, n \in \mathbb{N}$ are such that $m < n$, then

$$\operatorname{Res}^{U_n}_{U_m}(\xi_\pi) = \sum_{\pi' \in \mathcal{SP}(m)} \dim(\pi', \pi) \xi_{\pi'}$$

for all $\pi \in \mathcal{SP}(n)$; in particular, since $U_0 = \{1\}$ and $\xi_\varnothing$ is the trivial character of U_0, we have $\operatorname{Res}^{U_n}_{U_0}(\xi_\pi) = \xi_\pi(1)\xi_\varnothing$, and thus $\dim(\pi) = \xi_\pi(1)$ for all $\pi \in \mathcal{SP}$.

An infinite path $(\lambda_1, \ldots, \lambda_n, \ldots)$ in a branching graph $(\Gamma, \varkappa)$ is said to be *regular* if the limit

$$\lim_{n \to \infty} \frac{\dim(\lambda, \lambda_n)}{\dim(\lambda_n)}$$

exists for all $\lambda \in \Gamma$. We note that, for every $n \in \mathbb{N}$, the function $\varphi_n \colon \Gamma_n \to \mathbb{R}^+$, defined by

$$\varphi_n(\lambda) = \frac{\dim(\lambda, \lambda_n)}{\dim(\lambda_n)}, \qquad \lambda \in \Gamma,$$

is almost harmonic: the harmonicity condition holds for all $\lambda \in \Gamma$, $\lambda \neq \lambda_n$. Therefore, the limit above (if it exists) defines a normalized nonnegative harmonic function on Γ, and thus there exists an harmonic function φ_τ associated with every regular path τ in Γ. The following result asserts that every extreme harmonic function $\varphi \in \operatorname{Ex}(\Gamma)$ can be obtained in this way; it is an easy consequence of a (slightly) variation of Birkhoff's pointwise ergodic theorem, and a proof can be found in [30, Appendix 3] (see also [27]).

Theorem 3.3. *Let $(\Gamma, \varkappa)$ be a branching graph. For every $\varphi \in \operatorname{Ex}(\Gamma)$, there exists a regular path $(\lambda_1, \ldots, \lambda_n, \ldots)$ in Γ such that*

$$\varphi(\lambda) = \lim_{n \to \infty} \frac{\dim(\lambda, \lambda_n)}{\dim(\lambda_n)}$$

for all $\lambda \in \Gamma$.

As a consequence, we obtain the following result.

Theorem 3.4. *If ξ is an indecomposable supercharacter of U_∞, then there exists a sequence $\xi_1, \xi_2, \ldots, \xi_n, \ldots$ of supercharacters of $U_1, U_2, \ldots, U_n, \ldots$, respectively, such that the corresponding sequence $\{\widehat{\xi}_n\}_{n=1}^\infty$ of normalized supercharacters is pointwise convergent and*

$$\xi(g) = \lim_{n \to \infty} \frac{\xi_n(g)}{\xi_n(1)}$$

for all $g \in U_\infty$.

Proof. Let $\xi \in \mathrm{SCh}(U_\infty)$ be arbitrary, and consider the superbranching graph $(\mathcal{SP}, \varkappa)$ of U_∞. By Theorem 3.2, there exists $\varphi \in \mathcal{H}_1^+(\mathcal{SP})$ such that $\xi = \xi_\varphi$, and the extremality of ξ implies that $\varphi \in \mathrm{Ex}(\mathcal{SP})$. By the previous theorem, there exists a regular path $(\pi_1, \ldots, \pi_n, \ldots)$ in $\mathcal{SP}$ such that

$$\varphi(\pi) = \lim_{n\to\infty} \frac{\dim(\pi, \pi_n)}{\dim(\pi_n)}$$

for all $\pi \in \mathcal{SP}$. Then, for every $m \in \mathbb{N}$ and every $g \in U_m$, we deduce that

$$\begin{aligned}\xi(g) = \xi_\varphi(g) &= \sum_{\pi\in\mathcal{SP}(m)} \varphi(\pi)\xi_\pi(g) = \sum_{\pi\in\mathcal{SP}(m)} \left(\lim_{n\to\infty} \frac{\dim(\pi,\pi_n)}{\dim(\pi_n)}\right)\xi_\pi(g) \\ &= \lim_{n\to\infty} \frac{1}{\dim(\pi_n)}\left(\sum_{\pi\in\mathcal{SP}(m)} \dim(\pi,\pi_n)\xi_\pi(g)\right) = \lim_{n\to\infty} \frac{\xi_{\pi_n}(g)}{\xi_{\pi_n}(1)},\end{aligned}$$

because $\mathrm{Res}^{U_n}_{U_m}(\xi_{\pi_n}) = \sum_{\pi\in\mathcal{SP}(m)} \dim(\pi,\pi_n)\xi_\pi$ for all $n \in \mathbb{N}$, $n \geq m$. □

Example 1. As a first example, let I be an arbitrary subset of $\mathbb{N}$. For every $n \in \mathbb{N}$, let $1 \leq i_1 < \cdots < i_r \leq n$ be such that $I \cap [n] = \{i_1, \ldots, i_r\}$, and let $\pi_n \in \mathcal{SP}(n)$ be unique partition of $[n]$ such that

$$\mathcal{D}(\pi_n) = \{(i_s, n-s+1) \colon 1 \leq s \leq r\} \cap [[n]].$$

Then, the sequence $\{\widehat{\xi}_{\pi_n}\}_{n=1}^\infty$ of normalized supercharacters is pointwise convergent; indeed, if $n \in \mathbb{N}$ for $\sigma \in \mathcal{SP}(n)$, then

$$\widehat{\xi}_{\pi_n}(g) = \begin{cases} \dfrac{(-1)^{|\mathcal{D}(\pi_n)\cap\mathcal{D}(\sigma)|}}{q^{\mathrm{nst}_{\pi_n}(\sigma)}(q-1)^{|\mathcal{D}(\pi_n)\cap\mathcal{D}(\sigma)|}}, & \text{if } \mathcal{D}(\sigma) \subseteq \mathrm{Reg}(\pi_n), \\ 0, & \text{otherwise,} \end{cases}$$

and thus

$$\lim_{n\to\infty} \widehat{\xi}_{\pi_n}(g) = \begin{cases} \dfrac{1}{q^{\mathrm{nst}_I(\sigma)}}, & \text{if } \mathcal{D}(\sigma) \subseteq I^c \times \mathbb{N}, \\ 0, & \text{otherwise,} \end{cases}$$

where $I^c = \mathbb{N} \setminus I$ and $\mathrm{nst}_I(\sigma) = \sum_{i\in I} \mathrm{nst}_i(\sigma)$ (notice that this sum is finite). We denote by ξ_I the supercharacter of U_∞ defined by

$$\xi_I(g) = \lim_{n\to\infty} \widehat{\xi}_{\pi_n}(g)$$

for all $g \in U_\infty$, and observe that

$$\xi_I(g) = \prod_{i\in I} \xi_i(g) = \prod_{i\in I} \left(\lim_{n\to\infty} \widehat{\xi}_{i,n}(g)\right);$$

notice that, for every $g \in U_\infty$, the product $\prod_{i\in I} \xi_i(g)$ is finite. We will prove later that ξ_I is an indecomposable supercharacter of U_∞.

Example 2. As a second example, let π be an arbitrary set partition of $\mathbb{N}$ and define the set of arcs $\mathcal{D}(\pi)$ as in the finite case; we will denote by $\mathcal{SP}(\mathbb{N})$ the set consisting of all set partitions of $\mathbb{N}$. With π we associate a supercharacter $\xi_\pi \in \mathrm{SCh}(U_\infty)$ as follows. For every $n \in \mathbb{N}$, let $\pi_n \in \mathcal{SP}(n)$ be the unique set partition such that $\mathcal{D}(\pi_n) = \mathcal{D}(\pi) \cap [[n]]$. Then, the sequence $\{\widehat{\xi}_{\pi_n}\}_{n=1}^{\infty}$ is pointwise convergent; indeed, for every $g \in U_\infty$, the sequence $\{\widehat{\xi}_{\pi_n}(g)\}_{n=1}^{\infty}$ is eventually constant. We thus define

$$\xi_\pi(g) = \lim_{n\to\infty} \widehat{\xi}_{\pi_n}(g)$$

for all $g \in U_\infty$. We also note that ξ_π admits the representation the infinite product

$$\xi_\pi = \prod_{(i,j)\in\mathcal{D}(\pi)} \widehat{\xi}_{i,j}.$$

As in the previous example, we will prove that ξ_π is an indecomposable supercharacter of U_∞.

We are now able to prove the following result.

Theorem 3.5. *For every $n \in \mathbb{N}$, let ξ_n be a supercharacter of U_n, and assume that the sequence $\{\widehat{\xi}_n\}_{n=1}^{\infty}$ is pointwise convergent. If $\xi \in \mathrm{SCh}(U_\infty)$ is given by $\xi(g) = \lim_{n\to\infty} \widehat{\xi}_n(g)$ for all $g \in U_\infty$, then $\xi = \xi_I \xi_\pi$ for some subset $I \subseteq \mathbb{N}$ and some partition $\pi \in \mathcal{SP}(\mathbb{N})$ such that $\mathcal{D}(\pi) \cap (I \times \mathbb{N}) = \emptyset$.*

Proof. For every $n \in \mathbb{N}$, let $\pi_n \in \mathcal{SP}(n)$ be such that $\xi_n = \xi_{\pi_n}$; moreover, for every $i \in \mathbb{N}$, let $\pi_n^{(i)} \in \mathcal{SP}(n)$ be defined by $\mathcal{D}(\pi_n^{(i)}) = \mathcal{D}(\pi_n) \cap (\{i\} \times [n])$, let $\mathcal{D}^{(i)} = \bigcup_{n\in\mathbb{N}} \mathcal{D}(\pi_n^{(i)})$ and, in the case where $\mathcal{D}^{(i)}$ is finite and nonempty, let $j_i = \max\{j \in \mathbb{N} \colon (i,j) \in \mathcal{D}^{(i)}\}$. It is easy to see that, for every $i \in \mathbb{N}$, the sequence $\{\widehat{\xi}_{\pi_n^{(i)}}\}_{n=1}^{\infty}$ is pointwise convergent; indeed, for every $g \in U_\infty$, we have

$$\lim_{n\to\infty} \widehat{\xi}_{\pi_n^{(i)}}(g) = \begin{cases} \xi_i(g), & \text{if } \mathcal{D}^{(i)} \text{ is infinite,} \\ \widehat{\xi}_{i,j_i}(g), & \text{if } \mathcal{D}^{(i)} \text{ is finite and} \\ & \{n \in \mathbb{N} \colon \mathcal{D}(\pi_n^{(i)}) \neq \emptyset\} \text{ is infinite,} \\ 1, & \text{if } \{n \in \mathbb{N} \colon \mathcal{D}(\pi_n^{(i)}) \neq \emptyset\} \text{ is finite.} \end{cases}$$

Since

$$\xi_{\pi_n} = \prod_{(i,j)\in\mathcal{D}(\pi_n)} \widehat{\xi}_{i,j} = \prod_{i\in\mathbb{N}} \widehat{\xi}_{\pi_n^{(i)}}$$

for all $n \in \mathbb{N}$, we conclude that

$$\begin{aligned} \xi(g) &= \lim_{n\to\infty} \Big(\prod_{i\in\mathbb{N}} \widehat{\xi}_{\pi_n^{(i)}}(g) \Big) = \prod_{i\in\mathbb{N}} \Big(\lim_{n\to\infty} \widehat{\xi}_{\pi_n^{(i)}}(g) \Big) \\ &= \xi_I(g) \Big(\lim_{n\to\infty} \prod_{i\in I^c} \widehat{\xi}_{\pi_n^{(i)}}(g) \Big) \end{aligned}$$

where $I = \{i \in \mathbb{N} \colon \mathcal{D}^{(i)} \text{ is infinite}\}$. Let $\xi' \in \mathrm{SCh}(U_\infty)$ be defined by

$$\xi'(g) = \lim_{n\to\infty} \prod_{i\in I^c} \widehat{\xi}_{\pi_n^{(i)}}(g)$$

for all $g \in U_\infty$; by the above, it is clear that, if $g \in U_\infty$, then

$$\xi'(g) = \prod_{i\in I'} \widehat{\xi}_{i,j_i}(g)$$

where I' is the subset of I^c consisting of all $i \in I^c$ for which $\mathcal{D}^{(i)}$ is finite and $\{n \in \mathbb{N} \colon \mathcal{D}(\pi_n^{(i)}) \neq \emptyset\}$ is infinite.

We claim that there exists a set partition $\pi \in \mathcal{SP}(\mathbb{N})$ such that $\mathcal{D}(\pi) = \{(i, j_i) \colon i \in I'\}$. Indeed, assume that this is not the case; hence, there must exist $i, i' \in \mathbb{N}$ with $i < i'$ and such that $j_i = j_{i'}$. Then, by the definition, there exists an increasing sequence $n_1, n_2, n_3, \ldots \in \mathbb{N}$ such that $(i, j_i) \in \mathcal{D}(\pi_{n_k})$ for all $k \in \mathbb{N}$; similarly, there exists an increasing sequence $n'_1, n'_2, n'_3, \ldots \in \mathbb{N}$ such that $(i', j_i) \in \mathcal{D}(\pi_{n'_k})$ for all $k \in \mathbb{N}$. By equation (2.3), we have $\widehat{\xi}_{\pi_{n_k}}(1+e_{i,j_i}) = -1$ and $\widehat{\xi}_{\pi_{n'_k}}(1+e_{i,j_i}) = 1$ for all $k \in \mathbb{N}$, and thus

$$\lim_{k\to\infty} \widehat{\xi}_{\pi_{n_k}}(1+e_{i,j_i}) = -1 \neq 1 = \lim_{k\to\infty} \widehat{\xi}_{\pi_{n'_k}}(1+e_{i,j_i}).$$

This is impossible because the sequence $\{\widehat{\xi}_{\pi_n}(1+e_{i,j_i})\}_{n=1}^{\infty}$ is convergent, and so our claim is true.

The proof is complete. □

Corollary 3.6. *If ξ is an indecomposable supercharacter of U_∞, then there exist a subset $I \subseteq \mathbb{N}$ and a partition $\pi \in \mathcal{SP}(\mathbb{N})$ such that $\mathcal{D}(\pi) \cap (I \times \mathbb{N}) = \emptyset$ and $\xi = \xi_I \xi_\pi$.*

Proof. It is enough to apply Theorem 3.4 (and the previous theorem). □

4. Multiplicative graphs

A branching graph $(\Gamma, \varkappa)$ is said to be *multiplicative* if there is a unital associative graded $\mathbb{R}$-algebra

$$A = \bigoplus_{n\in\mathbb{N}_0} A_n$$

and an (injective) map $\iota \colon \Gamma \to A$ such that the following properties are satisfied:

1. $\iota(\varnothing) = 1$.
2. For every $n \in \mathbb{N}_0$, the set $\{\iota(\lambda) \colon \lambda \in \Gamma_n\}$ is an $\mathbb{R}$-basis of A_n.
3. The element $\iota(\star) = \sum_{\lambda\in\Gamma_1} \varkappa(\varnothing, \lambda)\iota(\lambda) \in A_1$ satisfies

$$\iota(\lambda)\iota(\star) = \sum_{\mu\in\Gamma_{n+1}} \varkappa(\lambda, \mu)\iota(\mu)$$

for all $\lambda \in \Gamma_n$ and all $n \in \mathbb{N}$.

If this is the situation, we let[2]

$$K = \sum_{\lambda \in \Gamma} \mathbb{R}_0^+ \iota(\lambda)$$

be the cone in A spanned by $\{\iota(\lambda) : \lambda \in \Gamma\}$, and let $\mathrm{Mult}_1^+(A)$ be the subset of the dual vector space A^* consisting of all multiplicative linear functions $\phi: A \to \mathbb{R}$ which are non-negative on K and satisfy $\phi(\iota(\star)) = 1$. We note that the inclusion $\iota: \Gamma \to A$ induces naturally a bijective map $\iota^*: A^* \to \mathcal{F}(\Gamma, \mathbb{R})$ where $\mathcal{F}(\Gamma, \mathbb{R})$ denotes the set consisting of all functions $\Gamma \to \mathbb{R}$; indeed, $\iota^*(\phi)(\lambda) = \phi(\iota(\lambda))$ for all $\phi \in A^*$ and all $\lambda \in \Gamma$.

Proposition 4.1. *The superbranching graph $(\mathcal{SP}, \varkappa)$ of U_∞ is multiplicative.*

Proof. For every $n \in \mathbb{N}$, let $\mathbf{SC}_n$ denote the vector space over $\mathbb{R}$ spanned by the set $\mathcal{Y}(n) = \{\xi_\pi : \pi \in \mathcal{SP}(n)\}$; hence, $\mathcal{Y}(n)$ is an $\mathbb{R}$-basis of $\mathbf{SC}_n$. Then, we define

$$\mathbf{SC} = \bigoplus_{n \in \mathbb{N}_0} \mathbf{SC}_n$$

where we agree that $\mathbf{SC}_0 = \mathbb{R}$. We define an (associative) multiplication on $\mathbf{SC}$ as follows. (We should mention that, if we define $\mathbf{SC}_n$, for $n \in \mathbb{N}$, to be the vector space over $\mathbb{R}$ spanned by the set of the (standard) supercharacters of the algebra group U_n, then the vector space $\mathbf{SC} = \bigoplus_{n \in \mathbb{N}_0} \mathbf{SC}_n$ has a Hopf algebra structure studied in [1] (see also [7]); the algebra structure considered in the present work is closely related to the algebra structure of the Hopf dual of the one studied in that paper.)

For every $m, n \in \mathbb{N}$, we naturally identify the direct product $U_m \times U_n$ with the subgroup

$$\widetilde{U}_{m,n} = \left\{ \begin{bmatrix} g & 0 \\ 0 & h \end{bmatrix} : g \in U_m,\ h \in U_n \right\}$$

of U_{m+n}. On the other hand, let $\xi \in \mathbf{SC}_m$ and $\zeta \in \mathbf{SC}_n$ be arbitrary, let $\xi \times \zeta : U_m \times U_n \to \mathbb{C}$ be the function defined by $(\xi \times \zeta)(g, h) = \xi(g)\zeta(h)$ for all $g \in U_m$ and all $h \in U_n$, and let $(\xi \times \zeta)^\circ : U_{m+n} \to \mathbb{C}$ be the extension by zero of $\xi \times \zeta$ to U_{m+n}. Notice that, if $\xi = \xi_\sigma$ is a supercharacter of U_m (corresponding to a set partition $\sigma \in \mathcal{SP}(m)$) and $\zeta = \xi_\tau$ is a supercharacter of U_n (corresponding to a set partition $\tau \in \mathcal{SP}(n)$), then $\xi \times \zeta = \xi_\sigma \times \xi_\tau$ is a supercharacter of $U_m \times U_n$ which corresponds (under the similar natural construction) to the $(B_m \times B_n)$-orbit of the element $\vartheta_\sigma \times \vartheta_\tau \in (\mathfrak{u}_m \times \mathfrak{u}_n)^\circ$. As in [14], we define the *superinduced function* $\mathrm{SInd}_{U_m \times U_n}^{U_{m+n}}(\xi \times \zeta) : U_{m+n} \to \mathbb{C}$ by the formula

$$\mathrm{SInd}_{U_m \times U_n}^{U_{m+n}}(\xi \times \zeta)(g) = \frac{1}{|B_{m+n}||B_m||B_n|} \sum_{h,k \in B_{m+n}} (\xi \times \zeta)^\circ(1 + h(g-1)k)$$

[2]Throughout the paper, we define $\mathbb{R}_0^+ = \mathbb{R}^+ \sqcup \{0\}$.

for all $g \in U_{m+n}$. By the definition, it is clear that $\mathrm{SInd}_{U_m \times U_n}^{U_{m+n}}(\xi \times \zeta)$ is a superclass function of U_{m+n}, and thus

$$\mathrm{SInd}_{U_m \times U_n}^{U_{m+n}}(\xi \times \zeta) = \sum_{\pi \in \mathcal{SP}(m+n)} \varkappa_{\xi,\zeta}(\pi)\, \xi_\pi$$

where $\varkappa_{\xi,\zeta}(\pi) \in \mathbb{C}$ for all $\pi \in \mathcal{SP}(m+n)$. The multiplicities $\varkappa_{\xi,\zeta}(\pi)$ may be calculated using a version for supercharacters of the well-known Frobenius reciprocity. Indeed, as in [14, Lemma 6.7], it is not difficult to check that

$$\big\langle \mathrm{SInd}_{U_m \times U_n}^{U_{m+n}}(\xi \times \zeta), \xi_\pi \big\rangle = \big\langle \xi \times \zeta, \mathrm{Res}_{U_m \times U_n}^{U_{m+n}}(\xi_\pi) \big\rangle$$

for all $\pi \in \mathcal{SP}(m+n)$; as usual, for any finite group G, we denote by $\langle \cdot, \cdot \rangle$ the Frobenius scalar product of class functions of G. Therefore, since $\mathcal{Y}(m+n)$ is as orhogonal basis of $\mathbf{SC}_{m+n}$ (with respect to the Frobenius scalar product), we deduce that

$$\varkappa_{\xi,\zeta}(\pi)\langle \xi_\pi, \xi_\pi \rangle = \big\langle \mathrm{SInd}_{U_m \times U_n}^{U_{m+n}}(\xi \times \zeta), \xi_\pi \big\rangle = \big\langle \xi \times \zeta, \mathrm{Res}_{U_m \times U_n}^{U_{m+n}}(\xi_\pi) \big\rangle,$$

and thus Proposition 3.1) implies that $\varkappa_{\xi,\zeta}(\pi)$ is a nonnegative real number for all $\pi \in \mathcal{SP}(m+n)$; in fact, if both ξ and ζ are supercharacters of U_m and U_n, respectively, then $\varkappa_{\xi,\zeta}(\pi)$ is a nonnegative rational number for all $\pi \in \mathcal{SP}(m+n)$. In particular, we conclude that $\mathrm{SInd}_{U_m \times U_n}^{U_{m+n}}(\xi \times \zeta) \in \mathbf{SC}_{m+n}$, and this allows us to define

$$\xi \cdot \zeta = \mathrm{SInd}_{U_m \times U_n}^{U_{m+n}}(\xi \times \zeta).$$

It is routine to check that the mapping $(\xi, \zeta) \mapsto \xi \cdot \zeta$ defines an associative product on $\mathbf{SC}^+ = \bigoplus_{n \in \mathbb{N}} \mathbf{SC}_n$; we extend this associative product to $\mathbf{SC}$ by imposing $1 \in \mathbb{R} = \mathbf{SC}_0$ to be the identity of $\mathbf{SC}$. Thus, $\mathbf{SC} = \bigoplus_{n \in \mathbb{N}_0} \mathbf{SC}_n$ becomes an associative graded $\mathbb{R}$-algebra with a countable $\mathbb{R}$-basis $\{\xi_\pi \colon \pi \in \mathcal{SP}\}$; we recall that $\xi_\varnothing = 1_{U_0}$ is the unit character of the trivial group $U_0 = \{1\}$.

We now define the (injective) map $\iota \colon \Gamma \to \mathbf{SC}$ by the rule

$$\iota(\pi) = \langle \xi_\pi, \xi_\pi \rangle^{-1} \xi_\pi$$

for all $\pi \in \mathcal{SP}$; notice that $\iota(\varnothing) = \xi_\varnothing = 1 \in \mathbf{SC}_0$, that $\{\iota(\pi) \colon \pi \in \mathcal{SP}\}$ is an $\mathbb{R}$-basis of $\mathbf{SC}$, and that $\iota(\{1\}) = \xi_{\{1\}} = 1_{U_1}$ is the unique supercharacter of $U_1 = \{1\}$ (indeed, $\pi = \{1\}$ is the unique set partition of $\{1\}$). Let $n \in \mathbb{N}$ be arbitrary, and recall that $U_n = U_n \times U_1$ (as subgroups of U_{n+1}). Thus, it is obvious that $\xi \times \xi_{\{1\}} = \xi$, and thus

$$\xi \cdot \xi_{\{1\}} = \mathrm{SInd}_{U_n}^{U_{n+1}}(\xi)$$

for all $\xi \in \mathbf{SC}_n$. Since

$$\big\langle \mathrm{SInd}_{U_n}^{U_{n+1}}(\xi_\pi), \xi_{\pi'} \big\rangle = \big\langle \xi_\pi, \mathrm{Res}_{U_n}^{U_{n+1}}(\xi_{\pi'}) \big\rangle = \varkappa(\pi, \pi')\langle \xi_\pi, \xi_\pi \rangle$$

for all $\pi \in \mathcal{SP}(n)$ and all $\pi' \in \mathcal{SP}(n+1)$, we conclude that

$$\xi_\pi \cdot \xi_{\{1\}} = \mathrm{SInd}_{U_n}^{U_{n+1}}(\xi_\pi) = \sum_{\pi' \in \mathcal{SP}(n+1)} \frac{\varkappa(\pi, \pi')\langle \xi_\pi, \xi_\pi \rangle}{\langle \xi_{\pi'}, \xi_{\pi'} \rangle}\, \xi_{\pi'}$$

for all $\pi \in \mathcal{SP}(n)$. Therefore, if we define $\iota(\star) = \xi_{\{1\}}$, then

$$\iota(\pi) \cdot \iota(\star) = \sum_{\pi' \in \mathcal{SP}(n+1)} \varkappa(\pi, \pi')\iota(\pi')$$

for all $\pi \in \mathcal{SP}(n)$, and this completes the proof. □

The main goal of this section is to obtain a parametrisation of the indecomposable supercharacters of U_∞ in terms of certain ring homomorphisms $\phi\colon \mathbf{SC} \to \mathbb{R}$. To this end, we will refer to the following Kerov–Vershik "ring theorem"; a proof can be found in [17, Proposition 8.4] (we should mention that the proof does not depend on the commutativity of the $\mathbb{R}$-algebra A (as it is required there)).

Theorem 4.2 (Kerov–Vershik). *Let A be a unital associative algebra over $\mathbb{R}$ and let $K \subseteq A$ be a convex cone satisfying the following conditions* [3]*:*

1. *$K - K = A$ (that is, K generates A).*
2. *$KK \subseteq K$ (that is, K is stable under multiplication).*
3. *K is spanned by a countable set of elements.*
4. *For every $a \in A$, there exists $\varepsilon \in \mathbb{R}^+$ such that $1 - \varepsilon a \in K$ (in particular, $1 \in K$).*

If $\mathcal{L}$ denotes the convex set consisting of all linear forms $\phi \in A^*$ which are non-negative on K and satisfy $\phi(1) = 1$, then a linear form $\phi \in \mathcal{L}$ is an extreme point of $\mathcal{L}$ if and only if ϕ is multiplicative (which means that $\phi(ab) = \phi(a)\phi(b)$ for all $a, b \in A$).

The following result is a non-commutative version of a theorem proved by Kerov–Vershik in [28]; for convenience of the reader, we give here a detailed proof (see also [29, Theorem 2.5]).

Theorem 4.3. *Let $(\Gamma, \varkappa)$ be a multiplicative branching graph associated with a graded $\mathbb{R}$-algebra $A = \bigoplus_{n \in \mathbb{N}_0} A_n$ via an injective map $\iota\colon \Gamma \to A$. Let K denote the cone in A spanned by $\{\iota(\lambda)\colon \lambda \in \Gamma\}$, and assume that K is stable under multiplication. Then, the natural map $\iota^*\colon A^* \to \mathcal{F}(\Gamma, \mathbb{R})$ induces an homeomorphism between $\mathrm{Mult}_1^+(A)$ and the subset $\mathrm{Ex}(\Gamma)$ consisting of all extreme harmonic functions defined on Γ.*

Proof. Let $\mathcal{J}$ be the ideal of A generated by the relation $\iota(\star) - 1$, let $\widetilde{A} = A/\mathcal{J}$ be the quotient $\mathbb{R}$-algebra, and let $\nu\colon A \to \widetilde{A}$ be the natural epimorphism; for every $a \in A$, we will denote by $\widetilde{a}$ the image $\nu(a) \in \widetilde{A}$.

Let $\widetilde{K} = \nu(K)$, and note that $\widetilde{K}$ is a cone in $\widetilde{A}$ satisfying $1 = \nu(1) = \nu(\iota(\star)) \in \widetilde{K}$. Since $KK \subseteq K$, it is clear that $\widetilde{K}\widetilde{K} \subseteq \widetilde{K}$; moreover, since K spans A, it is also obvious that $\widetilde{K}$ spans $\widetilde{A}$. On the other hand, since Γ has a countable set of vertices, $\widetilde{K}$ has a countable set of generators; indeed, $\widetilde{K}$ is spanned by the set $\widetilde{\iota(\Gamma)} = \nu(\iota(\Gamma))$. Finally, let $\widetilde{a} \in \widetilde{A}$ be arbitrary; we claim that there exists $\varepsilon \in \mathbb{R}^+$

[3]For arbitrary subsets $X, Y \subseteq A$, we define $X - Y = \{x - y\colon x \in X,\ y \in Y\}$ and $XY = \{xy\colon x \in X,\ y \in Y\}$

such that $1-\varepsilon\widetilde{a} \in \widetilde{K}$. Without loss of generality, we may assume that $\widetilde{a} \in \widetilde{K}$; indeed, since $\widetilde{K}-\widetilde{K}=\widetilde{A}$, there exist $\widetilde{b},\widetilde{c} \in \widetilde{K}$ such that $\widetilde{a}=\widetilde{b}-\widetilde{c}$, and thus, if $\varepsilon \in \mathbb{R}^+$ is such that $1-\varepsilon\widetilde{b} \in \widetilde{K}$, then $1-\varepsilon\widetilde{a} = 1-\varepsilon\widetilde{b}+\varepsilon\widetilde{c} \in \widetilde{K}$. In fact, since $\widetilde{K}$ is spanned by $\widetilde{\iota(\Gamma)}$, it is enough to show that, for every $\lambda \in \Gamma$, there exists $\varepsilon \in \mathbb{R}^+$ such that $1-\varepsilon\widetilde{\iota(\lambda)} \in \widetilde{K}$. Let $n \in \mathbb{N}$, and let $\lambda \in \Gamma_n$. Then,

$$\iota(\star)^n = \sum_{\mu\in\Gamma_n} \dim(\mu)\iota(\mu),$$

and thus $\iota(\star)^n - \dim(\lambda)\iota(\lambda) \in K$. Since $\nu(\iota(\star)^n) = \nu(\iota(\star))^n = 1$, we conclude that

$$1-\dim(\lambda)\widetilde{\iota(\lambda)} = \nu\big(\iota(\star)^n - \dim(\lambda)\iota(\lambda)\big) \in \nu(K) = \widetilde{K}.$$

Therefore, the pair $(\widetilde{A},\widetilde{K})$ satisfies the conditions of Theorem 4.2, and so, if $\widetilde{\mathcal{L}}$ denotes the convex set consisting of all linear forms $\widetilde{\phi} \in \widetilde{A}^*$ which are nonnegative on $\widetilde{K}$ and satisfy $\widetilde{\phi}(1)=1$, then a linear function $\widetilde{\phi} \in \widetilde{\mathcal{L}}$ is extreme if and only if $\widetilde{\phi}$ is multiplicative.

Now, consider the subset $\mathrm{Mult}_1^+(A)$ of A^*, and let $\phi \in \mathrm{Mult}_1^+(A)$ be arbitrary. Then, $\phi(\iota(\star))=1$; since $\phi(1)=1$ (because ϕ is multiplicative), we conclude that $\phi(\iota(\star)-1)=0$. Since ϕ is multiplicative, it follows that $\phi(a)=0$ for all $a \in \mathcal{I}$, and thus ϕ induces naturally a linear map $\widetilde{\phi} \in \widetilde{A}^*$ (by the rule $\widetilde{\phi}(\nu(a)) = \phi(a)$ for all $a \in A$). It is clear that $\widetilde{\phi}$ is multiplicative, nonnegative on $\widetilde{K}$ and satisfies $\widetilde{\phi}(1)=1$. Therefore, $\widetilde{\phi} \in \mathrm{Ex}(\widetilde{\mathcal{L}})$, and thus the mapping $\phi \mapsto \widetilde{\phi}$ defines a map $\Lambda\colon \mathrm{Mult}_1^+(A) \to \mathrm{Ex}(\widetilde{\mathcal{L}})$ which is clearly a homeomorphism.

Finally, we observe that there is a homeomorphism $\Lambda'\colon \mathrm{Ex}(\widetilde{\mathcal{L}}) \to \mathrm{Ex}(\Gamma)$. For every $\widetilde{\phi} \in A^*$, we define the function $\iota^*(\widetilde{\phi})\colon \Gamma \to \mathbb{R}$ by

$$\iota^*(\widetilde{\phi})(\lambda) = \widetilde{\phi}\big(\widetilde{\iota(\lambda)}\big)$$

for all $\lambda \in \Gamma$. Since $\widetilde{\iota(\star)}=1$, we deduce that

$$\begin{aligned}\iota^*(\widetilde{\phi})(\lambda) &= \widetilde{\phi}\big(\widetilde{\iota(\lambda)}\big) = \widetilde{\phi}\big(\widetilde{\iota(\star)}\widetilde{\iota(\lambda)}\big) = \widetilde{\phi}\Big(\sum_{\mu\in\Gamma_{n+1}} \varkappa(\lambda,\mu)\widetilde{\iota(\mu)}\Big)\\ &= \sum_{\mu\in\Gamma_{n+1}} \varkappa(\lambda,\mu)\widetilde{\phi}\big(\widetilde{\iota(\mu)}\big) = \sum_{\mu\in\Gamma_{n+1}} \varkappa(\lambda,\mu)\iota^*(\widetilde{\phi})(\mu)\end{aligned}$$

for all $\lambda \in \Gamma_n$ and all $n \in \mathbb{N}$, which means that $\iota^*(\widetilde{\phi}) \in \mathcal{H}(\Gamma)$. Therefore, the mapping $\widetilde{\phi} \mapsto \iota^*(\widetilde{\phi})$ defines a map $\iota^*\colon A^* \to \mathcal{H}(\Gamma)$; since $\{\iota(\lambda)\colon \lambda \in \Gamma\}$ is an $\mathbb{R}$-basis of A, this map is bijective, and it is clearly a homeomorphism. It is also clear that $\iota^*(\widetilde{\mathcal{L}}) = \mathcal{H}_1^+(\Gamma)$ and that $\iota^*(\mathrm{Ex}(\widetilde{\mathcal{L}})) = \mathrm{Ex}(\mathcal{H}_1^+(\Gamma)) = \mathrm{Ex}(\Gamma)$, Therefore, ι^* induces the required homeomorphism $\Lambda'\colon \mathrm{Ex}(\widetilde{\mathcal{L}}) \to \mathrm{Ex}(\Gamma)$, and this completes the proof. □

By Theorem 3.2, a supercharacter $\xi \in \mathrm{SCh}(U_\infty)$ corresponds uniquely to the harmonic function $\varphi \in \mathcal{H}_1^+(\mathcal{SP})$ defined by the rule

$$\xi(g) = \sum_{\pi \in \mathcal{SP}(n)} \varphi(\pi)\xi_\pi(g)$$

for all $g \in U_n$ and all $n \in \mathbb{N}$; hence, if we set $\widetilde{\xi}_\pi = \langle \xi_\pi, \xi_\pi \rangle^{-1} \xi_\pi$, then we have

$$\varphi(\pi) = \frac{\langle \mathrm{Res}_{U_n}^{U_\infty}(\xi), \xi_\pi \rangle}{\langle \xi_\pi, \xi_\pi \rangle} = \langle \mathrm{Res}_{U_n}^{U_\infty}(\xi), \widetilde{\xi}_\pi \rangle$$

for all $\pi \in \mathcal{SP}(n)$ and all $n \in \mathbb{N}$. On the other hand, by the above, the harmonic function φ corresponds uniquely to the linear function $\phi \in \mathbf{SC}^*$ defined on the basis $\{\widetilde{\xi}_\pi : \pi \in \mathcal{SP}\}$ of $\mathbf{SC}$ by the rule

$$\phi(\widetilde{\xi}_\pi) = \varphi(\pi) = \langle \mathrm{Res}_{U_n}^{U_\infty}(\xi), \widetilde{\xi}_\pi \rangle$$

for all $\pi \in \mathcal{SP}(n)$ and all $n \in \mathbb{N}$. Therefore, if $\mathcal{L}(\mathbf{SC})$ denotes the convex subset of $\mathbf{SC}^*$ consisting of all normalized linear functions which are nonnegative on the cone $K = \sum_{\pi \in \mathcal{SP}} \mathbb{R}_0^+ \xi_\pi$ in $\mathbf{SC}$, then we obtain a homeomorphism $\Lambda\colon \mathrm{SCh}(U_\infty) \to \mathcal{L}(\mathbf{SC})$ where, for every $\xi \in \mathrm{SCh}(U_\infty)$, the linear map $\Lambda(\xi) \in \mathbf{SC}^*$ is defined by

$$\Lambda(\xi)(\xi_\pi) = \langle \mathrm{Res}_{U_n}^{U_\infty}(\xi), \xi_\pi \rangle = \frac{1}{|U_n|} \sum_{g \in U_n} \xi(g)\xi_\pi(g)$$

for all $\pi \in \mathcal{SP}(n)$ and all $n \in \mathbb{N}$.

Besides the supercharacter basis, the $\mathbb{R}$-algebra $\mathbf{SC}$ also has a distinguished basis $\{\kappa_\pi : \pi \in \mathcal{SP}\}$ consisting of all *superclass characteristic functions*; for every $\pi \in \mathcal{SP}$, we define $\kappa_\pi : U_\infty \to \mathbb{C}$ by

$$\kappa_\pi(g) = \begin{cases} 1, & \text{if } g \in \mathcal{K}_\pi(\infty), \\ 0, & \text{otherwise,} \end{cases}$$

for all $g \in U_\infty$. Then, for every $\xi \in \mathrm{SCh}(U_\infty)$, we get

$$\Lambda(\xi)(\kappa_\pi) = \frac{1}{|U_n|} \sum_{g \in U_n} \xi(g)\kappa_\pi(g) = \frac{|\mathcal{K}_\pi|}{|U_n|} \xi(g_\pi)$$

for all $\pi \in \mathcal{SP}(n)$ and all $n \in \mathbb{N}$. In particular, if we define

$$\widetilde{\kappa}_\pi = \frac{|U_n|}{|\mathcal{K}_\pi|} \kappa_\pi$$

for all $\pi \in \mathcal{SP}(n)$ and all $n \in \mathbb{N}$, then we obtain

$$\Lambda(\kappa_\pi)(\widetilde{\kappa}_{\pi'}) = \delta_{\pi,\pi'}$$

for all $\pi, \pi' \in \mathcal{SP}$. Therefore, $\{\Lambda(\kappa_\pi) : \pi \in \mathcal{SP}\}$ is a basis of the graded dual vector space $\mathbf{SC}^*_{\mathrm{gr}} = \bigoplus_{n \in \mathbb{N}} \mathbf{SC}^*_n$, which is dual to the basis $\{\widetilde{\kappa}_\pi : \pi \in \Gamma\}$ of $\mathbf{SC}$; we note that $\mathbf{SC}^*_{\mathrm{gr}}$ should not be confused with the dual vector space $\mathbf{SC}^*$.

The multiplication in $\mathbf{SC}$ is easy to describe in terms of the superclass characteristic functions; indeed, if we define $\pi/\pi' \in \mathcal{SP}$ to the unique set partition such that $\mathcal{D}(\pi/\pi') = \mathcal{D}(\pi) \sqcup \{(i+n, j+n) \colon (i,j) \in \mathcal{D}(\pi')\}$, then it follows easily that from the definition of superinduction (see also [7, Lemma 4.7]) that

$$\widetilde{\kappa}_{\pi} \cdot \widetilde{\kappa}_{\pi'} = \widetilde{\kappa}_{\pi/\pi'}$$

for all $\pi, \pi' \in \mathcal{SP}$. Therefore, a linear map $\phi \in \mathbf{SC}^*$ is multiplicative if and only if

$$\phi(\widetilde{\kappa}_{\pi/\pi'}) = \phi(\widetilde{\kappa}_{\pi})\phi(\widetilde{\kappa}_{\pi'})$$

for all $\pi, \pi' \in \mathcal{SP}$, and thus $\xi \in \mathrm{SCh}(U_\infty)$ is an indecomposable supercharacter if and only if

$$\xi(g_{\pi/\pi'}) = \xi(g_{\pi})\xi(g_{\pi'}) \tag{4.1}$$

for all $\pi, \pi' \in \mathcal{SP}$. (Recall that, for every $\pi \in \mathcal{SP}$, $g_\pi \in U_\infty$ is the "normalized" element $g_\pi = 1 + e_\pi$ in the superclass $\mathcal{K}_\pi$.)

In particular, we conclude that every elementary supercharacter of U_∞ is indecomposable; by an *elementary supercharacter* of U_∞ we mean, either a supercharacter of the form ξ_i for $i \in \mathbb{N}$, or a supercharacter of the form $\widehat{\xi}_{i,j}$ for $i, j \in \mathbb{N}$ with $i < j$.

Theorem 4.4. *A supercharacter $\xi \in \mathrm{SCh}(U_\infty)$ is indecomposable if and only if $\xi = \xi_I \xi_\pi$ for some subset $I \subseteq \mathbb{N}$ and some partition $\pi \in \mathcal{SP}(\mathbb{N})$ such that $\mathcal{D}(\pi) \cap (I \times \mathbb{N}) = \emptyset$.*

Proof. By virtue of Corollary 3.6, it remains to prove that, for every $I \subseteq \mathbb{N}$ and every $\pi \in \mathcal{SP}(\mathbb{N})$ such that $\mathcal{D}(\pi) \cap (I \times \mathbb{N}) = \emptyset$, the supercharacter $\xi = \xi_I \xi_\pi$ is indecomposable. Since $\xi_I = \prod_{i \in I} \xi_i$ and $\xi_\pi = \prod_{(i,j) \in \mathcal{D}(\pi)} \widehat{\xi}_{i,j}$, this follows immediately because every elementary supercharacter is indecomposable. □

Acknowledgment

Many thanks to Eric Marberg for the very useful discussions we had during his 2015 visit to the Department of Mathematics of the Faculty of Sciences, University of Lisbon. The authors also thank the anonymous referees for the very useful suggestions (and corrections) which certainly helped us to improve this final version of the paper.

References

[1] M. Aguiar, C.A.M. André, C. Benedetti, N. Bergeron, Z. Chen, P. Diaconis, A. Hendrickson, S. Hsiao, I.M. Isaacs, A. Jedwab, K. Johnson, G. Karaali, A. Lauve, T. Le, S. Lewis, H. Li, K. Magaard, E. Marberg, J.-C. Novelli, A. Pang, F. Saliola, L. Tevlin, J.-Y. Thibon, N. Thiem, V. Venkateswaran, C.R. Vinroot, N. Yan, M. Zabrocki, *Supercharacters, symmetric functions in noncommuting variables, and related Hopf algebras.* Adv. Math. **229** (2012), no. 4, 2310–2337.

[2] C.A.M. André, *Basic characters of the unitriangular group.* J. Algebra **175** (1995), 287–319.

[3] C.A.M. André, *Basic sums of coadjoint orbits of the unitriangular group.* J. Algebra **176** (1995), 959–1000.

[4] C.A.M. André, *Irreducible characters of finite algebra groups.* In Matrices and Group Representations, Coimbra, 1998. Textos Mat. Sér. B **19** (1999), 65–80.

[5] C.A.M. André, *The basic character table of the unitriangular group.* J. Algebra **241** (2001), 437–471.

[6] C.A.M. André, *Basic characters of the unitriangular group (for arbitrary primes).* Proc. Amer. Math Soc. **130** (2002), 1943–1954.

[7] C.A.M. André, *Supercharacters of unitriangular groups and set partition combinatorics.* CIMPA school: "Modern Methods in Combinatorics ECOS2013", 2da Escuela Puntana de Combinatoria: Escuela de Combinatoria del Sur, Universidad Nacional de San Luis, Argentina, 2013.

[8] N. Bergeron and N. Thiem, *A supercharacter table decomposition via power-sum symmetric functions.* Internat. J. Algebra Comput. **23** (2013), 763–778.

[9] R.P. Boyer, *Characters and factor representations of the unitary group of the CAR-algebra.* J. Operator Theory **30** (1993), no. 2, 315–328.

[10] D. Bragg and N. Thiem, *Restrictions of rainbow supercharacters.* J. Algebra **451** (2016), 357–400.

[11] B. Chern, P. Diaconis, D.M. Kane, and R.C. Rhoades, *Closed expressions for averages of set partition statistics.* Res. Math. Sci. **1** (2014), no. 1, 1–32.

[12] B. Chern, P. Diaconis, D.M. Kane, and R.C. Rhoades, *Central limit theorems for some set partition statistics.* Adv. in Appl. Math. **70** (2015), 92–105.

[13] D. De Stavola, *A Plancherel measure associated to set partitions and its limit.* Adv. in Appl. Math. **92** (2018), 73–98.

[14] P. Diaconis and I.M. Isaacs, *Supercharacters and Superclasses for Algebra Groups.* Trans. Amer. Math. Soc. **360** (2007), no. 5, 2359–2392.

[15] A. Dudko, *Characters on the full group of an ergodic hyperfinite equivalence relation.* J. Funct. Anal. **261** (2011), 1401–1414.

[16] A. Dudko and K. Medynets, *On characters of inductive limits of symmetric groups.* Available online at `https://arxiv.org/pdf/1105.6325.pdf`.

[17] A. Gnedin and G. Olshanski, *Coherent permutations with descent statistic and the boundary problem for the graph of zigzag diagrams.* Int. Math. Res. Not. 2006, Art. ID 51968.

[18] K.R. Goodearl, *Partially ordered abelian groups with interpolation.* Mathematical Surveys and Monographs 20, Amer. Math. Soc., Providence, RI, 1986.

[19] E. Hirai and T. Hirai, *Positive definite class functions on a topological group and characters of factor representations.* J. Math. Kyoto Univ. **45** (2005), no. 2, 355–379.

[20] E. Hirai and T. Hirai, *Characters of wreath products of finite groups with the infinite symmetric group.* J. Math. Kyoto Univ. **45** (2005), no. 3, 547–597.

[21] E. Hirai, T. Hirai, and A. Hora, *Limits of characters of wreath products $\mathfrak{S}_n(T)$ of a compact group T with the symmetric groups and characters of $\mathfrak{S}_\infty(T)$*. I. Nagoya Math. J. 193 (2009), 1–93.

[22] E. Hirai, T. Hirai, and A. Hora, *Limits of characters of wreath products $\mathfrak{S}_n(T)$ of a compact group T with the symmetric groups and characters of $\mathfrak{S}_\infty(T)$. II. From a viewpoint of probability theory.* J. Math. Soc. Japan **60** (2008), no. 4, 1187–1217.

[23] I.M. Isaacs, *Characters of groups associated with finite algebras.* J. Algebra **177** (1995), 708–730.

[24] J. Keller, *Generalized Supercharacter Theories and Schur Rings for Hopf Algebras.* Ph.D. thesis, University of Colorado at Boulder, 2014.

[25] S.V. Kerov, *The boundary of Young lattice and random Young tableaux.* Formal Power Series and Algebraic Combinatorics (New Brunswick, NJ, 1994). DIMACS Ser. Discrete Math. Theoret. Comput. Sc., vol. 24, Amer. Math. Soc., Providence, RI, 1996, pp. 133–158.

[26] S.V. Kerov, *Asymptotic representation theory of the symmetric group and its applications in analysis. Translated from the Russian manuscript by N.V. Tsilevich. With a foreword by A. Vershik and comments by G. Olshanski.* Translations of Mathematical Monographs 219, Amer. Math. Soc., Providence, RI, 2003.

[27] S.V. Kerov and A.M. Vershik, *Asymptotic character theory of the symmetric group.* Funct. Analysis and its Applications **15** (1981), 15–27.

[28] S.V. Kerov and A.M. Vershik, *Characters and factor representations of the infinite symmetric group.* Dokl. Akad. Nauk SSSR **257** (1981), no. 5, 1037–1040. English translation: Soviet Math. Dokl. **23** (1981), no. 2, 389–392.

[29] C. Lecouvey and P. Tarrago, *Central measures on multiplicative graphs, representations of Lie algebras and weight polytopes.*
Available online at `https://arxiv.org/pdf/1609.00138.pdf`.

[30] G. Olshanski and A.M. Vershik, *Ergodic unitarily invariant measures on the space of infinite Hermitian matrices.* Contemporary Mathematical Physics: F.A. Berezin Volume, Amer. Math. Soc. Transl. **175**(2) (1996), 137–175.

[31] R.R. Phelps, *Lectures on Choquet's theorem.* Van Nostrand Mathematical Studies, Vol. 7, Princeton, NJ, 1966.

[32] H.L. Skudlarek, *Die unzerlegbaren Charaktere einiger diskreter Gruppen.* Math. Annal. **233** (1976), 213–231.

[33] N. Thiem, *Supercharacter theories of type A unipotent radicals and unipotent polytopes.* Available online at `https://arxiv.org/pdf/1512.03446.pdf`.

[34] E. Thoma, *Die unzerlegbaren, positiv-definiten Klassenfunktionen der abzahlbar unendlichen, symmetrischen Gruppe.* Math. Z. **85** (1964), 40–61.

[35] N. Yan, *Representation theory of the finite unipotent linear groups.* Unpublished Ph.D. Thesis, Department of Mathematics, University of Pennsylvania, 2001.

Carlos A.M. André, Filipe Gomes and Jocelyn Lochon
Centro de Análise Funcional, Estruturas Lineares e Aplicações (CEAFEL)
Departamento de Matemática
Faculdade de Ciências da Universidade de Lisboa
Campo Grande, Edifício C6, Piso 2
1749-016 Lisboa, Portugal
e-mail: `caandre@ciencias.ulisboa.pt`
`filipe13_gomes@hotmail.com`
`jocelyn.lochon@gmail.com`

Operator Theory:
Advances and Applications, Vol. 267, 25–64

A C^*-algebra of Singular Integral Operators with Shifts and Piecewise Quasicontinuous Coefficients

M. Amélia Bastos, Cláudio A. Fernandes and Yuri I. Karlovich

Abstract. The C^*-algebra $\mathfrak{B}$ of bounded linear operators on the space $L^2(\mathbb{T})$, which is generated by all multiplication operators by piecewise quasicontinuous functions, by the Cauchy singular integral operator and by the range of a unitary representation of a group G of orientation-preserving diffeomorphisms of $\mathbb{T}$ onto itself that have the same finite set of fixed points for all $g \in G \setminus \{e\}$, is studied. A Fredholm symbol calculus for the C^*-algebra $\mathfrak{B}$ and a Fredholm criterion for the operators $B \in \mathfrak{B}$ are established by using spectral measures and the local-trajectory method for studying C^*-algebras associated with C^*-dynamical systems.

Mathematics Subject Classification (2010). 45E05, 47A53, 47A67, 47B33, 47G10, 47L15.

Keywords. Singular integral operator with shifts, piecewise quasicontinuous function, C^*-algebra, amenable group, local-trajectory method, spectral measure, representation of a C^*-algebra, symbol calculus, Fredholmness.

1. Introduction

Given a Hilbert space $\mathcal{H}$, we denote by $\mathcal{B}(\mathcal{H})$ the C^*-algebra of all bounded linear operators on $\mathcal{H}$, by $\mathcal{K}(\mathcal{H})$ the ideal of all compact operators in $\mathcal{B}(\mathcal{H})$, and by $I \in \mathcal{B}(\mathcal{H})$ the identity operator on $\mathcal{H}$. If $S, T \in \mathcal{B}(\mathcal{H})$ and $S - T \in \mathcal{K}(\mathcal{H})$, we say that the operators S and T are equivalent and write $S \simeq T$. For an operator

---

This work was partially supported by the Fundação para a Ciência e a Tecnologia (Portuguese Foundation for Science and Technology) through the projects UID/MAT/04721/2013 (Centro de Análise Funcional, Estruturas Lineares e Aplicações) and UID/MAT/00297/2013 (Centro de Matemática e Aplicações). The third author was also supported by the SEP-CONACYT Project No. 168104 (México).

$A \in \mathcal{B}(\mathcal{H})$, we denote by $A^\pi := A + \mathcal{K}(\mathcal{H})$ the coset of A in the Calkin algebra $\mathcal{B}^\pi(\mathcal{H}) := \mathcal{B}(\mathcal{H})/\mathcal{K}(\mathcal{H})$ and by $|A|$ the essential norm of A, that is,

$$|A| = \|A^\pi\| = \inf\{\|A + K\| : K \in \mathcal{K}(\mathcal{H})\}.$$

An operator $A \in \mathcal{B}(\mathcal{H})$ is Fredholm on a Hilbert space $\mathcal{H}$ if and only if the coset A^π is invertible in the C^*-algebra $\mathcal{B}^\pi(\mathcal{H})$. Given two C^*-algebras $\mathcal{A}$ and $\mathcal{B}$, we write $\mathcal{A} \cong \mathcal{B}$ if these C^*-algebras are *-isomorphic and hence isometric.

Let $\mathcal{K} := \mathcal{K}(L^2(\mathbb{T}))$ be the ideal of compact operators acting on the Lebesgue space $L^2(\mathbb{T})$, where $\mathbb{T}$ is the unit circle in $\mathbb{C}$ with the length measure and anticlockwise orientation, and let $PQC(\mathbb{T})$ be the C^*-subalgebra of $L^\infty(\mathbb{T})$ generated by the quasicontinuous functions, $QC(\mathbb{T})$, and by the piecewise continuous functions, $PC(\mathbb{T})$, on $\mathbb{T}$. Consider the C^*-algebra

$$\mathfrak{A} := \mathrm{alg}(PQC(\mathbb{T}), S_{\mathbb{T}}) \subset \mathcal{B}(L^2(\mathbb{T})) \tag{1.1}$$

generated by all the multiplication operators aI with $a \in PQC(\mathbb{T})$ and by the Cauchy singular integral operator $S_{\mathbb{T}}$ defined by

$$(S_{\mathbb{T}}\varphi)(t) := \lim_{\varepsilon \to 0} \frac{1}{\pi i} \int_{\mathbb{T}\setminus\mathbb{T}(t,\varepsilon)} \frac{\varphi(\tau)}{\tau - t}\, d\tau,$$

where $\mathbb{T}(t,\varepsilon) = \{\tau \in \mathbb{T} : |\tau - t| < \varepsilon\}$, $t \in \mathbb{T}$.

Let G be a discrete group of orientation-preserving diffeomorphisms of $\mathbb{T}$ onto itself that have the same finite set $\Lambda \subset \mathbb{T}$ of fixed points for all $g \in G \setminus \{e\}$, where e is the unit of G and the group operation is given by $(gh)(t) = h(g(t))$ for all $g, h \in G$ and all $t \in \mathbb{T}$. In that case the group G is commutative (see [24, p. 284] and [10, § 2]) and therefore amenable [17]. Then the group G acts on the contour $\mathbb{T}$ *topologically freely* [1], that is, for each finite set $F \subset G$ and each open arc $\gamma \subset \mathbb{T}$ there exists a point $t \in \gamma$ such that the points $g(t)$ for $g \in F$ are pairwise distinct.

The aim of this paper is to construct a Fredholm symbol calculus (in other words, a faithful representation of the quotient C^*-algebra $\mathfrak{B}^\pi := \mathfrak{B}/\mathcal{K}$ in a Hilbert space) and to establish a Fredholm criterion for the operators in the nonlocal C^*-subalgebra

$$\mathfrak{B} := \mathrm{alg}(\mathfrak{A}, U_G) = \mathrm{alg}(PQC(\mathbb{T}), S_{\mathbb{T}}, U_G) \tag{1.2}$$

of $\mathcal{B}(L^2(\mathbb{T}))$ generated by all singular integral operators $A \in \mathfrak{A}$ and by all unitary weighted shift operators in the group $U_G := \{U_g : g \in G\}$, where

$$(U_g\varphi)(t) := |g'(t)|^{1/2}\varphi(g(t)) \quad \text{for } t \in \mathbb{T}. \tag{1.3}$$

To study the C^*-algebra $\mathfrak{B}$, we apply the local-trajectory method combined with using suitable spectral measures (see [18], [19] and [5]). Other approaches for studying nonlocal operator algebras see in [1]–[3].

The C^*-algebra $\mathfrak{B}$ in the case of piecewise slowly oscillating coefficients, a subset of the $PQC(\mathbb{T})$ coefficients, was studied in [5] (see also [6]–[9] for other classes of groups G with more complicate actions). The C^*-algebra $\mathfrak{B}$ with $PQC(\mathbb{T})$ coefficients was studied in [11] in the case of groups G acting freely on $\mathbb{T}$, making use of the local-trajectory method. In the present paper we assume that the shifts $g \in G$

have the same finite set Λ of fixed points, which essentially complicates the study of the C^*-algebra (1.2) because the action of the group G on the maximal ideal space of a central subalgebra $\mathcal{Z}^\pi$ of the quotient C^*-algebra $\mathfrak{A}^\pi := \mathfrak{A}/\mathcal{K}$ of $\mathfrak{A}$ given by (1.1) stops to be topologically free. To overcome arising difficulties, we need to apply the technique of spectral measures combined with the local-trajectory method and to use deep properties of quasicontinuous functions investigated in [25] and [26].

The paper is organized as follows. In Section 2, following [25]–[26], we consider the C^*-algebra $PQC(\mathbb{T})$ of piecewise quasicontinuous functions, collect their properties and describe the fibers and the Gelfand topology on the maximal ideal space $M(PQC(\mathbb{T}))$ of $PQC(\mathbb{T})$. Section 3 contains main results of the paper: the Fredholm symbol calculus for the C^*-algebra $\mathfrak{B}$ and a Fredholm criterion for the operators $B \in \mathfrak{B}$. In Section 4 we recall the local-trajectory method and its generalization on the basis of spectral measures. In Section 5, following [26] and [14], we describe the Fredholm symbol calculus for the C^*-algebra (1.1) and study the central C^*-algebra $\mathcal{Z}^\pi$ of $\mathfrak{A}^\pi$ generated by the cosets $(aI)^\pi$ with $a \in QC(\mathbb{T})$ and by the cosets $H^\pi_{P,t}$ related to singular integral operators with fixed singularities at points $t \in \mathbb{T}$. We describe here the maximal ideal space $M(\mathcal{Z}^\pi)$ of $\mathcal{Z}^\pi$ and its Gelfand transform, and calculate symbols of important operators $U_g V_t \in \mathfrak{A}$ for $t \in \mathbb{T}$, where the operators U_g are related to shifts $g \in G$ with a fixed point at t and the operator V_t has a fixed singularity at t.

In Section 6 we construct a spectral decomposition of the quotient C^*-algebra $\mathfrak{B}^\pi$ into the orthogonal sum of three operator C^*-subalgebras $\mathfrak{B}_{\rm arc}$, $\mathfrak{B}^\circ_\Lambda$ and $\mathfrak{B}^\diamond_\Lambda$ invariant with respect to the action of the group G on $M(\mathcal{Z}^\pi)$, and present an abstract Fredholm criterion for the operators $B \in \mathfrak{B}$ in terms of invertibility of their images $B_{\rm arc}$, B°_Λ and $B^\diamond_\Lambda$ in the C^*-algebras $\mathfrak{B}_{\rm arc}$, $\mathfrak{B}^\circ_\Lambda$ and $\mathfrak{B}^\diamond_\Lambda$, respectively. In Section 7 we study the invertibility in the C^*-algebra $\mathfrak{A}_{\rm arc} \subset \mathfrak{B}_{\rm arc}$. Applying the local-trajectory method and the results of Section 7, in Section 8 we obtain the invertibility criterion for the operators $B_{\rm arc} \in \mathfrak{B}_{\rm arc}$. In Section 9 we elaborate the invertibility criterion for the operators $B^\circ_\Lambda \in \mathfrak{B}^\circ_\Lambda$. Section 10 is devoted to studying the invertibility of functional operators with piecewise quasicontinuous coefficients and shifts in the group G. Finally, in Section 11 we prove that the invertibility of the operators $B_{\rm arc} \in \mathfrak{B}_{\rm arc}$ implies the invertibility of the operators $B^\diamond_\Lambda \in \mathfrak{B}^\diamond_\Lambda$. Combining the results of Sections 8–11, we obtain the Fredholm criterion for the operators $B \in \mathfrak{B}$ in terms of their Fredholm symbols (see Section 3).

2. The C^*-algebra $PQC(\mathbb{T})$

Let $L^\infty(\mathbb{T})$ be the C^*-algebra of all bounded measurable functions on the unit circle $\mathbb{T} := \{z \in \mathbb{C} : |z| = 1\}$. Let $C(\mathbb{T})$ and $PC(\mathbb{T})$ denote the C^*-subalgebras of $L^\infty(\mathbb{T})$ consisting, respectively, of all continuous functions on $\mathbb{T}$ and all piecewise continuous functions on $\mathbb{T}$, that is, the functions having finite one-sided limits at each point $t \in \mathbb{T}$.

For a Lebesgue measurable set $E \subset \mathbb{T}$, we will denote by $|E|$ its measure, $|E| := \int_E dm$. For each subarc I of $\mathbb{T}$ and each function $f \in L^1(\mathbb{T})$, the *average* of f over I is given by

$$I(f) := \frac{1}{|I|} \int_I f \, dm.$$

Taking $\delta > 0$ and considering arcs $I \subset \mathbb{T}$, we let

$$M_\delta(f) := \sup_{I \subset \mathbb{T},\, |I| \leq \delta} \frac{1}{|I|} \int_I |f(t) - I(f)| \, dm(t).$$

Obviously, $0 \leq M_{\delta_1}(f) \leq M_{\delta_2}(f)$ if $0 < \delta_1 \leq \delta_2 \leq 2\pi$. Let

$$M_0(f) := \lim_{\delta \to 0} M_\delta(f).$$

According to [25], a function $f \in L^1(\mathbb{T})$ is said to have *vanishing mean oscillation* on $\mathbb{T}$ if $M_0(f) = 0$. The set of functions of vanishing mean oscillation on $\mathbb{T}$ is denoted by $VMO(\mathbb{T})$.

Let $H^\infty(\mathbb{D})$ denote the Hardy space of bounded analytic functions on the unit disc $\mathbb{D} := \{z \in \mathbb{C} : |z| < 1\}$, and let H^∞ denote the set of all functions in $L^\infty(\mathbb{T})$ that are non-tangential limits on $\mathbb{T}$ of functions in $H^\infty(\mathbb{D})$. Equivalently, H^∞ consists of the functions $f \in L^\infty(\mathbb{T})$ for which the Fourier coefficients f_n defined for $n \in \mathbb{Z}$ by

$$f_n = \frac{1}{2\pi} \int_0^{2\pi} f(e^{i\theta}) e^{-in\theta} d\theta$$

equal zero for all $n < 0$.

The C^*-subalgebra of $L^\infty(\mathbb{T})$ defined by

$$QC(\mathbb{T}) := (C(\mathbb{T}) + H^\infty) \cap (C(\mathbb{T}) + \overline{H^\infty})$$

is referred to as the C^*-algebra of quasicontinuous functions. By [25],

$$QC(\mathbb{T}) = VMO(\mathbb{T}) \cap L^\infty(\mathbb{T}).$$

Let $PQC(\mathbb{T}) := \mathrm{alg}(QC(\mathbb{T}), PC(\mathbb{T}))$ be the C^*-subalgebra of $L^\infty(\mathbb{T})$ generated by the C^*-algebras $QC(\mathbb{T})$ and $PC(\mathbb{T})$. The functions in $PQC(\mathbb{T})$ are referred to as the piecewise quasicontinuous functions.

It is known that the maximal ideal space of $C(\mathbb{T})$ and $PC(\mathbb{T})$ can be identified, respectively, with $\mathbb{T}$ and $\mathbb{T} \times \{0, 1\}$,

$$M(C(\mathbb{T})) = \mathbb{T}, \quad M(PC(\mathbb{T})) = \mathbb{T} \times \{0, 1\},$$

where the points $t \in \mathbb{T}$ are identified with the evaluation functionals δ_t given for $f \in C(\mathbb{T})$ by $\delta_t(f) = f(t)$, and the pairs $(t, 0)$ and $(t, 1)$ are the multiplicative linear functionals defined for $a \in PC(\mathbb{T})$ by $(t, 0)a = a(t-0)$ and $(t, 1)a = a(t+0)$, where $a(t-0)$ and $a(t+0)$ are the left and right one-sided limits of a at the point $t \in \mathbb{T}$. The base of open sets on $\mathbb{T} \times \{0, 1\}$ consists of all sets of the form

$$(t, \tau) \times \{0, 1\}, \quad ((t, \tau] \times \{0\}) \cup ((t, \tau) \times \{1\}), \quad ((t, \tau) \times \{0\}) \cup ([t, \tau) \times \{1\}),$$

where $t, \tau \in \mathbb{T}$. Since $C(\mathbb{T}) \subset QC(\mathbb{T}) \subset PQC(\mathbb{T})$, it follows from [26] that

$$M(QC(\mathbb{T})) = \bigcup_{t\in\mathbb{T}} M_t(QC(\mathbb{T})), \quad M(PQC(\mathbb{T})) = \bigcup_{\xi\in M(QC(\mathbb{T}))} M_\xi(PQC(\mathbb{T})), \tag{2.1}$$

where the corresponding fibers are given for $t \in \mathbb{T}$ and $\xi \in M(QC(\mathbb{T}))$ by

$$M_t(QC(\mathbb{T})) = \{\xi \in M(QC(\mathbb{T})) : \xi|_{C(\mathbb{T})} = t\},$$
$$M_\xi(PQC(\mathbb{T})) = \{y \in M(PQC(\mathbb{T})) : y|_{QC(\mathbb{T})} = \xi\}.$$

For a function $f \in L^1(\mathbb{T})$ and points $t = e^{i\theta} \in \mathbb{T}$ and $\lambda \in (1, \infty)$, put

$$\delta_{(\lambda,t)}(f) = \frac{\lambda}{2\pi}\int_{\theta-\frac{\pi}{\lambda}}^{\theta+\frac{\pi}{\lambda}} f(e^{ix})dx.$$

Observe that, for each $(\lambda, t) \in (1, \infty) \times \mathbb{T}$, the map

$$\delta_{(\lambda,t)} : QC(\mathbb{T}) \to \mathbb{C}, \quad a \mapsto \delta_{(\lambda,t)}(a),$$

defines a linear functional in $QC(\mathbb{T})^*$ which allows one to identify $(\lambda, t) \in (1, \infty) \times \mathbb{T}$ with a subset of $QC(\mathbb{T})^*$. Let $M_t^0(QC(\mathbb{T}))$ denote the set of functionals in $M_t(QC(\mathbb{T}))$ that lie in the weak-star closure of $(1, \infty) \times \{t\}$,

$$M_t^0(QC(\mathbb{T})) := M_t(QC(\mathbb{T})) \cap \mathrm{clos}_{QC(\mathbb{T})^*}((1, \infty) \times \{t\}).$$

For each $t \in \mathbb{T}$, we also consider the sets

$$M_t^+(QC(\mathbb{T})) := \{\xi \in M_t(QC(\mathbb{T})) : \xi(f) = 0 \text{ whenever } f \in QC \text{ and } \limsup_{z\to t^+}|f(z)| = 0\},$$
$$M_t^-(QC(\mathbb{T})) := \{\xi \in M_t(QC(\mathbb{T})) : \xi(f) = 0 \text{ whenever } f \in QC \text{ and } \limsup_{z\to t^-}|f(z)| = 0\}.$$

According to [26, Lemma 8], we have the following.

Lemma 2.1. *For each $t \in \mathbb{T}$,*

$$M_t^+(QC(\mathbb{T})) \cap M_t^-(QC(\mathbb{T})) = M_t^0(QC(\mathbb{T})),$$
$$M_t^+(QC(\mathbb{T})) \cup M_t^-(QC(\mathbb{T})) = M_t(QC(\mathbb{T})).$$

By Lemma 2.1, for each $t \in \mathbb{T}$, we can split the fiber $M_t(QC(\mathbb{T}))$ into the three disjoint sets: $M_t^0(QC(\mathbb{T}))$ and

$$\widetilde{M}_t^+(QC(\mathbb{T})) := M_t^+(QC(\mathbb{T})) \setminus M_t^0(QC(\mathbb{T})),$$
$$\widetilde{M}_t^-(QC(\mathbb{T})) := M_t^-(QC(\mathbb{T})) \setminus M_t^0(QC(\mathbb{T})). \tag{2.2}$$

Hence, letting

$$M^\pm(QC(\mathbb{T})) := \bigcup_{t\in\mathbb{T}} M_t^\pm(QC(\mathbb{T})), \quad M^0(QC(\mathbb{T})) := \bigcup_{t\in\mathbb{T}} M_t^0(QC(\mathbb{T})),$$
$$\widetilde{M}^\pm(QC(\mathbb{T})) := \bigcup_{t\in\mathbb{T}} \widetilde{M}_t^\pm(QC(\mathbb{T})), \tag{2.3}$$

we obtain from (2.1) the following partition of $M(QC(\mathbb{T}))$:

$$M(QC(\mathbb{T})) = \widetilde{M}^-(QC(\mathbb{T})) \cup M^0(QC(\mathbb{T})) \cup \widetilde{M}^+(QC(\mathbb{T})). \tag{2.4}$$

There is a natural one-to-one mapping

$$w : M(PQC(\mathbb{T})) \to M(QC(\mathbb{T})) \times \{0,1\}$$

into a subset of $M(QC(\mathbb{T})) \times \{0,1\}$, which is given as follows: for every $y \in M(PQC(\mathbb{T}))$, defining $\xi = y|_{QC(\mathbb{T})}$, $t = y|_{C(\mathbb{T})}$ and $v = y|_{PC(\mathbb{T})}$, we conclude that $w(y) = (\xi, 0)$ if $v = (t, 0)$ and $w(y) = (\xi, 1)$ if $v = (t, 1)$.

We get the following characterization of fibers of $M(PQC(\mathbb{T}))$ from [26, Lemmas 8, 13].

Lemma 2.2. *Let $t \in \mathbb{T}$ and $\xi \in M_t(QC(\mathbb{T}))$. Then*

(i) $M_\xi(PQC(\mathbb{T})) = \{(\xi, 1)\}$ *whenever* $\xi \in \widetilde{M}_t^+(QC(\mathbb{T}))$;

(ii) $M_\xi(PQC(\mathbb{T})) = \{(\xi, 0)\}$ *whenever* $\xi \in \widetilde{M}_t^-(QC(\mathbb{T}))$;

(iii) $M_\xi(PQC(\mathbb{T})) = \{(\xi, 0), (\xi, 1)\}$ *whenever* $\xi \in M_t^0(QC(\mathbb{T}))$. *In this case, if* $\{\lambda_n\} \subset (1, \infty)$ *is such that* $(\lambda_n, t) \to \xi$ *in the weak-star topology on* $QC(\mathbb{T})^*$, *then*

$$(\xi, 1)f = \lim_{n\to\infty} \frac{\lambda_n}{\pi} \int_\theta^{\theta+\frac{\pi}{\lambda_n}} f(e^{ix})dx, \quad (\xi, 0)f = \lim_{n\to\infty} \frac{\lambda_n}{\pi} \int_{\theta-\frac{\pi}{\lambda_n}}^{\theta} f(e^{ix})dx$$

for every $f \in PQC(\mathbb{T})$, *where* $t = e^{i\theta}$.

The Gelfand topology on $M(PQC(\mathbb{T}))$ can be described as follows. If $\xi \in M_t(QC(\mathbb{T}))$ with $t \in \mathbb{T}$, then a base of neighborhoods for $(\xi, \mu) \in M(PQC(\mathbb{T}))$ consists of all open sets of the form

$$W_{(\xi,\mu)} = \begin{cases} \big[(U_{\xi,t} \times \{0\}) \cup (U_{\xi,t}^- \times \{0,1\})\big] \cap M(PQC(\mathbb{T})) & \text{if } \mu = 0, \\ \big[(U_{\xi,t} \times \{1\}) \cup (U_{\xi,t}^+ \times \{0,1\})\big] \cap M(PQC(\mathbb{T})) & \text{if } \mu = 1, \end{cases} \tag{2.5}$$

where $U_{\xi,t} = U_\xi \cap M_t(QC(\mathbb{T}))$, U_ξ is an open neighborhood of $\xi \in M(QC(\mathbb{T}))$, and $U_{\xi,t}^-$, $U_{\xi,t}^+$ consists of all $\zeta \in U_\xi$ such that $\tau = \zeta|_{C(\mathbb{T})}$ belong, respectively, to the open arcs $(te^{-i\varepsilon}, t)$ and $(t, te^{i\varepsilon})$ of $\mathbb{T}$ for some $\varepsilon \in (0, 2\pi)$.

In order to study the C^*-algebra $\mathfrak{B}$ defined in (1.2), we need to know how the operators of multiplication aI ($a \in PQC(\mathbb{T})$), the Cauchy singular integral operator $S_\mathbb{T}$ and the unitary shift operators U_g ($g \in G$) interact modulo compact operators in $\mathcal{B}(L^2(\mathbb{T}))$.

It follows from the Hartman theorem (see, e.g., [13, Theorem 2.18]) that

$$aS_\mathbb{T} \simeq S_\mathbb{T} aI \quad \text{for all } a \in QC(\mathbb{T}). \tag{2.6}$$

As all shifts $g \in G$ are orientation-preserving diffeomorphisms, we get

$$U_g S_\mathbb{T} \simeq S_\mathbb{T} U_g \quad \text{for all } g \in G,$$

which allows us to conclude that

$$U_g S_\mathbb{T} U_g^* \simeq S_\mathbb{T} \quad \text{for all } g \in G, \tag{2.7}$$

(see, e.g., [22, Theorem 4.1]). An easy computation shows that

$$U_g a U_g^* = (a \circ g)I \quad \text{for all} \quad a \in L^\infty(\mathbb{T}). \tag{2.8}$$

Obviously, for every $a \in L^\infty(\mathbb{T})$ the function $a \circ g$ is also in $L^\infty(\mathbb{T})$. It was observed in [14] that $a \circ g \in QC(\mathbb{T})$ whenever $a \in QC(\mathbb{T})$. Consequently, we have the following fact.

Lemma 2.3. *If g is a diffeomorphism of $\mathbb{T}$ onto itself, $a \in QC(\mathbb{T})$ and $b \in PQC(\mathbb{T})$, then $a \circ g \in QC(\mathbb{T})$ and $b \circ g \in PQC(\mathbb{T})$.*

As a consequence of (2.7)–(2.8) and Lemma 2.3, for every shift $g \in G$ the mapping

$$\alpha_g : A^\pi \mapsto U_g^\pi A^\pi (U_g^\pi)^{-1} \tag{2.9}$$

defines a *-automorphism of the C^*-algebra $\mathfrak{A}^\pi = \mathfrak{A}/\mathcal{K}$, where $\mathcal{K} = \mathcal{K}(L^2(\mathbb{T}))$ is the ideal of all compact operators on $L^2(\mathbb{T})$.

3. Main results: Fredholmness in the C^*-algebra $\mathfrak{B}$

Let G be a group of orientation-preserving diffeomorphisms of $\mathbb{T}$ onto itself such that all $g \in G \setminus \{e\}$ have the same finite set Λ of fixed points on $\mathbb{T}$. Consider the C^*-algebra

$$\mathfrak{B} = \text{alg}(PQC(\mathbb{T}), S_\mathbb{T}, U_G) \subset \mathcal{B}(L^2(\mathbb{T}))$$

generated by all multiplication operators by the piecewise quasicontinuous functions on $\mathbb{T}$, by the Cauchy singular integral operator $S_\mathbb{T}$ and by all shift operators U_g $(g \in G)$. Let

$$\mathfrak{N}_{\text{arc}} := \bigcup_{\tau \in \mathcal{O}_{\text{arc}}} M_\tau(QC(\mathbb{T})) \times \overline{\mathbb{R}}, \quad \mathfrak{N}_\Lambda := \bigcup_{\tau \in \Lambda} M_\tau^0(QC(\mathbb{T})) \times \mathbb{R}, \tag{3.1}$$

where $\mathcal{O}_{\text{arc}}$ is a subset of $\mathbb{T} \setminus \Lambda$ that contains exactly one point in each G-orbit $G(\tau) = \{g(\tau) : g \in G\}$ defined for $\tau \in \mathbb{T} \setminus \Lambda$ by the group of shifts G. Thus, $\mathbb{T} \setminus \Lambda = \bigcup_{\tau \in \mathcal{O}_{\text{arc}}} G(\tau)$. Along with $\mathfrak{N}_{\text{arc}}$, we define the sets

$$\mathfrak{N}_{\text{arc}}^\pm := \bigcup_{\tau \in \mathcal{O}_{\text{arc}}} \widetilde{M}_\tau^\pm(QC(\mathbb{T})) \times \overline{\mathbb{R}}, \quad \mathfrak{N}_{\text{arc}}^0 := \bigcup_{\tau \in \mathcal{O}_{\text{arc}}} M_\tau^0(QC(\mathbb{T})) \times \overline{\mathbb{R}},$$

where the sets $\widetilde{M}_\tau^\pm(QC(\mathbb{T}))$, $M_t^0(QC(\mathbb{T}))$ are given by (2.3) and Lemma 2.1.

For each $(\xi, x) \in \mathfrak{N}_{\text{arc}}$, we introduce the representation

$$\Phi_{\xi,x} : \mathfrak{B} \to \mathcal{B}(l^2(G, \mathbb{C}^2)), \quad B \mapsto \Phi_{\xi,x}(B), \tag{3.2}$$

given on the generators of the C^*-algebra $\mathfrak{B}$ for all $g \in G$ by

$$\begin{aligned}
[\Phi_{\xi,x}(aI)f](g) &= \begin{cases} \operatorname{diag}\{(a\circ g)(\xi,0),(a\circ g)(\xi,0)\}f(g) & \text{if } (\xi,x)\in\mathfrak{N}_{\rm arc}^-, \\ \operatorname{diag}\{(a\circ g)(\xi,1),(a\circ g)(\xi,0)\}f(g) & \text{if } (\xi,x)\in\mathfrak{N}_{\rm arc}^0, \\ \operatorname{diag}\{(a\circ g)(\xi,1),(a\circ g)(\xi,1)\}f(g) & \text{if } (\xi,x)\in\mathfrak{N}_{\rm arc}^+, \end{cases} \\
[\Phi_{\xi,x}(S_{\mathbb{T}})f](g) &= \begin{bmatrix} \tanh(\pi x) & 1/\cosh(\pi x) \\ 1/\cosh(\pi x) & -\tanh(\pi x) \end{bmatrix} f(g), \\
[\Phi_{\xi,x}(U_h)f](g) &= f(gh),
\end{aligned} \tag{3.3}$$

where $a \in PQC(\mathbb{T})$, $a(\xi,\mu)$ is the value of the Gelfand transform of a at the point $(\xi,\mu) \in M(PQC(\mathbb{T}))$, $h \in G$, and $f \in l^2(G,\mathbb{C}^2)$.

For each $(\xi,x) \in \mathfrak{N}_\Lambda$, we introduce the representation

$$\Phi_{\xi,x} : \mathfrak{B} \to \mathcal{B}(\mathbb{C}^2), \quad B \mapsto \Phi_{\xi,x}(B), \tag{3.4}$$

given on the generators of $\mathfrak{B}$ by

$$\begin{aligned}
[\Phi_{\xi,x}(aI)]f &= \operatorname{diag}\{a(\xi,1),a(\xi,0)\}f, \\
[\Phi_{\xi,x}(S_{\mathbb{T}})]f &= \begin{bmatrix} \tanh(\pi x) & 1/\cosh(\pi x) \\ 1/\cosh(\pi x) & -\tanh(\pi x) \end{bmatrix} f, \\
[\Phi_{\xi,x}(U_h)]f &= \operatorname{diag}\{e^{ix\ln h'(\xi)}, e^{ix\ln h'(\xi)}\}f,
\end{aligned} \tag{3.5}$$

where $a \in PQC(\mathbb{T})$, $a(\xi,\mu)$ is the value of the Gelfand transform of a at the point $(\xi,\mu) \in M(PQC(\mathbb{T}))$, $h \in G$, and $f \in \mathbb{C}^2$.

We establish below the following Fredholm criterion for the operators B in the C^*-algebra $\mathfrak{B}$.

Theorem 3.1. *An operator $B \in \mathfrak{B}$ is Fredholm on the space $L^2(\mathbb{T})$ if and only if the following two conditions are satisfied:*

(i) *for all $(\xi,x) \in \mathfrak{N}_{\rm arc}$ the operators $\Phi_{\xi,x}(B)$ are invertible on the space $l^2(G,\mathbb{C}^2)$ and*

$$\sup_{(\xi,x)\in\mathfrak{N}_{\rm arc}} \big\|(\Phi_{\xi,x}(B))^{-1}\big\| < \infty;$$

(ii) *for all $(\xi,x) \in \mathfrak{N}_\Lambda$ the operators $\Phi_{\xi,x}(B)$ are invertible on the space $\mathbb{C}^2$ and*

$$\inf_{(\xi,x)\in\mathfrak{N}_\Lambda} \big\|(\Phi_{\xi,x}(B))^{-1}\big\| < \infty$$

or, equivalently,

$$\inf_{(\xi,x)\in\mathfrak{N}_\Lambda} |\det(\Phi_{\xi,x}(B))| > 0,$$

where the operators $\Phi_{\xi,x}(B)$ are identified with their 2×2 matrices.

The Fredholm criterion presented in Theorem 3.1 does not depend on the concrete choice of $\mathfrak{N}_{\rm arc}$ because, for every operator $B \in \mathfrak{B}$, its representations associated with different points of the same orbit $G(\tau)$ are unitarily equivalent, and therefore it suffices to use representations related to only one point in each orbit. Moreover, one can easily see that assertion (i) of Theorem 3.1 remains true with $\mathfrak{N}_{\rm arc}$ replaced by

$$\mathfrak{N}_{\mathbb{T}\setminus\Lambda} := \bigcup_{\tau\in\mathbb{T}\setminus\Lambda} M_\tau(QC(\mathbb{T})) \times \overline{\mathbb{R}}.$$

Consider the Hilbert space

$$\mathcal{H} := \Big(\bigoplus_{(\xi,x)\in\mathfrak{N}_{\rm arc}} l^2(G,\mathbb{C}^2)\Big) \oplus \Big(\bigoplus_{(\xi,x)\in\mathfrak{N}_\Lambda} \mathbb{C}^2\Big) \tag{3.6}$$

and the operator function $\Phi(B) : (\xi, x) \mapsto \Phi_{\xi,x}(B)$ defined on $\mathfrak{N}_{\rm arc} \cup \mathfrak{N}_\Lambda$ by (3.2)–(3.5) and equipped with the norm

$$\|\Phi(B)\|_{\mathcal{B}(\mathcal{H})} = \max\Big\{ \sup_{(\xi,x)\in\mathfrak{N}_{\rm arc}} \big\|\Phi_{\xi,x}(B)\big\|_{\mathcal{B}(l^2(G,\mathbb{C}^2))}, \sup_{(\xi,x)\in\mathfrak{N}_\Lambda} \big\|\Phi_{\xi,x}(B)\big\|_{\mathcal{B}(\mathbb{C}^2)}\Big\}.$$

The operator function $\Phi(B)$ is referred to as the *Fredholm symbol* of an operator $B \in \mathfrak{B}$. Clearly, the set $\Phi(\mathfrak{B}) := \{\Phi(B) : B \in \mathfrak{B}\}$ is a C^*-algebra, and the mapping

$$\Phi : \mathfrak{B} \to \mathcal{B}(\mathcal{H}), \quad B \mapsto \Phi(B) \tag{3.7}$$

is a C^*-algebra homomorphism of the C^*-algebra $\mathfrak{B}$ onto the C^*-algebra $\Phi(\mathfrak{B})$ with kernel $\operatorname{Ker}\Phi = \mathcal{K}$. Hence, for the quotient C^*-algebra $\mathfrak{B}^\pi = \mathfrak{B}/\mathcal{K}$, it follows that $\mathfrak{B}^\pi \cong \Phi(\mathfrak{B})$, that is, the map $\mathfrak{B}^\pi \to \mathcal{B}(\mathcal{H})$ is a faithful representation in the Hilbert space (3.6). Making use of this symbol calculus, Theorem 3.1 can be rewritten in the following form.

Theorem 3.2. *An operator $B \in \mathfrak{B}$ is Fredholm on the space $L^2(\mathbb{T})$ if and only if its symbol $\Phi(B)$ is invertible.*

4. The local-trajectory method and spectral measures

In order to proceed to the study of the C^*-algebra $\mathfrak{B}$ of singular integral operators with $PQC(\mathbb{T})$ coefficients and shifts $g \in G$, we recall here the local-trajectory method and its generalization on the basis of spectral measures (cf. [18], [19], [5]).

Let $\mathcal{A}$ be a unital C^*-algebra, $\mathcal{Z}$ a central C^*-subalgebra of $\mathcal{A}$ with the same unit I, G a discrete group with unit e, $U : g \mapsto U_g$ a homomorphism of the group G onto a group $U_G = \{U_g : g \in G\}$ of unitary elements such that $U_{g_1g_2} = U_{g_1}U_{g_2}$ and $U_e = I$. Suppose that $\mathcal{A}$ and U_G are contained in a C^*-algebra $\mathcal{D}$. Let

$$\mathcal{B} := \operatorname{alg}(\mathcal{A}, U_G) \tag{4.1}$$

be the minimal C^*-subalgebra of $\mathcal{D}$ containing the C^*-algebra $\mathcal{A}$ and the group U_G. Assume that

(A1) *for every $g \in G$, the mappings $\alpha_g : a \mapsto U_g\, a\, U_g^*$ are *-automorphisms of the C^*-algebras $\mathcal{A}$ and $\mathcal{Z}$;*

(A2) *G is an amenable discrete group.*

According to [17, § 1.2], a discrete group G is called *amenable* if the C^*-algebra $l^\infty(G)$ of all bounded complex-valued functions on G with sup-norm has an invariant mean, that is, a positive linear functional ρ of norm 1 such that

$$\rho(f) = \rho({}_sf) = \rho(f_s) \quad \text{for all } s \in G \text{ and all } f \in l^\infty(G),$$

where $({}_sf)(g) = f(s^{-1}g)$, $(f_s)(g) = f(gs)$, $g \in G$.

By virtue of (A1), the C^*-algebra $\mathcal{B} = \text{alg}(\mathcal{A}, U_G)$ is the closure of the set $\mathcal{B}^0$ of elements $b = \sum a_g U_g$, where $a_g \in \mathcal{A}$ and g runs through finite subsets of G.

Let $M(\mathcal{Z})$ be the maximal ideal space of the commutative C^*-algebra $\mathcal{Z}$. By the Gelfand– Naimark theorem [23, § 16], $\mathcal{Z} \cong C(M(\mathcal{Z}))$, where $C(M(\mathcal{Z}))$ is the C^*-algebra of all continuous complex-valued functions on $M(\mathcal{Z})$. Under assumption (A1), identifying the characters φ_m of the commutative C^*-algebra $\mathcal{Z}$ and the maximal ideals $m = \text{Ker}\,\varphi_m \in M(\mathcal{Z})$, we see that each *-automorphism $\alpha_g : \mathcal{Z} \to \mathcal{Z}$ induces a homeomorphism $\beta_g : M(\mathcal{Z}) \to M(\mathcal{Z})$ given by the rule

$$z(\beta_g(m)) = (\alpha_g(z))(m), \quad z \in \mathcal{Z},\ m \in M(\mathcal{Z}),\ g \in G, \tag{4.2}$$

where $z(\cdot) \in C(M(\mathcal{Z}))$ is the Gelfand transform of the element $z \in \mathcal{Z}$. The set $G(m) := \{\beta_g(m) : g \in G\}$ is called the G-orbit of a point $m \in M(\mathcal{Z})$.

Let $\mathcal{P}_\mathcal{A}$ denote the set of all pure states of the C^*-algebra $\mathcal{A}$ with the induced weak* topology, and let J_m be the closed two-sided ideal of $\mathcal{A}$ generated by the maximal ideal $m \in M(\mathcal{Z})$ of the central C^*-algebra $\mathcal{Z} \subset \mathcal{A}$. If $\mu \in \mathcal{P}_\mathcal{A}$, then $\text{Ker}\,\mu \supset J_m$ where $m := \mathcal{Z} \cap \text{Ker}\,\mu \in M(\mathcal{Z})$ (see, e.g., [12, Lemma 4.1]), and hence $\mathcal{P}_\mathcal{A} = \bigcup_{m\in M}\{\nu \in \mathcal{P}_\mathcal{A} : \text{Ker}\,\nu \supset J_m\}$.

Let the following version of *topologically free* action of the group G hold:

(A3) *there is a set $M_0 \subset M(\mathcal{Z})$ such that for every finite set $G_0 \subset G \setminus \{e\}$ and every nonempty open set $W \subset \mathcal{P}_\mathcal{A}$ there exists a state $\nu \in W$ such that $\beta_g(m_\nu) \neq m_\nu$ for all $g \in G_0$, where $m_\nu = \mathcal{Z} \cap \text{Ker}\,\nu$ belongs to the G-orbit $G(M_0) := \{\beta_g(m) :\ g \in G,\ m \in M_0\}$ of the set M_0.*

For every $m \in M(\mathcal{Z})$, let $\widetilde{\pi}_m$ be an isometric representation

$$\widetilde{\pi}_m :\ \mathcal{A}/J_m \to \mathcal{B}(\mathcal{H}_m) \tag{4.3}$$

of the quotient C^*-algebra $\mathcal{A}/J_m$ in a Hilbert space $\mathcal{H}_m$. Consider the canonical *-homomorphism $\varrho_m : \mathcal{A} \to \mathcal{A}/J_m$ and the representation

$$\pi'_m :\ \mathcal{A} \to \mathcal{B}(\mathcal{H}_m),\quad A \mapsto (\widetilde{\pi}_m \circ \varrho_m)(A). \tag{4.4}$$

Since $\alpha_g(J_{\beta_g(m)}) = J_m$ for all $g \in G$ and all $m \in M(\mathcal{Z})$ in view of (A1), it follows that the quotient C^*-algebras $\mathcal{A}/J_{\beta_g(m)}$ and $\mathcal{A}/J_m$ are *-isomorphic, and therefore the spaces $\mathcal{H}_{\beta_g(m)}$ can be chosen equal for all $g \in G$.

Let $\Omega(M_0)$ be the set of G-orbits of all points $m \in M_0$. For each G-orbit $\omega \in \Omega(M_0)$, fix a point $m = m_\omega \in \omega$, put $\mathcal{H}_\omega = \mathcal{H}_m$, and let $l^2(G, \mathcal{H}_\omega)$ be the Hilbert space of all functions $f : G \mapsto \mathcal{H}_\omega$ such that $f(g) \neq 0$ for at most countable set of points $g \in G$ and $\|f\| := (\sum_g \|f(g)\|^2)^{1/2} < \infty$. For every $\omega \in \Omega(M_0)$ we consider the representation $\pi_\omega : \mathcal{B} \to \mathcal{B}(l^2(G, \mathcal{H}_\omega))$ defined by

$$[\pi_\omega(a)f](g) = \pi'_{m_\omega}(\alpha_g(a))f(g), \quad [\pi_\omega(U_h)f](g) = f(gh) \tag{4.5}$$

for all $a \in \mathfrak{A}$, all $g, h \in G$ and all $f \in l^2(G, \mathcal{H}_\omega)$.

Under the above three conditions we have the following result ([19, Theorems 4.1, 4.12], [5, Theorem 3.1]).

Theorem 4.1. *If assumptions* (A1)–(A3) *are satisfied, then an element* $b \in \mathcal{B}$ *is invertible in* $\mathcal{B}$ *if and only if for every orbit* $\omega \in \Omega(M_0)$ *the operator* $\pi_\omega(b)$ *is invertible on the space* $l^2(G, \mathcal{H}_\omega)$ *and, in the case of infinite set* $\Omega(M_0)$,

$$\sup\big\{\|(\pi_\omega(b))^{-1}\|_{\mathcal{B}(l^2(G,\mathcal{H}_\omega))} : \ \omega \in \Omega(M_0)\big\} < \infty.$$

Note that Theorem 4.1 remains true with $M_0 = M(\mathcal{Z})$, and then $\Omega(M_0) = \Omega(M(\mathcal{Z}))$ is the set of all G-orbits on $M(\mathcal{Z})$.

We will present now a generalization of the local-trajectory method for the case when condition (A3) is not fulfilled. Such generalization based on the notion of spectral measures was developed in [19] and [5].

Let M be a compact Hausdorff space and H a Hilbert space. By [23, p. 249], a *spectral measure* $P(\cdot)$ is a map from the σ-algebra of all Borel sets of M into the set of orthogonal projections in $\mathcal{B}(H)$ such that for every $\xi \in H$ the function $\Delta \to (P(\Delta)\xi, \xi)$ is the restriction to Borel sets of a measure on M defined by some integral on $C(M)$. Hence, $P(\emptyset) = 0$, $P(M) = I$, $P(\Delta_1 \cap \Delta_2) = P(\Delta_1)P(\Delta_2)$ for all Borel sets $\Delta_1, \Delta_2 \subset M$, and $P(\Delta_1 \cup \Delta_2) = P(\Delta_1) + P(\Delta_2)$ if $\Delta_1 \cap \Delta_2 = \emptyset$.

Suppose the C^*-algebra $\mathcal{B} = \text{alg}(\mathcal{A}, U_G)$ satisfies only the conditions (A1)–(A2) of the local-trajectory method. Let $\mathfrak{R}(M(\mathcal{Z}))$ denote the σ-algebra of all Borel subsets of $M(\mathcal{Z})$, and let

$$\mathfrak{R}_G(M(\mathcal{Z})) := \big\{\Delta \in \mathfrak{R}(M(\mathcal{Z})) : \ \beta_g(\Delta) = \Delta \ \text{ for all } \ g \in G\big\}, \tag{4.6}$$

where the homeomorphisms β_g are given by (4.2). As is well known (see, e.g., [15, Theorem 2.6.1]), there exists an isometric representation $\pi : \mathcal{B} \to \mathcal{B}(\mathcal{H})$ of the C^*-algebra $\mathcal{B}$ in a Hilbert space $\mathcal{H}$. According to [23, § 17], for the representation $\pi|_\mathcal{Z} : \mathcal{Z} \to \mathcal{B}(\mathcal{H})$ of the unital commutative C^*-algebra $\mathcal{Z}$, there is a unique spectral measure $P_\pi : \mathfrak{R}(M(\mathcal{Z})) \to \mathcal{B}(\mathcal{H})$ that commutes with all operators in the C^*-algebra $\pi(\mathcal{Z})$ and in its commutant $\pi(\mathcal{Z})'$, and such that

$$\pi(z) = \int_{M(\mathcal{Z})} z(m)\, dP_\pi(m) \quad \text{for all} \ \ z \in \mathcal{Z}, \tag{4.7}$$

where $z(\cdot) \in C(M(\mathcal{Z}))$ is the Gelfand transform of the element $z \in \mathcal{Z}$.

The integral in (4.7) is defined as follows. Since

$$\int_{M(\mathcal{Z})} z(m)\, dP_\pi(m) = \int_{M(\mathcal{Z})} \operatorname{Re} z(m)\, dP_\pi(m) + i \int_{M(\mathcal{Z})} \operatorname{Im} z(m)\, dP_\pi(m),$$

it is sufficient to define such integral for real-valued functions $f \in C(M(\mathcal{Z}))$. Any real-valued function $f \in C(M(\mathcal{Z}))$ is the uniform limit in $L^\infty(M(\mathcal{Z}))$ of a sequence $\{f_n\}_{n\in\mathbb{N}}$ of simple Borel real-valued functions $f_n = \sum_{k=1}^n c_{k,n}\chi_{\Delta_{k,n}}$, where $\Delta_{1,n}, \dots, \Delta_{n,n}$ are pairwise disjoint Borel subsets of $M(\mathcal{Z})$ for every $n \in \mathbb{N}$, $c_{k,n}$ are real constants that can be chosen equal to values of f at some points

$m_{k,n} \in \Delta_{k,n}$, and $\chi_{\Delta_{k,n}}$ are the characteristic functions of $\Delta_{k,n}$. Then

$$\int_{M(\mathcal{Z})} f(m)\, dP_\pi(m) = \lim_{n\to\infty} \sum_{k=1}^{n} f(m_{k,n}) P_\pi(\Delta_{k,n}) \in \mathcal{B}(\mathcal{H}),$$

where the convergence is uniform in $\mathcal{B}(\mathcal{H})$ because $\|f - f_n\|_{L^\infty(M(\mathcal{Z}))} \to 0$ as $n \to \infty$. Hence, for every $\xi \in \mathcal{H}$,

$$\Big(\Big(\int_{M(\mathcal{Z})} f(m)\, dP_\pi(m)\Big)\xi, \xi\Big) = \int_{M(\mathcal{Z})} f(m)\, d\mu_\xi(m),$$

where $\mu_\xi(\cdot) := (P_\pi(\cdot)\xi, \xi)$ is a Borel measure, and

$$\mu_\xi(\Delta) = (P_\pi(\Delta)\xi, \xi) = \int_{M(\mathcal{Z})} \chi_\Delta(m) d\mu_\xi(m) \quad \text{for all} \quad \Delta \in \mathfrak{R}(M(\mathcal{Z})).$$

Since (A1) holds and since $az = za$ for all $a \in \mathcal{A}$ and all $z \in \mathcal{Z}$, we deduce from [19, Lemma 4.6] and (4.6)–(4.7) that

$$\pi(b)P_\pi(\Delta) = P_\pi(\Delta)\pi(b) \quad \text{for all} \quad b \in \mathcal{B} \quad \text{and all} \quad \Delta \in \mathfrak{R}_G(M(\mathcal{Z})). \tag{4.8}$$

Given $\Delta \in \mathfrak{R}_G(M(\mathcal{Z}))$ such that $P_\pi(\Delta) \neq 0$, let us define the Hilbert space $\mathcal{H}_\Delta := P_\pi(\Delta)\mathcal{H} = \{P_\pi(\Delta)\xi : \xi \in \mathcal{H}\}$ and introduce the following three C^*-subalgebras of $\mathcal{B}(\mathcal{H}_\Delta)$:

$$\begin{aligned} \mathcal{B}_\Delta &:= \{P_\pi(\Delta)\pi(b) : b \in \mathcal{B}\}, \\ \mathcal{A}_\Delta &:= \{P_\pi(\Delta)\pi(a) : a \in \mathcal{A}\}, \\ \mathcal{Z}_\Delta &:= \{P_\pi(\Delta)\pi(z) : z \in \mathcal{Z}\}. \end{aligned} \tag{4.9}$$

Since $\mathcal{Z}$ is a central C^*-subalgebra of $\mathcal{A}$, it follows from (4.8) that $\mathcal{Z}_\Delta$ is a central C^*-subalgebra of $\mathcal{A}_\Delta$, where $\mathcal{A}_\Delta \subset \mathcal{B}_\Delta$.

For each Borel set $\Delta \in \mathfrak{R}(M(\mathcal{Z}))$, let $\operatorname{Int}\Delta$ and $\overline{\Delta}$ denote the interior and the closure of Δ, respectively, and let $\widetilde{\Delta}$ be the closed subset of $\overline{\Delta}$ given by

$$\widetilde{\Delta} := \{m \in M(\mathcal{Z}) \colon P_\pi(W_m \cap \Delta) \neq 0 \text{ for each open neighborhood } W_m \text{ of } m\}.$$

The next lemma summarizes some important properties of sets $\widetilde{\Delta}$.

Lemma 4.2. [19, Lemmas 5.1–5.2] *If* $\Delta \in \mathfrak{R}(M(\mathcal{Z}))$ *and* $\operatorname{Int}\Delta \neq 0$, *then*

(i) $P_\pi(\Delta) \neq 0$,
(ii) $\mathcal{Z}_\Delta \cong C(\widetilde{\Delta})$,
(iii) $\widetilde{\Delta} = \overline{\operatorname{Int}\Delta}$ *in case* $P_\pi(\Delta \setminus \overline{\operatorname{Int}\Delta}) = 0$.

Fix $\Delta \in \mathfrak{R}_G(M(\mathcal{Z}))$. For each $g \in G$, consider the unitary operator $U_{g,\Delta} := P_\pi(\Delta)\pi(U_g)$ on $\mathcal{H}_\Delta$. As condition (A1) holds, the mappings

$$\alpha_{g,\Delta} : P_\pi(\Delta)\pi(a) \mapsto U_{g,\Delta}P_\pi(\Delta)\pi(a)U_{g,\Delta}^* = P_\pi(\Delta)\pi(U_g a U_g^*) \quad (g \in G),$$

are *-automorphisms of the C^*-algebras $\mathcal{Z}_\Delta$ and $\mathcal{A}_\Delta$ defined by (4.9). Since $\mathcal{Z}_\Delta \cong C(\widetilde{\Delta})$, where $\widetilde{\Delta} \in \mathfrak{R}_G(M(\mathcal{Z}))$ and the isomorphism $\mathcal{Z}_\Delta \to C(\widetilde{\Delta})$ is given by $P_\pi(\Delta)\pi(z) \mapsto z(\cdot)|_{\widetilde{\Delta}}$, with restriction $z(\cdot)|_{\widetilde{\Delta}}$ of the Gelfand transform of $z \in \mathcal{Z}$, we conclude that each *-automorphism $\alpha_{g,\Delta}$ induces on $\widetilde{\Delta}$ the homeomorphism $\beta_{g,\Delta} := \beta_g|_{\widetilde{\Delta}}$, where β_g is defined by (4.2).

We will need below the following decomposition result.

Theorem 4.3. [6, Proposition 3.3] *Let $\pi : \mathcal{B} \to \mathcal{B}(\mathcal{H})$ be an isometric representation of the C^*-algebra $\mathcal{B} = \mathrm{alg}(\mathcal{A}, U_G)$ in a Hilbert space $\mathcal{H}$ and let $\{\Delta_i\}$ be an at most countable family of disjoint Borel sets in $\mathfrak{R}_G(M(\mathcal{Z}))$ such that $P_\pi(\Delta_i) \neq 0$ for all i and $P_\pi(M(\mathcal{Z}) \setminus \bigcup_i \Delta_i) = 0$. If condition* (A1) *holds, then the mapping*

$$\Theta : \mathcal{B} \to \bigoplus_i \mathcal{B}_{\Delta_i}, \quad b \mapsto \bigoplus_i P_\pi(\Delta_i)\pi(b),$$

is an isometric C^*-algebra homomorphism from the C^*-algebra $\mathcal{B}$ into the C^*-algebra $\widetilde{\mathcal{B}} := \bigoplus_i \mathcal{B}_{\Delta_i}$. Then an element $b \in \mathcal{B}$ is invertible if and only if for each i the operator $P_\pi(\Delta_i)\pi(b)$ is invertible on the Hilbert space $\mathcal{H}_{\Delta_i}$ and

$$\sup_i \left\|(P_\pi(\Delta_i)\pi(b))^{-1}\right\| < \infty \quad \textit{in case} \quad \{\Delta_i\} \quad \textit{is countable.}$$

It should be noted that, for the C^*-algebra $\mathcal{B} = \mathrm{alg}(\mathcal{A}, U_G)$, a G-invariant decomposition considered in Theorem 4.3 does not always exist. On the other hand, such decompositions always exist for the C^*-algebras $\mathcal{B} = \mathrm{alg}(\mathcal{A}, U_G)$ with groups G containing nonempty sets of common fixed points for all $g \in G$ (see Theorem 6.3 below).

5. The C^*-algebra $\mathfrak{A}$

Consider the C^*-algebra

$$\mathfrak{A} = \mathrm{alg}(PQC(\mathbb{T}), S_{\mathbb{T}}) \subset \mathcal{B}(L^2(\mathbb{T}))$$

of singular integral operators on $L^2(\mathbb{T})$ with $PQC(\mathbb{T})$ coefficients. With the C^*-algebra $\mathfrak{A}$ we associate the set

$$\mathfrak{M} := M(QC(\mathbb{T})) \times \overline{\mathbb{R}}, \tag{5.1}$$

where $\overline{\mathbb{R}} = [-\infty, +\infty]$ is the two-point compactification of the real line $\mathbb{R} = (-\infty, +\infty)$. Let $B(\mathfrak{M}, \mathbb{C}^{2\times 2})$ stand for the C^*-algebra of all bounded matrix functions $f : \mathfrak{M} \to \mathbb{C}^{2\times 2}$. Put

$$\mathfrak{M}^\pm := \widetilde{M}^\pm(QC(\mathbb{T})) \times \overline{\mathbb{R}}, \quad \mathfrak{M}^0 := M^0(QC(\mathbb{T})) \times \overline{\mathbb{R}}, \tag{5.2}$$

where the sets $\widetilde{M}^\pm(QC(\mathbb{T}))$ and $M^0(QC(\mathbb{T}))$ are defined by (2.3).

According to [14, Section 7.4] and [5, Theorem 5.1], we obtain the following symbol calculus for the C^*-algebra $\mathfrak{A}$.

Theorem 5.1. *The map* $\mathrm{Sym} : \{aI : a \in PQC(\mathbb{T})\} \cup \{S_{\mathbb{T}}\} \to B(\mathfrak{M}, \mathbb{C}^{2\times 2})$ *given by the matrix functions*

$$(\mathrm{Sym}\, aI)(\xi, x) := \begin{cases} \mathrm{diag}\{a(\xi,0), a(\xi,0)\} & \textit{for all } (\xi,x) \in \mathfrak{M}^-, \\ \mathrm{diag}\{a(\xi,1), a(\xi,0)\} & \textit{for all } (\xi,x) \in \mathfrak{M}^0, \\ \mathrm{diag}\{a(\xi,1), a(\xi,1)\} & \textit{for all } (\xi,x) \in \mathfrak{M}^+, \end{cases} \tag{5.3}$$

$$(\mathrm{Sym}\, S_{\mathbb{T}})(\xi, x) := \begin{bmatrix} u(x) & w(x) \\ w(x) & -u(x) \end{bmatrix} \quad \textit{for all } (\xi,x) \in \mathfrak{M},$$

where $a(\xi,\mu)$ *is the Gelfand transform of a function* $a \in PQC(\mathbb{T})$ *at the point* $(\xi,\mu) \in M(PQC(\mathbb{T}))$, *the sets* $\mathfrak{M}^{\pm}$ *and* $\mathfrak{M}^0$ *are given by* (5.2), *and*

$$u(x) := \tanh(\pi x), \quad w(x) := 1/\cosh(\pi x) \quad \textit{for all } x \in \overline{\mathbb{R}}, \tag{5.4}$$

extends to a C^**-algebra homomorphism*

$$\mathrm{Sym} : \mathfrak{A} \to B(\mathfrak{M}, \mathbb{C}^{2\times 2}) \tag{5.5}$$

whose kernel consists of all compact operators on $L^2(\mathbb{T})$. *An operator* $A \in \mathfrak{A}$ *is Fredholm on the space* $L^2(\mathbb{T})$ *if and only if*

$$\det((\mathrm{Sym}\, A)(\xi,x)) \neq 0 \quad \textit{for all } (\xi,x) \in \mathfrak{M}.$$

Proof. Let $M := \bigcup_{t\in\mathbb{T}} M_t$, where, with $\widetilde{M}_t^{\pm}(QC(\mathbb{T}))$ given by (2.2),

$$M_t := \widetilde{M}_t^-(QC(\mathbb{T})) \cup \big(M_t^0(QC(\mathbb{T})) \times [0,1]\big) \cup \widetilde{M}_t^+(QC(\mathbb{T})), \quad t \in \mathbb{T}.$$

For the C^*-algebra $\mathfrak{A}$, we know from [14, Section 7] that the mapping

$$\widetilde{\mathrm{Sym}} : \{aI : a \in PQC(\mathbb{T})\} \cup \{S_{\mathbb{T}}\} \to B(M, \mathbb{C}^{2\times 2}),$$

defined by the matrix functions

$$(\widetilde{\mathrm{Sym}}\, S_{\mathbb{T}})(\zeta) = \mathrm{diag}\{1, -1\} \quad \text{for all } \zeta \in M,$$

$$(\widetilde{\mathrm{Sym}}\, aI)(\zeta) = \begin{cases} \mathrm{diag}\{a(\xi,0), a(\xi,0)\} \quad \text{for all } \zeta = \xi \in \widetilde{M}^-(QC(\mathbb{T})), \\ \begin{bmatrix} a(\xi,1)\mu + a(\xi,0)(1-\mu) & [a(\xi,1) - a(\xi,0)]\varrho(\mu) \\ [a(\xi,1) - a(\xi,0)]\varrho(\mu) & a(\xi,1)(1-\mu) + a(\xi,0)\mu \end{bmatrix} \\ \qquad\qquad \text{for all } \zeta = (\xi,\mu) \in M^0(QC(\mathbb{T})) \times [0,1], \\ \mathrm{diag}\{a(\xi,1), a(\xi,1)\} \quad \text{for all } \zeta = \xi \in \widetilde{M}^+(QC(\mathbb{T})), \end{cases}$$

where $\varrho(\mu) = \sqrt{\mu(1-\mu)}$ for all $\mu \in [0,1]$ and the sets $\widetilde{M}^{\pm}(QC(\mathbb{T}))$ and $M^0(QC(\mathbb{T}))$ in the partition (2.4) of $M(QC(\mathbb{T}))$ are defined by (2.3), extends to a C^*-algebra homomorphism $\widetilde{\mathrm{Sym}} : \mathfrak{A} \to B(M, \mathbb{C}^{2\times 2})$ whose kernel consists of all compact operators in the C^*-algebra $\mathcal{B}(L^2(\mathbb{T}))$, and an operator $A \in \mathfrak{A}$ is Fredholm on space $L^2(\mathbb{T})$ if and only if

$$\det((\widetilde{\mathrm{Sym}}\, A)(\zeta)) \neq 0 \quad \text{for all } \zeta \in M.$$

Let $\mathcal{M} := M(QC(\mathbb{T})) \times [0,1]$ and consider the mapping

$$\mathrm{Sym}_{[0,1]} : \mathfrak{A} \to B(\mathcal{M}, \mathbb{C}^{2\times 2})$$

defined for all operator $A \in \mathfrak{A}$ by

$$(\mathrm{Sym}_{[0,1]}A)(\xi,\mu) := \begin{cases} (\widetilde{\mathrm{Sym}}\,A)(\xi,\mu) & \text{if } (\xi,\mu) \in M^0(QC(\mathbb{T})) \times [0,1], \\ (\widetilde{\mathrm{Sym}}\,A)(\xi) & \text{if } (\xi,\mu) \in \mathcal{M} \setminus (M^0(QC(\mathbb{T})) \times [0,1]). \end{cases}$$

Clearly, $\mathrm{Sym}_{[0,1]}$ is a C^*-algebra homomorphism with the same kernel as $\widetilde{\mathrm{Sym}}$, and such that an operator $A \in \mathfrak{A}$ is Fredholm on space $L^2(\mathbb{T})$ if and only if

$$\det((\mathrm{Sym}_{[0,1]}A)(\xi,\mu)) \neq 0 \quad \text{for all } (\xi,\mu) \in \mathcal{M}.$$

Since $\mu(x) = (1+u(x))/2 \in [0,1]$ for all $x \in \overline{\mathbb{R}}$, it remains to see that

$$(\mathrm{Sym}\,A)(\xi,x) =$$
$$\begin{bmatrix} \sqrt{\mu(x)} & \sqrt{1-\mu(x)} \\ \sqrt{1-\mu(x)} & -\sqrt{\mu(x)} \end{bmatrix} (\mathrm{Sym}_{[0,1]}A)(\xi,\mu(x)) \begin{bmatrix} \sqrt{\mu(x)} & \sqrt{1-\mu(x)} \\ \sqrt{1-\mu(x)} & -\sqrt{\mu(x)} \end{bmatrix}^{-1}$$

for all $(\xi,x) \in \mathfrak{M}$. □

By Theorem 5.1,

$$|A| = \|(\mathrm{Sym}\,A)I\|_{\mathcal{B}(l^2(\mathfrak{M},\mathbb{C}^2))} \quad \text{for all } A \in \mathfrak{A}.$$

For each point $t \in \mathbb{T}$, we consider the operator $V_t \in \mathcal{B}(L^2(\mathbb{T}))$, with fixed singularity at t, given by

$$(V_t\varphi)(z) := \frac{\chi_t^+(z)}{\pi i}\int_{\mathbb{T}} \frac{\varphi(y)\chi_t^+(y)}{y+z-2t}\,dy - \frac{\chi_t^-(z)}{\pi i}\int_{\mathbb{T}} \frac{\varphi(y)\chi_t^-(y)}{y+z-2t}\,dy \quad \text{for } z \in \mathbb{T}, \tag{5.6}$$

where $\chi_t^\pm$ are the characteristic functions of arcs $\gamma_t^\pm$ such that $\gamma_t := \gamma_t^+ \cup \gamma_t^-$ is a neighborhood of t, $\gamma_t^+ \cap \gamma_t^- = \{t\}$, γ_t is separated from $-t$, and $\gamma_t^+ \cap (-t,t) = \varnothing$, $\gamma_t^- \cap (t,-t) = \varnothing$. Let

$$v(x) := -iw(x) = -i/\cosh(\pi x) \quad \text{for all } x \in \overline{\mathbb{R}}. \tag{5.7}$$

Lemma 5.2. *For each $t \in \mathbb{T}$, the operator V_t given by* (5.6) *belongs to the C^*-algebra $\mathfrak{A}$ and its symbol is given by*

$$(\mathrm{Sym}V_t)(\xi,x) := \begin{cases} \mathrm{diag}\{v(x),v(x)\} & \text{if } (\xi,x) \in M_t^0(QC(\mathbb{T})) \times \overline{\mathbb{R}}, \\ 0_{2\times 2} & \text{if } (\xi,x) \in \mathfrak{M} \setminus (M_t^0(QC(\mathbb{T})) \times \overline{\mathbb{R}}), \end{cases} \tag{5.8}$$

where the function v and the set $\mathfrak{M}$ are defined by (5.7) *and* (5.1), *respectively.*

Proof. For every $t \in \mathbb{T}$, the operator V_t belongs to the C^*-algebra

$$\mathfrak{S} := \mathrm{alg}(PC(\mathbb{T}), S_{\mathbb{T}}) \subset \mathfrak{A}$$

(cf. [5, Lemma 5.3]). Fix a sequence of polynomials $P_n(u) = \sum_{k=0}^n a_{k,n}u^k$ with coefficients $a_{k,n} \in \mathbb{C}$ that uniformly converges on $[-1,1]$ to the function $\sqrt{u^2-1} \in$

$C([-1,1])$. Hence, because $S_{\mathbb{T}}^2 = I$ and therefore the spectrum of the operator $S_{\mathbb{T}}$ consists of the points ± 1, we conclude that

$$\lim_{n\to\infty} P_n(S_{\mathbb{T}}) = \sqrt{S_{\mathbb{T}}^2 - I} = 0. \tag{5.9}$$

By the proof of [5, Lemma 5.3],

$$(V_t)^\pi = \lim_{n\to\infty} \left(\chi_{t,n} P_n(\chi_t^+ S_{\mathbb{T}} \chi_t^+ I - \chi_t^- S_{\mathbb{T}} \chi_t^- I)\right)^\pi, \tag{5.10}$$

where $\chi_{t,n} \in PC(\mathbb{T})$ are characteristic functions of open neighborhoods of the point t, $\operatorname{supp}\chi_{t,n+1} \subset \operatorname{supp}\chi_{t,n}$ and $\bigcap_{n\in\mathbb{N}} \operatorname{supp}\chi_{t,n} = \{t\}$. Since

$$\left(\mathrm{Sym}(\chi_t^+ S_{\mathbb{T}} \chi_t^+ I - \chi_t^- S_{\mathbb{T}} \chi_t^- I)\right)(\xi, x) = \operatorname{diag}\{u(x), u(x)\} \tag{5.11}$$

for all $(\xi, x) \in M_t^0(SO(\mathbb{T})) \times \overline{\mathbb{R}}$, and since the sequence $\{P_n(u(x))\}$ converges uniformly on $\overline{\mathbb{R}}$ to the function $v(x) = \sqrt{u^2(x) - 1}$, where the functions u and v are given by (5.4) and (5.7), respectively, we conclude from (5.10) and (5.11) similarly to [5, Lemma 5.3] that

$$(\mathrm{Sym} V_t)(\xi, x) = \operatorname{diag}\{v(x), v(x)\} \quad \text{for all} \quad (\xi, x) \in M_t^0(QC(\mathbb{T})) \times \overline{\mathbb{R}}. \tag{5.12}$$

On the other hand, if $(\xi, x) \in \widetilde{M}_t^{\pm}(QC(\mathbb{T})) \times \overline{\mathbb{R}}$, then, respectively,

$$\left(\mathrm{Sym}(\chi_t^+ S_{\mathbb{T}} \chi_t^+ I - \chi_t^- S_{\mathbb{T}} \chi_t^- I)\right)(\xi, x) = \left(\mathrm{Sym}(\pm S_{\mathbb{T}})\right)(\xi, x).$$

Hence, taking into account (5.10), (5.3) and (5.9), we infer that

$$(\mathrm{Sym} V_t)(\xi, x) = \lim_{n\to\infty} \left(\mathrm{Sym}(\chi_{t,n} P_n(\pm S_{\mathbb{T}}))\right)(\xi, x) = 0 \tag{5.13}$$

for all $(\xi, x) \in \widetilde{M}_t^{\pm}(QC(\mathbb{T})) \times \overline{\mathbb{R}}$. Finally, since $\bigcap_{n\in\mathbb{N}} \operatorname{supp}\chi_{t,n} = \{t\}$, we obtain (5.8) from (5.12), (5.13) and (5.3). □

We now study the product $U_g V_t$ of the shift operator U_g given by (1.3) and the operator V_t with a fixed singularity at a fixed point $t \in \mathbb{T}$ of g.

Lemma 5.3. *Let g be an orientation-preserving diffeomorphism of $\mathbb{T}$ onto itself and let $t \in \mathbb{T}$ be a fixed point of g. Then $U_g V_t \in \mathfrak{A}$, $U_g V_t \simeq V_t U_g$ and*

$$(\mathrm{Sym}(U_g V_t))(\xi, x) := \begin{cases} \operatorname{diag}\{e^{ix \ln g'(t)} v(x), e^{ix \ln g'(t)} v(x)\} & \text{if } (\xi, x) \in M_t^0(QC(\mathbb{T})) \times \overline{\mathbb{R}}, \\ 0_{2\times 2} & \text{if } (\xi, x) \in \mathfrak{M} \setminus (M_t^0(QC(\mathbb{T})) \times \overline{\mathbb{R}}). \end{cases} \tag{5.14}$$

Proof. Let $t \in \mathbb{T}$ be a fixed point of a $g \in G$. By the proof of [5, Lemma 5.4], $U_g V_t \in \mathfrak{A}$, $U_g V_t \simeq V_t U_g$ and

$$(U_g V_t)^\pi = \lim_{n\to\infty} \left(\chi_{t,n} P_n(\chi_t^+ S_{\mathbb{T}} \chi_t^+ I - \chi_t^- S_{\mathbb{T}} \chi_t^- I)\right)^\pi, \tag{5.15}$$

where the sequence $\{P_n(u(x))\}$ of functions $P_n(u(x)) = \sum_{k=0}^n a_{k,n} u^k(x)$ with coefficients $a_{k,n} \in \mathbb{C}$ converges uniformly on $\overline{\mathbb{R}}$ to the continuous function $x \mapsto$

$e^{ix\ln g'(t)}v(x)$, $u(x)=\tanh(\pi x)$ and $v(x)=-i/\cosh(\pi x)$. Repeating now the arguments used in the proof of Lemma 5.2, we establish (5.14) on the basis of (5.15) and Theorem 5.1. □

We know from [5, (4.10)] that

$$aV_t \simeq V_t aI, \quad S_{\mathbb{T}}V_t \simeq V_t S_{\mathbb{T}} \quad \text{for all } a\in PC(\mathbb{T}) \text{ and all } t\in\mathbb{T}. \tag{5.16}$$

Since $V_t\in\mathfrak{S}=\operatorname{alg}(PC(\mathbb{T}),S_{\mathbb{T}})$, we infer from (2.6) that

$$aV_t \simeq V_t aI \quad \text{for all } a\in QC(\mathbb{T}) \text{ and all } t\in\mathbb{T}. \tag{5.17}$$

Hence, relations (5.16)–(5.17) imply

$$aV_t \simeq V_t aI \quad \text{for all } a\in PQC(\mathbb{T}) \text{ and all } t\in\mathbb{T}. \tag{5.18}$$

We now introduce the C^*-algebra

$$\mathcal{Z}:=\operatorname{alg}\big\{aI,\, H_{P,t}:\ a\in QC(\mathbb{T}),\ P\in\mathcal{P},\ t\in\mathbb{T}\big\}\subset\mathfrak{A}\subset\mathcal{B}(L^2(\mathbb{T})) \tag{5.19}$$

generated by all multiplication operators aI $(a\in QC(\mathbb{T}))$ and by all operators

$$H_{P,t}:=P(\chi_t^+S_{\mathbb{T}}\chi_t^+I-\chi_t^-S_{\mathbb{T}}\chi_t^-I)V_t\in\mathfrak{S}\subset\mathfrak{A} \quad (P\in\mathcal{P},\ t\in\mathbb{T}), \tag{5.20}$$

where

$$\mathcal{P}:=\Big\{P_n(u)=\sum_{k=0}^{n}a_k u^k:\ a_k\in\mathbb{C},\ n\in\mathbb{N}\Big\}. \tag{5.21}$$

By (5.16) and (5.18), we also obtain the relations

$$aH_{P,t}\simeq H_{P,t}aI, \quad S_{\mathbb{T}}H_{P,t}\simeq H_{P,t}S_{\mathbb{T}} \quad \text{for all } a\in PQC(\mathbb{T}),\ P\in\mathcal{P},\ t\in\mathbb{T}. \tag{5.22}$$

Moreover, due to (2.6) and (5.22), $\mathcal{Z}^\pi:=(\mathcal{Z}+\mathcal{K})/\mathcal{K}$ is a commutative central C^*-subalgebra of the C^*-algebra $\mathfrak{A}^\pi:=\mathfrak{A}/\mathcal{K}$.

Applying Lemma 5.2, we infer the following result by analogy with [5, Lemma 6.1].

Lemma 5.4. *For each $t\in\mathbb{T}$ and each polynomial $P\in\mathcal{P}$, the symbol of the operator $H_{P,t}\in\mathfrak{A}\subset\mathcal{B}(L^2(\mathbb{T}))$ is given by*

$$(\operatorname{Sym}H_{P,t})(\xi,x)$$
$$=\begin{cases}\operatorname{diag}\{P(u(x))v(x),P(u(x))v(x)\} & \text{if } (\xi,x)\in M_t^0(QC(\mathbb{T}))\times\overline{\mathbb{R}},\\ 0_{2\times2} & \text{if } (\xi,x)\in\mathfrak{M}\setminus(M_t^0(QC(\mathbb{T}))\times\overline{\mathbb{R}}),\end{cases}$$

where the functions u, v and the set $\mathfrak{M}$ are defined by (5.4), (5.7) *and* (5.1), *respectively.*

Let $\mathfrak{A}^0$ be the non-closed subalgebra of $\mathfrak{A}$ generated by the operators aI $(a\in PQC^0(\mathbb{T}))$ and $S_{\mathbb{T}}$, where $PQC^0(\mathbb{T})$ is the dense subalgebra of $PQC(\mathbb{T})$ consisting of the functions $\sum_{k=1}^n a_kp_k$, where $a_k\in QC(\mathbb{T})$ and $p_k\in PC^0(\mathbb{T})$ for $k=1,2,\dots,n$, and $PC^0(\mathbb{T})$ consists of all functions in $PC(\mathbb{T})$ with finite sets of discontinuities.

Theorem 5.5. *Every operator $A \in \mathfrak{A}$ is uniquely represented in the form*

$$A = a_+ P_{\mathbb{T}}^+ + a_- P_{\mathbb{T}}^- + H_A, \tag{5.23}$$

where $a_\pm \in PQC(\mathbb{T})$, $P_{\mathbb{T}}^\pm = (I \pm S_{\mathbb{T}})/2$ and $(\mathrm{Sym}\, H_A)(\xi, x) = 0_{2\times 2}$ for all $(\xi, x) \in \mathfrak{M} \setminus (M^0(QC(\mathbb{T})) \times \mathbb{R})$. Moreover, the mappings $A \mapsto a_\pm$ are C^*-algebra homomorphisms of the C^*-algebra $\mathfrak{A}$ onto the C^*-algebra $PQC(\mathbb{T})$, and

$$\|a_\pm\|_{L^\infty(\mathbb{T})} \le |A| := \inf_{K \in \mathcal{K}} \|A + K\|. \tag{5.24}$$

Proof. Obviously, every operator $A \in \mathfrak{A}^0$ is of the form (5.23), where $a_\pm \in PQC^0(\mathbb{T})$ and $(\mathrm{Sym}\, H_A)(\xi, \pm\infty) = 0_{2\times 2}$ for all $\xi \in M(QC(\mathbb{T}))$.

Moreover, $(\mathrm{Sym}\, H_A)(\xi, x) = 0_{2\times 2}$ for all $(\xi, x) \in \widetilde{M}^\pm(QC(\mathbb{T})) \times \overline{\mathbb{R}}$. Therefore, by (5.3) and (5.4), $(\mathrm{Sym}\, A)(\xi, \pm\infty)$ is a diagonal matrix for every $\xi \in M(QC(\mathbb{T}))$ and its diagonal entries $(\mathrm{Sym}\, A)_{11}(\xi, \pm\infty)$ and $(\mathrm{Sym}\, A)_{22}(\xi, \pm\infty)$ satisfy the relations

$$\begin{aligned}
(\mathrm{Sym}\, A)_{11}(\xi, +\infty) &= \begin{cases} a_+(\xi, 1) & \text{if } \xi \in M^+(QC(\mathbb{T})), \\ a_+(\xi, 0) & \text{if } \xi \in \widetilde{M}^-(QC(\mathbb{T})), \end{cases} \\
(\mathrm{Sym}\, A)_{22}(\xi, -\infty) &= \begin{cases} a_+(\xi, 1) & \text{if } \xi \in \widetilde{M}^+(QC(\mathbb{T})), \\ a_+(\xi, 0) & \text{if } \xi \in M^-(QC(\mathbb{T})), \end{cases} \\
(\mathrm{Sym}\, A)_{11}(\xi, -\infty) &= \begin{cases} a_-(\xi, 1) & \text{if } \xi \in M^+(QC(\mathbb{T})), \\ a_-(\xi, 0) & \text{if } \xi \in \widetilde{M}^-(QC(\mathbb{T})), \end{cases} \\
(\mathrm{Sym}\, A)_{22}(\xi, +\infty) &= \begin{cases} a_-(\xi, 1) & \text{if } \xi \in \widetilde{M}^+(QC(\mathbb{T})), \\ a_-(\xi, 0) & \text{if } \xi \in M^-(QC(\mathbb{T})). \end{cases}
\end{aligned} \tag{5.25}$$

Since $\mathfrak{A}^\pi \cong \mathrm{Sym}\, \mathfrak{A}$ according to Theorem 5.1, we deduce from (5.25) in view of topology (2.5) that

$$\begin{aligned}
\|a_\pm\|_{L^\infty(\mathbb{T})} &= \max\Big\{ \max_{\xi \in M^+(QC(\mathbb{T}))} |a_\pm(\xi, 1)|, \max_{\xi \in M^-(QC(\mathbb{T}))} |a_\pm(\xi, 0)| \Big\} \\
&\le \max_{\xi \in M(QC(\mathbb{T}))} \max\big\{ |(\mathrm{Sym}\, A)_{11}(\xi, \pm\infty)|, |(\mathrm{Sym}\, A)_{22}(\xi, \mp\infty)| \big\} \le |A|,
\end{aligned}$$

respectively. Consequently, the mappings $A \mapsto a_\pm$ extends by continuity to C^*-algebra homomorphisms of the C^*-algebra $\mathfrak{A}$ onto $PQC(\mathbb{T})$ that preserve (5.24). This implies the unique representation (5.23) for every $A \in \mathfrak{A}$. □

Let $\dot{\mathbb{R}} := \mathbb{R} \cup \{\infty\}$. Using reasonings similar to those in the proof of [5, Theorem 6.3], we easily infer from (5.19) that

$$\mathcal{Z}^\pi \cong C(\dot{\mathfrak{M}}), \quad \dot{\mathfrak{M}} := M(QC(\mathbb{T})) \times \dot{\mathbb{R}}, \tag{5.26}$$

where $\dot{\mathfrak{M}}$ is the compact Hausdorff space equipped with the Gelfand topology whose neighborhood base of a point $(\xi, x) \in \dot{\mathfrak{M}}$ consists of all open sets of the form

$$W_{\xi,x} = \begin{cases} U^0_{\xi,t} \times (x-\varepsilon, x+\varepsilon) & \text{if } (\xi,x) \in M^0(QC(\mathbb{T})) \times \mathbb{R}, \\ (U_\xi \times \dot{\mathbb{R}}) \setminus (U^0_{\xi,t} \times [-\varepsilon,\varepsilon]) & \text{if } (\xi,x) \in M^0(QC(\mathbb{T})) \times \{\infty\}, \\ (\widetilde{U}^\pm_{\xi,t} \cup U^\pm_{\xi,t}) \times \dot{\mathbb{R}} & \text{if } (\xi,x) \in \widetilde{M}^\pm(QC(\mathbb{T})) \times \dot{\mathbb{R}}, \end{cases} \tag{5.27}$$

where $\varepsilon > 0$, U_ξ is an open neighborhood of a point $\xi \in M(QC(\mathbb{T}))$ and, for $t = \xi|_{C(\mathbb{T})} \in \mathbb{T}$, $U^0_{\xi,t} := U_\xi \cap M^0_t(QC(\mathbb{T}))$, $\widetilde{U}^\pm_{\xi,t} := U_\xi \cap \widetilde{M}^\pm_t(QC(\mathbb{T}))$ and $U^-_{\xi,t}$, $U^+_{\xi,t}$ consists of all $\zeta \in U_\xi$ such that $\tau = \zeta|_{C(\mathbb{T})}$ belong, respectively, to the open arcs $(te^{-i\varepsilon}, t)$ and $(t, te^{i\varepsilon})$ of $\mathbb{T}$ for some $\varepsilon \in (0, 2\pi)$.

Theorem 5.1 and Lemma 5.4 imply the following.

Theorem 5.6. *The maximal ideal space $M(\mathcal{Z}^\pi)$ of the C^*-algebra $\mathcal{Z}^\pi$ is homeomorphic to the compact Hausdorff set $\dot{\mathfrak{M}}$ given by (5.26), and the Gelfand transform of $\mathcal{Z}^\pi$ is defined by*

$$\Gamma: \ \mathcal{Z}^\pi \to C(\dot{\mathfrak{M}}), \quad Z^\pi \mapsto z(\xi, x),$$

where

$$z(\xi,x) = \begin{cases} (\operatorname{Sym} Z)_{11}(\xi,x) & \textit{if} \quad (\xi,x) \in M(QC(\mathbb{T})) \times \mathbb{R}, \\ (\operatorname{Sym} Z)_{11}(\xi,\pm x) & \textit{if} \quad (\xi,x) \in M(QC(\mathbb{T})) \times \{\infty\}, \end{cases}$$

and $(\operatorname{Sym} Z)(\xi,x)$ *given by* (5.3)–(5.5) *for each operator* $Z \in \mathcal{Z}$ *has the form*

$$(\operatorname{Sym} Z)(\xi,x) = \operatorname{diag}\{(\operatorname{Sym} Z)_{11}(\xi,x), (\operatorname{Sym} Z)_{22}(\xi,x)\} \quad \textit{for all } (\xi,x) \in \dot{\mathfrak{M}}$$

and possesses the properties

$$(\operatorname{Sym} Z)_{11}(\xi,x) = (\operatorname{Sym} Z)_{22}(\xi,x), \quad (\operatorname{Sym} Z)(\xi,+\infty) = (\operatorname{Sym} Z)(\xi,-\infty).$$

It follows from [21, (5.24)] that

$$U_g V_t \simeq V_{g^{-1}(t)} U_g \quad \text{for } \ t \in \mathbb{T}, \ g \in G. \tag{5.28}$$

Taking into account (5.20), (2.7)–(2.8) and Lemma 2.3, we infer that

$$U_g H_{P,t} \simeq H_{P,g^{-1}(t)} U_g \quad \text{for all } \ t \in \mathbb{T}, \ g \in G, \ P \in \mathcal{P}.$$

Hence, for every $g \in G$, the mapping $\alpha_g : A^\pi \mapsto U^\pi_g A^\pi (U^\pi_g)^{-1}$ introduced in (2.9) is also a *-automorphism of the central C^*-subalgebra $\mathcal{Z}^\pi$ of $\mathfrak{A}^\pi$ that induces on the set $M(\mathcal{Z}^\pi) = \dot{\mathfrak{M}}$ given by (5.26) the homeomorphism

$$\beta_g : \ \dot{\mathfrak{M}} \to \dot{\mathfrak{M}}, \quad (\xi,x) \mapsto (g(\xi), x), \tag{5.29}$$

where $\xi \mapsto g(\xi)$ is the homeomorphism on $M(QC(\mathbb{T}))$ given by

$$a(g(\xi)) = (a \circ g)(\xi) \quad \text{for all } \ a \in QC(\mathbb{T}) \ \text{ and all } \ \xi \in M(QC(\mathbb{T})) \tag{5.30}$$

(as usual $a(\xi) := \xi(a)$).

6. Representations of the C^*-algebra $\mathfrak{B}^\pi$

Consider the C^*-algebra $\mathfrak{B}$ given by (1.2) and fix an isometric representation

$$\varphi : \mathfrak{B}^\pi \to \mathcal{B}(\mathcal{H}_\varphi) \tag{6.1}$$

of the C^*-algebra $\mathfrak{B}^\pi := \mathfrak{B}/\mathcal{K}$ in an abstract Hilbert space $\mathcal{H}_\varphi$.

For each point $t \in \mathbb{T}$, we consider the following subsets of $\dot{\mathfrak{M}}$ and $\mathfrak{M}$ (see (5.26) and (5.1)):

$$\begin{gathered} \mathfrak{M}_t^\circ := M_t^0(QC(\mathbb{T})) \times \mathbb{R}, \quad \overline{\mathfrak{M}_t^\circ} := M_t^0(QC(\mathbb{T})) \times \overline{\mathbb{R}}, \\ \dot{\mathfrak{M}}_t^\infty := M_t(QC(\mathbb{T})) \times \{\infty\}, \quad \mathfrak{M}_t^\infty := M_t(QC(\mathbb{T})) \times \{\pm\infty\}, \\ \dot{\mathfrak{M}}_t := M_t(QC(\mathbb{T})) \times \dot{\mathbb{R}}, \quad \mathfrak{M}_t := M_t(QC(\mathbb{T})) \times \overline{\mathbb{R}}. \end{gathered} \tag{6.2}$$

Let $\Lambda \subset \mathbb{T}$ be the finite set of all fixed points for all shifts $g \in G \setminus \{e\}$. We also introduce the sets

$$\begin{gathered} \dot{\mathfrak{M}}_\Lambda^\infty := \bigcup_{t\in\Lambda} \dot{\mathfrak{M}}_t^\infty, \quad \mathfrak{M}_\Lambda^\infty := \bigcup_{t\in\Lambda} \mathfrak{M}_t^\infty, \quad \mathfrak{M}_\Lambda := \bigcup_{t\in\Lambda} \mathfrak{M}_t, \quad \mathfrak{M}_\Lambda^\circ := \bigcup_{t\in\Lambda} \mathfrak{M}_t^\circ, \\ \dot{\mathfrak{M}}_{\mathrm{arc}} := \bigcup_{t\in\mathbb{T}\setminus\Lambda} \dot{\mathfrak{M}}_t, \quad \mathfrak{M}_{\mathrm{arc}} := \bigcup_{t\in\mathbb{T}\setminus\Lambda} \mathfrak{M}_t, \quad \mathfrak{M}_\Lambda^\diamond := \bigcup_{t\in\Lambda} (\dot{\mathfrak{M}}_t \setminus \mathfrak{M}_t^\circ), \end{gathered} \tag{6.3}$$

where the sets $\dot{\mathfrak{M}}_{\mathrm{arc}}$ and $\mathfrak{M}_\Lambda^\circ$ are open in $\dot{\mathfrak{M}}$, while the sets $\dot{\mathfrak{M}}_\Lambda^\infty$ and $\mathfrak{M}_\Lambda^\diamond$ are closed in $\dot{\mathfrak{M}}$. As a result, we get the following partition of $\dot{\mathfrak{M}}$:

$$\dot{\mathfrak{M}} = \dot{\mathfrak{M}}_{\mathrm{arc}} \cup \mathfrak{M}_\Lambda^\circ \cup \mathfrak{M}_\Lambda^\diamond. \tag{6.4}$$

The following simple assertion contains properties of sets (6.2)–(6.3).

Lemma 6.1. *For each $g \in G$ and each $t \in \mathbb{T}$, the homeomorphism $\beta_g : \dot{\mathfrak{M}} \to \dot{\mathfrak{M}}$ defined by (5.29)–(5.30) sends the sets $\dot{\mathfrak{M}}_t$ and $\mathfrak{M}_t^\circ$ onto the sets $\dot{\mathfrak{M}}_{g(t)}$ and $\mathfrak{M}_{g(t)}^\circ$, respectively, whence the sets $\dot{\mathfrak{M}}_{\mathrm{arc}}$, $\mathfrak{M}_\Lambda^\circ$, $\mathfrak{M}_\Lambda^\diamond$ and $\dot{\mathfrak{M}}_\Lambda^\infty$ belong to the set*

$$\mathfrak{R}_G(\dot{\mathfrak{M}}) := \big\{\Delta \in \mathfrak{R}(\dot{\mathfrak{M}}) : \ \beta_g(\Delta) = \Delta \ \text{ for all } \ g \in G\big\}, \tag{6.5}$$

where $\mathfrak{R}(\dot{\mathfrak{M}})$ denotes the σ-algebra of all Borel subsets of $\dot{\mathfrak{M}} = M(\mathcal{Z}^\pi)$.

Let $\mathcal{H}_\phi$ denote the Hilbert space

$$\mathcal{H}_\phi := l^2(\mathfrak{M}_{\mathrm{arc}}, \mathbb{C}^2) \oplus l^2(M_\Lambda^0(QC(\mathbb{T})), L_2^2(\mathbb{R})), \tag{6.6}$$

where

$$M_\Lambda^0(QC(\mathbb{T})) := \bigcup_{t\in\Lambda} M_t^0(QC(\mathbb{T})), \tag{6.7}$$

Consider the C^*-subalgebra $\phi(\mathfrak{A}^\pi)$ of $\mathcal{B}(\mathcal{H}_\phi)$ consisting of the operators

$$\phi(A^\pi) := \Big(\bigoplus_{(\xi,x)\in\mathfrak{M}_{\mathrm{arc}}} (\mathrm{Sym}\,A)(\xi,x)I\Big) \oplus \Big(\bigoplus_{\xi\in M_\Lambda^0(QC(\mathbb{T}))} (\mathrm{Sym}\,A)(\xi,\cdot)I\Big), \tag{6.8}$$

where $A \in \mathfrak{A}$ and $(\mathrm{Sym}\,A)(\xi,\cdot)$ for every $\xi \in M_\Lambda^0(QC(\mathbb{T}))$ is the matrix function $x \mapsto (\mathrm{Sym}\,A)(\xi,x)$ defined for all $x \in \overline{\mathbb{R}}$.

Consider the C^*-algebras $(\mathrm{Sym}\,\mathfrak{A})|_{\mathfrak{M}_{\mathrm{arc}}}$ and $(\mathrm{Sym}\,\mathfrak{A})|_{\overline{\mathfrak{M}}_{\mathrm{arc}}}$ consisting, respectively, of the restrictions $(\mathrm{Sym}\,A)|_{\mathfrak{M}_{\mathrm{arc}}}$ and $(\mathrm{Sym}\,A)|_{\overline{\mathfrak{M}}_{\mathrm{arc}}}$ of $\mathrm{Sym}\,A$ for all operators $A \in \mathfrak{A}$ to the sets $\mathfrak{M}_{\mathrm{arc}}$ and

$$\overline{\mathfrak{M}}_{\mathrm{arc}} := \mathfrak{M} \setminus \mathfrak{M}^{\circ}_{\Lambda} = \mathfrak{M}_{\mathrm{arc}} \cup (\mathfrak{M}_{\Lambda} \setminus \mathfrak{M}^{\circ}_{\Lambda}), \tag{6.9}$$

where the sets $\mathfrak{M}_{\mathrm{arc}}$, $\mathfrak{M}_{\Lambda}$ and $\mathfrak{M}^{\circ}_{\Lambda}$ are given by (6.2)–(6.3).

Lemma 6.2. $(\mathrm{Sym}\,\mathfrak{A})|_{\mathfrak{M}_{\mathrm{arc}}} \cong (\mathrm{Sym}\,\mathfrak{A})|_{\overline{\mathfrak{M}}_{\mathrm{arc}}}$.

Proof. By (5.3) and (5.4), $(\mathrm{Sym}\,A)(\xi, \pm\infty)$ is a diagonal matrix for every $\xi \in M(QC(\mathbb{T}))$ and every $A \in \mathfrak{A}$, and its entries

$$(\mathrm{Sym}\,A)_{11}(\xi, \pm\infty) \quad \text{and} \quad (\mathrm{Sym}\,A)_{22}(\xi, \pm\infty)$$

for all $\xi \in M_t(QC(\mathbb{T}))$ with $t \in \Lambda$ are values of two $PQC(\mathbb{T})$ functions due to (5.25). Since the action of the group G on the maximal ideal space $M(QC(\mathbb{T}))$ is topologically free according to [11, Lemma 4.2], the mentioned entries for $\xi \in M_t(QC(\mathbb{T}))$ with $t \in \Lambda$ can be approximated, in view of (5.25) and the Gelfand topology (2.5) on $M(PQC(\mathbb{T}))$, by the corresponding entries of the matrices

$$(\mathrm{Sym}\,A)(\zeta, \pm\infty),$$

where $\zeta \in \bigcup_{\tau \in \mathbb{T}\setminus\Lambda} M_\tau(QC(\mathbb{T}))$ and, for $(\mathrm{Sym}\,A)_{11}(\xi, \pm\infty)$, τ belong to the right semi-neighborhood of t if $\xi \in M^+(QC(\mathbb{T}))$ and τ are in the left semi-neighborhood of t if $\xi \in \widetilde{M}^-(QC(\mathbb{T}))$, while for $(\mathrm{Sym}\,A)_{22}(\xi, \pm\infty)$, τ belong to the right semi-neighborhood of t if $\xi \in \widetilde{M}^+(QC(\mathbb{T}))$ and τ belong to the left semi-neighborhood of t if $\xi \in M^-(QC(\mathbb{T}))$. Hence, for every $A \in \mathfrak{A}$,

$$\sup_{(\xi,x)\in\mathfrak{M}_{\mathrm{arc}}} \|(\mathrm{Sym}\,A)(\xi, x)\|_{sp} = \sup_{(\xi,x)\in\mathfrak{M}_{\mathrm{arc}}\cup\mathfrak{M}^{\infty}_{\Lambda}} \|(\mathrm{Sym}\,A)(\xi, x)\|_{sp}, \tag{6.10}$$

where $\|\cdot\|_{sp}$ is the spectral norm, and the set $\mathfrak{M}^{\infty}_{\Lambda}$ is given by (6.2)–(6.3).

On the other hand, the singular values of the matrices $(\mathrm{Sym}\,A)(\xi, x)$ for all $(\xi, x) \in \mathfrak{M}^- \cup \mathfrak{M}^+$ coincide with the singular values

$$|(\mathrm{Sym}\,A)_{11}(\xi, \pm\infty)| \quad \text{and} \quad |(\mathrm{Sym}\,A)_{22}(\xi, \pm\infty)|$$

of the matrices $(\mathrm{Sym}\,A)(\xi, \pm\infty)$, respectively, where

$$(\mathrm{Sym}\,A)_{11}(\xi, \pm\infty) = (\mathrm{Sym}\,A)_{22}(\xi, \mp\infty)$$

and the sets $\mathfrak{M}^{\pm}$ are given by (5.2). Hence,

$$\sup_{(\xi,x)\in\mathfrak{M}^{\infty}_{\Lambda}} \|(\mathrm{Sym}\,A)(\xi, x)\|_{sp} = \sup_{(\xi,x)\in\mathfrak{M}_{\Lambda}\setminus\mathfrak{M}^{\circ}_{\Lambda}} \|(\mathrm{Sym}\,A)(\xi, x)\|_{sp}.$$

This together with (6.9) and (6.10) implies the equality

$$\sup_{(\xi,x)\in\mathfrak{M}_{\mathrm{arc}}} \|(\mathrm{Sym}\,A)(\xi, x)\|_{sp} = \sup_{(\xi,x)\in\overline{\mathfrak{M}}_{\mathrm{arc}}} \|(\mathrm{Sym}\,A)(\xi, x)\|_{sp},$$

which completes the proof. ∎

Since $\mathrm{Sym}\,\mathfrak{A} \cong (\mathrm{Sym}\,\mathfrak{A})|_{\mathfrak{M}_{\mathrm{arc}}\cup\mathfrak{M}_\Lambda^\circ}$ according to Lemma 6.2, we infer from (6.6)–(6.8) and Theorem 5.1 that the homomorphism

$$\phi : \mathfrak{A}^\pi \to \mathcal{B}(\mathcal{H}_\phi), \quad A^\pi \mapsto \phi(A^\pi), \tag{6.11}$$

is an isometric representation of $\mathfrak{A}^\pi$ in the Hilbert space $\mathcal{H}_\phi$. Let

$$P_\varphi : \mathfrak{R}(\dot{\mathfrak{M}}) \to \mathcal{B}(\mathcal{H}_\varphi), \quad P_\phi : \mathfrak{R}(\dot{\mathfrak{M}}) \to \mathcal{B}(\mathcal{H}_\phi) \tag{6.12}$$

be the unique spectral measures associated to the representations (6.1) and (6.11) in the Hilbert spaces $\mathcal{H}_\varphi$ and $\mathcal{H}_\phi$, respectively, which are restricted to the unital commutative C^*-algebra $\mathcal{Z}^\pi$.

For the representation (6.11) given by (6.6)–(6.8) we get the equalities

$$P_\phi(\dot{\mathfrak{M}}_{\mathrm{arc}}) = I_{\mathrm{arc}} \oplus O_\Lambda, \quad P_\phi(\dot{\mathfrak{M}}_\Lambda) = O_{\mathrm{arc}} \oplus I_\Lambda, \tag{6.13}$$

where O_{arc} and I_{arc} are, respectively, the zero and the identity operators on the Hilbert space $l^2(\mathfrak{M}_{\mathrm{arc}}, \mathbb{C}^2)$, while O_Λ and I_Λ are the zero and identity operators on Hilbert space $l^2(M_\Lambda^0(QC(\mathbb{T})), L_2^2(\mathbb{R}))$.

With partition (6.4) we associate the following C^*-subalgebras of $\varphi(\mathfrak{B}^\pi)$:

$$\mathfrak{B}_{\mathrm{arc}} := \mathrm{alg}\{P_\varphi(\dot{\mathfrak{M}}_{\mathrm{arc}})\varphi(A^\pi),\ P_\varphi(\dot{\mathfrak{M}}_{\mathrm{arc}})\varphi(U_g^\pi) : \ A \in \mathfrak{A},\ g \in G\}; \tag{6.14}$$

$$\mathfrak{B}_\Lambda^\circ := \mathrm{alg}\{P_\varphi(\mathfrak{M}_\Lambda^\circ)\varphi(A^\pi),\ P_\varphi(\mathfrak{M}_\Lambda^\circ)\varphi(U_g^\pi) : \ A \in \mathfrak{A},\ g \in G\}; \tag{6.15}$$

$$\mathfrak{B}_\Lambda^\diamond := \mathrm{alg}\{P_\varphi(\mathfrak{M}_\Lambda^\diamond)\varphi(A^\pi),\ P_\varphi(\mathfrak{M}_\Lambda^\diamond)\varphi(U_g^\pi) : \ A \in \mathfrak{A},\ g \in G\}. \tag{6.16}$$

Since the sets $\dot{\mathfrak{M}}_{\mathrm{arc}}$, $\mathfrak{M}_\Lambda^\circ$ and $\mathfrak{M}_\Lambda^\diamond$ belong to $\mathfrak{R}_G(\dot{\mathfrak{M}})$, we infer from Theorem 4.3 and partition (6.4) the following abstract Fredholm criterion.

Theorem 6.3. *An operator $B \in \mathfrak{B}$ is Fredholm on the Lebesgue space $L^2(\mathbb{T})$ if and only if*

(i) *the operator $B_{\mathrm{arc}} := P_\varphi(\dot{\mathfrak{M}}_{\mathrm{arc}})\varphi(B^\pi)$ is invertible on the Hilbert space*
$$P_\varphi(\dot{\mathfrak{M}}_{\mathrm{arc}})\mathcal{H}_\varphi;$$
(ii) *the operator $B_\Lambda^\circ := P_\varphi(\mathfrak{M}_\Lambda^\circ)\varphi(B^\pi)$ is invertible on the Hilbert space*
$$P_\varphi(\mathfrak{M}_\Lambda^\circ)\mathcal{H}_\varphi;$$
(iii) *the operator $B_\Lambda^\diamond := P_\varphi(\mathfrak{M}_\Lambda^\diamond)\varphi(B^\pi)$ is invertible on the Hilbert space*
$$P_\varphi(\mathfrak{M}_\Lambda^\diamond)\mathcal{H}_\varphi.$$

7. An invertibility symbol calculus for the C^*-algebra $\mathfrak{A}_{\mathrm{arc}}$

Let us study the invertibility in the C^*-subalgebra

$$\mathfrak{A}_{\mathrm{arc}} := P_\varphi(\dot{\mathfrak{M}}_{\mathrm{arc}})\varphi(\mathfrak{A}^\pi) \subset \mathcal{B}(P_\varphi(\dot{\mathfrak{M}}_{\mathrm{arc}})\mathcal{H}_\varphi) \tag{7.1}$$

consisting of the operators $P_\varphi(\dot{\mathfrak{M}}_{\mathrm{arc}})\varphi(A^\pi)$ for all $A \in \mathfrak{A}$. The C^*-algebra $\mathcal{Z}_{\mathrm{arc}} := P_\varphi(\dot{\mathfrak{M}}_{\mathrm{arc}})\varphi(\mathcal{Z}^\pi)$, which consists of the operators $P_\varphi(\dot{\mathfrak{M}}_{\mathrm{arc}})\varphi(Z^\pi)$ with $Z \in \mathcal{Z}$, is a central C^*-subalgebra of $\mathfrak{A}_{\mathrm{arc}}$. Since the set $\dot{\mathfrak{M}}_{\mathrm{arc}}$ given in (6.3) is open in the

Gelfand topology of $\mathfrak{M}$ (see (5.27)), we infer from Lemma 4.2 that $\mathcal{Z}_{\rm arc} \cong C(\widetilde{\mathfrak{M}}_{\rm arc})$, where $\widetilde{\mathfrak{M}}_{\rm arc}$ is the closure of $\dot{\mathfrak{M}}_{\rm arc}$ in $\mathfrak{M}$,

$$\widetilde{\mathfrak{M}}_{\rm arc} := \mathfrak{M} \setminus \mathfrak{M}_\Lambda^\circ = \dot{\mathfrak{M}}_{\rm arc} \cup \mathfrak{M}_\Lambda^\diamond. \tag{7.2}$$

We now consider the Hilbert space $\mathcal{H}_\phi$ given by (6.6) and its subspace $P_\phi(\dot{\mathfrak{M}}_{\rm arc})\mathcal{H}_\phi$, which is isometrically isomorphic to the Hilbert space $l^2(\mathfrak{M}_{\rm arc}, \mathbb{C}^2)$ by (6.13). Along with the C^*-algebra $\mathfrak{A}_{\rm arc}$, we consider the C^*-algebra

$$\widetilde{\mathfrak{A}}_{\rm arc} := P_\phi(\dot{\mathfrak{M}}_{\rm arc})\phi(\mathfrak{A}^\pi) \subset \mathcal{B}(l^2(\mathfrak{M}_{\rm arc}, \mathbb{C}^2)),$$

which consists of the operators $P_\phi(\dot{\mathfrak{M}}_{\rm arc})\phi(A^\pi)$ with $A \in \mathfrak{A}$. Comparing the images of spectral measures (6.12), we obtain the following.

Theorem 7.1. *The mapping given by*

$$P_\varphi(\dot{\mathfrak{M}}_{\rm arc})\varphi(A^\pi) \mapsto P_\phi(\dot{\mathfrak{M}}_{\rm arc})\phi(A^\pi) \quad \textit{for all} \;\; A \in \mathfrak{A} \tag{7.3}$$

is a C^*-algebra isomorphism of the C^*-algebra $\mathfrak{A}_{\rm arc}$ onto the C^*-algebra $\widetilde{\mathfrak{A}}_{\rm arc}$.

Proof. According to [5, Lemma 3.5], for the open Borel set $\dot{\mathfrak{M}}_{\rm arc} \subset \mathfrak{M}$ and each operator $A \in \mathfrak{A}$, we have the equalities

$$\|P_\varphi(\dot{\mathfrak{M}}_{\rm arc})\varphi(A^\pi)\|_{\mathcal{B}(\mathcal{H}_\varphi)} = \sup_{Z \in \mathcal{Z}(\dot{\mathfrak{M}}_{\rm arc})} \|\varphi(Z^\pi A^\pi)\|_{\mathcal{B}(\mathcal{H}_\varphi)}, \tag{7.4}$$

$$\|P_\phi(\dot{\mathfrak{M}}_{\rm arc})\phi(A^\pi)\|_{\mathcal{B}(\mathcal{H}_\phi)} = \sup_{Z \in \mathcal{Z}(\dot{\mathfrak{M}}_{\rm arc})} \|\phi(Z^\pi A^\pi)\|_{\mathcal{B}(\mathcal{H}_\phi)}, \tag{7.5}$$

where the set $\mathcal{Z}(\dot{\mathfrak{M}}_{\rm arc})$ consists of the operators $Z \in \mathcal{Z}$ for which the Gelfand transform of the coset Z^π is a function $z(\cdot,\cdot) \in C(\mathfrak{M})$ with values in $[0,1]$ and with support contained in $\widetilde{\mathfrak{M}}_{\rm arc}$. Since φ and ϕ are isometric representations of the C^*-algebra $\mathfrak{A}^\pi$, we conclude that the right parts of (7.4) and (7.5) are equal, and therefore

$$\|P_\varphi(\dot{\mathfrak{M}}_{\rm arc})\varphi(A^\pi)\|_{\mathcal{B}(\mathcal{H}_\varphi)} = \|P_\phi(\dot{\mathfrak{M}}_{\rm arc})\phi(A^\pi)\|_{\mathcal{B}(\mathcal{H}_\phi)} \quad \text{for all} \;\; A \in \mathfrak{A} \tag{7.6}$$

or, equivalently, for all $A \in \mathfrak{A}$,

$$\|P_\varphi(\dot{\mathfrak{M}}_{\rm arc})\varphi(A^\pi)\|_{\mathcal{B}(P_\varphi(\dot{\mathfrak{M}}_{\rm arc})\mathcal{H}_\varphi)} = \|P_\phi(\dot{\mathfrak{M}}_{\rm arc})\phi(A^\pi)\|_{\mathcal{B}(P_\phi(\dot{\mathfrak{M}}_{\rm arc})\mathcal{H}_\phi)}$$

because $\dot{\mathfrak{M}}_{\rm arc} \in \mathfrak{R}(\dot{\mathfrak{M}})$. This implies that the mapping (7.3) is a well-defined isometric *-isomorphism of the C^*-algebra $\mathfrak{A}_{\rm arc}$ onto the C^*-algebra $\widetilde{\mathfrak{A}}_{\rm arc}$. □

Combining Theorem 7.1 and Lemma 6.2, we obtain an invertibility symbol calculus for the abstract C^*-algebra $\mathfrak{A}_{\rm arc}$ given by (7.1).

Theorem 7.2. *The mapping*

$$\mathrm{Sym}_{\rm arc} : \mathfrak{A}_{\rm arc} \to B(\overline{\mathfrak{M}}_{\rm arc}, \mathbb{C}^{2\times 2}), \quad P_\varphi(\dot{\mathfrak{M}}_{\rm arc})\varphi(A^\pi) \mapsto (\mathrm{Sym}\, A)|_{\overline{\mathfrak{M}}_{\rm arc}}, \tag{7.7}$$

is an isometric C^*-algebra homomorphism.

For every $A \in \mathfrak{A}$ the operator $P_\varphi(\dot{\mathfrak{M}}_{\rm arc})\varphi(A^\pi)$ is invertible on the space $P_\varphi(\dot{\mathfrak{M}}_{\rm arc})\mathcal{H}_\varphi$ if and only if

$$\det((\mathrm{Sym}\,A)(\xi,x)) \neq 0 \quad \textit{for all} \quad (\xi,x) \in \overline{\mathfrak{M}}_{\rm arc}.$$

Proof. By (6.8) and (6.13), the C^*-algebra $\widetilde{\mathfrak{A}}_{\rm arc}$ is isometrically *-isomorphic to the C^*-algebra of all matrix functions $(\mathrm{Sym}\,\mathfrak{A})|_{\mathfrak{M}_{\rm arc}}$, which, in its turn, is isometrically *-isomorphic to the C^*-algebra $(\mathrm{Sym}\,\mathfrak{A})|_{\overline{\mathfrak{M}}_{\rm arc}}$ according to Lemma 6.2. Hence, the mapping

$$\widetilde{\mathfrak{A}}_{\rm arc} \to B(\overline{\mathfrak{M}}_{\rm arc}, \mathbb{C}^{2\times 2}), \quad P_\phi(\dot{\mathfrak{M}}_{\rm arc})\phi(A^\pi) \mapsto (\mathrm{Sym}\,A)|_{\overline{\mathfrak{M}}_{\rm arc}} \tag{7.8}$$

is an isometric C^*-algebra homomorphism. Since $\mathfrak{A}_{\rm arc} \cong \widetilde{\mathfrak{A}}_{\rm arc}$ in virtue of Theorem 7.1, we conclude from (7.8) that the map (7.7) taken in the form

$$P_\varphi(\dot{\mathfrak{M}}_{\rm arc})\varphi(A^\pi) \mapsto P_\phi(\dot{\mathfrak{M}}_{\rm arc})\phi(A^\pi) \mapsto (\mathrm{Sym}\,A)|_{\overline{\mathfrak{M}}_{\rm arc}}$$

is an isometric C^*-algebra homomorphism of $\mathfrak{A}_{\rm arc}$ into $B(\overline{\mathfrak{M}}_{\rm arc}, \mathbb{C}^{2\times 2})$, and its image coincides with $(\mathrm{Sym}\,\mathfrak{A})|_{\overline{\mathfrak{M}}_{\rm arc}}$. Then the invertibility criterion for elements in $\mathfrak{A}_{\rm arc}$ follows immediately. □

8. The C^*-algebra $\mathfrak{B}_{\rm arc}$

The local-trajectory method will allow us to establish in this section an invertibility criterion for the C^*-algebra $\mathfrak{B}_{\rm arc}$ given by (6.14). Observe that C^*-algebra $\mathfrak{B}_{\rm arc}$ can be viewed similarly to (4.1) as

$$\mathfrak{B}_{\rm arc} := \mathrm{alg}(\mathfrak{A}_{\rm arc}, (U_G)_{\rm arc}) \subset \mathcal{B}(P_\varphi(\dot{\mathfrak{M}}_{\rm arc})\mathcal{H}_\varphi),$$

the C^*-algebra generated by all operators $P_\varphi(\dot{\mathfrak{M}}_{\rm arc})\varphi(A^\pi) \in \mathfrak{A}_{\rm arc}$ and by all unitary operators $U_{g,\,{\rm arc}} = P_\varphi(\dot{\mathfrak{M}}_{\rm arc})\varphi(U_g^\pi)$, with $g \in G$, in the commutative (and therefore amenable) group $(U_G)_{\rm arc}$.

For each $g \in G$, the mapping

$$\alpha_{g,\,{\rm arc}} : P_\varphi(\dot{\mathfrak{M}}_{\rm arc})\varphi(\mathfrak{A}^\pi) \mapsto U_{g,\,{\rm arc}}(P_\varphi(\dot{\mathfrak{M}}_{\rm arc})\varphi(\mathfrak{A}^\pi))U_{g,\,{\rm arc}}^*$$

is a *-automorphism of the C^*-algebra $\mathfrak{A}_{\rm arc}$ and its central C^*-subalgebra $\mathcal{Z}_{\rm arc}$ because

$$U_{g,\,{\rm arc}}(P_\varphi(\dot{\mathfrak{M}}_{\rm arc})\varphi(\mathfrak{A}^\pi))U_{g,\,{\rm arc}}^* = P_\varphi(\dot{\mathfrak{M}}_{\rm arc})\varphi(U_g^\pi\mathfrak{A}^\pi(U_g^\pi)^*)$$

and mapping (2.9) is a *-automorphism of the C^*-algebras $\mathcal{Z}^\pi$ and $\mathfrak{A}^\pi$.

Thus, conditions (A1)–(A2) of the local-trajectory method are satisfied for the C^*-algebra $\mathfrak{B}_{\rm arc}$. For every $g \in G$, the *-automorphism $\alpha_{g,\,{\rm arc}}$ induces on the maximal ideal space $M(\mathcal{Z}_{\rm arc}) = \widetilde{\mathfrak{M}}_{\rm arc}$ of $\mathcal{Z}_{\rm arc}$ the homeomorphism

$$\beta_{g,\,{\rm arc}} : \widetilde{\dot{\mathfrak{M}}}_{\rm arc} \to \widetilde{\dot{\mathfrak{M}}}_{\rm arc}, \quad (\xi,x) \mapsto \beta_g(\xi,x), \tag{8.1}$$

where β_g and $\widetilde{\mathfrak{M}}_{\rm arc}$ are given by (5.29) and (7.2), respectively. It follows from (5.29)–(5.30) that for every $g \in G\setminus\{e\}$ the set of fixed points of $\beta_{g,\,\rm arc}$ is contained in the set $\dot{\mathfrak{M}}_\Lambda := \bigcup_{t\in\Lambda}\dot{\mathfrak{M}}_t$.

Let us verify the fulfillment of condition (A3) of the local-trajectory method for the C^*-algebra $\mathfrak{B}_{\rm arc}$. For each $(\xi,x)\in\widetilde{\mathfrak{M}}_{\rm arc}$, let $J_{(\xi,x)}$ be the smallest closed two-sided ideal of $\mathfrak{A}_{\rm arc}$ that contains the set

$$\{P_\varphi(\dot{\mathfrak{M}}_{\rm arc})\varphi(Z^\pi):\ Z\in\mathcal{Z},\ (\mathrm{Sym}\,Z)(\xi,x)=0_{2\times2}\}.$$

The set $\mathcal{P}_{\rm arc}$ of all pure states of the C^*-algebra $\mathfrak{A}_{\rm arc}$ can be written as

$$\mathcal{P}_{\rm arc}=\bigcup_{(\xi,x)\in\widetilde{\mathfrak{M}}_{\rm arc}}\mathcal{P}_{(\xi,x)},\quad \mathcal{P}_{(\xi,x)}:=\{\rho\in\mathcal{P}_{\rm arc}:\ \mathrm{Ker}\,\rho\supset J_{(\xi,x)}\}.$$

According to [15, Theorem 2.11.8(i)] (see also [19]), the set $\mathcal{P}_{(\xi,x)}$ can be identified with the set of all pure states of the quotient C^*-algebra $\mathfrak{A}_{\rm arc}/J_{(\xi,x)}$.

By Theorem 7.2, for every $(\xi,x)\in\mathfrak{M}^\circ_{\rm arc}:=\bigcup_{t\in\mathbb{T}\setminus\Lambda}\mathfrak{M}^\circ_t$, the quotient C^*-algebra $\mathfrak{A}_{\rm arc}/J_{(\xi,x)}$ is isometrically *-isomorphic to the C^*-subalgebra $(\mathrm{Sym}\,\mathfrak{A})(\xi,x)$ of $\mathbb{C}^{2\times2}$, and this isomorphism is given by

$$\widetilde{\pi}_{(\xi,x)}:P_\varphi(\dot{\mathfrak{M}}_{\rm arc})\varphi(A^\pi)+J_{(\xi,x)}\mapsto(\mathrm{Sym}\,A)(\xi,x)\quad\text{for all }\ A\in\mathfrak{A}.\tag{8.2}$$

For each $(\xi,x)\in\widetilde{\mathfrak{M}}_{\rm arc}\setminus\mathfrak{M}^\circ_{\rm arc}$, we infer from Theorems 5.1, 7.2 and the proof of Lemma 6.2 that the C^*-algebra $\mathfrak{A}_{\rm arc}/J_{(\xi,x)}$ is isometrically *-isomorphic to the C^*-subalgebra $\mathrm{diag}\{(\mathrm{Sym}\,\mathfrak{A})(\xi,+\infty),\ (\mathrm{Sym}\,\mathfrak{A})(\xi,-\infty)\}$ of $\mathbb{C}^{4\times4}$ if $(\xi,x)\in\mathfrak{M}^{0,\infty}:=M^0(QC(\mathbb{T}))\times\{\infty\}$, and to the C^*-algebra $(\mathrm{Sym}\,\mathfrak{A})(\xi,+\infty)\subset\mathbb{C}^{2\times2}$ if $(\xi,x)\in\dot{\mathfrak{M}}^\pm:=\widetilde{M}^\pm(QC(\mathbb{T}))\times\dot{\mathbb{R}}$, where the isomorphism is defined for all $A\in\mathfrak{A}$ by

$$\widetilde{\pi}_{(\xi,x)}:P_\varphi(\dot{\mathfrak{M}}_{\rm arc})\varphi(A^\pi)+J_{(\xi,x)}\mapsto\begin{cases}\mathrm{diag}\{(\mathrm{Sym}\,A)(\xi,+\infty),\ (\mathrm{Sym}\,A)(\xi,-\infty)\}\\ \qquad\qquad\qquad\qquad\text{if }\ (\xi,x)\in\mathfrak{M}^{0,\infty},\\ (\mathrm{Sym}\,A)(\xi,+\infty)\quad\text{if }\ (\xi,x)\in\dot{\mathfrak{M}}^\pm.\end{cases}\tag{8.3}$$

Since the matrices $(\mathrm{Sym}\,A)(\xi,\pm\infty)$ are diagonal for all $A\in\mathfrak{A}$ and all $\xi\in M(QC(\mathbb{T}))$, we conclude from (8.3) that the C^*-algebras $\mathfrak{A}_{\rm arc}/J_{(\xi,x)}$ are commutative, and therefore for every $(\xi,x)\in\widetilde{\mathfrak{M}}_{\rm arc}\setminus\mathfrak{M}^\circ_{\rm arc}$ the set $\mathcal{P}_{(\xi,x)}=\mathcal{P}_{\mathfrak{A}_{\rm arc}/J_{(\xi,x)}}$ of pure states of the C^*-algebra $\mathfrak{A}_{\rm arc}/J_{(\xi,x)}$ consists of four multiplicative linear functionals whose values coincide with the diagonal entries of matrices $(\mathrm{Sym}\,A)(\xi,\pm\infty)$ if $(\xi,x)\in\mathfrak{M}^{0,\infty}$, and consists of two multiplicative linear functionals whose values coincide with the diagonal entries of the matrix $(\mathrm{Sym}\,A)(\xi,+\infty)$ if $(\xi,x)\in\dot{\mathfrak{M}}^\pm$.

Hence, for every $\xi\in M(QC(\mathbb{T}))$ and every $(\xi,x)\in\widetilde{\mathfrak{M}}_{\rm arc}\setminus\mathfrak{M}^\circ_{\rm arc}$,

$$\mathcal{P}_{(\xi,x)}=\mathcal{P}_{(\xi,\infty)}:=\{\rho^{(1)}_{\xi,+\infty},\rho^{(2)}_{\xi,+\infty},\rho^{(1)}_{\xi,-\infty},\rho^{(2)}_{\xi,-\infty}\},$$

where the pure states $\rho^{(j)}_{\xi,\pm\infty}$ for $j=1,2$ are given by

$$\rho^{(j)}_{\xi,\pm\infty}:\mathfrak{A}_{\rm arc}\to\mathbb{C},\quad P_\varphi(\dot{\mathfrak{M}}_{\rm arc})\varphi(A^\pi)\mapsto(\mathrm{Sym}\,A)_{jj}(\xi,\pm\infty),\tag{8.4}$$

$(\mathrm{Sym}\,A)_{jj}(\xi,\pm\infty)$ is the (j,j)-entry of the matrix $(\mathrm{Sym}\,A)(\xi,\pm\infty)$, and the set $\mathcal{P}_{(\xi,\infty)}$ consists of two elements for all $(\xi,x)\in\dot{\mathfrak{M}}^{\pm}$ because in that case $\rho^{(1)}_{\xi,+\infty}=\rho^{(2)}_{\xi,-\infty}$ and $\rho^{(2)}_{\xi,+\infty}=\rho^{(1)}_{\xi,-\infty}$.

Since $\beta_g(\xi,x)\neq(\xi,x)$ for all $(\xi,x)\in\dot{\mathfrak{M}}_{\mathrm{arc}}$, to verify the fulfillment of condition (A3) of the local-trajectory method for the C^*-algebra $\mathfrak{B}_{\mathrm{arc}}$, it only remains to prove the approximability (in the weak* topology) of the pure states $\rho\in\mathcal{P}_{(\xi,\infty)}$, with $\xi\in\dot{\mathfrak{M}}^{\infty}_{\Lambda}$, by pure states in $\mathcal{P}_{(\xi,x)}$, with $(\xi,x)\in\dot{\mathfrak{M}}_{\mathrm{arc}}$. In that case $M_0=\dot{\mathfrak{M}}_{\mathrm{arc}}$ in condition (A3).

Fix $\xi\in M_t(QC(\mathbb{T}))$ for $t\in\Lambda$. By (8.4) and the proof of Lemma 6.2, every open neighborhood of $\rho^{(1)}_{\xi,\pm\infty}$ and $\rho^{(2)}_{\xi,\pm\infty}$ in the weak* topology contains, respectively, pure states $\rho^{(1)}_{\zeta,\pm\infty}$ and $\rho^{(2)}_{\zeta,\pm\infty}$, where $\zeta\in\bigcup_{\tau\in\mathbb{T}\setminus\Lambda}M_\tau(QC(\mathbb{T}))$ and τ is on the right of t for $\rho^{(1)}_{\xi,\pm\infty}$ and $\rho^{(2)}_{\xi,\pm\infty}$ if, respectively, $\xi\in M_t^+(QC(\mathbb{T}))$ and $\xi\in\widetilde{M}_t^+(QC(\mathbb{T}))$, and τ is on the left of t for $\rho^{(1)}_{\xi,\pm\infty}$ and $\rho^{(2)}_{\xi,\pm\infty}$ if, respectively, $\xi\in\widetilde{M}_t^-(QC(\mathbb{T}))$ and $\xi\in M_t^-(QC(\mathbb{T}))$. Thus, condition (A3) is also fulfilled for the C^*-algebra $\mathfrak{B}_{\mathrm{arc}}$.

For each $(\xi,x)\in\mathfrak{M}_{\mathrm{arc}}$, we consider the representation

$$\pi_{(\xi,x)}:\mathfrak{B}_{\mathrm{arc}}\to\mathcal{B}(l^2(G,\mathbb{C}^2)) \tag{8.5}$$

given on the generators of the C^*-algebra $\mathfrak{B}_{\mathrm{arc}}$ by

$$\begin{aligned}\big[\pi_{(\xi,x)}\big(P_\varphi(\dot{\mathfrak{M}}_{\mathrm{arc}})\varphi((aI)^\pi)\big)f\big](g)&=\big[(\mathrm{Sym}\,((a\circ g)I))(\xi,x)\big]f(g),\\ \big[\pi_{(\xi,x)}\big(P_\varphi(\dot{\mathfrak{M}}_{\mathrm{arc}})\varphi(S^\pi_{\mathbb{T}})\big)f\big](g)&=\big[(\mathrm{Sym}\,S_{\mathbb{T}})(\xi,x)\big]f(g),\\ \big[\pi_{(\xi,x)}\big(P_\varphi(\dot{\mathfrak{M}}_{\mathrm{arc}})\varphi(U^\pi_h)\big)f\big](g)&=f(gh),\end{aligned} \tag{8.6}$$

where $a\in PQC(\mathbb{T})$, $g,h\in G$ and $f\in l^2(G,\mathbb{C}^2)$.

Fix a set $\mathcal{O}_{\mathrm{arc}}\subset\mathbb{T}\setminus\Lambda$ which contains exactly one point in each orbit defined by the group of shifts G on $\mathbb{T}\setminus\Lambda$, and consider the sets

$$\dot{\mathfrak{N}}_{\mathrm{arc}}:=\bigcup_{\tau\in\mathcal{O}_{\mathrm{arc}}}M_\tau(QC(\mathbb{T}))\times\dot{\mathbb{R}},\quad \mathfrak{N}_{\mathrm{arc}}=\bigcup_{\tau\in\mathcal{O}_{\mathrm{arc}}}M_\tau(QC(\mathbb{T}))\times\overline{\mathbb{R}}. \tag{8.7}$$

Theorem 8.1. *For each $B\in\mathfrak{B}$, the operator $B_{\mathrm{arc}}=P_\varphi(\dot{\mathfrak{M}}_{\mathrm{arc}})\varphi(B^\pi)\in\mathfrak{B}_{\mathrm{arc}}$ is invertible on the space $P_\varphi(\dot{\mathfrak{M}}_{\mathrm{arc}})\mathcal{H}_\varphi$ if and only if for all $(\xi,x)\in\mathfrak{N}_{\mathrm{arc}}$ the operators $\pi_{(\xi,x)}(B_{\mathrm{arc}})$ are invertible on the space $l^2(G,\mathbb{C}^2)$ and*

$$\sup_{(\xi,x)\in\mathfrak{N}_{\mathrm{arc}}}\big\|(\pi_{(\xi,x)}(B_{\mathrm{arc}}))^{-1}\big\|<\infty. \tag{8.8}$$

Proof. The set $\dot{\mathfrak{N}}_{\mathrm{arc}}$ given by (8.7) contains exactly one point in each G-orbit defined on the set $\dot{\mathfrak{M}}_{\mathrm{arc}}\subset\widetilde{\dot{\mathfrak{M}}}_{\mathrm{arc}}$ by the group $\{\beta_{g,\,\mathrm{arc}}:g\in G\}$ of homeomorphisms given by (8.1). Thus, for each $(\xi,x)\in\dot{\mathfrak{N}}_{\mathrm{arc}}$, starting with the representations (8.2)–(8.3) and following (4.3)–(4.5), we obtain the family of representations (8.5) defined by (8.6) and indexed by the points of the set $\mathfrak{N}_{\mathrm{arc}}$, where we take into account the fact that the Hilbert spaces $l^2(G,\mathbb{C}^4)$ and $l^2(G,\mathbb{C}^2)\oplus l^2(G,\mathbb{C}^2)$ are isometrically isomorphic. Since assumptions (A1)–(A3) are fulfilled for the C^*-algebra $\mathfrak{B}_{\rm arc}$, Theorem 4.1 implies the assertion of the theorem. □

For each $(\xi,x)\in\mathfrak{N}_{\rm arc}$, we now consider the representation

$$\mathfrak{B}\to\mathcal{B}(l^2(G,\mathbb{C}^2)),\quad B\mapsto\pi_{(\xi,x)}(B_{\rm arc}). \tag{8.9}$$

It is easily seen from (8.5)–(8.6) that representation (8.9) coincides with the representation $\Phi_{\xi,x}$ given by (3.2)–(3.3), that is,

$$\Phi_{\xi,x}(B)=\pi_{(\xi,x)}(B_{\rm arc})\quad\text{for all}\quad B\in\mathfrak{B}\quad\text{and all}\quad(\xi,x)\in\mathfrak{N}_{\rm arc}.$$

Consequently, Theorem 8.1 is equivalent to part (i) of Theorem 3.1.

9. The C^*-algebra $\mathfrak{B}_\Lambda^\circ$

In this section we establish an invertibility criterion for the operators in the C^*-algebra $\mathfrak{B}_\Lambda^\circ=P_\varphi(\mathfrak{M}_\Lambda^\circ)\varphi(\mathfrak{B}^\pi)$ represented in the form (6.15).

Theorem 9.1. *The mapping given by*

$$P_\varphi(\mathfrak{M}_\Lambda^\circ)\varphi(A^\pi)\mapsto P_\phi(\mathfrak{M}_\Lambda^\circ)\phi(A^\pi)\quad\textit{for all}\quad A\in\mathfrak{A} \tag{9.1}$$

is a C^*-algebra isomorphism of the C^*-algebra $\mathfrak{A}_\Lambda^\circ:=P_\varphi(\mathfrak{M}_\Lambda^\circ)\varphi(\mathfrak{A}^\pi)$ onto the C^*-algebra $\widetilde{\mathfrak{A}}_\Lambda^\circ:=P_\phi(\mathfrak{M}_\Lambda^\circ)\phi(\mathfrak{A}^\pi)$.

Proof. Since $\mathfrak{M}_\Lambda^\circ$ is an open subset of $\dot{\mathfrak{M}}$ and the C^*-algebras $\varphi(\mathfrak{A}^\pi)$ and $\phi(\mathfrak{A}^\pi)$ are isometrically *-isomorphic, applying [5, Lemma 3.5], we infer that

$$\|P_\varphi(\mathfrak{M}_\Lambda^\circ)\varphi(A^\pi)\|_{\mathcal{B}(\mathcal{H}_\varphi)}=\|P_\phi(\mathfrak{M}_\Lambda^\circ)\phi(A^\pi)\|_{\mathcal{B}(\mathcal{H}_\phi)}\quad\text{for all}\quad A\in\mathfrak{A} \tag{9.2}$$

by analogy with (7.6). This implies that (9.1) is a well-defined isometric *-isomorphism of the C^*-algebra $\mathfrak{A}_\Lambda^\circ$ onto the C^*-algebra $\widetilde{\mathfrak{A}}_\Lambda^\circ$. □

Let $\overline{\mathfrak{M}_\Lambda^\circ}:=\bigcup_{t\in\Lambda}\overline{\mathfrak{M}_t^\circ}$, where $\overline{\mathfrak{M}_t^\circ}$ is given by (6.2). Since the Hilbert space $P_\phi(\mathfrak{M}_\Lambda^\circ)\mathcal{H}_\phi$ is isometrically isomorphic to the Hilbert space

$$\mathcal{H}_\Lambda^\circ:=l^2(M_\Lambda^0(QC(\mathbb{T})),L_2^2(\mathbb{R})), \tag{9.3}$$

where $M_\Lambda^0(QC(\mathbb{T}))$ is given by (6.7), we immediately obtain from (6.8) and (6.13) the following.

Lemma 9.2. *The mapping*

$$\widetilde{\mathfrak{A}}_\Lambda^\circ\to\mathcal{B}(\mathcal{H}_\Lambda^\circ),\quad P_\phi(\mathfrak{M}_\Lambda^\circ)\phi(A^\pi)\mapsto\bigoplus_{\xi\in M_\Lambda^0(QC(\mathbb{T}))}(\operatorname{Sym}A)(\xi,\cdot)I$$

is an isometric C^*-algebra homomorphism. An operator $P_\phi(\mathfrak{M}_\Lambda^\circ)\phi(A^\pi)$ for $A\in\mathfrak{A}$ is invertible on the space $P_\phi(\mathfrak{M}_\Lambda^\circ)\mathcal{H}_\phi$ if and only if

$$\det((\operatorname{Sym}A)(\xi,x))\neq0\quad\textit{for all}\quad(\xi,x)\in\overline{\mathfrak{M}_\Lambda^\circ}. \tag{9.4}$$

Theorem 9.1 and Lemma 9.2 imply the following.

Theorem 9.3. *The map* $\mathrm{Sym}_\Lambda^\circ : \mathfrak{A}_\Lambda^\circ \to \mathcal{B}(\mathcal{H}_\Lambda^\circ)$*, defined by*

$$P_\varphi(\mathfrak{M}_\Lambda^\circ)\varphi(A^\pi) \mapsto \bigoplus_{\xi \in M_\Lambda^0(QC(\mathbb{T}))} (\mathrm{Sym}\, A)(\xi,\cdot)I \quad \textit{for} \quad A \in \mathfrak{A},$$

is an isometric C^**-algebra homomorphism. An operator* $P_\varphi(\mathfrak{M}_\Lambda^\circ)\varphi(A^\pi)$ *for* $A \in \mathfrak{A}$ *is invertible on the space* $P_\varphi(\mathfrak{M}_\Lambda^\circ)\mathcal{H}_\varphi$ *if and only if* (9.4) *holds.*

Given $t \in \Lambda$, let $\widetilde{\mathfrak{H}}_t^\pi$ be the closed two-sided ideal of the C^*-algebra $\mathcal{Z}^\pi$ generated by all the cosets $H_{P,t}^\pi = H_{P,t} + \mathcal{K}$ with $P \in \mathcal{P}$ (see (5.20) and (5.21)). Since $\mathcal{Z}^\pi$ is a commutative C^*-algebra, we conclude that

$$\widetilde{\mathfrak{H}}_t^\pi = \mathrm{clos}\Big\{ \sum_{j=1}^m Z_j^\pi H_{P_j,t}^\pi : Z_j \in \mathcal{Z},\ P_j \in \mathcal{P},\ m \in \mathbb{N} \Big\}, \tag{9.5}$$

where the closure is taken in the essential norm $\|\cdot\|$. Let $\widetilde{\mathfrak{H}}_\Lambda^\pi$ be the closed two-sided ideal of $\mathcal{Z}^\pi$ generated by all ideals $\widetilde{\mathfrak{H}}_t^\pi$ $(t \in \Lambda)$.

Following [5, Lemma 3.5], we define the set

$$\mathcal{Z}(\mathfrak{M}_\Lambda^\circ) := \Big\{ Z^\pi \in \mathcal{Z}^\pi : \mathrm{supp}\, z(\cdot,\cdot) \subset \dot{\mathfrak{M}}_\Lambda^\circ,\ z(\xi,x) \in [0,1] \text{ for all } (\xi,x) \in \dot{\mathfrak{M}} \Big\}, \tag{9.6}$$

where $z(\cdot,\cdot) \in C(\dot{\mathfrak{M}})$ is the Gelfand transform of the coset Z^π, $\mathrm{supp}\, z(\cdot,\cdot)$ is the support of $z(\cdot,\cdot)$, and $\dot{\mathfrak{M}}_\Lambda^\circ := \bigcup_{t\in\Lambda} M_t^0(QC(\mathbb{T})) \times \dot{\mathbb{R}}$ is the closure in $\dot{\mathfrak{M}}$ of the set $\mathfrak{M}_\Lambda^\circ$ given by (6.3).

Lemma 9.4. *For the ideal* $\widetilde{\mathfrak{H}}_\Lambda^\pi$*, the following assertions hold:*

(i) *for every* $B^\pi \in \mathfrak{B}^\pi$ *and every* $H_\Lambda^\pi \in \widetilde{\mathfrak{H}}_\Lambda^\pi$*,*

$$B^\pi H_\Lambda^\pi = H_\Lambda^\pi B^\pi \in \mathfrak{A}^\pi; \tag{9.7}$$

(ii) $\widetilde{\mathfrak{H}}_\Lambda^\pi = \mathcal{Z}(\mathfrak{M}_\Lambda^\circ)$.

Proof. (i) For $B^\pi \in \mathfrak{A}^\pi$, the property (9.7) follows from the fact that $\mathcal{Z}^\pi$ is a central subalgebra of $\mathfrak{A}^\pi$. Since every coset $H^\pi \in \widetilde{\mathfrak{H}}_\Lambda^\pi$ is of the form $H^\pi = \sum_{t\in\Lambda} H_t^\pi$, where $H_t^\pi \in \widetilde{\mathfrak{H}}_t^\pi$, we can see that relation (9.7) for $B^\pi = U_g^\pi$ follows from Lemma 5.3 and (5.20). Hence, part (i) is proved for every $B^\pi \in \mathfrak{B}^\pi$.

(ii) Fix $t \in \Lambda$. The symbols of operators $H_{P,t}$ $(P \in \mathcal{P})$ are equal to $0_{2\times2}$ for all $(\xi,x) \in \mathfrak{M} \setminus \mathfrak{M}_t^\circ$ (see Lemma 5.4). Hence, by Theorem 5.6, the Gelfand transform of the coset $H_{P,t}^\pi$ is a continuous function on $\dot{\mathfrak{M}}$ with support contained in the closure $\dot{\mathfrak{M}}_t^\circ = M_t^0(QC(\mathbb{T})) \times \dot{\mathbb{R}}$ of the set $\mathfrak{M}_t^\circ$ in $\dot{\mathfrak{M}}$. By (9.5), this property also holds for the Gelfand transform of any element H_t^π of the ideal $\widetilde{\mathfrak{H}}_t^\pi$. Thus, $\widetilde{\mathfrak{H}}_t^\pi \subset \mathcal{Z}(\mathfrak{M}_t^\circ)$, where

$$\mathcal{Z}(\mathfrak{M}_t^\circ) := \Big\{ Z^\pi \in \mathcal{Z}^\pi : \mathrm{supp}\, z(\cdot,\cdot) \subset \dot{\mathfrak{M}}_t^\circ,\ z(\xi,x) \in [0,1] \text{ for all } (\xi,x) \in \dot{\mathfrak{M}} \Big\}.$$

On the other hand, we infer from [5, Lemma 6.2] and (9.5) that any element $Z^\pi \in \mathcal{Z}(\mathfrak{M}_t^\circ)$ can be approximated in the C^*-algebra $\mathcal{Z}^\pi$ by cosets $H_t^\pi \in \widetilde{\mathfrak{H}}_t^\pi$, which implies that $\mathcal{Z}(\mathfrak{M}_t^\circ) \subset \widetilde{\mathfrak{H}}_t^\pi$. Moreover, Theorem 5.6 and [5, Lemma 6.2] imply that the ideal $\widetilde{\mathfrak{H}}_t^\pi$ is isomorphic to the ideal of all continuous functions on the compact $\dot{\mathfrak{M}}$ that vanish on $\mathfrak{M} \setminus \mathfrak{M}_t^\circ$. Hence, $\widetilde{\mathfrak{H}}_t^\pi = \mathcal{Z}(\mathfrak{M}_t^\circ)$, which together with (9.6) implies part (ii). □

Consider the Hilbert space $\mathcal{H}_\Lambda^\circ$ given by (9.3), with $M_\Lambda^0(QC(\mathbb{T}))$ given by (6.7), and introduce the C^*-algebra

$$\Psi(\mathfrak{B}_\Lambda^\circ) := \operatorname{alg}\{\Psi_\Lambda^\circ(A^\pi),\ \Psi_\Lambda^\circ(U_g^\pi) :\ A \in \mathfrak{A},\ g \in G\} \subset \mathcal{B}(\mathcal{H}_\Lambda^\circ) \tag{9.8}$$

generated by the operators $\Psi_\Lambda^\circ(A^\pi)$ $(A \in \mathfrak{A})$ and $\Psi_\Lambda^\circ(U_g^\pi)$ $(g \in G)$, where

$$\Psi_\Lambda^\circ(A^\pi) := \bigoplus_{\xi \in M_\Lambda^0(QC(\mathbb{T}))} (\operatorname{Sym} A)(\xi, \cdot)I, \quad \Psi_\Lambda^\circ(U_g^\pi) := \bigoplus_{\xi \in M_\Lambda^0(QC(\mathbb{T}))} e_{\ln g'(\xi)}(\cdot)I, \tag{9.9}$$

and $e_{\ln g'(\xi)}(x) := e^{ix \ln g'(\xi)}$ for all $(\xi, x) \in \mathfrak{M}_\Lambda^\circ$.

The mapping $g \mapsto \Psi_\Lambda^\circ(U_g^\pi)$ is a unitary representation of the group G in the Hilbert space $\mathcal{H}_\Lambda^\circ$, the adjoint operator $\Psi_\Lambda^\circ(U_g^\pi)^*$ equals $\Psi_\Lambda^\circ(U_{g^{-1}}^\pi)$, and

$$\Psi_\Lambda^\circ(U_g^\pi)\Psi_\Lambda^\circ(A^\pi)\Psi_\Lambda^\circ(U_g^\pi)^* = \Psi_\Lambda^\circ(A^\pi)$$

for all $g \in G$ and all $A \in \mathfrak{A}$ due to (9.9). Consequently, the C^*-algebra $\Psi(\mathfrak{B}_\Lambda^\circ)$ is the closure of the algebra $\Psi(\mathfrak{B}_\Lambda^\circ)^0$ composed by the finite sums of the form $\sum_g \Psi_\Lambda^\circ(A_g^\pi)\Psi_\Lambda^\circ(U_g^\pi)$, where $A_g \in \mathfrak{A}$.

Theorem 9.5. *The mapping*

$$P_\varphi(\mathfrak{M}_\Lambda^\circ)\varphi\Big(\sum_{g\in F} A_g^\pi U_g^\pi\Big) \mapsto \sum_{g \in F} \Psi_\Lambda^\circ(A_g^\pi)\Psi_\Lambda^\circ(U_g^\pi), \tag{9.10}$$

where F is a finite subset of G and $A_g \in \mathfrak{A}$ for $g \in F$, extends to a C^*-algebra isomorphism of the C^*-algebra $\mathfrak{B}_\Lambda^\circ$ onto the C^*-algebra $\Psi(\mathfrak{B}_\Lambda^\circ)$ given by (9.8).

Proof. Fix an operator $B = \sum_{g\in F} A_g U_g \subset \mathfrak{B}$, where F is a finite subset of G and $A_g \in \mathfrak{A}$ for $g \in F$, and put $\Psi_\Lambda^\circ(B^\pi) := \sum_{g \in F} \Psi_\Lambda^\circ(A_g^\pi)\Psi_\Lambda^\circ(U_g^\pi)$. Since the set $\mathfrak{M}_\Lambda^\circ$ is open and since $P_\varphi(\mathfrak{M}_\Lambda^\circ)\varphi(B^\pi) = \varphi(B^\pi)P_\varphi(\mathfrak{M}_\Lambda^\circ)$, we infer similarly to [5, Lemma 3.5] that

$$\|P_\varphi(\mathfrak{M}_\Lambda^\circ)\varphi(B^\pi)\|_{\mathcal{B}(\mathcal{H}_\varphi)} = \sup_{Z^\pi \in \mathcal{Z}(\mathfrak{M}_\Lambda^\circ)} \|\varphi(B^\pi Z^\pi)\|_{\mathcal{B}(\mathcal{H}_\varphi)}, \tag{9.11}$$

where the set $\mathcal{Z}(\mathfrak{M}_\Lambda^\circ)$ is given by (9.6).

For each coset $H_\Lambda^\pi \in \widetilde{\mathfrak{H}}_\Lambda^\pi$, it follows from Lemma 9.4(i) that $B^\pi H_\Lambda^\pi \in \mathfrak{A}^\pi$. Hence, by (6.8), (5.3), (9.9) and by Lemmas 5.3–5.4, we obtain $\phi_\Lambda(B^\pi H_\Lambda^\pi) = \Psi_\Lambda^\circ(B^\pi)\phi_\Lambda(H_\Lambda^\pi)$ where $\phi_\Lambda : \mathfrak{A}^\pi \to \mathcal{B}(\mathcal{H}_\Lambda^\circ)$ is the restriction of the representation ϕ (see (6.6), (6.8) and (6.11)) to the space $\mathcal{H}_\Lambda^\circ$ considered as an invariant Hilbert subspace of $\mathcal{H}_\phi$. Therefore, applying (9.2) or Theorem 9.3, we get

$$\begin{aligned}\|P_\varphi(\mathfrak{M}_\Lambda^\circ)\varphi(B^\pi H_\Lambda^\pi)\|_{\mathcal{B}(\mathcal{H}_\varphi)} &= \|P_\phi(\mathfrak{M}_\Lambda^\circ)\phi(B^\pi H_\Lambda^\pi)\|_{\mathcal{B}(\mathcal{H}_\phi)} \\ &= \|\Psi_\Lambda^\circ(B^\pi)\phi_\Lambda(H_\Lambda^\pi)\|_{\mathcal{B}(\mathcal{H}_\Lambda^\circ)}.\end{aligned} \tag{9.12}$$

Since $\mathcal{Z}(\mathfrak{M}_\Lambda^\circ) = \widetilde{\mathfrak{H}}_\Lambda^\pi$ due to Lemma 9.4(ii) and since

$$P_\varphi(\mathfrak{M}_\Lambda^\circ)\varphi(B^\pi Z^\pi) = \varphi(B^\pi Z^\pi) \quad \text{for all} \quad Z^\pi \in \mathcal{Z}(\mathfrak{M}_\Lambda^\circ),$$

we deduce from (9.5) and the equalities (9.11)–(9.12) that

$$\begin{aligned} \sup_{Z^\pi \in \mathcal{Z}(\mathfrak{M}_\Lambda^\circ)} \|\Psi_\Lambda^\circ(B^\pi)\phi_\Lambda(Z^\pi)\|_{\mathcal{B}(\mathcal{H}_\Lambda^\circ)} &= \sup_{Z^\pi \in \mathcal{Z}(\mathfrak{M}_\Lambda^\circ)} \|P_\varphi(\mathfrak{M}_\Lambda^\circ)\varphi(B^\pi Z^\pi)\|_{\mathcal{B}(\mathcal{H}_\varphi)} \\ = \sup_{Z^\pi \in \mathcal{Z}(\mathfrak{M}_\Lambda^\circ)} \|\varphi(B^\pi Z^\pi)\|_{\mathcal{B}(\mathcal{H}_\varphi)} &= \|P_\varphi(\mathfrak{M}_\Lambda^\circ)\varphi(B^\pi)\|_{\mathcal{B}(\mathcal{H}_\varphi)}. \end{aligned} \tag{9.13}$$

We now consider the identical representation π of the unital C^*-algebra $\Psi(\mathfrak{B}_\Lambda^\circ)$ in the Hilbert space $\mathcal{H}_\Lambda^\circ$. By (9.9), $\phi_\Lambda(\mathcal{Z}^\pi)$ is a central C^*-subalgebra of $\Psi(\mathfrak{B}_\Lambda^\circ)$ with the same unit. Clearly, the maximal ideal space of $\phi_\Lambda(\mathcal{Z}^\pi)$ coincides with $\mathfrak{M}_\Lambda^\circ$. Since the set $\mathfrak{M}_\Lambda^\circ$ is an open subset of $\dot{\mathfrak{M}}_\Lambda$ and since the corresponding spectral projection $P_\pi(\mathfrak{M}_\Lambda^\circ)$ is the identity operator on Hilbert space $\mathcal{H}_\Lambda^\circ$, we conclude from [5, Lemma 3.5] that

$$\begin{aligned} \|\Psi_\Lambda^\circ(B^\pi)\|_{\mathcal{B}(\mathcal{H}_\Lambda^\circ)} &= \|P_\phi(\mathfrak{M}_\Lambda^\circ)\Psi_\Lambda^\circ(B^\pi)\|_{\mathcal{B}(\mathcal{H}_\Lambda^\circ)} \\ &= \sup_{Z^\pi \in \mathcal{Z}(\mathfrak{M}_\Lambda^\circ)} \|\Psi_\Lambda^\circ(B^\pi)\phi_\Lambda(Z^\pi)\|_{\mathcal{B}(\mathcal{H}_\Lambda^\circ)}, \end{aligned}$$

which together with (9.13) implies that

$$\|P_\varphi(\mathfrak{M}_\Lambda^\circ)\varphi(B^\pi)\|_{\mathcal{B}(\mathcal{H}_\varphi)} = \|\Psi_\Lambda^\circ(B^\pi)\|_{\mathcal{B}(\mathcal{H}_\Lambda^\circ)} \tag{9.14}$$

for all finite sums $B^\pi = \sum_{g \in F} A_g^\pi U_g^\pi \in \mathfrak{B}^\pi$ with $A_g^\pi \in \mathfrak{A}^\pi$. Since the set of such finite sums is dense in $\mathfrak{B}^\pi$ and since (9.14) holds, the mapping (9.10) uniquely extends to a C^*-algebra isomorphism of $\mathfrak{B}_\Lambda^\circ$ onto $\Psi(\mathfrak{B}_\Lambda^\circ)$. □

Every coset B^π of the C^*-algebra $\mathfrak{B}^\pi$ is the limit of a sequence of cosets of the form $B_n^\pi = \sum_{g \in F_n} A_{g,n}^\pi U_g^\pi$ where $A_{g,n}^\pi \in \mathfrak{A}^\pi$ and g runs through finite subsets F_n of G ($n \in \mathbb{N}$). Then, according to Theorem 9.5, the operator $\Psi_\Lambda(B^\pi)$ in the C^*-algebra $\Psi(\mathfrak{B}_\Lambda^\circ)$ has the form

$$\Psi_\Lambda^\circ(B^\pi) = \lim_{n\to\infty} \sum_{g \in F_n} \Psi_\Lambda^\circ(A_{g,n}^\pi)\Psi_\Lambda^\circ(U_g^\pi),$$

where the *-homomorphism $\Psi_\Lambda^\circ : \mathfrak{B}^\pi \to \mathcal{B}(\mathcal{H}_\Lambda^\circ)$ is an extension of the *-homomorphism $\phi_\Lambda : \mathfrak{A}^\pi \to \mathcal{B}(\mathcal{H}_\Lambda^\circ)$ to the C^*-algebra $\mathfrak{B}^\pi$ in view of (6.8) and (9.9). Thus,

$$\begin{gathered} \Psi_\Lambda^\circ(B^\pi) = \bigoplus_{\xi \in M_\Lambda^\circ(QC(\mathbb{T}))} B_\Lambda^\circ(\xi,\cdot)I \in \mathcal{B}(\mathcal{H}_\Lambda^\circ), \\ B_\Lambda^\circ(\xi,\cdot) : \mathbb{R} \to \mathbb{C}^{2\times 2}, \quad x \mapsto \lim_{n\to\infty} \sum_{g \in F_n} (\mathrm{Sym}\, A_{g,n})(\xi, x)\, e_{\ln g'(\xi)}(x). \end{gathered} \tag{9.15}$$

Taking into account the inverse closedness of the C^*-algebras $\mathfrak{B}_\Lambda^\circ$ and $\Psi(\mathfrak{B}_\Lambda^\circ)$ in the C^*-algebras $\mathcal{B}(P_\varphi(\mathfrak{M}_\Lambda^\circ)\mathcal{H}_\varphi)$ and $\mathcal{B}(\mathcal{H}_\Lambda^\circ)$, respectively, we immediately obtain an invertibility criterion for the operators in the C^*-algebra $\mathfrak{B}_\Lambda^\circ$ from Theorem 9.5 and (9.15).

Theorem 9.6. *For each $B \in \mathfrak{B}$, the operator $B_\Lambda^\circ = P_\varphi(\mathfrak{M}_\Lambda^\circ)\varphi(B^\pi) \in \mathfrak{B}_\Lambda^\circ$ is invertible on the space $P_\varphi(\mathfrak{M}_\Lambda^\circ)\mathcal{H}_\varphi$ if and only if the operator $\Psi_\Lambda^\circ(B^\pi) \in \Psi(\mathfrak{B}_\Lambda^\circ)$ is invertible on the space $\mathcal{H}_\Lambda^\circ = l^2(M_\Lambda^0(QC(\mathbb{T})), L_2^2(\mathbb{R}))$, that is, if*

$$\min_{\xi \in M_\Lambda^0(QC(\mathbb{T}))} \inf_{x \in \mathbb{R}} \big|\det(B_\Lambda^\circ(\xi, x))\big| > 0.$$

For each $(\xi, x) \in \mathfrak{N}_\Lambda$, where $\mathfrak{N}_\Lambda$ is given by (3.1), we introduce the representation

$$\mathfrak{B} \to \mathcal{B}(\mathbb{C}^2), \quad B \mapsto B_\Lambda^\circ(\xi, x)I, \tag{9.16}$$

given on the generators of the C^*-algebra $\mathfrak{B}$ by the following 2×2 matrix functions $B_\Lambda^\circ(\cdot,\cdot)$ according to (5.3), (5.4) and (9.15):

$$\begin{aligned}
(aI)_\Lambda^\circ(\xi, x) &= \operatorname{diag}\big\{a(\xi, 1), a(\xi, 0)\big\},\\
(S_\mathbb{T})_\Lambda^\circ(\xi, x) &= \begin{bmatrix} \tanh(\pi x) & 1/\cosh(\pi x) \\ 1/\cosh(\pi x) & -\tanh(\pi x) \end{bmatrix},\\
(U_h)_\Lambda^\circ(\xi, x) &= \operatorname{diag}\big\{e^{ix \ln h'(\xi)}, e^{ix \ln h'(\xi)}\big\},
\end{aligned} \tag{9.17}$$

where $a \in PQC(\mathbb{T})$, $a(\xi, \mu)$ is the value of the Gelfand transform of a at the point $(\xi, \mu) \in M(PQC(\mathbb{T}))$ and $h \in G$. It is easily seen from (9.17) that for each $(\xi, x) \in \mathfrak{N}_\Lambda$ representation (9.16) coincides with the representation $\Phi_{\xi,x}$ defined by (3.4)–(3.5), that is,

$$\Phi_{\xi,x}(B) = B_\Lambda^\circ(\xi, x)I \quad \text{for all} \quad B \in \mathfrak{B} \quad \text{and all} \quad (\xi, x) \in \mathfrak{N}_\Lambda.$$

Hence, identifying the operators $\Phi_{\xi,x}(B)$ and the matrices $B_\Lambda^\circ(\xi, x)$, we conclude that Theorem 9.6 is equivalent to part (ii) of Theorem 3.1.

10. Invertibility of functional operators

Let us study the invertibility of functional operators in the C^*-algebra

$$\mathcal{A} := \operatorname{alg}(PQC(\mathbb{T}), U_G) \subset \mathcal{B}(L^2(\mathbb{T}))$$

generated by the multiplication operators by piecewise quasicontinuous functions on $\mathbb{T}$ and by the isometric shift operators U_g $(g \in G)$.

The C^*-algebra $\mathcal{A}$ is the closure of the algebra $\mathcal{A}^0 \subset \mathcal{A}$ consisting of the functional operators $A = \sum_{g\in F} a_g U_g$, where $a_g \in PQC^0(\mathbb{T})$ and F runs through the finite subsets of G. In order to obtain an invertibility criterion for the operators $A \in \mathcal{A}$, we will apply the local-trajectory method.

Let $\widetilde{\mathfrak{A}} := \widetilde{\mathcal{Z}} := \{aI : a \in PQC(\mathbb{T})\}$. As $\widetilde{\mathcal{Z}} \cong PQC(\mathbb{T})$, we get $M(\widetilde{\mathcal{Z}}) = M(PQC(\mathbb{T}))$. Let us check the fulfillment of assumptions made in Section 4.

Lemma 2.3 implies that for every $g \in G$ and all $a \in PQC(\mathbb{T})$ the mapping $\widetilde{\alpha}_g : aI \mapsto U_g a U_g^{-1} = (a \circ g)I$ is a *-automorphism of the commutative C^*-algebras $\widetilde{\mathfrak{A}} = \widetilde{\mathcal{Z}} \subset \mathcal{B}(L^2(\mathbb{T}))$. Since G is a commutative and therefore an amenable group, we see that conditions (A1)–(A2) for the C^*-algebra $\mathcal{A}$ are satisfied.

For every $g \in G$, the *-automorphism $\widetilde{\alpha}_g$ induces the homeomorphism

$$\widetilde{\beta}_g : M(PQC(\mathbb{T})) \to M(PQC(\mathbb{T})), \quad (\xi,\mu) \mapsto (g(\xi),\mu), \tag{10.1}$$

where $g(\xi) \in M(QC(\mathbb{T}))$ is given by (5.30). If the ideal $\xi \in M(QC(\mathbb{T}))$ belongs to the fiber $M_t(SO(\mathbb{T}))$, then $g(\xi) \in M_{g(t)}(QC(\mathbb{T}))$. Moreover, if ξ belongs to $M^0(QC(\mathbb{T}))$ or $\widetilde{M}^{\pm}(QC(\mathbb{T}))$, respectively, then so is $g(\xi)$. Hence, taking into account the topologically free action of the group G on $M(QC(\mathbb{T}))$ (see [11, Lemma 4.2]) and the Gelfand topology (2.5) on $M(PQC(\mathbb{T}))$, we easily conclude that condition (A3) for the C^*-algebra $\mathcal{A}$ also holds, with $M_0 := M_{\mathbb{T}\setminus\Lambda}(PQC(\mathbb{T}))$, where $M_{\mathbb{T}\setminus\Lambda}(PQC(\mathbb{T})) := \bigcup_{\tau\in\mathbb{T}\setminus\Lambda} M_\tau(PQC(\mathbb{T}))$.

With each maximal ideal $(\xi,\mu) \in M(PQC(\mathbb{T}))$ we associate the representation

$$\Pi_{(\xi,\mu)} : \mathcal{A} \to \mathcal{B}(l^2(G)), \quad A \mapsto A_{(\xi,\mu)} \tag{10.2}$$

given for the operators $A = \sum_{g\in F} a_g U_g \in \mathcal{A}^0$ by

$$(A_{(\xi,\mu)}f)(h) = \sum_{g\in F} [(a_g \circ h)(\xi,\mu)] f(hg) \quad (h \in G,\ f \in l^2(G)). \tag{10.3}$$

Following Section 8, we fix a set $\mathcal{O}_{\rm arc} \subset \mathbb{T} \setminus \Lambda$ containing exactly one point in each orbit defined by the group of shifts G on $\mathbb{T} \setminus \Lambda$ and consider the set

$$\mathfrak{R}_{\rm arc} := \bigcup_{\tau\in\mathcal{O}_{\rm arc}} M_\tau(PQC(\mathbb{T})). \tag{10.4}$$

The set $\mathfrak{R}_{\rm arc}$ contains exactly one point in each G-orbit defined by the action of the group G on $M_{\mathbb{T}\setminus\Lambda}(PQC(\mathbb{T}))$ by means of the homeomorphisms $\widetilde{\beta}_g$ $(g \in G)$ given by (10.1).

Since conditions (A1)–(A3) are fulfilled, we get the following.

Theorem 10.1. *A functional operator $A \in \mathcal{A}$ is invertible on the space $L^2(\mathbb{T})$ if and only if for all $(\xi,\mu) \in \mathfrak{R}_{\rm arc}$ the operators $A_{(\xi,\mu)}$ are invertible on the space $l^2(G)$ and*

$$\sup_{(\xi,\mu)\in\mathfrak{R}_{\rm arc}} \left\| (A_{(\xi,\mu)})^{-1} \right\| < \infty. \tag{10.5}$$

Proof. Take the maximal ideal $J_{(\xi,\mu)} := \{aI : a \in PQC(\mathbb{T}),\ a(\xi,\mu) = 0\}$ of $\widetilde{\mathcal{Z}}$ associated to each functional $(\xi,\mu) \in M(PQC(\mathbb{T}))$. As $\widetilde{\mathfrak{A}} = \widetilde{\mathcal{Z}}$, the mapping

$$\widetilde{\Pi}_{(\xi,\mu)} : \widetilde{\mathfrak{A}}/J_{(\xi,\mu)} \to \mathbb{C}, \quad aI + J_{(\xi,\mu)} \mapsto a(\xi,\mu),$$

is an isometric representation of the C^*-algebra $\widetilde{\mathfrak{A}}/J_{(\xi,\mu)}$ in $\mathbb{C}$. Following (4.3)–(4.5), we construct representations of the C^*-algebra $\mathcal{A}$ in the Hilbert space $l^2(G)$ by formulas (10.2) and (10.3). Since $\mathcal{A}$ satisfies conditions (A1)–(A3) of the local-trajectory method, Theorem 4.1 immediately implies the statement of the theorem. □

Applying Theorems 10.1 and 8.1 we can prove the following result.

Theorem 10.2. *For any functional operator $A \in \mathcal{A}$, its Fredholmness on the space $L^2(\mathbb{T})$ is equivalent to its invertibility on this space.*

Proof. Obviously, we only need to prove that the Fredholmness of each functional operator A implies its invertibility. Suppose that an operator $A \in \mathcal{A}$ is Fredholm on the space $L^2(\mathbb{T})$. Then the operator $A_{\rm arc} := P_\varphi(\dot{\mathfrak{M}}_{\rm arc})\varphi(A^\pi)$ is invertible on the Hilbert space $P_\varphi(\dot{\mathfrak{M}}_{\rm arc})\mathcal{H}_\varphi$. Consequently, by Theorem 8.1, for all $(\xi, x) \in \mathfrak{N}_{\rm arc}$ the operators $\pi_{(\xi,x)}(A_{\rm arc})$ are invertible on the space $l^2(G, \mathbb{C}^2)$ and

$$\sup_{(\xi,x)\in\mathfrak{N}_{\rm arc}} \big\|(\pi_{(\xi,x)}(A_{\rm arc}))^{-1}\big\| < \infty. \tag{10.6}$$

By (8.5)–(8.6), (5.3) and (10.2)–(10.3), we infer that, for all $x \in \overline{\mathbb{R}}$,

$$\pi_{(\xi,x)}(A_{\rm arc}) = \begin{cases} {\rm diag}\{A_{(\xi,0)}, A_{(\xi,0)}\} & \text{if } \xi \in \widetilde{M}^-_{\mathbb{T}\setminus\Lambda}(QC(\mathbb{T})), \\ {\rm diag}\{A_{(\xi,1)}, A_{(\xi,0)}\} & \text{if } \xi \in M^0_{\mathbb{T}\setminus\Lambda}(QC(\mathbb{T})), \\ {\rm diag}\{A_{(\xi,1)}, A_{(\xi,1)}\} & \text{if } \xi \in \widetilde{M}^+_{\mathbb{T}\setminus\Lambda}(QC(\mathbb{T})), \end{cases} \tag{10.7}$$

where

$$\widetilde{M}^\pm_{\mathbb{T}\setminus\Lambda}(QC(\mathbb{T})) := \bigcup_{\tau\in\mathbb{T}\setminus\Lambda} \widetilde{M}^\pm_\tau(QC(\mathbb{T})), \quad M^0_{\mathbb{T}\setminus\Lambda}(QC(\mathbb{T})) := \bigcup_{\tau\in\mathbb{T}\setminus\Lambda} M^0_\tau(QC(\mathbb{T})). \tag{10.8}$$

Comparing the sets $\mathfrak{N}_{\rm arc}$ and $\mathfrak{R}_{\rm arc}$ given by (3.1) and (10.4), respectively, we conclude from (10.7) and (10.6) that for every $(\xi, \mu) \in \mathfrak{R}_{\rm arc}$ the operator $A_{(\xi,\mu)}$ is invertible on the space $l^2(G)$ and condition (10.5) is fulfilled. Hence, by Theorem 10.1, the operator A is invertible on the space $L^2(\mathbb{T})$. □

11. The C^*-algebra $\mathfrak{B}^\diamond_\Lambda$

In this section we will show that for every $B \in \mathfrak{B}$ the invertibility of the operator $B_{\rm arc} = P_\varphi(\dot{\mathfrak{M}}_{\rm arc})\varphi(B^\pi)$ on the Hilbert space $P_\varphi(\dot{\mathfrak{M}}_{\rm arc})\mathcal{H}_\varphi$ implies the invertibility of the operators $B^\diamond_\Lambda = P_\varphi(\mathfrak{M}^\diamond_\Lambda)\varphi(B^\pi)$ on the Hilbert space $P_\varphi(\mathfrak{M}^\diamond_\Lambda)\mathcal{H}_\varphi$. This means that condition (iii) in Theorem 6.3 is superfluous.

Observe that the C^*-algebra $\mathfrak{B} = {\rm alg}(PQC(\mathbb{T}), S_\mathbb{T}, U_G)$ can be viewed as the C^*-algebra $\mathfrak{B} = {\rm alg}(\mathcal{A}, S_\mathbb{T})$ generated by the C^*-algebra $\mathcal{A}$ studied in Section 10 and by the Cauchy singular integral operator $S_\mathbb{T}$. By analogy with Theorem 5.5, we can get another general form of operators $B \in \mathfrak{B}$.

Let $\mathfrak{B}^0$ denote the dense non-closed subalgebra of the C^*-algebra $\mathfrak{B}$ consisting of all operators of the form $\sum_{i=1}^n T_{i1}T_{i2}\dots T_{ij_i}$ where $n, j_i \in \mathbb{N}$ and $T_{i,k} \in \mathcal{A}^0 \cup \{S_\mathbb{T}\}$. We denote by $\mathfrak{H}$ the closed two-sided ideal of $\mathfrak{B}$ generated by all the commutators $[aI, S_\mathbb{T}] := aS_\mathbb{T} - S_\mathbb{T}aI$ with $a \in PC(\mathbb{T})$, that is, the closure of the set

$$\mathfrak{H}^0 := \Big\{ \sum_{i=1}^n B_iH_iC_i : B_i, C_i \in \mathfrak{B}^0,\ H_i = [a_iI, S_\mathbb{T}],\ a_i \in PC^0(\mathbb{T}),\ n \in \mathbb{N} \Big\}.$$

The ideal $\mathcal{K}$ of all compact operators on $L^2(\mathbb{T})$ is contained in $\mathfrak{H}$ (see, e.g., [16]). Hence, the commutators $[aI, S_{\mathbb{T}}]$ $(a \in QC(\mathbb{T}))$ and $[U_g, S_{\mathbb{T}}]$ $(g \in G)$ belong to the ideal $\mathfrak{H}$ because these operators are compact on the space $L^2(\mathbb{T})$ according to (2.6) and (2.7). As a result, for every $A \in \mathcal{A}$ the commutators $[A, S_{\mathbb{T}}]$ also are in $\mathfrak{H}$.

Let $\widetilde{\mathcal{A}}$ be the C^*-algebra of the 2×2 diagonal matrices with $\mathcal{A}$-valued entries. We now obtain an analogue of Theorem 5.5 for the C^*-algebra $\mathfrak{B}$ (cf. [20], [21]).

Theorem 11.1. *Every operator $B \in \mathfrak{B}$ is uniquely represented in the form*

$$B = A^+ P_{\mathbb{T}}^+ + A^- P_{\mathbb{T}}^- + H_B, \tag{11.1}$$

where $A^\pm$ are functional operators in the C^*-algebra $\mathcal{A}$, $P_{\mathbb{T}}^\pm = (I \pm S_{\mathbb{T}})/2$ are the orthogonal projections associated with the Cauchy singular integral operator $S_{\mathbb{T}}$, $H_B \in \mathfrak{H}$, the map $B \mapsto \operatorname{diag}\{A^+, A^-\}$ is a C^*-algebra homomorphism of the C^*-algebra $\mathfrak{B}$ onto the C^*-algebra $\widetilde{\mathcal{A}}$ with kernel $\mathfrak{H}$, and

$$\|A^\pm\| \le \inf_{H\in\mathfrak{H}} \|B + H\| \le |B| = \inf_{K\in\mathcal{K}} \|B + K\|. \tag{11.2}$$

Proof. Obviously, every operator $B \in \mathfrak{B}^0$ is represented in the form (11.1), and the mapping $B \mapsto \operatorname{diag}\{A^+, A^-\}$ is an algebraic homomorphism of the non-closed algebra $\mathfrak{B}^0$ into $\widetilde{\mathcal{A}}$ with the kernel contained in $\mathfrak{H}^0$. This mapping is defined for the generators of the algebra $\mathfrak{B}$ by

$$aI \mapsto \operatorname{diag}\{aI, aI\}, \quad U_g \mapsto \operatorname{diag}\{U_g, U_g\}, \quad S_{\mathbb{T}} \mapsto \operatorname{diag}\{I, -I\}.$$

Obviously, it only remains to prove (11.2) for all $B \in \mathfrak{B}^0$.

Taking into account the facts that the ideal $\mathfrak{H} \subset \mathfrak{B}$ is generated by all the commutators $[aI, S_{\mathbb{T}}]$ $(a \in PC(\mathbb{T}))$ and that

$$(\operatorname{Sym}([aI, S_{\mathbb{T}}]))(\xi, \pm\infty) = 0_{2\times 2} \quad \text{for all } a \in PC(\mathbb{T}) \text{ and all } \xi \in M(QC(\mathbb{T}))$$

according to (5.3), we infer from (8.6) that for any operator $H \in \mathfrak{H}$,

$$\pi_{(\xi,\pm\infty)}(H_{\mathrm{arc}}) = 0 \quad \text{for all } \xi \in M(QC(\mathbb{T})), \tag{11.3}$$

where $H_{\mathrm{arc}} := P_\varphi(\dot{\mathfrak{M}}_{\mathrm{arc}})\varphi(H^\pi)$. Analogously, by (5.3), for all $\xi \in M(QC(\mathbb{T}))$,

$$(\operatorname{Sym} P_{\mathbb{T}}^\pm)(\xi, \pm\infty) = \operatorname{diag}\{1, 0\}, \quad (\operatorname{Sym} P_{\mathbb{T}}^\pm)(\xi, \mp\infty) = \operatorname{diag}\{0, 1\}. \tag{11.4}$$

Further, from (10.7) it follows that for $x \in \{+\infty, -\infty\}$,

$$\pi_{(\xi,x)}(A^\pm_{\mathrm{arc}}) = \begin{cases} \operatorname{diag}\{A^\pm_{(\xi,0)}, A^\pm_{(\xi,0)}\} & \text{if } \xi \in \widetilde{M}^-_{\mathbb{T}\setminus\Lambda}(QC(\mathbb{T})), \\ \operatorname{diag}\{A^\pm_{(\xi,1)}, A^\pm_{(\xi,0)}\} & \text{if } \xi \in M^0_{\mathbb{T}\setminus\Lambda}(QC(\mathbb{T})), \\ \operatorname{diag}\{A^\pm_{(\xi,1)}, A^\pm_{(\xi,1)}\} & \text{if } \xi \in \widetilde{M}^+_{\mathbb{T}\setminus\Lambda}(QC(\mathbb{T})), \end{cases} \tag{11.5}$$

where the sets on the right of (11.5) are given by (10.8). Hence, for every operator $B = A^+P_{\mathbb{T}}^+ + A^-P_{\mathbb{T}}^- + H_B \in \mathfrak{B}^0$ we deduce from (11.3)–(11.5) that

$$\pi_{(\xi,\pm\infty)}(B_{\mathrm{arc}}) = \operatorname{diag} \begin{cases} \{A^\pm_{(\xi,0)}, A^\mp_{(\xi,0)}\} & \text{if } \xi \in \widetilde{M}^-_{\mathbb{T}\setminus\Lambda}(QC(\mathbb{T})), \\ \{A^\pm_{(\xi,1)}, A^\mp_{(\xi,0)}\} & \text{if } \xi \in M^0_{\mathbb{T}\setminus\Lambda}(QC(\mathbb{T})), \\ \{A^\pm_{(\xi,1)}, A^\mp_{(\xi,1)}\} & \text{if } \xi \in \widetilde{M}^+_{\mathbb{T}\setminus\Lambda}(QC(\mathbb{T})). \end{cases} \tag{11.6}$$

Therefore, letting $M^{\pm}_{\mathbb{T}\setminus\Lambda}(QC(\mathbb{T})) := \bigcup_{\tau\in\mathbb{T}\setminus\Lambda} M^{\pm}_{\tau}(QC(\mathbb{T}))$, we infer that

$$\max\Big\{\sup_{\xi\in M^{-}_{\mathbb{T}\setminus\Lambda}(QC(\mathbb{T}))}\|A^{\pm}_{(\xi,0)}\|,\ \sup_{\xi\in M^{+}_{\mathbb{T}\setminus\Lambda}(QC(\mathbb{T}))}\|A^{\pm}_{(\xi,1)}\|\Big\} \le \|P_\varphi(\dot{\mathfrak{M}}_{\mathrm{arc}})\varphi(B^\pi+H^\pi)\| \le \|B+H\| \quad \text{for all } H\in\mathfrak{H}. \tag{11.7}$$

By Theorem 10.1, the C^*-algebra $\mathcal{A}$ is isometrically *-isomorphic to the C^*-algebra

$$\Big\{\bigoplus_{(\xi,\mu)\in\mathfrak{R}_{\mathrm{arc}}} A_{(\xi,\mu)} : A\in\mathcal{A}\Big\} \subset \bigoplus_{(\xi,\mu)\in\mathfrak{R}_{\mathrm{arc}}} \mathcal{B}(l^2(G))$$

equipped with the norm $\sup_{(\xi,\mu)\in\mathfrak{R}_{\mathrm{arc}}}\|A_{(\xi,\mu)}\|$. Therefore, for every $A\in\mathcal{A}$,

$$\|A\| = \sup_{(\xi,\mu)\in\mathfrak{R}_{\mathrm{arc}}}\|A_{(\xi,\mu)}\|. \tag{11.8}$$

Finally, from (11.7) and (11.8) it follows that $\|A^{\pm}\| \le \|B+H\|$ for all $B\in\mathfrak{B}^0$ and all $H\in\mathfrak{H}$, which immediately implies (11.2). □

Let $G(\mathbb{T})$ denote the set of the G-orbits $G(t) := \{g(t) : g\in G\}\subset\mathbb{T}$ for all $t\in\mathbb{T}$. For each $w\in G(\mathbb{T})$ we denote by $\mathfrak{H}_w$ the closed two-sided ideal of $\mathfrak{B}$ generated by the operator V_t given by (5.6), where t is an arbitrary point of the G-orbit w. Observe that $V_tV_\tau \simeq 0$ for $t,\tau\in\mathbb{T}$ and $t\ne\tau$. Hence, $H_wH_v\simeq 0$ if $w,v\in G(\mathbb{T})$ and $w\ne v$. Similarly to [21, Lemma 5.4] we get the following result.

Lemma 11.2. *Every operator $H\in\mathfrak{H}^0$ is of the form $H\simeq\sum_{w\in G_H}H_w$, where G_H is an at most finite subset of $G(\mathbb{T})$ and $H_w\in\mathfrak{H}_w$.*

Consider now the C^*-algebra $\mathfrak{B}^\diamond_\Lambda = P_\varphi(\mathfrak{M}^\diamond_\Lambda)\varphi(\mathfrak{B}^\pi)$ (see (6.16)). Given $t\in\Lambda$, let $\mathfrak{H}_t$ be the closed two-sided ideal of $\mathfrak{B}$ generated by the operator V_t given by (5.6). By Lemma 9.4(i), the ideal $\mathfrak{H}_t$ is contained in the C*-algebra $\mathfrak{A}$. Clearly, the ideal $\mathfrak{H}$ is generated by all ideals $\mathfrak{H}_t$ $(t\in\mathbb{T})$.

Lemma 11.3. *If $H\in\mathfrak{H}$, then*

$$P_\varphi(\mathfrak{M}^\diamond_\Lambda)\varphi(H^\pi) = 0. \tag{11.9}$$

Proof. First, let us prove that, for the set $\mathfrak{M}^\diamond_\Lambda$ closed in $\dot{\mathfrak{M}}$,

$$P_\varphi(\mathfrak{M}^\diamond_\Lambda)\varphi(V^\pi_\tau) = 0 \quad \text{for all } \tau\in\mathbb{T}. \tag{11.10}$$

Fix $\tau\in\mathbb{T}$. Let Δ_n be a sequence of open sets of $\dot{\mathfrak{M}}$ such that $\bigcap_n\Delta_n = \mathfrak{M}^\diamond_\Lambda$. Analogously to Theorem 7.1 one can establish that

$$\|P_\varphi(\Delta_n)\varphi(V^\pi_\tau)\|_{\mathcal{B}(\mathcal{H}_\varphi)} = \|P_\phi(\Delta_n)\phi(V^\pi_\tau)\|_{\mathcal{B}(\mathcal{H}_\phi)} \quad \text{for all } n\in\mathbb{N}. \tag{11.11}$$

If $\tau\in\mathbb{T}\setminus\Lambda$, then for the Hilbert space $\mathcal{H}_\phi$ we have the equality

$$\lim_{n\to\infty}\|P_\phi(\Delta_n)\phi(V^\pi_\tau)\|_{\mathcal{B}(\mathcal{H}_\phi)} = 0 \tag{11.12}$$

because $\tau \notin \overline{\Delta_n}$ for all sufficiently large n, where $\overline{\Delta_n}$ is the closure of Δ_n in $\dot{\mathfrak{M}}$. Hence, we derive from (11.11) that, for the abstract Hilbert space $\mathcal{H}_\varphi$,

$$\lim_{n\to\infty} \|P_\varphi(\Delta_n)\varphi(V_\tau^\pi)\|_{\mathcal{B}(\mathcal{H}_\varphi)} = 0 \tag{11.13}$$

for all $\tau \in \mathbb{T} \setminus \Lambda$. Let now $\tau \in \Lambda$. Then we conclude from Lemma 5.2 that again (11.12) holds, which in view of (11.11) gives (11.13) for all $\tau \in \Lambda$. Thus,

$$\lim_{n\to\infty} \|P_\varphi(\Delta_n)\varphi(V_\tau^\pi)\|_{\mathcal{B}(\mathcal{H}_\varphi)} = 0 \quad \text{for all} \quad \tau \in \mathbb{T}. \tag{11.14}$$

Let $\varphi|_{\mathcal{Z}^\pi}$ be the restriction of the representation φ to the central C^*-subalgebra $\mathcal{Z}^\pi$. Consider the decomposition of the representation $\varphi|_{\mathcal{Z}^\pi}$ into the direct sum of cyclic representations φ_α in the mutually orthogonal subspaces $\mathcal{H}_{\varphi,\alpha}$ of the space $\mathcal{H}_\varphi$. Then, by [5, Lemma 3.4], for each subspace $\mathcal{H}'_\varphi = \bigoplus_{\alpha\in Q} \mathcal{H}_{\varphi,\alpha} \subset \mathcal{H}_\varphi$ with a finite set Q, we have

$$P_\varphi(\mathfrak{M}_\Lambda^\diamond)\varphi(V_\tau^\pi) = \operatorname*{s-lim}_{n\to\infty} P_\varphi(\Delta_n)\varphi(V_\tau^\pi).$$

Hence,

$$\begin{aligned} 0 \le \|P_\varphi(\mathfrak{M}_\Lambda^\diamond)\varphi(V_\tau^\pi)\|_{\mathcal{B}(\mathcal{H}'_\varphi)} &\le \liminf_{n\to\infty} \|P_\varphi(\Delta_n)\varphi(V_\tau^\pi)\|_{\mathcal{B}(\mathcal{H}'_\varphi)} \\ &\le \liminf_{n\to\infty} \|P_\varphi(\Delta_n)\varphi(V_\tau^\pi)\|_{\mathcal{B}(\mathcal{H}_\varphi)}, \end{aligned}$$

which implies according to (11.14) that for all considered subspaces $\mathcal{H}'_\varphi \subset \mathcal{H}_\varphi$,

$$\|P_\varphi(\mathfrak{M}_\Lambda^\diamond)\varphi(V_\tau^\pi)\|_{\mathcal{B}(\mathcal{H}'_\varphi)} = 0. \tag{11.15}$$

Since the set of vectors $\xi = \{\xi_\alpha\}_\alpha \in \mathcal{H}_\varphi$ with a finite number of non-zero entries $\xi_\alpha \in \mathcal{H}_{\varphi,\alpha}$ is dense in $\mathcal{H}_\varphi$, the relations (11.15) imply (11.10).

Consider now an operator $H_w \in \mathfrak{H}_w$ with $w \in G(\mathbb{T})$. It follows from (5.16), (5.18) and (5.28) that

$$H_w^\pi = \lim_{n\to\infty} \sum_{\tau\in G_{n,w}} B_{n,\tau}^\pi V_\tau^\pi, \tag{11.16}$$

where $G_{n,w}$ are finite subsets of the G-orbit w and $B_{n,\tau}$ are operators in the C^*-algebra $\mathfrak{B}$. Since the set $\mathfrak{M}_\Lambda^\diamond$ belongs to the family $\mathfrak{R}_G(\dot{\mathfrak{M}})$ defined in (6.5), we infer from (4.8) that for all $B \in \mathfrak{B}$,

$$P_\varphi(\mathfrak{M}_\Lambda^\diamond)\varphi(B^\pi) = P_\varphi(\mathfrak{M}_\Lambda^\diamond)\varphi(B^\pi)P_\varphi(\mathfrak{M}_\Lambda^\diamond).$$

Therefore, from (11.16) and (11.10) it follows that

$$P_\varphi(\mathfrak{M}_\Lambda^\diamond)\varphi(H_w^\pi) = \lim_{n\to\infty} \sum_{\tau\in G_{n,w}(\mathbb{T})} P_\varphi(\mathfrak{M}_\Lambda^\diamond)\varphi(B_{n,\tau}^\pi)P_\varphi(\mathfrak{M}_\Lambda^\diamond)\varphi(V_\tau^\pi) = 0. \tag{11.17}$$

Finally, by Lemma 11.2, for any operator $H \in \mathfrak{H}^0$ there are a finite subset G_H of $G(\mathbb{T})$ and operators $H_w \in \mathfrak{H}_w$ such that $H^\pi = \sum_{w\in G_H} H_w^\pi$. Hence, by (11.17), for $H \in \mathfrak{H}^0$ we obtain

$$P_\varphi(\mathfrak{M}_\Lambda^\diamond)\varphi(H^\pi) = \sum_{w\in G_H} P_\varphi(\mathfrak{M}_\Lambda^\diamond)\varphi(H_w^\pi) = 0,$$

which implies (11.9) because the set $\mathfrak{H}^0$ is dense in $\mathfrak{H}$. □

Lemma 11.3 shows that the operators $H \in \mathfrak{H}$ do not have influence on the operators in the C^*-algebra $\mathfrak{B}^\circ_\Lambda$ and, in particular, implies the following.

Corollary 11.4. *If A is a functional operator in $\mathcal{A}$, then*

$$P_\varphi(\mathfrak{M}^\circ_\Lambda)\varphi([A, S_\mathbb{T}]^\pi) = 0.$$

Theorem 11.5. *If $B \in \mathfrak{B}$ and the operator $B_{\rm arc} = P_\varphi(\dot{\mathfrak{M}}_{\rm arc})\varphi(B^\pi)$ is invertible on the Hilbert space $P_\varphi(\dot{\mathfrak{M}}_{\rm arc})\mathcal{H}_\varphi$, then the operator $B^\circ_\Lambda := P_\varphi(\mathfrak{M}^\circ_\Lambda)\varphi(B^\pi)$ is invertible on the Hilbert space $P_\varphi(\mathfrak{M}^\circ_\Lambda)\mathcal{H}_\varphi$.*

Proof. Fix an operator $B \in \mathfrak{B}$. By Theorem 11.1, the operator B has the form $B = A^+P^+_\mathbb{T} + A^-P^-_\mathbb{T} + H_B$, with $A^\pm \in \mathcal{A}$ and $H_B \in \mathfrak{H}$. We then infer from Lemma 11.3 that

$$B^\circ_\Lambda = P_\varphi(\mathfrak{M}^\circ_\Lambda)\varphi((A^+P^+_\mathbb{T} + A^-P^-_\mathbb{T})^\pi). \tag{11.18}$$

If the operator $B_{\rm arc}$ is invertible on the space $P_\varphi(\dot{\mathfrak{M}}_{\rm arc})\mathcal{H}_\varphi$, then the operators $(A^\pm)^\circ_\Lambda := P_\varphi(\mathfrak{M}^\circ_\Lambda)\varphi((A^\pm)^\pi)$ are invertible on the space $P_\varphi(\mathfrak{M}^\circ_\Lambda)\mathcal{H}_\varphi$. Indeed, if the operator $B_{\rm arc}$ is invertible on the Hilbert space $P_\varphi(\dot{\mathfrak{M}}_{\rm arc})\mathcal{H}_\varphi$, then, by Theorem 8.1, the operators $\pi_{(\xi,x)}(B_{\rm arc})$ for all $(\xi, x) \in \mathfrak{N}_{\rm arc}$ are invertible on the Hilbert space $l^2(G, \mathbb{C}^2)$ and condition (8.8) is fulfilled. In particular, the operators $\pi_{(\xi,\pm\infty)}(B_{\rm arc})$ are invertible on the space $l^2(G, \mathbb{C}^2)$ for all $(\xi, \pm\infty) \in \mathfrak{R}_{\rm arc}$, where $\mathfrak{R}_{\rm arc}$ is defined by (10.4). Then it follows from (11.6) that all the operators $A^\pm_{(\xi,\mu)}$ for $(\xi, \mu) \in \mathfrak{R}_{\rm arc}$ are invertible on the space $l^2(G)$ and

$$\sup_{(\xi,\mu)\in\mathfrak{R}_{\rm arc}} \left\|\left(A^\pm_{(\xi,\mu)}\right)^{-1}\right\| < \infty.$$

Hence, by Theorem 10.1, the functional operators $A^\pm$ are invertible on the space $L^2(\mathbb{T})$, which implies the invertibility of both the operators $(A^\pm)^\circ_\Lambda$ on the Hilbert space $P_\varphi(\mathfrak{M}^\circ_\Lambda)\mathcal{H}_\varphi$.

Corollary 11.4 implies that for every functional operator $A \in \mathcal{A}$,

$$P_\varphi(\mathfrak{M}^\circ_\Lambda)\varphi(A^\pi)P_\varphi(\mathfrak{M}^\circ_\Lambda)\varphi((P^\pm_\mathbb{T})^\pi) = P_\varphi(\mathfrak{M}^\circ_\Lambda)\varphi((P^\pm_\mathbb{T})^\pi)P_\varphi(\mathfrak{M}^\circ_\Lambda)\varphi(A^\pi).$$

Therefore, taking into account the invertibility of the operators $(A^\pm)^\circ_\Lambda$ and the equalities

$$P_\varphi(\mathfrak{M}^\circ_\Lambda)\varphi((P^+_\mathbb{T})^\pi)P_\varphi(\mathfrak{M}^\circ_\Lambda)\varphi((P^-_\mathbb{T})^\pi) = P_\varphi(\mathfrak{M}^\circ_\Lambda)\varphi((P^+_\mathbb{T}P^-_\mathbb{T})^\pi) = 0,$$

we conclude from (11.18) that the operator

$$(B^\circ_\Lambda)^{-1} := ((A^+)^\circ_\Lambda)^{-1}P_\varphi(\mathfrak{M}^\circ_\Lambda)\varphi((P^+_\mathbb{T})^\pi) + ((A^-)^\circ_\Lambda)^{-1}P_\varphi(\mathfrak{M}^\circ_\Lambda)\varphi((P^-_\mathbb{T})^\pi),$$

where $((A^\pm)^\circ_\Lambda)^{-1}$ are the inverses for the operators $(A^\pm)^\circ_\Lambda$, is the inverse of the operator B°_Λ on the space $P_\varphi(\mathfrak{M}^\circ_\Lambda)\mathcal{H}_\varphi$. □

Finally, to complete the proof of Theorem 3.1, which gives the Fredholm criterion for the operators B in the C^*-algebra $\mathfrak{B}$, it remains to combine Theorems 6.3, 8.1, 9.6 and 11.5. This also gives the Fredholm symbol calculus for the C^*-algebra $\mathfrak{B}$ (see (3.7)).

Acknowledgment

We are grateful to the referee for the useful comments and suggestions.

References

[1] A. Antonevich, *Linear Functional Equations. Operator Approach.* Operator Theory: Advances and Applications, vol. 83. Birkhäuser, Basel, 1996; Russian original: University Press, Minsk, 1988.

[2] A. Antonevich, M. Belousov, and A. Lebedev, *Functional Differential Equations: II. C^*-Applications. Part 2 Equations with Discontinuous Coefficients and Boundary Value Problems.* Pitman Monogr. Surveys Pure Appl. Math., vol. 95, Longman, Harlow, 1998.

[3] A. Antonevich and A. Lebedev, *Functional Differential Equations: I. C^*-Theory.* Pitman Monogr. Surveys Pure Appl. Math., vol. 70. Longman, Harlow, 1994.

[4] M.A. Bastos, C.A. Fernandes, and Yu.I. Karlovich, *C^*-algebras of integral operators with piecewise slowly oscillating coefficients and shifts acting freely.* Integr. Equ. Oper. Theory **55** (2006) 19–67.

[5] M.A. Bastos, C.A. Fernandes, and Yu.I. Karlovich, *Spectral measures in C^*-algebras of singular integral operators with shifts.* J. Funct. Anal. **242** (2007) 86–126.

[6] M.A. Bastos, C.A. Fernandes, and Yu.I. Karlovich, *C^*-algebras of singular integral operators with shifts having the same nonempty set of fixed points.* Compl. Anal. Oper. Theory **2** (2008) 241–272.

[7] M.A. Bastos, C.A. Fernandes, and Yu.I. Karlovich, *A nonlocal C^*-algebra of singular integral operators with shifts having periodic points.* Integr. Equ. Oper. Theory **71** (2011) 509–534.

[8] M.A. Bastos, C.A. Fernandes, and Yu.I. Karlovich, *A C^*-algebra of singular integral operators with shifts admitting distinct fixed points.* J. Math. Anal. Appl. **432** (2014) 502–524.

[9] M.A. Bastos, C.A. Fernandes, and Yu.I. Karlovich, *A C^*-algebras of singular integral operators with shifts similar to affine mappings.* In "Operator Theory, Operator Algebras and Applications", eds. M.A. Bastos et al. Operator Theory: Advances and Applications, vol. 242, 2014, pp. 53–79.

[10] L.A. Beklaryan, *Groups of homeomorphisms of the line and the circle. Topological characteristics and metric invariants.* Russian Math. Surveys **59** (2004) 599–660.

[11] A. Böttcher, Yu.I. Karlovich, and B. Silbermann, *Singular integral equations with PQC coefficients and freely transformed argument.* Math. Nachr. **166** (1994) 113–133.

[12] A. Böttcher, Yu.I. Karlovich, and I.M. Spitkovsky, *The C^*-algebra of singular integral operators with semi-almost periodic coefficients.* J. Funct. Anal. **204** (2003) 445–484.

[13] A. Böttcher, Yu.I. Karlovich, and I.M. Spitkovsky, *Convolution Operators and Factorization of Almost Periodic Matrix Functions.* Operator Theory: Advances and Applications, vol. 131. Birkhäuser, Basel, 2002.

[14] A. Böttcher, S. Roch, B. Silbermann, and I.M. Spitkovsky, *A Gohberg–Krupnik–Sarason symbol calculus for algebras of Toeplitz, Hankel, Cauchy, and Carleman*

operators. In "Topics in Operator Theory. Ernst D. Hellinger Memorial Volume", eds. L. de Branges et al. Operator Theory: Advances and Applications, vol. 48, 1990, pp. 189–234.

[15] J. Dixmier, *C^*-Algebras.* North-Holland, Amsterdam, 1977.

[16] I. Gohberg and N. Krupnik, *On the algebra generated by the one-dimensional singular integral operators with piecewise continuous coefficients.* Funct. Anal. Appl. **4** (1970) 193–201.

[17] F.P. Greenleaf, *Invariant Means on Topological Groups and Their Representations.* Van Nostrand-Reinhold, New York, 1969.

[18] Yu.I. Karlovich, *The local-trajectory method of studying invertibility in C^*-algebras of operators with discrete groups of shifts.* Soviet. Math. Dokl. **37** (1988) 407–411.

[19] Yu.I. Karlovich, *A local-trajectory method and isomorphism theorems for nonlocal C^*-algebras.* In "Modern Operator Theory and Applications. The Igor Borisovich Simonenko Anniversary Volume", eds. Ja.M. Erusalimsky et al. Operator Theory: Advances and Applications, vol. 170, 2007, pp. 137–166.

[20] Yu.I. Karlovich and V.G. Kravchenko, *An algebra of singular integral operators with piecewise-continuous coefficients and piecewise-smooth shift on a composite contour.* Math. USSR-Izv. **23** (1984) 307–352.

[21] Yu.I. Karlovich and B. Silbermann, *Fredholmness of singular integral operators with discrete subexponential groups of shifts on Lebesgue spaces.* Math. Nachr. **272** (2004) 55–94.

[22] P.S. Muhly and J. Xia, *Calderón–Zygmund operators, local mean oscillation and certain automorphisms of the Toeplitz algebra.* Amer. J. Math. **117** (1995) 1157–1201.

[23] M.A. Naimark, *Normed Algebras.* Wolters-Noordhoff Publishing, Groningen, 1972.

[24] S.P. Novikov, *Topology of foliations.* Trans. Mosc. Math. Soc. **14** (1965) 268–304.

[25] D. Sarason, *Functions of vanishing mean oscillation.* Trans. Amer. Math. Soc. **207** (1975) 391–405.

[26] D. Sarason, *Toeplitz operators with piecewise qasicontinuous symbols.* Indiana Univ. Math. J. **26** (1977) 817–838.

M. Amélia Bastos
Departamento de Matemática
Instituto Superior Técnico
Universidad de Lisboa
Av. Rovisco Pais
1049–001 Lisboa, Portugal
e-mail: abastos@math.ist.utl.pt

Cláudio A. Fernandes
Centro de Matemática e Aplicações
Departamento de Matemática
Faculdade de Ciências e Tecnologia
Universidade Nova de Lisboa
Quinta da Torre
2829–516 Caparica, Portugal
e-mail: caf@fct.unl.pt

Yuri I. Karlovich
Centro de Investigación en Ciencias
Instituto de Investigación en Ciencias Básicas y Aplicadas
Universidad Autónoma del Estado de Morelos
Av. Universidad 1001, Col. Chamilpa
C.P. 62209 Cuernavaca, Morelos, México
e-mail: karlovich@uaem.mx

Operator Theory:
Advances and Applications, Vol. 267, 65–78

Non-Hermitian Quantum Mechanics of Bosonic Operators

Natália Bebiano, João da Providência and J.P. da Providência

Abstract. The spectral analysis of a family of non-Hermitian operators appearing in quantum physics is our main concern. The properties of such operators are essentially different from those of Hermitian Hamiltonians, namely due to spectral instabilities. We demonstrate that the considered operators and their adjoints can be diagonalized when expressed in terms of certain conveniently constructed operators. We show that their eigenfunctions constitute complete systems, but do not form Riesz bases. Attempts to overcome this difficulty in the quantum mechanical set up are pointed out.

Mathematics Subject Classification (2010). 47A10, 81Q12, 34L10, 35A27, 47B44.

Keywords. Bosonic operator, non-Hermitian Hamiltonian, spectral analysis, Riesz basis, metric.

1. Introduction

The main motivation for this article is the following. In conventional formulations of non-relativistic quantum mechanics, the Hamiltonian operator is Hermitian (synonymously, self-adjoint), and so it has real eigenvalues and an orthonormal set of eigenfunctions. These fundamental issues are in the heart of von Neumann quantum paradigm for physical observability and dynamical evolution. Certain relativistic extensions of quantum mechanics lead to non-Hermitian Hamiltonian operators, $H \neq H^*$, for H^* the adjoint of H (e.g., see [5, Chapter VIII]). Extending the set of allowed operators by including $\mathcal{PT}$-symmetric ones, $\mathcal{P}$ being the *reflection* operator, $\mathcal{P}f(x) = f(-x)$ and $\mathcal{T}$ the *time reflection* operator, $\mathcal{T}f(x) = \overline{f(x)}$,

This work was partially supported by the Centro de Matemática da Universidade de Coimbra (CMUC), funded by the European Regional Development Fund through the program COMPETE and by the Portuguese Government through the FCT – Fundação para a Ciência e a Tecnologia under the project PEst-C/MAT/UI0324/2011.

yields spectra with reflection symmetry with respect to the real axis. In this context, non-Hermitian Hamiltonians with a purely discrete spectrum have been the object of intense research activity [1, 3, 4, 8, 11, 12, 14, 15], attempting to build quantum mechanical theories with physical observables described by these operators. The subtleties of non self-adjoint Hamiltonians deserved the attention of physicists and mathematicians, as they may originate new and unexpected phenomena, and the non self-adjoint theory has revealed difficulties and challenging problems.

We focus on the spectral analysis of non-Hermitian operators, a field with applications in many areas of physics, most remarkably in quantum mechanics. We summarise theoretical classical and recent developments on this topic, and the theory is illustrated with a concrete example. We shall be particularly concerned with $L^2(\mathbb{R}^2)$, the Hilbert space of square-integrable complex valued functions in two real variables x, y, equipped with the inner product

$$\langle \Phi, \Psi \rangle = \int_{-\infty}^{+\infty} \int_{-\infty}^{+\infty} \Phi(x,y)\overline{\Psi(x,y)}\mathrm{d}x\mathrm{d}y,$$

and corresponding norm $\|\cdot\|$. The main topic of the present paper is the study of the spectral properties of the operator in $L^2(\mathbb{R}^2)$,

$$H := -\frac{1}{4}\left(\frac{\partial^2}{\partial x^2} + \frac{\partial^2}{\partial y^2}\right) + \gamma\left(y\frac{\partial}{\partial x} + x\frac{\partial}{\partial y}\right) + (x^2+y^2), \quad \gamma \in \mathbb{R}, \tag{1}$$

which is obviously non self-adjoint for $\gamma \neq 0$. We explicitly determine the eigenfunctions and eigenvalues of H, which can be rewritten as a quadratic function of two interacting bosonic operators and their adjoints.

Let $\mathcal{H}$ be an infinite-dimensional separable Hilbert space endowed with the inner product $\langle \cdot, \cdot \rangle$, and corresponding norm $\|\cdot\|$. For H a non-Hermitian operator in $\mathcal{H}$ with a purely discrete spectrum, the strategy of finding a Hermitian operator H_0 and an invertible operator P such that

$$H = PH_0P^{-1}, \tag{2}$$

has been exploited. This idea refers that such a non-Hermitian H can be viewed essentially as an alternative representation of a Hermitian operator with the same eigenvalues and multiplicities, whenever P, P^{-1} are both bounded (see [10]).

The relation (2) is closely related to the *quasi-Hermicity* of H

$$QH = H^*Q, \tag{3}$$

where $Q = (P^{-1})^*P^{-1}$ is a positive definite operator called a *metric operator*. (The relation (3) means that QH and H^*Q have the same domain and act in the same manner.) In fact, H with property (2), is formally Hermitian with respect to the modified inner product $\langle Q\,\cdot, \cdot \rangle$,

$$\langle QHx, y \rangle = \langle x, H^*Qy \rangle = \langle x, QHy \rangle, \quad x, y \in \mathcal{H}.$$

The metric is said to be *non singular* if it is bounded, invertible and boundedly invertible, otherwise it is *singular*. The pathologies of non-Hermitian operators with singular metric may be serious, since the metric can transform a basis into a set without any reasonable basicity properties. Moreover, the spectral stability with respect to small perturbations is not ensured and complex eigenvalues can appear distant to the original ones [13].

The rest of this note is organized as follows. In Section 2 we recall some issues used subsequently. In Section 3 we investigate spectral properties of H in (1). It is shown that the obtained eigenfunctions are complete systems but do not form Riesz bases, and so the validity of useful properties of Hermitian operators is not guaranteed. In fact, non-Hermitian operators behave in an essentially different way, due in particular to spectral instabilities and to the lack of basicity of the Hamiltonian eigenfunctions. In Section 5 we discuss incidences of non-Hermiticity in quantum physics.

2. Preliminaries

Many important operators in quantum physics are unbounded, which restricts their domains of definition to adequate subsets of the Hilbert space where they live. For instance, these domains are considered to be dense.

The set of eigenfunctions of an operator H, $\{\Psi_n\}_{n=1}^{\infty}$, forms a *complete* family in $\mathcal{H}$ if the span of Ψ_n is dense in this space, or, equivalently, if the orthogonal complement of this linear span in the Hilbert space reduces to the zero function.

Eigenfunctions of non-Hermitian operators are in general not orthogonal or even do not form a complete family. It should be stressed that the completeness of an orthogonal set $\{\Psi_n\}_{n=1}^{\infty}$, adequately normalized, does not imply, by its own, basicity. An orthonormal set $\{\Psi_n\}_{n=1}^{\infty}$ does not guarantee that any $\psi \in \mathcal{H}$ admits a unique expansion of the form

$$\psi = \sum_{n=1}^{\infty} c_n \Psi_n,$$

where, in this event, it should be $c_n = \langle \psi, \Psi_n \rangle, \; n \geq 1$.

The concept of eigenbasis is very important in quantum mechanics. In contrast to the case of eigenfunctions of Hermitian operators, which are basis, for non-Hermitian operators this property fails in general, and so the notion of *Riesz basis* may be of interest. We say that $\{\Psi_n\}_{n=1}^{\infty}$ is a Riesz basis if for any $\psi \in \mathcal{H}$, there exists a positive constant C independent of ψ such that

$$C^{-1}\|\psi\|^2 \leq \sum_{n=1}^{\infty} |\langle \psi, \Psi_n \rangle|^2 \leq C\|\psi\|^2.$$

The eigenfunctions of an operator H with purely discrete spectrum constitute a Riesz basis if and only if H is quasi-Hermitian via a nonsingular bounded metric Q (see [2]). The eigenfunctions of a non-Hermitian H, despite possibly being a complete family, may not form a Riesz basis, as occurs frequently with several models [10]. It should be noticed that Riesz basicity, like the spectrum, is not preserved by unbounded metrics.

We introduce, as a first example of familiar operators in quantum mechanics, the *multiplication* operators x and y

$$f(x,y) \to xf(x,y), \quad f(x,y) \to yf(x,y),$$

defined in their maximal domains $\mathcal{D}(x) = \{\psi \in L^2(\mathbb{R}^2) : x\psi \in L^2(\mathbb{R}^2)\}$ and $\mathcal{D}(y) = \{\psi \in L^2(\mathbb{R}^2) : y\psi \in L^2(\mathbb{R}^2)\}$.

We consider the partial differential operators $\partial/\partial x$ and $\partial/\partial y$ defined by

$$f(x,y) \to \frac{\partial f(x,y)}{\partial x}, \quad f(x,y) \to \frac{\partial f(x,y)}{\partial y},$$

with domains $\mathcal{D}(\partial/\partial x) = \mathcal{D}(\partial/\partial y) = W^{1,2}(\mathbb{R}^2)$, where $W^{1,2}(\mathbb{R}^2)$ stands for the usual Lebesgue space of functions in $L^2(\mathbb{R}^2)$, whose first partial derivatives belong to $L^2(\mathbb{R}^2)$. These operators are known as *momentum* operators. The partial differential operators $\partial^2/\partial x^2$ and $\partial^2/\partial y^2$ are defined by

$$f(x,y) \to \frac{\partial^2 f(x,y)}{\partial x^2}, \quad f(x,y) \to \frac{\partial^2 f(x,y)}{\partial y^2},$$

with domains $\mathcal{D}(\partial^2/\partial x^2) = \mathcal{D}(\partial^2/\partial y^2) = W^{2,2}(\mathbb{R}^2)$, the Lebesgue space of functions in $L^2(\mathbb{R}^2)$ whose first and second derivatives belong to $L^2(\mathbb{R}^2)$.

The (*standard*) *bosonic operators*

$$a := x + \frac{1}{2}\frac{\partial}{\partial x}, \quad b := y + \frac{1}{2}\frac{\partial}{\partial y},$$

with domains

$$\mathcal{D}(a) = \mathcal{D}(b) = \{\Psi \in W^{2,2}(\mathbb{R}^2) : x\Psi, y\Psi \in L^2(\mathbb{R}^2)\}$$

are useful in our discussion. The operators a and b are known to be densely defined, closed, and their adjoints read

$$a^* = x - \frac{1}{2}\frac{\partial}{\partial x}, \quad b^* = y - \frac{1}{2}\frac{\partial}{\partial y}.$$

We recall that, conventionally, a, b are said to be *annihilation operators*, while a^*, b^* are *creation operators*. It is worth noticing that these operators are unbounded, and they satisfy the commutation rules (CR's),

$$[a, a^*] = [b, b^*] = 1, \tag{4}$$

where 1 is the identity operator on $L^2(\mathbb{R}^2)$. (This means that $aa^*f - a^*af = bb^*f - b^*bf = f$ for any f in $\mathcal{D}(a)$ and $\mathcal{D}(b)$). Furthermore,

$$[a, b^*] = [b, a^*] = [a^*, b^*] = [a, b] = 0. \tag{5}$$

As it is well known, the canonical commutation relations (4) and (5) characterize an algebra of Weil–Heisenberg (W-H). Moreover, the following holds,

$$a\Phi_0 = b\Phi_0 = 0,$$

for $\Phi_0 = \mathrm{e}^{-(x^2+y^2)}$ in $\mathcal{D}(a)$ and $\mathcal{D}(b)$, a so-called *vacuum state*. The set of functions

$$\{\Phi_{m,n} = a^{*m} b^{*n} \Phi_0 : \ m, n \geq 0\},$$

constitutes a basis of $\mathcal{H}$, that is, every vector in $L^2(\mathbb{R}^2)$ can be uniquely expressed in terms of this system, which is *complete*, since 0 is the only vector orthogonal to all its elements.

3. Non self-adjoint Hamiltonian H describing two interacting bosonic operators

3.1. Model

The closedness of the operator H in (1) is the essential starting point for the study of its spectrum. In the perspective of a convenient domain of H, aiming at defining $\mathcal{D}(H)$, we regard the cofactor of γ in the non-self-adjoint term,

$$V = \left(y\frac{\partial}{\partial x} + x\frac{\partial}{\partial y} \right),$$

as a perturbation of the 2D harmonic oscillator,

$$\Re(H) = -\frac{1}{4}\left(\frac{\partial^2}{\partial x^2} + \frac{\partial^2}{\partial y^2} \right) + (x^2 + y^2).$$

It is easily seen that this term is relatively bounded with the relative bound less than one with respect to $\Re(H)$, provided that γ is less than one. Under this condition, H is a closed operator on the domain of the harmonic oscillator:

$$W^{2,2}(\mathbb{R}^2) \cap L^2(\mathbb{R}^2, (x^4 + y^4)\mathrm{d}x\mathrm{d}y).$$

Here, we have used the standard perturbation result that states the following. If $\Re(H)$ is closed and V is relatively bounded with respect to $\Re(H)$, with the relative bound smaller than 1, then $\Re(H) + \lambda V$, with $\lambda < 1$, is closed [9, Theorem 3.3].

We notice that $\mathcal{D}(H)$ contains the subspace $\mathfrak{S}(\mathbb{R}^2)$ constituted by functions $f(x, y)$ such that $e^{-2\gamma xy} f(x, y) \in \mathfrak{S}(\mathbb{R}^2)$ and, in turn, this one contains $C_0^\infty(\mathbb{R}^2)$.

In terms of creation and annihilation operators, H is equivalently defined as

$$H = a^*a + bb^* + \gamma(a^*b^* - ab), \quad \gamma \in \mathbb{R}, \tag{6}$$

and so H is quadratic in the bosonic operators a, b and in their adjoints.

3.2. Spectrum

As it is well known, the *resolvent set* of H, denoted by $\rho(H)$, is constituted by all the complex numbers for which the *resolvent operator* $\lambda \in \rho(H) \to (H - \lambda)^{-1}$ exists as a bounded operator on $\mathcal{H}$. The complement

$$\sigma(H) = \mathbb{C} \backslash \rho(H)$$

is called the *spectrum* of H. The set of all eigenvalues of H is called the *point spectrum*, denoted by $\sigma_p(H)$, and formed by complex numbers λ for which $H - \lambda :$ $\mathcal{D}(H) \to \mathcal{H}$ is not injective. The spectrum of an operator on a finite-dimensional

Hilbert space is exhausted by the eigenvalues, but, in the infinite-dimensional setting, there are additional parts to be considered. Those λ which are not eigenvalues but $H-\lambda$ is not bijective constitute the *continuous* or *residual spectrum*, depending on the range $\mathrm{Ran}(H-\lambda)$ being, respectively, dense or not. The spectrum $\sigma(H)$ is the union of these three disjoint spectra. The spectrum of self-adjoint operators is nonempty, real, and the residual spectrum is empty, while the spectrum of non self-adjoint operators can be empty or coincide with the whole complex plane (see, e.g., refs. [13, 15]).

4. Accretivity

It may be advantageous to estimate the spectrum in terms of the *numerical range*:

$$W(H) := \{\langle H\psi, \psi\rangle \,:\, \psi \in \mathcal{D}(H),\ \|\psi\| = 1\}.$$

In general $W(H)$ is neither open nor closed, even when H is a closed operator. However, it is always convex and for H bounded,

$$\sigma(H) \subset \overline{W(H)}.$$

Proposition 4.1. *The numerical range of H is bounded by the hyperbola*

$$y^2 + \gamma^2(1-x^2) = 0, \quad x \geq 1.$$

Proof. The numerical range of H is determined as follows. Recalling that $W(H)$ is convex, let us consider the supporting line of $W(H)$ perpendicular to the direction θ. Recall that the distance of this line to the origin is the lowest eigenvalue of

$$\Re(\mathrm{e}^{-i\theta}H) = (a^*a + b^*b)\cos\theta - i\gamma(a^*b^* - ab)\sin\theta,$$

provided this operator is bounded from below, which occurs for $-\pi/2 \leq \theta < \pi/2$. The eigenvalues of $\Re(\mathrm{e}^{-i\theta}H)$ are readily determined by the EMM [5], and they are found to be

$$E_\theta(1+n+m), \quad n, m \geq 0,$$

where

$$E_\theta = \frac{\sqrt{1-\gamma^2+\cos(2\theta)+\gamma^2\cos(2\theta)}}{\sqrt{2}}.$$

Thus, the supporting line under consideration is given by

$$x\cos\theta + y\sin\theta = E_\theta. \tag{7}$$

As it is well known, the equation of the boundary of $W(H)$ is found eliminating θ between (7) and

$$-x\sin\theta + y\cos\theta = E'_\theta,$$

being easily obtained as the branch of the hyperbola

$$y = \pm\gamma\sqrt{x^2-1}, \quad x \geq 1. \qquad \square$$

Sectorial operators are defined by the property that their numerical range is the subset of a sector

$$S_{w,\theta} = \{z \in \mathbb{C} : |\arg(z - w)| \leq \theta\}$$

with $w \in \mathbb{R}$ and $0 \leq \theta < \pi/2$, called, respectively, the *vertex* and *semi-angle* of H. An operator is said to be *accretive* if the vertex can be chosen at the origin, i.e., $W(H) \subset S_{0,\pi/2}$. An operator H is *m-accretive* if its numerical range is contained in the right closed half-plane and the *resolvent bound* for any λ with $\Re\lambda < 0$, holds:

$$\forall \lambda \in \mathbb{C}, \quad \Re\lambda < 0, \quad \|(H - \lambda)^{-1}\| \leq 1/|\Re\lambda|.$$

Proposition 4.2. *The operator H is m-accretive.*

Proof. We show that for any $z \in \mathbb{C}$, with $\Re z < 0$, the resolvent bound holds. We have

$$\operatorname{dist}(z, \overline{W(H)}) \leq |\langle H\psi, \psi\rangle - z| = |\langle (H - z)\psi, \psi\rangle| \leq \|(H - z)\psi\|.$$

As $\operatorname{dist}(z, \overline{W(H)}) \geq |\Re z|$, the result follows, having in mind Proposition 4.1. □

Any m-accretive operator is closed and densely defined. In fact, by [2, Proposition 5.2.1, p. 246], if H is not closed, then $\sigma(H) = \mathbb{C}$.

A closed operator H in $\mathcal{H}$ has a *compact resolvent* if $\rho(H) \neq \emptyset$ and $(H - \lambda)^{-1}$, for some $\lambda \in \rho(H)$, is a compact operator.

It is known that [9, Theorem IX, 2.3], if H has a compact resolvent, then $\sigma(H) = \sigma_p(H)$. The operator H has a compact resolvent, as $\Re(H)$ is an m-accretive operator (since $\Re(H)$ is Hermitian and $W(\Re(H))$ lies on the positive real axis) with compact resolvent and, moreover, V is relatively bounded with respect to $\Re(H)$ with relative bound smaller than 1. Then, $\Re(H) + \lambda V$ has a compact resolvent [9, Theorem 5.4.1].

4.1. Eigenvectors and eigenvalues of H

To obtain the eigenvalues of H, we firstly consider the selfadjoint operator in $L^2(\mathbb{R}^2)$,

$$H_0 = -\frac{1}{4}\left(\frac{\partial^2}{\partial x^2} + \frac{\partial^2}{\partial y^2}\right) + (1 + \gamma^2)(x^2 + y^2), \quad \gamma < 1. \tag{8}$$

We know *a priori* that the eigenfunctions of H_0 (after normalization) form an orthonormal family in $L^2(\mathbb{R}^2)$ and the corresponding eigenvalues are real. The eigenvalues of H_0 are easily determined as

$$E_{m,n} = (1 + m + n)\sqrt{1 + \gamma^2}, \quad m, n \geq 0,$$

and the associated eigenfunctions $\Phi_{m,n}$ are

$$\Phi_{m,n} = K H_m((1 + \gamma^2)^{1/4}x)\mathrm{e}^{-\sqrt{1+\gamma^2}x^2} \times H_n((1 + \gamma^2)^{1/4}y)\mathrm{e}^{-\sqrt{1+\gamma^2}y^2},$$

where $K = (2(1 + \gamma^2)/\pi)^{1/2}$ and $H_n(t)$ is the nth Hermite polynomial in t. That is, $\Phi_{m,n}$ is factorized as follows

$$\Phi_{m,n}(x, y) = \Phi_m(x)\Phi_n(y),$$

where $\Phi_n(t)$ are the usual eigenfunctions of the famous harmonic oscillator [2]. For the sake of completeness, we show how to obtain $\Phi_{m,n}$. Indeed, let us consider the differential operators

$$g = \frac{1}{2(1+\gamma^2)^{1/4}}\frac{\partial}{\partial x} + (1+\gamma^2)^{1/4}x, \qquad h = \frac{1}{2(1+\gamma^2)^{1/4}}\frac{\partial}{\partial y} + (1+\gamma^2)^{1/4}y,$$
$$g^* = -\frac{1}{2(1+\gamma^2)^{1/4}}\frac{\partial}{\partial x} + (1+\gamma^2)^{1/4}x, \ h^* = -\frac{1}{2(1+\gamma^2)^{1/4}}\frac{\partial}{\partial y} + (1+\gamma^2)^{1/4}y,$$

which satisfy the Weil–Heisenberg commutation rules,

$$[g, g^*] = [h, h^*] = 1, [g, h] = [g^*, h^*] = [g^*, h] = [g, h^*] = 0.$$

We easily find that

$$H_0 = \sqrt{1+\gamma^2}(g^*g + h^*h + 1).$$

The *groundstate* eigenfunction of H_0, which is the vacuum of the operators g, h, is

$$\Phi_0(x, y) = \kappa_{0,0}\mathrm{e}^{-\sqrt{1+\gamma^2}(x^2+y^2)},$$

being the remaining eigenfunctions,

$$\Phi_{m,n}(x, y) = \kappa_{m,n}g^{*m}h^{*n}\Phi_0(x, y),$$

where $\kappa_{m,n}$ are normalization factors.

The eigenfunctions $\Phi_{m,n}(x, y)$ constitute a complete set in $L^2(\mathbb{R}^2)$, as can be easily verified by adapting the standard proof of completeness of Hermite functions. The functions are orthogonal, and the system forms a basis of $L^2(\mathbb{R}^2)$. Details can be found, for instance, in [5, 6].

To find the eigenvalues of H we notice that H is *formally* similar to H_0

$$H_0 = \mathrm{e}^{-2\gamma xy}H\mathrm{e}^{2\gamma xy}.$$

The word "formally" refers to the fact that the operator $P = \mathrm{e}^{2\gamma xy}$ is unbounded. Nevertheless, the similarity relation is well defined on the eigenfunctions of H_0.

For $\Phi(x, y)$ an arbitrary differentiable function in the domain of H_0, we have

$$\left(-\frac{1}{4}\left(\frac{\partial^2}{\partial x^2} + \frac{\partial^2}{\partial y^2}\right) + \gamma\left(y\frac{\partial}{\partial x} + x\frac{\partial}{\partial y}\right) + (x^2+y^2)\right)\mathrm{e}^{2\gamma xy}\Phi(x, y)$$
$$= \mathrm{e}^{2\gamma xy}\left(-\frac{1}{4}\left(\frac{\partial^2}{\partial x^2} + \frac{\partial^2}{\partial y^2}\right) + (1+\gamma^2)(x^2+y^2)\right)\Phi(x, y),$$

that is

$$H\mathrm{e}^{2\gamma xy}\Phi = \mathrm{e}^{2\gamma xy}H_0\Phi,$$

and the operator equality

$$H\mathrm{e}^{2\gamma xy} = \mathrm{e}^{2\gamma xy}H_0$$

holds.

Since

$$H_0\Phi_{m,n} = E_{m,n}\Phi_{m,n},$$

we easily get

$$H\mathrm{e}^{2\gamma xy}\Phi_{m,n} = E_{m,n}\mathrm{e}^{2\gamma xy}\Phi_{m,n}.$$

The eigenfunctions of H are expressed as

$$\Psi_{m,n} = \mathrm{e}^{2\gamma xy}\Phi_{m,n},$$

and the eigenvalues of H are those of H_0. Thus, the eigenvalues of H are real positive and $\Psi_{m,n} \in \mathcal{D}(H)$, as it should be.

4.2. Eigenfunctions and eigenvalues of H^*

Following the procedure in Subsection 4.1, it can be shown that, formally, we may write,

$$H_0 = \mathrm{e}^{2\gamma xy} H^* \mathrm{e}^{-2\gamma xy},$$

implying that eigenfunctions and associated eigenvalues of H^* are given respectively by

$$\tilde{\Psi}_{m,n} = \mathrm{e}^{-2\gamma xy}\Phi_{m,n} \in \mathcal{D}(H^*), \quad E_{m,n} = (m+n+1)\sqrt{1+\gamma^2}, \quad m,n \geq 0.$$

4.3. Biorthogonality of the eigenfunctions

It is straightforward to see that the set $\{\Psi_{m,n} : m,n \geq 0\}$ is not constituted by orthogonal functions. However, the vector system $\{\tilde{\Psi}_{m,n} : m,n \geq 0\}$, formed by the eigenfunctions of H^*, is biorthogonal to the eigensystem of H, $\{\Psi_{m,n} : m,n \geq 0\}$. Indeed, we have

$$\langle \Psi_{m,n}, \tilde{\Psi}_{p,q}\rangle = \langle \mathrm{e}^{2\gamma xy}\Phi_{m,n}, \mathrm{e}^{-2\gamma xy}\Phi_{p,q}\rangle = \langle \Phi_{m,n}, \Phi_{p,q}\rangle = m!n!\delta_{mp}\delta_{nq},$$

where $\delta_{ij} = 1$ for $i = j$, otherwise $\delta_{ij} = 0$, represents the Kronecker symbol.

4.4. Completeness of the eigenfunctions

The completeness of both eigensystems of H and H^* can be shown to hold. The operator is m-accretive as its numerical range lies a hyperbolical region limited by the branch of hyperbola $y = \sqrt{x^2 - 1}$, $x \geq 1$. Moreover, it can be easily seen that the imaginary part of the resolvent of $-iH$ at $\delta < 0$, is non-negative Hermitian

$$\frac{1}{2i}\left((-iH - \delta)^{-1} - (iH^* - \delta)^{-1}\right) \geq 0.$$

As the resolvent is a trace class function, by the completeness theorem [7, Theorem VII.8.1], we may conclude that the eigenfunctions of H form a complete system. Analogous arguments are valid for the eigensystem of H^*.

We observe that completeness does not imply that any $\psi \in L^2(\mathbb{R}^2)$ has a unique expansion

$$\psi = \sum_{m,n=0}^{\infty} c_{m,n}\Psi_{m,n},$$

a fundamental issue in quantum mechanics.

4.5. Asymptotic behavior of the eigenfunctions

We wish to discuss the asymptotic behavior of the Hamiltonian eigenfunctions. To this end, it is convenient to introduce the Planck constant $\hbar$, explicitly. We also change the notation slightly. We replace the notation x, y, used for the particle coordinates, by x_1, x_2, and we denote the respective momenta by

$$p_1 = -i\hbar\frac{\partial}{\partial x_1}, \quad p_2 = -i\hbar\frac{\partial}{\partial x_2}.$$

For simplicity, we also consider $\gamma = 1$, so that (6) becomes

$$H = \frac{1}{4}(p_1^2 + p_2^2) + x_1^2 + x_2^2 + i(x_1p_2 + x_2p_1) - 1.$$

Next, we introduce the change of variables, which constitutes a canonical transformation,

$$X = \frac{1}{2}(x_1 + x_2), \quad x = x_1 - x_2, \quad P = p_1 + p_2, \quad p = \frac{1}{2}(p_1 - p_2).$$

In terms of the new variables, the Hamiltonian is the sum of two summands, each one involving only one type of variable, and becomes

$$H = \frac{1}{8}P^2 + 2X^2 + iXP + \frac{1}{2}p^2 + \frac{1}{2}x^2 + ixp - 1.$$

We observe that it is equivalent to consider $m, n \to \infty$ or $\hbar \to 0$. We firstly concentrate on the summand

$$H_X = \frac{1}{8}P^2 + 2X^2 + iXP.$$

In order to characterize the asymptotic behavior of the eigenfunctions we use the well-known Wentzel–Kramers–Brillouin [WKB] approximation [5, Chapter III] in leading order. Indeed, we express the eigenfunction of the energy operator as

$$\Psi(X) = e^{i\mathrm{S}(X)/\hbar}.$$

The function $\mathrm{S}(X)$, in leading order, is determined by the Jacobi equation

$$\frac{1}{8}\left(\frac{\mathrm{dS}}{\mathrm{d}X}\right)^2 + 2X^2 + iX\frac{\mathrm{dS}}{\mathrm{d}X} = E,$$

where E denotes the associated eigenvalue of the energy operator. Thus, we readily obtain,

$$\frac{\mathrm{dS}}{\mathrm{d}X} = 2\sqrt{2E - 8X^2} - 4iX,$$

and

$$S = X\sqrt{2E - 8X^2} + \frac{E}{\sqrt{2}}\arctan\frac{2X}{\sqrt{E - 4X^2}} - 2iX^2.$$

Similarly, the eigenfunction of H^* is

$$\tilde{\Psi}(X) = e^{i\tilde{\mathrm{S}}(X)/\hbar},$$

with

$$\tilde{S} = X\sqrt{2E - 8X^2} + \frac{E}{\sqrt{2}} \arctan \frac{2X}{\sqrt{E - 4X^2}} + 2iX^2.$$

Therefore, the integral

$$\int_{-\sqrt{E/2}}^{\sqrt{E/2}} \overline{\Psi}\Psi \, dX$$

approaches $+\infty$ as $\hbar \to 0$, while the integral

$$\int_{-\sqrt{E/2}}^{\sqrt{E/2}} \overline{\tilde{\Psi}}\tilde{\Psi} \, dX$$

approaches 0 as $\hbar \to 0$.

On the other hand, the integral

$$\int_{-\sqrt{E/2}}^{\sqrt{E/2}} \overline{\Psi}\tilde{\Psi} \, dX$$

remains finite as $\hbar \to 0$.

The summand $H_x = \frac{1}{2}p^2 + \frac{1}{2}x^2 + ixp$ may be similarly treated.

As a consequence, we get

$$\lim_{m,n\to\infty} \|\Psi_{m,n}\| = \infty.$$

Thus, the system of eigenvectors of $\{\Psi_{m,n}\}$ does not form a Riesz basis for $L^2(\mathbb{R}^2)$. Analogous conclusion holds for $\{\tilde{\Psi}_{m,n}\}$.

4.6. Metric operator

The existence of a positive definite Q satisfying (3) is equivalent to the fact that the resolvent of H satisfies

$$Q(H - z_0)^{-1} = (H^* - z_0)^{-1}Q$$

for $z_0 \in \rho(H) \cap \rho(H^*) \cap \mathbb{R}^2$.

As a consequence of the reality of the discrete spectrum of H and of the completeness of the corresponding eigenfunctions, it can be shown, using the procedure in [13], that there exists a bounded metric for H. The proof relies on the fact that the existence of bounded metric for an unbounded H can be transferred to the same problem for its bounded resolvent.

Nevertheless, there does not exist a non singular metric operator ensuring quasi-Hermiticity, because the eigenfunctions of H do not form a Riesz basis. As a consequence, in the similarity of H to a self-adjoint operator, the basicity properties of these operators may be very different (see [1] and [2]). However, despite these negative features, in the subspace $\mathcal{S} = \text{span}\{\Psi_{m,n}\}$, a new inner product may be meaningfully defined with the help of a metric operator (non-singular in $\mathcal{S}$), with respect to which the restriction of H to $\mathcal{S}$, say h, should represent H.

4.7. Physical Hilbert space

In order to specify the *physical Hilbert space* associated with the Hamiltonian (1), we chose its domain to be the function space

$$\mathrm{Dom}(H) = \{\Psi : \mathrm{e}^{-2\gamma xy}\Psi \in W^{2,2}(\mathbb{R}^2) \cap L^2(\mathbb{R}^2, (x^4 + y^4)\mathrm{d}x\mathrm{d}y)\},$$

endowed with the inner product,

$$\langle\langle \Xi_1, \Xi_2 \rangle\rangle = \int_{-\infty}^{\infty} \mathrm{d}x \int_{-\infty}^{\infty} \mathrm{d}y \mathrm{e}^{-4\gamma xy} \Xi_1(x,y)\overline{\Xi_2(x,y)}, \quad \Xi_1, \Xi_2 \in \mathrm{Dom}(H),$$

which involves the weight function $\mathrm{e}^{-4\gamma xy}$. Since H_0 in (8) is a closed operator on the domain of the harmonic oscillator

$$\mathrm{Dom}(H_0) = W^{2,2}(\mathbb{R}^2) \cap L^2(\mathbb{R}^2, (x^4 + y^4)\mathrm{d}x\mathrm{d}y),$$

and

$$H\mathrm{e}^{2\gamma xy}\Phi = \mathrm{e}^{2\gamma xy} H_0 \Phi, \quad \forall \Phi \in \mathrm{Dom}(H_0),$$

while

$$\mathrm{e}^{-2\gamma xy} H\Psi = H_0 \mathrm{e}^{-2\gamma xy}\Psi, \quad \forall \Psi \in \mathrm{Dom}(H),$$

it follows that H is a closed operator on $\mathrm{Dom}(H)$.

Let

$$\mathcal{S} = \mathrm{span}\{\Psi_{m,n}\}, \quad \tilde{\mathcal{S}} = \mathrm{span}\{\tilde{\Psi}_{m,n}\},$$

and $\mathcal{D} = \mathrm{span}\{\Phi_{m,n}\}$.

In terms of the linear operators $\mathrm{e}^{2\gamma xy} : \mathcal{D} \to \mathcal{S}$ and $\mathrm{e}^{-2\gamma xy} : \mathcal{D} \to \tilde{\mathcal{S}}$ we may write

$$\Psi_{m,n} = \mathrm{e}^{2\gamma xy}\Phi_{m,n}, \quad \tilde{\Psi}_{m,n} = \mathrm{e}^{-2\gamma xy}\Phi_{m,n}. \tag{9}$$

From (9) it follows that

$$\mathcal{S} = \mathrm{e}^{S}(\mathcal{D}), \quad \tilde{\mathcal{S}} = \mathrm{e}^{-S}(\mathcal{D}).$$

Since

$$\langle \mathrm{e}^{-S}\Psi_{m,n}, \mathrm{e}^{-S}\Psi_{p,q} \rangle = \delta_{mp}\delta_{nq}, \quad S = 2\gamma xy,$$

it follows that the eigenfunctions $\Psi_{m,n}$ are orthogonal for this inner product,

$$\langle\langle \Psi_{m,n}, \Psi_{p,q} \rangle\rangle = \delta_{mp}\delta_{nq}.$$

Next we show that the linear space $\mathcal{S}$, equipped with the inner product $\langle\langle\cdot,\cdot\rangle\rangle$, allows the probabilistic interpretation of quantum mechanics. The symmetry is obviously satisfied. It may be pointed out that, from the point of view of physics, only the action of the metric operator $\exp(-2S)$ on $\mathcal{S}$ is significant and, in this event, it is unambiguously defined. We verify that $\exp(-2S)$ is a metric. For $\phi, \psi \in \mathcal{S}$, $\langle \exp(-2S)\phi, \psi \rangle$ is finite, and so the operator $\exp(-2S)$ is bounded in $\mathcal{S}$. (Similarly, $\exp(2S)$ is bounded in $\tilde{\mathcal{S}}$.) Moreover,

$$\langle\langle \psi, \psi \rangle\rangle, \ \psi \in \mathcal{S},$$

is nonnegative. (We notice that H leaves $\mathcal{S}$ invariant.)

The vectors $\Psi_{m,n}$ are orthogonal with respect to the new inner product and any $\Psi \in \mathcal{S}$ may be expanded as a unique finite linear combination of $\Psi_{n,m}$,

$$\Psi = \sum\nolimits_{n,m} c_{nm} \Psi_{n,m}, \quad \text{with} \quad \langle\!\langle \Psi, \Psi_{n,m} \rangle\!\rangle = c_{nm}.$$

In order to ensure the probabilistic interpretation of quantum mechanics, we impose the normalization $\langle\!\langle \Psi, \Psi \rangle\!\rangle = 1$, so that

$$\langle\!\langle \Psi, \Psi \rangle\!\rangle = \sum\nolimits_{n,m} |c_{nm}|^2 = 1.$$

The probability amplitude is given by c_{nm} and satisfies $\sum_{n,m} |c_{nm}|^2 = 1$.

5. Conclusions

We have initially considered a non self-adjoint Hamiltonian H whose eigenvalues and corresponding eigenfunctions have been explicitly determined. The investigated Hamiltonian and its adjoint have real eigenvalues and systems of biorthogonal eigenvectors. They have infinite diagonal matrix representations in the respective eigensystems, which are complete. Nevertheless, they do not form Riesz bases.

Viewing H as the Hamiltonian of a physical model, problems arise from non Hermiticity. The original inner product defined in $\mathcal{H}$ is not adequate for the physical interpretation of the model. A new Q-metric, which is appropriate for that purpose, may be introduced (see Subsection 4.7). Following Mostafazadeh [11], one can define a subspace of the Hilbert space, and the restriction of the Hamiltonian operator to that subspace, so that it has the same spectrum and eigenfunctions as the original one. The referred subspace remains invariant under the action of H. Remarkably, stating that this Hermitian operator represents in a reasonable sense the non-Hermitian operator may be controversial, since relevant information on the Hamiltonian may not be captured in the mentioned subspace.

Non-Hermitian operators have typically non-trivial pseudospectra. It is known that the relation (2) holds via a bounded and boundedly invertible positive transformation if and only if (3) holds with a positive bounded and boundedly invertible metric [10]. Further, if (3) holds with a positive bounded and boundedly invertible metric, then the pseudospectrum of H is trivial. The concept of pseudospectrum is of great relevance for the description of non-Hermitian operators in the context of quantum mechanics. A non-trivial pseudospectrum ensures the non-existence of a bounded metric.

References

[1] F. Bagarello, *Construction of pseudo-bosons systems.* J. Math. Phys. **51** (2010), 023531.

[2] F. Bagarello, J.-P. Gazeau, F.H. Szafraniec, and M. Znojil (eds.), *Non-Selfadjoint Operators in Quantum Physics: Mathematical Aspects.* John Wiley & Sons, Hoboken, NJ, 2015.

[3] N. Bebiano, J. da Providência, and J.P. da Providência, *Mathematical aspects of quantum systems with a pseudo-Hermitian Hamiltonian*. Brazilian Journal of Physics **46** (2016), 152–156.

[4] C.M. Bender and S. Boettcher, *Real spectra in non-Hermitian Hamiltonians having PT symmetry*. Phys. Rev. Lett. **80** (1998), 5243–5246.

[5] A.S. Davydov, *Quantum Mechanics*. Pergamon Press, 1965.

[6] P.A.M. Dirac, *The Principles of Quantum Mechanics*. 4th ed. Oxford University Press, Oxford, Oxford, 1958.

[7] I. Gohberg, S. Goldberg, and M. Kaashoek, *Classes of Linear Operators*. Vol. I. Birkäuser, Basel, 1990.

[8] A. Gonzaléz Lopéz and T. Tanaka, *Nonlinear pseudo-supersymmetry in the framework of N-fold supersymmetry*. J. Phys. A: Math. Gen. **39** (2006), 3715–3723.

[9] D. Krejcirik and P. Siegl, *Elements of spectral theory without the spectral theorem*. In: "Non-Selfadjoint Operators in Quantum Physics: Mathematical Aspects". John Wiley & Sons, Hoboken, NJ, 2015, pp. 241–291.

[10] D. Krejcirik, P. Siegl, M. Tater, and J. Viola, *Pseudospectra in non-Hermitian quantum mechanics*. J. Math. Phys. **56** (2015), 103513.

[11] A. Mostafazadeh, *Pseudo-Hermitian quantum mechanics with unbounded metric operators*. Phil. Trans. R. Soc. A **371** (2013), 20120050.

[12] J. da Providência, N. Bebiano, and J.P. da Providência, *Non Hermitian operators with real spectra in quantum mechanics*. Brazilian Journal of Physics **41** (2011), 78–85.

[13] P. Siegl and D. Krejcirik, *The metric operator for the imaginary cubic oscillator*. Phys. Rev. D **86** (2012), 11702(R).

[14] F.G. Scholtz, H.B. Geyer, and F.J.W. Hahne, *Quasi-Hermitian operators in quantum mechanics and the variational principle*. Ann. Phys. NY **213** (1992), 74.

[15] M. Znojil, *Should PT symmetric quantum mechanics be interpreted as nonlinear?* J. Nonlin. Math. Phys. **9** (2002), 122–123.

Natália Bebiano
CMUC, Departament of Mathematics
Universidade de Coimbra
3001-454 Coimbra, Portugal
e-mail: `bebiano@mat.uc.pt`

João da Providência
Departamento de Física
Universidade de Coimbra
3001-454 Coimbra, Portugal
e-mail: `providencia@fis.uc.pt`

J.P. da Providência
Department of Physics
University of Beira. Interior
P-6201-001 Covilhã, Portugal
e-mail: `joaodaprovidencia@daad-alumni.de`

Operator Theory:
Advances and Applications, Vol. 267, 79–122

Fredholm Conditions on Non-compact Manifolds: Theory and Examples

Catarina Carvalho, Victor Nistor and Yu Qiao

Abstract. We give explicit Fredholm conditions for classes of pseudodifferential operators on suitable singular and non-compact spaces. In particular, we include a "user's guide" to Fredholm conditions on particular classes of manifolds including asymptotically hyperbolic manifolds, asymptotically Euclidean (or conic) manifolds, and manifolds with poly-cylindrical ends. The reader interested in applications should be able to read right away the results related to those examples, beginning with Section 5. Our general, theoretical results are that an operator adapted to the geometry is Fredholm if, and only if, it is elliptic and all its *limit operators* (in a sense to be made precise) are invertible. Central to our theoretical results is the concept of a "Fredholm groupoid." By definition, a *Fredholm groupoid* is one for which this characterization of the Fredholm condition is valid. We use the notions of *exhaustive* and *strictly spectral families of representations* to obtain a general characterization of Fredholm groupoids. In particular, we introduce the class of the so-called *groupoids with Exel's property* as the groupoids for which the regular representations are exhaustive. We show that the class of "stratified submersion groupoids" has Exel's property, where stratified submersion groupoids are defined by gluing fibered pull-backs of bundles of Lie groups. We prove that a stratified submersion groupoid is Fredholm whenever its isotropy groups are amenable. Many groupoids, and hence many pseudodifferential operators appearing in practice, fit into this framework. This fact is exploited to yield Fredholm conditions not only in the above-mentioned classes, but also on manifolds that are obtained by desingularization or by blow-up of singular sets.

Mathematics Subject Classification (2010). 58J40 (primary), 58H05, 47L80, 46L60.

Keywords. Pseudodifferential operator, differential operator, Fredholm operator, groupoid, Lie group, Sobolev space, C^*-algebras, compact operators, non-compact manifold, singular manifold.

Carvalho was partially supported by Fundação para a Ciência e Tecnologia UID/MAT/04721/2013 (Portugal). Nistor has been partially supported by ANR-14-CE25-0012-01. Qiao was partially supported by NSF of China (11301317, 11571211).
Manuscripts available from `http://iecl.univ-lorraine.fr/Victor.Nistor/`.

1. Introduction

We obtain in this paper necessary and sufficient conditions for classes of operators to be Fredholm. Our results specialize to yield Fredholm conditions for classical pseudodifferential operators on manifolds with cylindrical and poly-cylindrical ends, on manifolds that are asymptotically Euclidean, and on manifolds that are asymptotically hyperbolic. Other examples of non-compact manifolds covered by our results include natural operators on spaces obtained by desingularization of suitable singular spaces by successively blowing up the lowest-dimensional singular strata.

1.1. Background and main result

Let M_0 be a Riemannian manifold and let $P\colon H^s(M_0;E) \to H^{s-m}(M_0;F)$ be an order m classical pseudodifferential operator acting between Sobolev sections of two smooth, Hermitian vector bundles E, F. Recall that P is called *elliptic* if its principal symbol $\sigma_m(P)$ is invertible outside the zero section. We have the following well-known, classical result [19, 96, 97].

Theorem 1.1 (Classical). *Assume that M_0 is* compact *and smooth. We have that $P\colon H^s(M_0;E) \to H^{s-m}(M_0;F)$ is Fredholm if, and only if, it is elliptic.*

This classical Fredholm result has many applications, so a natural question to ask is to what extent it extends to (suitable) non-compact manifolds. The example of constant coefficient differential operators on $\mathbb{R}^n$ shows that this classical result might be no longer true as stated if M_0 is not compact.

On certain classes of manifolds, however, it is possible to reformulate Theorem 1.1 as follows. Let M_0 be a non-compact manifold with "amenable ends" (see Subsection 4.2 for the precise definition). Then we can associate to M_0 the following data:

(1) Smooth manifolds M_α, $\alpha \in I$, and
(2) Lie groups G_α acting freely and properly on M_α, $\alpha \in I$,

(for a suitable index set I), satisfying the following theorem.

Theorem 1.2. *Let P be an order m classical pseudodifferential operator on M_0 compatible with the geometry. Then one can associate to P certain G_α-invariant pseudodifferential operators P_α on M_α with the following property:*

$$P : H^s(M_0;E) \to H^{s-m}(M_0;F) \textit{ is Fredholm} \quad \Leftrightarrow \quad P \textit{ is elliptic and}$$
$$P_\alpha\colon H^s(M_\alpha;E) \to H^{s-m}(M_\alpha;F) \quad \textit{is invertible for every } \alpha \in I\,.$$

We stress here the appearance of the additional operators P_α, whose invertibility is necessary for the given operator to be Fredholm. These are "limit" or "boundary" operators and are, in a certain sense, limits at infinity of our operator P. This is consistent with the physical intuition behind the HVZ theorem for the N-body hamiltonian [32] and the references therein. See also [22, 31, 32, 50, 56, 82, 89] and the references therein.

Many results similar to Theorem 1.2 were obtained for psedudodifferential operators before. See for instance [25, 34, 55, 60, 62, 94, 95] and the references therein for a very small sample. Similar results for *differential* operators were also obtained earlier, see [24, 45, 46, 71, 80] and the references therein, again for a very small sample. Other related results appeared in the context of localization principles, often in relation to the N-body problem mentioned above. Inspired by these works, we shall call the operators P_α the *limit operators* of P. We shall refer to results similar to Theorem 1.2 as *non-local Fredholm conditions.*

The class of "manifolds with amenable ends" is introduced such that, almost by definition, it is close to being the largest class of manifolds for which Theorem 1.2 is valid. The challenge then becomes to provide a large enough class of manifolds with amenable ends, which we do in this paper. The term "amenable" comes from the fact that certain isotropy groups at infinity are assumed to be amenable. The manifolds M_α and the groups G_α are obtained from the orbits of certain natural vector fields acting on a *compactification* M of M_0 and from their isotropies. One can see this very clearly in the case of manifolds with cylindrical and polycylindrical ends, of manifolds that are asymptotically Euclidean, and of manifolds that are asymptotically hyperbolic, which are all particular cases of manifolds with amenable ends. See the last section of the paper for explicit statements covering in detail the case of these classes of manifolds. It is interesting to notice that the case of asymptotically hyperbolic manifolds leads to the study of certain operators on *non-commutative* solvable Lie groups.

Theorem 1.2 is essentially a consequence of the results in [48] or [49], but the results of those papers turned out to be difficult to use directly by non-specialists. Moreover, the approach in this paper is direct, and we do not use the results of [48, 49], but rather rederive them in the needed form. In particular, we have tried in the last section of the paper to provide a review of the applications such that the reader can *understand the main results of this paper on Fredholm operators without any knowledge of groupoids or C^*-algebras.*

1.2. Exhaustive families of representations

It was realized by some of the authors mentioned above that the non-local Fredholm conditions of Theorem 1.2 are related to the representation theory of certain C^*-algebra $\mathfrak{A}$ (a different algebra for each M_0). See also [18, 19, 80, 101] and the references therein. The C^*-algebra $\mathfrak{A}$ is such that they contain the operators of the form $a = (1 + \Delta)^{(s-m)/2} P(1 + \Delta)^{-s/2}$ acting on $L^2(M_0)$ (where Δ is the positive Laplacian and P is an order m pseudodifferential operator as in Theorem 1.2). The reason for considering the operator a is that P is Fredholm if, and only if, a is Fredholm; moreover, a is bounded (unlike P, in general). In most practical applications, the C^*-algebra $\mathfrak{A}$ can be chosen to be a groupoid C^*-algebra. Properties of the operator P can then be inferred from the structure of $\mathfrak{A}$, taking advantage of the fact that one has a well-developed theory of groupoid C^*-algebras, see Section 2 for more details and references. In particular, these ideas are used in the proof of Theorem 4.12, which contains Theorem 1.2 as a particular case. The operators

P_α in Theorem 1.2 then can be obtained as homomorphic images of the operator P, that is $P_\alpha = \pi_\alpha(P)$, for suitable representations $\{\pi_\alpha\}$ of the algebra $\mathfrak{A}$.

The relevant families of representations in this setting are those of *strictly spectral* (*and strictly norming*) families of representations, introduced by Roch in [88] and recently studied in detail by Nistor and Prudhon in [76]. In that paper, the concept of an *exhaustive family* of representations has emerged as a useful concept to study strictly spectral families of representations. From a technical point of view, these families of representations form the backbone of the theoretical results in this paper.

1.3. Fredholm Lie groupoids

The key point of our approach is to start with a general study of Fredholm conditions for pseudodifferential operators in the framework of "Fredholm Lie groupoids." All pseudodifferential operators considered in this paper will be classical. To each Lie groupoid $\mathcal{G}$ with units a manifold with corners M, one can associate a class $\Psi^m(\mathcal{G}; E, F)$ of (classical) pseudodifferential operators on $\mathcal{G}$ acting between sections of vector bundles $E, F \to M$, whose closure contains the groupoid C^*-algebra $C^*_r(\mathcal{G})$ as an essential ideal if $m = 0$ and $E = F = \mathbb{C}$ (see Section 4.2 and the references therein). Let ∂M denote the union of the proper faces of M (the *boundary* of M) and $M_0 := M \smallsetminus \partial M$ be the *interior* of M. For some results, we shall need that M be compact, so, for simplicity, we shall assume that this is the case, unless otherwise stated. The smooth manifold M_0 and its compactification M have the same meaning as in Subsection 1.1. Often one chooses for the algebra $\mathfrak{A}$ considered in Subsection 1.2 the algebra $\mathfrak{A} := C^*_r(\mathcal{G})$, the "reduced C^*-algebra" of $\mathcal{G}$, whose definition we recall below. Alternatively, on can consider $\mathfrak{A} := \overline{\Psi}(\mathcal{G})$, the norm closure of $\Psi^0(\mathcal{G})$.

A *Fredholm Lie groupoid* $\mathcal{G}$ is, by definition, a Lie groupoid for which the Fredholm property of the associated operators is equivalent to the invertibility of the fiberwise boundary restrictions and of the principal symbol on M (the principal symbol has to be suitably defined at the boundary, using the extension of TM_0 to a Lie algebroid on M). More precisely, restricting to operators on negative order and assuming that $\mathcal{G}_{M_0} = M_0 \times M_0$ is the pair groupoid on M_0, we say that $\mathcal{G}$ is a *Fredholm Lie groupoid* if it has the following property:

$$\text{"}a \in 1 + C^*_r(\mathcal{G}) \text{ is Fredholm} \Leftrightarrow \pi_x(a) \text{ is invertible } \forall x \in \partial M,\text{"} \tag{1}$$

for the regular representations π_x of $C^*(\mathcal{G})$, where the Fredholmness is considered on $L^2(M_0)$. (See 4.3 for a more detailed formulation of the definition of a Fredholm groupoid.) The reader will recognize in condition (1) the type of Fredholm conditions that appeared in Theorem 1.2 (and which are typically used in practice). It turns out that if the defining condition (1) is satisfied for all $a \in 1 + C^*_r(\mathcal{G})$, then it will be satisfied for many other operators associated to $\mathcal{G}$, in particular it will be satisfied for suitable operators $a = P \in \Psi^m(\mathcal{G}; E, F)$ (if $m > 0$, one has to add the ellipticity condition and take into account the domain of definition).

The extension of the Fredholm conditions in Equation (1) from operators in $1 + C_r^*(\mathcal{G})$ to operators in $\Psi^m(\mathcal{G}; E, F)$ turns out to be almost automatic and is conveniently formalized through the use of strictly spectral and exhaustive families of representations of groupoid C^*-algebras (see Section 3 for definitions and references). Thus, in our paper, we shift the study of the Fredholm conditions from the study of a *single* operator to the study of suitable *algebras* containing it. In particular, it will be enough to study the properties and the representations of *regularizing* pseudodifferential operators. The Fredholm properties of higher-order pseudodifferential operators will then follow simply by including the ellipticity condition. Thus the Fredholm conditions established in this paper will be formulated in terms of representations of groupoid C^*-algebras.

To make our results applicable, we give various characterizations of Fredholm groupoids and provide methods to prove that individual groupoids are Fredholm. In particular, we introduce the class of "stratified submersion groupoids." A *stratified submersion groupoid* is one that is built out by gluing fibered pull-backs of bundles of Lie groups. We show that a stratified submersion groupoid with amenable stabilizers is Fredholm and hence that the interior of the set of its units yields a manifold with amenable ends, and hence Theorem 1.2 is valid for these manifolds.

1.4. Examples and applications

The main significance of the class of stratified submersion Lie groupoids mentioned in the last paragraph is that many of the groupoids that appear in practice are naturally stratified submersion Lie groupoids. Typically, using the notations as above, we have that M identifies with a compactification of M_0 to a manifold with corners, where the behavior at $F := \partial M := M \setminus M_0$ models the behavior at infinity ('at the ends'). Often the corresponding isotropy groups are amenable. A class of manifolds that fits this perspective is the class of Lie manifolds [48, 49], a few examples of which appear in Section 5. See [33].

Moreover, stratified submersion Lie groupoids are also tailored to applications to singular spaces obtained by (iterative) "desingularization procedures," the desingularization being the analog in the category of groupoids of the blow-up construction in the category of manifolds with corners. In fact, it can be seen that desingularization preserves Fredholm groupoids, and that the class of stratified submersion groupoids is closed under desingularization. While we do not pursue the applications to singular spaces in full detail here, we outline the key ideas involved and give the construction of such a groupoid yielding Fredholm conditions for the edge calculus.

Our results can be extended to products of such manifolds, or to manifolds that locally at infinity are products of such manifolds.

1.5. Contents of the paper

The paper consists of roughly two parts: the theoretical part and applications. The applications are included in the last two sections. The most concrete applications are contained in the last section, which we tried to write in such a way that it can, to a large extent, be read independently of the rest of the paper. The reader interested only in applications can thus start immediately with Section 5. At least the main results of that section should be understandable *without any knowledge of groupoids.*

We now describe in detail the main contents of the paper. We start with reviewing the relevant topics related to groupoids and groupoid C^*-algebras, so Section 2 consists mostly of background material on locally compact groupoids $\mathcal{G}$, on their Haar systems, and on their C^*-algebras. We then review manifolds with corners and tame submersions, and we recall the definitions of Lie groupoids in the framework that we need, that is, that of manifolds with corners. Note that all our Lie groupoids will be second countable and Hausdorff. We finish this section with a review of some examples of Lie groupoids that will play a role in what follows.

Section 3 contains preliminaries on exhaustive families of representations from [76] and some general results on groupoid C^*-algebras. We recall the notions of strictly spectral and strictly norming families [88] and the main results from [76]. We then introduce groupoids with "Exel's property," and, respectively, with "strong Exel's property." A groupoid satisfies *Exel's property* (respectively, the *strong Exel's property*) when its family of regular representations is exhaustive for the reduced C^*-algebra (respectively, for the full C^*-algebra). We prove that groupoids given by fibered pull-backs of bundles of amenable Lie groups always have Exel's strong property. Motivated by this result, we then introduce the class of *stratified submersion groupoids*, given essentially by gluing fibered pull-backs of bundles of Lie groups. We then prove that stratified submersion Lie groupoids with amenable isotropy groups always have Exel's strong property.

In Section 4, we define Fredholm Lie groupoids. Note that in this paper, we always work in the setting of Lie groupoids, except where otherwise explicitly stated. We provide a characterization of Fredholm Lie groupoids using strictly spectral families of regular representations and show that groupoids that have the strong Exel's property are Fredholm. It follows that stratified submersion Lie groupoids with amenable isotropy groups are Fredholm. We then specialize to algebras of pseudodifferential operators on Lie groupoids and obtain the crucial Theorems 4.12 and 4.17.

The last section of the paper, Section 5, contains examples and applications of our results, namely of Theorem 4.17. We start with group actions, define the associated transformation groupoid, and use this construction to describe Fredholm conditions related to several pseudodifferential calculi: the b-groupoid that models operators on manifolds with cylindrical and poly-cylindrical ends; the scattering groupoid that models operators on manifolds that are asymptotically Euclidean, and also operators on asymptotically hyperbolic spaces. We then consider the edge calculus and construct a suitable Fredholm stratified submersion groupoid that will recover Fredholm conditions. This is a particular case of a desingularization groupoid, whose construction is briefly recalled in the following subsection. ("Desingularization groupoids" were introduced in [75], where more details on this construction can be found.) We then give more explicit examples of the desingularization and blow-up process. A first example of the desingularization procedure deals with the blow-up of a smooth submanifold along another smooth, compact manifold. The second example extends this first example to manifolds with boundary. Finally, the last example deals with the iterated blow-up of a singular stratified subset of dimension one. (See the main body of the paper for specific references to the existing literature related to these results.)

For simplicity, in view of the applications considered in this paper, we shall work almost exclusively with *Lie groupoids* (and their reductions), although some definitions and results are valid in greater generality. For the most part, our manifolds will be Hausdorff and second countable.

2. Groupoids and their C^*-algebras

We recall in this section some basic definitions and properties of groupoids. Although our interest in applications lies mainly in Lie groupoids, we have found it convenient to consider occasionally also the general case of locally compact groupoids, so in the beginning we consider both cases. We refer to Mackenzie's books [53, 54] for more details and, in general, for a nice introduction to the subject of Lie groups and Lie groupoids, as well as to further references and historical comments. See also [12, 64, 85, 109] for the more specialized issues relating to analytic applications. We use the approach in [75], where most of the needed definitions and results can also be found.

2.1. Locally compact groupoids

Let us introduce groupoids as in [12, 54, 85].

Definition 2.1. A *groupoid* is a pair $\mathcal{G} = (\mathcal{G}^{(1)}, \mathcal{G}^{(0)})$, where $\mathcal{G}^{(i)}$, $i = 0, 1$, are sets, together with structural morphisms

(1) $d, r : \mathcal{G}^{(1)} \to \mathcal{G}^{(0)}$ (the "domain" and "range"),
(2) $\iota : \mathcal{G}^{(1)} \to \mathcal{G}^{(1)}$ (the "inverse"),
(3) $u : \mathcal{G}^{(0)} \to \mathcal{G}^{(1)}$ (the inclusion of units), and
(4) $\mu : \mathcal{G}^{(2)} := \{(g, h) \in \mathcal{G}^{(1)} \times \mathcal{G}^{(1)} \,|\, d(g) = r(h)\} \to \mathcal{G}^{(1)}$ (the product),

with the following properties (we write $gh := \mu(g, h)$, for simplicity):

(1) $d(gh) = d(h)$, $r(gh) = r(g)$ if $d(g) = r(h)$;
(2) $g_1(g_2g_3) = (g_1g_2)g_3$ for all $g_i \in \mathcal{G}$ such that $d(g_i) = r(g_{i+1})$;
(3) $d(u(x)) = x = r(u(x))$, for all $x \in M$.
(4) $gu(d(g)) = g$ and $u(r(g))g = g$ for all $g \in \mathcal{G}$.
(5) $g\iota(g) = u(r(g))$ and $\iota(g)g = u(d(g))$ for all $g \in \mathcal{G}$.

Moreover, we see that $\mathcal{G}^{(0)}$ can be regarded as the set of objects of a category with morphisms $\mathcal{G}^{(1)}$. The objects of $\mathcal{G}$ will also be called *units* and the morphisms of $\mathcal{G}$ will also be called *arrows*. Actually, *a groupoid $\mathcal{G}$ is simply a small category in which every morphism is invertible.* (A *small category* is one whose objects and morphisms are sets.) For convenience, we shall identify $\mathcal{G} = \mathcal{G}^{(1)}$ and denote $M := \mathcal{G}^{(0)}$. We shall write $\mathcal{G} \rightrightarrows M$ for a groupoid $\mathcal{G}$ with units M. In this paper, $\mathcal{G}$ always denotes a groupoid.

Definition 2.2. A *locally compact groupoid* is a groupoid $\mathcal{G} \rightrightarrows M$ such that:

(1) $\mathcal{G}$ and M are locally compact spaces, with M Hausdorff;
(2) the structural morphisms d, r, ι, u, and μ are continuous;
(3) d is surjective and open.

See Subsection 2.4 for examples of groupoids. Let us notice that the definition requires only the space of units M to be Hausdorff. In this paper, however, all spaces will be assumed or proved to be Hausdorff.

In what follows, we denote by $\mathcal{G}_A := d^{-1}(A)$, $\mathcal{G}^A = r^{-1}(A)$ and $\mathcal{G}_A^B := d^{-1}(A) \cap r^{-1}(B)$. We call $\mathcal{G}_A^A$ the *reduction* of $\mathcal{G}$ to A. If A is $\mathcal{G}$*-invariant*, in the sense that $\mathcal{G}_A^A = \mathcal{G}_A = \mathcal{G}^A = r^{-1}(A)$, then $\mathcal{G}_A$ is also a groupoid, called the *restriction* of $\mathcal{G}$ to A. For any $x \in M$, $\mathcal{G}_x^x = d^{-1}(x) \cap r^{-1}(x)$ is a group, called the *isotropy group* at x.

We will study several examples of groupoids in detail when we will introduce Lie groupoids.

2.2. Haar systems and C^*-algebras

We now recall the definition of a Haar system of a locally compact groupoid and we use this opportunity to fix some more notations to be used throughout the rest of the paper. We refer to [12, 43, 85] for more information on the topics discussed in this subsection.

If $\mathcal{G}$ is a locally compact groupoid, we shall denote by $\mathcal{C}_c(\mathcal{G})$ the space of continuous, complex-valued, compactly supported functions on $\mathcal{G}$.

Definition 2.3. A *right Haar system* for a locally compact groupoid $\mathcal{G}$ is a family $\lambda = \{\lambda_x\}_{x \in M}$, where λ_x is a Borel regular measure on $\mathcal{G}$ with support $\mathrm{supp}(\lambda_x) = d^{-1}(x) =: \mathcal{G}_x$ for every $x \in M = \mathcal{G}^{(0)}$, satisfying

(i) *The continuity condition:*

$$M \ni x \mapsto \lambda_x(\varphi) := \int_{\mathcal{G}_x} \varphi(g) \mathrm{d}\lambda_x(g) \in \mathbb{C}$$

is continuous for $\varphi \in \mathcal{C}_c(\mathcal{G})$.

(ii) *The invariance condition:*

$$\int_{\mathcal{G}_{r(g)}} \varphi(hg) \mathrm{d}\lambda_{r(g)}(h) = \int_{\mathcal{G}_{d(g)}} \varphi(h) \mathrm{d}\lambda_{d(g)}(h)$$

for all $g \in \mathcal{G}$ and $\varphi \in \mathcal{C}_c(\mathcal{G})$.

We shall assume from now on that all our locally compact groupoids are endowed with a (right) Haar system. Let thus $\mathcal{G}$ be a locally compact groupoid and $\{\lambda_x\}$ be the Haar system associated to it. The space $\mathcal{C}_c(\mathcal{G})$ has a natural product given by the formula

$$(\varphi_1 * \varphi_2)(g) := \int\limits_{d^{-1}(d(g))} \varphi_1(gh^{-1})\varphi_2(h)\mathrm{d}\lambda_{d(g)}(h).$$

This makes $\mathcal{C}_c(\mathcal{G})$ into an associative $*$-algebra with the involution

$$\varphi^*(g) := \overline{\varphi(g^{-1})}$$

for all $g \in \mathcal{G}$ and $\varphi \in \mathcal{C}_c(\mathcal{G})$. There also exists a natural algebra norm on $\mathcal{C}_c(\mathcal{G})$ defined by

$$\|\varphi\|_I := \max\Big\{ \sup_{x\in M} \int_{\mathcal{G}_x} |\varphi|\mathrm{d}\lambda_x,\ \sup_{x\in M} \int_{\mathcal{G}_x} |\varphi^*|\mathrm{d}\lambda_x \Big\}.$$

The completion of $\mathcal{C}_c(\mathcal{G})$ with respect to the norm $\|\cdot\|_I$ is denoted $L^1(\mathcal{G})$.

Recall [28] that a *C^*-algebra* is a complex algebra A together with a conjugate linear involution $*$ and a complete norm $\|\ \|$ such that $(ab)^* = b^*a^*$, $\|ab\| \leq \|a\|\|b\|$, and $\|a^*a\| = \|a\|^2$, for all $a, b \in A$. Let $\mathcal{H}$ be a Hilbert space and denote by $\mathcal{L}(\mathcal{H})$ the space of linear, bounded operators on $\mathcal{H}$. Then $\mathcal{L}(\mathcal{H})$ is a C^*-algebra. A *representation* of a C^*-algebra A on the Hilbert space $\mathcal{H}_\pi$ is a $*$-morphism $\pi : A \to \mathcal{L}(\mathcal{H}_\pi)$.

To a locally compact groupoid $\mathcal{G}$ (endowed with a Haar system), there are associated two basic C^*-algebras, the *full* and *reduced* C^*-algebras $C^*(\mathcal{G})$ and $C^*_r(\mathcal{G})$, whose definition we now recall.

Definition 2.4. The *full C^*-algebra* associated to $\mathcal{G}$, denoted $C^*(\mathcal{G})$, is defined as the completion of $\mathcal{C}_c(\mathcal{G})$ with respect to the norm

$$\|\varphi\| := \sup_\pi \|\pi(\varphi)\|,$$

where π ranges over all *contractive* $*$-representations of $\mathcal{C}_c(\mathcal{G})$. (Recall that a $*$-representation π of $\mathcal{C}_c(\mathcal{G})$ is called contractive if $\|\pi(\varphi)\| \leq \|\varphi\|_I$, for all $\varphi \in \mathcal{C}_c(\mathcal{G})$.) Let us define as usual for any $x \in M$ the *regular* representation $\pi_x : C^*(\mathcal{G}) \to \mathcal{L}(L^2(\mathcal{G}_x, \lambda_x))$ by the formula

$$(\pi_x(\varphi)\psi)(g) := \varphi * \psi(g) := \int_{\mathcal{G}_{d(g)}} \varphi(gh^{-1})\psi(h)d\lambda_{d(g)}(h), \quad \phi \in \mathcal{C}_c(\mathcal{G}).$$

We then define similarly the *reduced C^*-algebra* $C^*_r(\mathcal{G})$ as the completion of $\mathcal{C}_c(\mathcal{G})$ with respect to the norm

$$\|\varphi\|_r := \sup_{x\in M} \|\pi_x(\varphi)\|$$

which is well defined, as all representations π_x are contractive (of norm ≤ 1). The groupoid $\mathcal{G}$ is said to be *metrically amenable* if the canonical surjective $*$-homomorphism $C^*(\mathcal{G}) \to C^*_r(\mathcal{G})$, induced by the definitions above, is also injective.

Also, for further use, we note that if $\mathcal{G}$ is second countable, then $C^*(\mathcal{G})$ is a separable C^*-algebra.

Remark 2.5. For any $\mathcal{G}$-invariant, locally closed subset $A \subseteq M$, the *reduced* groupoid $\mathcal{G}_A = \mathcal{G}_A^A$ is locally compact and has a Haar system λ_A obtained by restricting the Haar system λ of $\mathcal{G}$ to $\mathcal{G}_A$. In particular, we can construct as above the corresponding C^*-algebra $C^*(\mathcal{G}_A)$ and the reduced C^*-algebra $C_r^*(\mathcal{G}_A)$. For any closed subset $A \subseteq M$, the subset $d^{-1}(A) \subseteq \mathcal{G}$ is also closed, so the restriction map $\mathcal{C}_c(\mathcal{G}) \to \mathcal{C}_c(d^{-1}(A))$ is well defined. If A is also $\mathcal{G}$-invariant then the restriction extends by continuity to both a $*$-homomorphism $\rho_A\colon C^*(\mathcal{G}) \to C^*(\mathcal{G}_A)$ and a $*$-homomorphism $(\rho_A)_r\colon C_r^*(\mathcal{G}) \to C_r^*(\mathcal{G}_A)$.

We have the following well-known, but important result [68, 69, 86] that we record for further reference.

Proposition 2.6. *Let $\mathcal{G} \rightrightarrows M$ be a second countable, locally compact groupoid with a Haar system. Let $U \subset M$ be an open $\mathcal{G}$-invariant subset, $F := M \smallsetminus U$.*

(i) *$C^*(\mathcal{G}_U)$ is a closed two-sided ideal of $C^*(\mathcal{G})$ that yields the short exact sequence*

$$0 \to C^*(\mathcal{G}_U) \to C^*(\mathcal{G}) \xrightarrow{\rho_F} C^*(\mathcal{G}_F) \to 0\,.$$

(ii) *If $\mathcal{G}_F$ is metrically amenable, then one has the exact sequence*

$$0 \to C_r^*(\mathcal{G}_U) \to C_r^*(\mathcal{G}) \xrightarrow{(\rho_F)_r} C_r^*(\mathcal{G}_F) \to 0\,.$$

(iii) *If the groupoids $\mathcal{G}_F$ and $\mathcal{G}_U$ (respectively, $\mathcal{G}$) are metrically amenable, then $\mathcal{G}$ (respectively, $\mathcal{G}_U$) is also metrically amenable.*

Proof. The first assertion is well known, see for instance [69, Lemma 2.10]. The second statement is in [86, Remark 4.10]. The last part follows from (ii) and the Five Lemma. □

Remark 2.7. We notice that the exact sequence of Proposition 2.6 corresponds to a disjoint union decomposition $\mathcal{G} = \mathcal{G}_F \sqcup \mathcal{G}_U$.

2.3. Manifolds with corners and Lie groupoids

Even if one is interested only in analysis on smooth manifolds (that is, in manifolds without corners or boundary), one important class of applications will be to spaces obtained as the blow-up with respect to suitable submanifolds, which leads to manifolds with corners, as each blow-up increases the highest codimension of corners (the rank of the manifolds) by one. To model the analysis on these spaces, we need Lie groupoids. The advantage of using Lie groupoids is that they have a distinguished class of Haar systems defined using the smooth structure.

We begin with some background material following [75]. It will be important to distinguish between smooth manifolds without corners (or boundaries) and (smooth) manifolds with corners. The former will be *smooth manifolds*, while the later will be simply *manifolds.* Thus, in this paper, a *manifold M* is a *second countable, Hausdorff* topological space locally modeled by open subsets of $[-1, 1]^n$ with smooth coordinate changes. In particular, our manifolds may have corners. By contrast, a *smooth manifold* will not have corners (or boundary). A point

$p \in M$ of a manifold (with corners) is called of *depth* k if it has a neighborhood V_p diffeomorphic to $[0, a)^k \times (-a, a)^{n-k}$, $a > 0$, by a diffeomorphism $\phi_p : V_p \to [0, a)^k \times (-a, a)^{n-k}$ with $\phi_p(p) = 0$. The maximal depth of points in M will be called the *rank* of M. Thus, a manifold is smooth if it has rank 0.

A function $f : M \to M_1$ between two manifolds with corners will be called *smooth* if its components are smooth in all coordinate charts. A little bit of extra care is needed here in defining the derivatives. This is illustrated clearly in one dimension, as one may have to consider side derivatives: if $f : [0, 1] \to \mathbb{R}$, then $f'(0) := \lim_{h\to 0, h>0} h^{-1}(f(h) - f(0))$, whereas $f'(1) := \lim_{h\to 0, h>0} h^{-1}(f(1) - f(1-h))$. Since $\lim_{h\to 0} h^{-1}(f(x+h)-f(x)) = \lim_{h\to 0} h^{-1}(f(x)-f(x-h)) =: f'(x)$, for $0 < x < 1$, this defines unambiguously $f'(x)$. A similar comment is in order in higher dimensions as well. In local coordinates, a smooth function $[0, 1]^n \to [0, 1]^k$ is one that is the restriction of a smooth function $\mathbb{R}^n \to \mathbb{R}^k$.

For $x \in M$, the set of *inward pointing tangent vectors* in $T_x(M)$ defines a closed cone denoted $T_x^+(M)$. Of course, if $x \in M_0 := M \smallsetminus \partial M$, then $T_x^+(M) = T_x(M)$. Let M and M_1 be manifolds with corners and $f : M_1 \to M$ be a smooth map. Then f induces a vector bundle map $df : TM_1 \to TM$ such that $df(T_x^+(M_1)) \subset T_{f(x)}^+ M$, $x \in \partial M$. If the smooth map $f : M_1 \to M$ is injective, has injective differential df, and has *locally closed range,* then we say that $f(M_1)$ is a *submanifold* of M. We stress the condition that submanifolds be locally closed (that is, the intersection of an open and a closed subset, or, equivalently, every point x of a submanifold has an open neighborhood U_x that intersects the submanifold in a closed subset of U_x). We hence rule out embedded submanifolds. Other than these conditions, our concept of submanifold is the most general possible.

Definition 2.8. A *tame submersion* h between two manifolds with corners M_1 and M is a smooth map $h : M_1 \to M$ such that its differential dh is surjective everywhere and

$$(dh_x)^{-1}(T_{h(x)}^+ M) = T_x^+ M_1 , \quad x \in M.$$

(That is, $dh(v)$ is an inward pointing vector of M if, and only if, v is an inward pointing vector of M_1.)

For instance, the maps $h_1, h_2 : [0, \infty) \times [0, \infty) \to [0, \infty)$ defined by $h_1(x, y) = x + y$, $h_2(x, y) = x$, are not tame submersions. Clearly, if $h : M_1 \to M$ is a tame submersion of manifolds with corners, then x and $h(x)$ will have the same depth. In particular, the rank of M_1 is at most the rank of M. We have the following well-known lemma (see for instance [75]).

Lemma 2.9. *Let $h : M_1 \to M$ be a tame submersion of manifolds with corners.*

(i) *For $m_1 \in M_1$, there exists an open neighborhood U of m_1 in M_1 such that $h(U)$ is open and the restriction of h to U is a $\mathcal{C}^\infty$ fibration with basis $h(U)$.*
(ii) *Let $L \subset M$ be a submanifold, then $L_1 := h^{-1}(L)$ is a submanifold of M_1 of rank $\leq$ the rank of L.*

We now define Lie groupoids roughly by replacing our spaces with manifolds with corners and the continuity conditions with smoothness conditions. The reason why Lie groupoids play an important role in applications is that they model the distribution kernels of many interesting classes of pseudodifferential operators.

Definition 2.10. A *Lie groupoid* is a groupoid $\mathcal{G} \rightrightarrows M$ such that

(i) $\mathcal{G}$ and M are manifolds with corners,
(ii) the structural morphisms d, r, ι, and u are smooth,
(iii) d is a tame submersion of manifolds with corners, and
(iv) μ is smooth.

In particular, the Lie groupoids used in this paper are second countable, Hausdorff, locally compact groupoids. In different contexts in other papers, it is sometimes useful *not to assume* the set of arrows $\mathcal{G}$ of a Lie groupoid to be Hausdorff, although the space of units M of a Lie groupoid $\mathcal{G}$ is always assumed to be Hausdorff. Lie groupoids were introduced by Ehresmann. A more general class also useful in applications is that of *continuous family groupoids,* where, roughly speaking, the structural maps are just continuous along the units, but keeping smoothness along the fibers. See [48, 79].

For the rest of this section, $\mathcal{G}$ will always be a *Lie groupoid.* In fact, most of the groupoids that we will consider in what follows will be Lie groupoids. However, one has to be careful since the restriction of a Lie groupoid is not necessarily a Lie groupoid itself. Moreover, we shall assume that all our Lie groupoids are Hausdorff.

Since d and r are tame submersions, it follows from Lemma 2.9 that the fibers $\mathcal{G}_x := d^{-1}(x)$ and $\mathcal{G}^x := r^{-1}(x)$, $x \in M$, are smooth manifolds (that is, they have no corners). Similarly, the same lemma implies that the set $\mathcal{G}^{(2)} \subset \mathcal{G} \times \mathcal{G}$ of composable units is a manifold as well (but it may have corners). In particular, it makes sense to consider smooth maps $\mathcal{G}^{(2)} \to \mathcal{G}^{(1)}$.

Remark 2.11. Let $\mathcal{G} \rightrightarrows M$ be a Lie groupoid and let $A(\mathcal{G}) := \cup_{x \in M} T_x\mathcal{G}_x$, where $\mathcal{G}_x := d^{-1}(x)$, as usual. Let us denote by $f_* : TM_1 \to TM_2$ the differential of a smooth map $f : M_1 \to M_2$. Then $A(\mathcal{G})$ is the restriction to units of the vector bundle $\ker(d_*)$, and hence it has a natural structure of vector bundle over M. It is called the *Lie algebroid* of $\mathcal{G}$. The differential r_* of the range map $r : \mathcal{G} \to M$ then induces a map $\varrho := r_*|_{A(\mathcal{G})} : A(\mathcal{G}) \to TM$, called the *anchor map* of $A(\mathcal{G})$. We let $\mathrm{Lie}(\mathcal{G}) := \varrho(\Gamma(M; A(\mathcal{G})))$ denote the image by ϱ of the space of sections of $A(\mathcal{G})$. It coincides with the image by r_* of the space d-vertical (i.e., tangent to the fibers $\mathcal{G}_x$ of d) vector fields on $\mathcal{G}$ that are invariant for the action of $\mathcal{G}$ on itself by right multiplication. Therefore both $\Gamma(M; A(\mathcal{G}))$ and $\mathrm{Lie}(\mathcal{G}) := \varrho(\Gamma(M; A(\mathcal{G})))$ are Lie algebras for the Lie bracket. Both Lie algebras will play an important role in the analysis of differential operators.

Remark 2.12. We can use the vector bundle $A(\mathcal{G})$ to define a Haar system. Let $D := |\Lambda^n A(\mathcal{G})|$, where n is the dimension of the Lie algebroid of $\mathcal{G}$. The pull-back vector bundle $r^*(D)$ is the bundle of 1-densities along the fibers of d. A trivialization of D will hence give rise to a right invariant set of measures on $\mathcal{G}_x$ and hence to a right Haar system.

We conclude that a Lie groupoid always has a (right) Haar system, which we shall always assume to be obtained as in the remark above.

2.4. Examples of Lie groupoids

We continue with various examples of constructions of Lie groupoids that will be needed in what follows. Recall that we only consider Hausdorff groupoids in this paper. Also, recall that a Lie groupoid $\mathcal{G}$ is metrically amenable if, by definition, the canonical morphism $C^*(\mathcal{G}) \to C^*_r(\mathcal{G})$ is an isomorphism.

Example 2.13. Any Lie group G gives rise to a Lie groupoid with set of units reduced to one point: the identity element of G. Thus $d, r : G \to M := \{e\}$ are constant, $\mu : G \times G \to G$ is the group multiplication, $\iota(g) := g^{-1}$ is the (usual) inverse, and $u : M \to G$ is the inclusion of the unit. We have that the Lie algebroid $A(G)$ of G coincides with the Lie algebra of G (the definition of a Lie algebroid was recalled in Remark 2.11). Hence $\mathrm{Lie}(G) := \Gamma(A(\mathcal{G}))$ is also the Lie algebra of G. We have that G is metrically amenable if, and only if, it is amenable in the usual sense of groups. In particular, if G is solvable, then G is amenable.

Example 2.14. Let M be a manifold with corners (hence *Hausdorff* by our conventions), let $\mathcal{G} = M$, and $d = r = u = \iota = id_M$. Then $\mathcal{G}$ is a Lie groupoid with only units. We shall call a Lie groupoid with these properties a *manifold.* We have that $A(M) = M \times \{0\}$, that is, the *zero* vector bundle. We have $\mathrm{Lie}(\mathcal{G}) = 0$.

Example 2.15. Let $\mathcal{G}_i \to M_i$, $i = 1, 2$, be two Lie groupoids. Then $\mathcal{G}_1 \times \mathcal{G}_2$ is a Lie groupoid with units $M_1 \times M_2$ and $A(\mathcal{G}_1 \times \mathcal{G}_2) = A(\mathcal{G}_1) \times A(\mathcal{G}_2)$. We have that $\mathrm{Lie}(\mathcal{G}_1 \times \mathcal{G}_2)$ is a suitable topological tensor product of $\mathrm{Lie}(\mathcal{G}_1)$ and $\mathrm{Lie}(\mathcal{G}_2)$.

The product of a manifold with a Lie group is thus again a Lie groupoid. We can "twist" this example to obtain a (smooth) "bundle of Lie groups."

Example 2.16. Let G be a Lie group with automorphism group $\mathrm{Aut}(G)$ and let $P \to M$ be a smooth, locally trivial, principal $\mathrm{Aut}(G)$-bundle, with M and P manifolds with corners (hence Hausdorff, by our conventions). Then the associated fiber bundle $\mathcal{G} := P \times_{\mathrm{Aut}(G)} G$ with fiber G is a Lie groupoid with units M. A Lie groupoid of this form will be called a *smooth bundle of Lie groups.* It satisfies $d = r$. We have $A(\mathcal{G}) \simeq P \times_{\mathrm{Aut}(G)} Lie(G)$ is a smooth bundle of Lie algebras and $\varrho : A(\mathcal{G}) \to TM$ is the zero map. The Lie algebra $\mathrm{Lie}(\mathcal{G})$ identifies thus with the family of sections of the bundle of Lie algebras $P \times_{\mathrm{Aut}(G)} Lie(G)$. If G is amenable, then $\mathcal{G}$ is metrically amenable. The fact that $\mathcal{G}$ is metrically amenable when G is amenable will be of crucial importance to us and will be used when the group G is a solvable Lie group.

The following simple example of the "pair groupoid" will be fundamental in what follows.

Example 2.17. Let M be a *smooth* manifold. (Thus M has no corners, according to our terminology.) Then the *pair groupoid* of M is $\mathcal{G} := M \times M$. It is a groupoid with units M. The structural morphisms are as follows: d is the second projection, r is the first projection, and the product μ is given by $(m_1, m_2)(m_2, m_3) = (m_1, m_3)$. This determines $u(m) = (m, m)$ and $\iota(m, m') = (m', m)$. We thus have that the product on $\mathcal{C}_c^\infty(\mathcal{G}) = \mathcal{C}_c^\infty(M \times M)$ is the product of operators with integral kernels. We have $A(M \times M) = TM$, with anchor map the identity map. One crucial feature of the pair groupoid is that, for any x, the regular representation π_x defines an isomorphism between $C^*(M \times M)$ and the ideal of compact operators in $\mathcal{L}(L^2(M))$. In particular, all pair groupoids are metrically amenable.

Example 2.18. Let M be a smooth manifold (so without corners) and $\widetilde{M}$ be it universal covering. Let $\pi_1(M)$ be the fundamental group of M associated to some fixed point of M. A related example to the pair groupoid is that of the *path groupoid* $\mathcal{P}(M) := (\widetilde{M} \times \widetilde{M})/\pi_1(M)$ of M, which will have the same Lie algebroid as the pair groupoid: $A(\mathcal{P}(M)) = TM$. The analysis associated to $\mathcal{P}(M)$ is that of $\pi_1(M)$-*invariant* operators on the covering space $\widetilde{M}$, thus quite different to that of the pair groupoid. This underscores the importance of choosing the right integrating groupoid when one is interested in Analysis.

We extend the example of the pair groupoid by defining *fibered pull-back groupoids* [40, 41] (we use the terminology in [75], however).

Example 2.19. Let again M and L be manifolds and $f : M \to L$ be *continuous* map. Let $\mathcal{H} \rightrightarrows L$ be a Lie groupoid. The *fibered pull-back groupoid* is then

$$f^{\downarrow\downarrow}(\mathcal{H}) := \{ (m, g, m') \in M \times \mathcal{H} \times M |\, f(m) = r(g),\, d(g) = f(m') \},$$

with units M and product $(m, g, m')(m', g', m'') = (m, gg', m'')$. We shall also sometimes write $M \times_f \mathcal{H} \times_f M = f^{\downarrow\downarrow}(\mathcal{H})$ for the fibered pull-back groupoid. Assume now that f is a tame submersion (in particular, that it is smooth). Then $f^{\downarrow\downarrow}(\mathcal{H})$ is a Lie groupoid and its Lie algebroid is

$$A(f^{\downarrow\downarrow}(\mathcal{H})) = \{(\xi, X) |\, \xi \in A(\mathcal{H}),\ X \in TM,\ \varrho(\xi) = f_*(X)\},$$

the *thick pull-back* Lie algebroid $f^{\downarrow\downarrow}A(\mathcal{H})$ of $A(\mathcal{H})$, see [53, 54].

The following particular example of a fibered pull-back will be useful for the study of the b-groupoid $\mathcal{G}_b$ in Section 5.

Example 2.20. Let us assume that M is a smooth manifold and let B be its set of connected components. Let G be a Lie group. We let $\mathcal{H} := B \times G$, the product of a manifold and a Lie group, and $f : M \to B$ be the map that associates to a point its connected component. Then $f^{\downarrow\downarrow}(\mathcal{H})$ is the topological disjoint union of the groupoids $(F \times F) \times G$ (product of the pair groupoid and a Lie group) for F ranging through the connected components of M.

3. Exhaustive families of representations and Exel's question

We next recall some basic facts on *exhaustive* families of representations following [76]. A groupoid has *Exel's property* when its set of regular representations is exhaustive. Exel [29] has asked which groupoids have this property. Establishing that a groupoid has Exel's property will turn out to be important in establishing Fredholm conditions in the next section. At the end of the section, we prove that groupoids given by fibered pull-backs of bundles of Lie groups always have Exel's property, which leads us to introduce, and extend this property to the more general class of *stratified submersion groupoids.*

3.1. Exhaustive families of representations of C^*-algebras

A two-sided ideal $I \subset A$ of a C^*-algebra A is called *primitive* if it is the kernel of an irreducible representation of A. We shall denote by $\mathrm{Prim}(A)$ the set of primitive ideals of A. The trivial ideal A (of A) is *not* considered a primitive ideal (of A), so $A \notin \mathrm{Prim}(A)$. For any representation ϕ of A, we define its *support* $\mathrm{supp}(\phi) \subset \mathrm{Prim}(A)$ as the set of primitive ideals of A *containing* $\ker(\phi)$. Let J be a closed, two-sided ideal of A. Then $\mathrm{Prim}(J) = \{I \in \mathrm{Prim}(A) | \, J \not\subset I\}$ and $\mathrm{Prim}(A/J)$ identifies with $\{I \in \mathrm{Prim}(A) | \, J \subset I\} = \mathrm{Prim}(J)^c$. We endow $\mathrm{Prim}(A)$ with the hull-kernel topology, which is the topology whose open sets are those of the form $\mathrm{Prim}(J)$, with J a two-sided ideal of A.

We shall need the following definition [76].

Definition 3.1. Let $\mathcal{F}$ be a set of representations of a C^*-algebra A. We say that $\mathcal{F}$ is *exhaustive* if $\mathrm{Prim}(A) = \bigcup_{\phi \in \mathcal{F}} \mathrm{supp}(\phi)$.

For applications, we shall need the following definition [88].

Definition 3.2. Let $\mathcal{F}$ be a set of representations of a unital C^*-algebra A.

(i) We say that $\mathcal{F}$ is *strictly norming* if, for any $a \in A$, there exists $\phi \in \mathcal{F}$ such that $\|\phi(a)\| = \|a\|$.

(ii) We say that $\mathcal{F}$ is *strictly spectral* if, for any $a \in A$, we have that a is invertible in A if, and only if, $\phi(a)$ is invertible for all $\phi \in \mathcal{F}$.

The two definitions above can be formulated for sets of morphisms or sets of primitive ideals. We also have that $\mathcal{F}_* := \{\pi \in \mathcal{F} | \, \pi \neq 0\}$ has the same properties ("exhaustive," "strictly norming," ...) as $\mathcal{F}$.

If A is non-unital, we modify the last definition as follows [76]. Let $A^+ := A \oplus \mathbb{C}$ and $\chi_0 \colon A^+ \to \mathbb{C}$ be the morphism defined by $\chi_0 = 0$ on A and $\chi_0(1) = 1$. We then replace A with A^+ and $\mathcal{F}$ with $\mathcal{F}^+ := \mathcal{F} \cup \{\chi_0\}$. This works also for exhaustive families since $\mathcal{F}$ is exhaustive for A if, and only if, $\mathcal{F}^+$ is exhaustive for A^+. Sometimes it is convenient to use the following alternative characterization of strictly spectral sets of representations. The set of representations $\mathcal{F}$ of A is strictly spectral if it satisfies the following property: $1 + a \in A^+$, $a \in A$, is invertible if, and only if, $1 + \phi(a)$ is invertible for any $\phi \in \mathcal{F}$.

A set of representations containing a representative in each equivalence classes of irreducible representations of a unital C^*-algebra is strictly norming (see [28,

29]). Therefore any exhaustive family is strictly norming [76]. In this regard, let us recall now the following results from [76, 88], which establish the relations between these concepts.

Theorem 3.3. *Let $\mathcal{F}$ be a set of non-degenerate representations of a C^*-algebra A. Then $\mathcal{F}$ is strictly norming if, and only if, it is strictly spectral. Every exhaustive set of representations is strictly spectral. If A is furthermore separable, then the converse is also true.*

Let A be a C^*-algebra and $I \subset A$ be a closed two-sided ideal. Recall that any representation $\pi\colon I \to \mathcal{L}(\mathcal{H})$ extends to a representation $\pi\colon A \to \mathcal{L}(\mathcal{H})$. If π is non-degenerate, this extension is unique. (See [28, Proposition 2.10.4].) This leads to the following results (see [76], Proposition 3.15 and Corollary 3.16).

Proposition 3.4. *Let $I \subset A$ be an ideal of a C^*-algebra. Let $\mathcal{F}_I$ be a set of non-degenerate representations of I and $\mathcal{F}_{A/I}$ be a set of representations of A/I. Let $\mathcal{F} := \mathcal{F}_I \cup \mathcal{F}_{A/I}$, regarded as a family of representations of A. If $\mathcal{F}_I$ and $\mathcal{F}_{A/I}$ are both exhaustive, then $\mathcal{F}$ is also exhaustive. The same result holds by replacing exhaustive with strictly norming.*

In the following corollary, one should think of the invertibility of a in A/I as an "ellipticity" condition.

Corollary 3.5. *Let $I \subset A$ be an ideal of a unital C^*-algebra A and let $\mathcal{F}_I$ be a strictly spectral set of nondegenerate representations of I. Let $a \in A$. Then a is invertible in A if, and only if, it is invertible in A/I and $\phi(a)$ is invertible for all $\phi \in \mathcal{F}_I$.*

We now address Morita equivalence [87] in a very simple form (see also [79, 85]). It is known that there exists a homeomorphism between the primitive ideal spectra of Morita equivalent C^*-algebras, so exhaustive families of representations will correspond to exhaustive families of representations under Morita equivalence. In particular, we have the following result.

Proposition 3.6. *Let $\mathcal{F}$ be a set of representations of a C^*-algebra A.*

(i) *If $\mathcal{F}_n := \{\pi \otimes 1 \,|\, \pi \in \mathcal{F}\}$ is the corresponding family of representations of $M_n(A) = A \otimes M_n(\mathbb{C})$, then $\mathcal{F}$ is exhaustive if, and only if, $\mathcal{F}_n$ is exhaustive.*
(ii) *Let $I \subset A$ be a closed, two-sided ideal. If $\mathcal{F}$ is exhaustive, then $\mathcal{F}_I := \{\pi|_I \,|\, \pi \in \mathcal{F}\}$ is an exhaustive family of representations of I.*
(iii) *Let $e \in M_n(A)$ be a projection. Then $\mathcal{F}_n$ defines, by restriction, an exhaustive family of representations of $eM_n(A)e$.*

Proof. The first part follows by Morita equivalence. The second part follows since $\operatorname{supp}(\phi|_I) = \operatorname{supp}(\phi) \cap \operatorname{Prim}(I)$. The last part is proved by combining the first two parts. Indeed, $\mathcal{F}_n$ defines an exhaustive set of representations of $M_n(A)$, and hence an exhaustive set of representations of $M_n(A)eM_n(A)$. Since the algebra $eM_n(A)e$ is Morita equivalent to $M_n(A)eM_n(A)$, by Proposition 4.27 in [35] (see also [87]), the result follows. □

3.2. Exel's question and properties

We will let $\mathcal{G}$ denote a Lie groupoid and $\mathcal{R}(\mathcal{G})$ denote its set of regular representations. The discussion below usually makes sense in the more general setting of locally compact groupoids, but we restrict, nevertheless to Lie groupoids, for simplicity.

In [29], Exel has asked for which groupoids $\mathcal{G}$ is the set $\mathcal{R}(\mathcal{G})$ a strictly norming set of representations of $C^*_r(\mathcal{G})$. An example due to Voiculescu, see [76], shows that not every locally compact groupoid has this property. (The question whether all Lie groupoids have Exel's property is still open.) It is easier to work with exhaustive families of representations, so we introduce the following definition.

Definition 3.7. We say that $\mathcal{G}$ has *Exel's property* if $\mathcal{R}(\mathcal{G})$, the set of regular representations of $\mathcal{G}$, is an exhaustive set of representations for $C^*_r(\mathcal{G})$. We say that $\mathcal{G}$ has *Exel's strong property* if $\mathcal{R}(\mathcal{G})$ is an exhaustive set of representations for $C^*(\mathcal{G})$.

We have that

$$\bigcup_{\phi\in\mathcal{F}} \operatorname{supp}(\phi) = \bigcup_{\phi\in\mathcal{F}} \operatorname{Prim}(A)\setminus \operatorname{Prim}(\ker(\phi)) = \operatorname{Prim}(A)\setminus \bigcap_{\phi\in\mathcal{F}} \operatorname{Prim}(\ker(\phi)).$$

Hence, if $\mathcal{F}$ is an exhaustive family of representations of a C^*-algebra A, then $\bigcap_{\phi\in\mathcal{F}} \operatorname{Prim}(\ker(\phi)) = \emptyset$, and hence $\mathcal{F}$ is a faithful family of representations of A. In particular, $\mathcal{G}$ has Exel's strong property if, and only if, $\mathcal{G}$ has Exel's property and is metrically amenable.

We shall need the following consequence of Propositions 2.6 and 3.4 of the previous subsection.

Corollary 3.8. *Let $U \subset M$ be an open, $\mathcal{G}$-invariant subset and assume $\mathcal{G}$ to be second countable. We let $F := M \smallsetminus U$. Let us assume that $\mathcal{G}_U$ and $\mathcal{G}_F$ have Exel's strong property. Then $\mathcal{G}$ also has Exel's strong property.*

Proof. Let $I := C^*(\mathcal{G}_U) = C^*_r(\mathcal{G}_U)$, $A := C^*(\mathcal{G})$. Then $A/I \simeq C^*(\mathcal{G}_F) \simeq C^*_r(\mathcal{G}_F)$ by Proposition 2.6(ii). Since $\mathcal{G}_F$ and $\mathcal{G}_U$ are metrically amenable, it follows that $\mathcal{G}$ is metrically amenable as well, by Proposition 2.6(iii). Proposition 3.4 then gives that $\mathcal{R}(\mathcal{G}) = \mathcal{R}(\mathcal{G}_U) \cup \mathcal{R}(\mathcal{G}_F)$ is an exhaustive family of representations of $C^*(\mathcal{G}) \simeq C^*_r(\mathcal{G})$, that is, $\mathcal{G}$ has Exel's strong property, as claimed. □

More generally, we can consider groupoids given by a filtration.

Corollary 3.9. *Let $\mathcal{G} \rightrightarrows M$ be a second countable groupoid and assume that U_i are open, $\mathcal{G}$-invariant subsets of M, such that $\emptyset =: U_{-1} \subset U_0 \subset \cdots \subset U_{i-1} \subset U_i \subset \cdots \subset U_N := M$, and for each $S := U_i \smallsetminus U_{i-1}$, $\mathcal{G}_S$ has Exel's strong property. Then $\mathcal{G}$ also has Exel's strong property.*

Proof. Corollary 3.8 yields the result by induction: If $N = 1$, it reduces to Corollary 3.8. Assuming that the result holds for a filtration in $N-1$ sets, then $\mathcal{G}_{U_{N-1}}$ has Exel's strong property. Since by assumption $\mathcal{G}_{M\setminus U_{N-1}}$ also has Exel's strong property, the result follows again by Corollary 3.8. □

Here is a simple, but very important example of a class of Lie groupoids $\mathcal{G}$ for which the set $\mathcal{R}(\mathcal{G})$ of regular representations is an exhaustive set of representations of $C^*(\mathcal{G})$. The idea of the proof below is that, locally, the C^*-algebra of $\mathcal{G}$ is of the form $C^*(G) \otimes \mathcal{K}$, where $\mathcal{K}$ are the compact operators on the (typical) fiber of f. We found it convenient to formalize the proof of this well-known result using Morita equivalence of groupoids.

Proposition 3.10. *Let $\mathcal{H} \rightrightarrows L$ be a bundle of Lie groups with fiber G and let $f : M \to L$ be a tame submersion. Define $\mathcal{G} := f^{\downarrow\downarrow}(\mathcal{H})$. Then $\mathcal{G}$ is a Lie groupoid with Exel's property. If G is amenable, then $\mathcal{G}$ has Exel's strong property.*

Proof. We have already seen that $\mathcal{G} := f^{\downarrow\downarrow}(\mathcal{H})$ is a Lie groupoid (because f is a tame submersion). We endow then $\mathcal{G}$ with one of the standard Haar measures coming from the Lie groupoid structure. Let

$$X = M \times_L \mathcal{H} := \{(m,\gamma) \in M \times \mathcal{H} |\, f(m) = r(\gamma)\}.$$

Then X defines a Morita equivalence between $\mathcal{G}$ and $\mathcal{H}$, and hence a Morita equivalence between their full and reduced C^*-algebras of these groupoids, by the classical results of [68, 99]. See also [103]. In particular, $\mathrm{Prim}(C^*(\mathcal{G}))$ and $\mathrm{Prim}(C^*(\mathcal{H}))$ are homeomorphic by natural homeomorphisms [87]. The same is true for $\mathrm{Prim}(C_r^*(\mathcal{G}))$ and $\mathrm{Prim}(C_r^*(\mathcal{H}))$.

All irreducible representations of $C_r^*(\mathcal{H})$ factor through an evaluation morphism $e_x : C_r^*(\mathcal{H}) \to C_r^*(G_x) \simeq C_r^*(G)$, for some $x \in M$, where G_x is the fiber of $\mathcal{H} \to M$ above x, since locally $\mathcal{H}$ is a product of a group and a space. We have that the regular representation π_x is obtained from the regular representation of G_x via e_x, the evaluation at x. Hence $\mathcal{G}$ has Exel's property.

Let us assume now that G is amenable. Hence the fibers of $\mathcal{H} \to M$ are amenable, and therefore $\mathcal{H}$ is metrically amenable. Hence $\mathcal{G}$ is also metrically amenable, by their Morita equivalence. Since $\mathcal{G}$ was already proved to have Exel's property, it follows that it has also Exel's strong property. □

We notice that if $\mathcal{G}$ is isomorphic to a groupoid as in the proposition above, then $\mathcal{H} \rightrightarrows L$ and $f : M \to L$ are uniquely determined, up to equivalence. Indeed, L is diffeomorphic to the set of orbits of $\mathcal{G}$ (acting on M), f becomes the quotient map $M \to M/\mathcal{G}$ and $\mathcal{H}$ is obtained from the isotropy groupoid of $\mathcal{G}$.

The above results suggest the following definition.

Definition 3.11. Let $\mathcal{H}$ be a groupoid with units M. We say that $\mathcal{H}$ is a *stratified submersion groupoid* with filtration (U_i) if U_i are open, $\mathcal{H}$-invariant subsets of M,

$$\emptyset =: U_{-1} \subset U_0 \subset \cdots \subset U_{i-1} \subset U_i \subset \cdots \subset U_N := M,$$

and each $S := U_i \smallsetminus U_{i-1}$ is a manifold, possibly with corners, such that there exist a Lie group bundle $G_S \to B_S$ and a tame submersion $f_S : S \to B_S$ of manifolds with corners such that

$$\mathcal{H}_S \simeq f_S^{\downarrow\downarrow}(G_S).$$

A groupoid $\mathcal{H}$ will be called a *stratified submersion* groupoid if it is a *stratified submersion* groupoid for some filtration.

Remark 3.12. It is clear from the definition that if $\mathcal{H}$ is a stratified submersion groupoid, then, with the notation of the definition, each $\mathcal{H}_{U_k \smallsetminus U_j}$, $j < k$, is also a stratified submersion groupoid. We have to notice, however, that $\mathcal{H}_{U_k \smallsetminus U_j}$ may not be a Lie groupoid, even if $\mathcal{H}$ is a Lie groupoid. This is because $U_k \smallsetminus U_j$ may not be a manifold (it is only a union of manifolds).

As we shall see, the class of stratified submersion groupoids is large enough to comprise many classes of groupoids and operators that arise in practice. Combining Corollary 3.9 with Proposition 3.10, we now obtain the following result.

Theorem 3.13. *Let $\mathcal{G} \rightrightarrows M$ be a stratified submersion Lie groupoid and assume that all the isotropy groups $\mathcal{G}_x^x$, $x \in M$, are amenable. Then the set $\mathcal{R}(\mathcal{G})$ of regular representations of $\mathcal{G}$ is an exhaustive family of representations of $C^*(\mathcal{G})$. In particular, $\mathcal{G}$ is metrically amenable and has Exel's property.*

Proof. Let us use the notation of Definition 3.11. Indeed then, the isotropy group $\mathcal{G}_x^x$ is isomorphic to the fiber of $G_S \to B_S$ above $f_S(x)$, for $x \in S$. Thus, by Proposition 3.10, all the groupoids $\mathcal{G}_S$ are metrically amenable and have Exel's property. Using Corollary 3.9, we obtain that $\mathcal{G}$ has Exel's strong property, that is, that it is metrically amenable and has Exel's property. □

In the next section, we will see how this property plays a role in establishing Fredholm conditions for operators on groupoids.

Remark 3.14. Let $\mathcal{H}$ be a stratified submersion Lie groupoid as in Definition 3.11. In view of the remark 2.7, we have then the following disjoint union decomposition:

$$\mathcal{H} = \sqcup_S f_S^{\downdownarrows}(G_S) = \sqcup_S S \times_B S \times_B G_S \,,$$

where $\times_B$ is the fibered product over the base (that is, over B_S). We notice also that two regular representations π_x and π_y are unitarily equivalent if there is $g \in \mathcal{G}$ such that $d(g) = x$ and $r(g) = y$. For a stratified submersion groupoid, this is the case exactly if $x, y \in S := U_{i-1} \smallsetminus U_i$ *and* $f_S(x) = f_S(y) = b \in B_s$, in which case, these two representations act, up to a unitary equivalence, on the set of square integrable functions on the space $Z_b := f_S^{-1}(b) \times (G_S)_b$. We denote by π_b the corresponding representation on $L^2(Z_b)$, $b \in B_S$.

4. Fredholm conditions

We now study Fredholm conditions for operators in algebras $\mathbf{\Psi}$ containing a reduced groupoid C^*-algebra $C_r^*(\mathcal{G})$ as an essential ideal. (Recall that an ideal of an algebra is *essential* if its annihilator vanishes.) The groupoids for which we obtain the kind of Fredholm conditions that we want (the kind that are typically used in practice) will be called "Fredholm groupoids." They are introduced and discussed next. Examples of Fredholm groupoids will be provided in the next section.

Throughout the rest of this paper, $\mathcal{G} \rightrightarrows M$ denote a second countable Lie groupoid, which is hence a locally compact groupoid. Let us fix a metric on $A(\mathcal{G})$ and hence a right Haar system λ_x on $\mathcal{G}$. Also, recall that all our Lie groupoids are assumed to be Hausdorff.

4.1. Fredholm groupoids and their characterization

In this subsection, we introduce Fredholm groupoids and give a first characterization of these groupoids. Recall that $\pi_x \colon C^*(\mathcal{G}) \to \mathcal{L}(L^2(\mathcal{G}_x, \lambda_x))$ denotes the regular representation on $\mathcal{G}_x$, given by left convolution, $\pi_x(\phi)\psi := \phi * \psi$ (Definition 2.4). The following notation will be used throughout the rest of the paper.

Notations 4.1. We assume that $U \subset M$ is an open, $\mathcal{G}$-invariant subset with $\mathcal{G}_U \simeq U \times U$ (the pair groupoid, see Example 2.17). For any $x_0 \in U$, the range map then defines a bijection $r : \mathcal{G}_{x_0} \to U$ and hence a measure μ on U corresponding to λ_{x_0}. This measure does not depend on the choice of $x_0 \in U$ and leads to isometries $L^2(\mathcal{G}_{x_0}, \lambda_{x_0}) \simeq L^2(U; \mu)$ that commute with the action of $\mathcal{G}$. We then denote by π_0 the corresponding representation of $C^*(\mathcal{G})$ on $L^2(U; \mu)$ and call it the *vector representation* of $C^*(\mathcal{G})$. We shall usually write $L^2(U) := L^2(U; \mu)$. The representation π_0 then defines an isomorphism $C^*_r(\mathcal{G}_U) \simeq \pi_0(C^*_r(\mathcal{G}_U)) = \mathcal{K}$, the algebra of compact operators on $L^2(U)$.

Remark 4.2. An important remark is that, since we only consider *Hausdorff* Lie groupoids, the vector representation π_0 is injective, by a result of Khoshkam and Skandalis [44]. In particular, we have that $1 + \pi_0(a)$ is invertible in $\pi_0(C^*_r(\mathcal{G}))/\mathcal{K}$ if, and only if, $1 + a$ is invertible in $C^*_r(\mathcal{G})/C^*_r(\mathcal{G}_U)$. We shall thus identify $C^*_r(\mathcal{G})$ with its image under π_0, that is, with a class of operators on $L^2(U)$, without further comment.

In the following definition, we shall use that $C^*_r(\mathcal{G}_U) \subset \ker \pi_x$ whenever $x \notin U$, by definition, and hence that π_x descends to a representation of the quotient algebra $C^*_r(\mathcal{G})/C^*_r(\mathcal{G}_U)$.

Definition 4.3. We say that $\mathcal{G}$ is a *Fredholm Lie groupoid* if:

(i) There is an open, dense, $\mathcal{G}$-invariant subset $U \subset M$ such that $\mathcal{G}_U \simeq U \times U$.

(ii) Let $a \in C^*_r(\mathcal{G})$. We have that $1 + a$ is Fredholm if, and only if, $1 + \pi_x(a)$ is invertible for each $x \in M \smallsetminus U$ (see 4.1 for notation).

The set $F := M \smallsetminus U$ will be called the set of *boundary units* of $\mathcal{G}$ and a set U as in this definition will be called a manifold with *amenable ends.*

For a Fredholm Lie groupoid $\mathcal{G} \rightrightarrows M$, we shall always denote by $U \subset M$ the $\mathcal{G}$-invariant open subset from the definition of a Lie Fredholm groupoid. Since U is a dense orbit, it is uniquely determined by $\mathcal{G}$. Hence, $F := M \smallsetminus U$ is closed and $\mathcal{G}$-invariant.

We shall need the following result for the matricial case.

Proposition 4.4. *Let us assume that $\mathcal{G} \rightrightarrows M$ is a Fredholm Lie groupoid, that M is compact, and that $e \in M_n(\mathcal{C}(M))$ is a projection.*

(i) *The set of representations* π_x, $x \in F$ *defines a strictly spectral family of representations of* $eM_n\big(C_r^*(\mathcal{G})/C_r^*(\mathcal{G}_U)\big)e$.
(ii) *For any* $a \in eM_n(C_r^*(\mathcal{G}))e$, *we have that* $1 + a$ *is Fredholm if, and only if,* $1 + \pi_x(a)$ *is invertible for each* $x \in F := M \smallsetminus U$.

Proof. By Atkinson's theorem (which states that an operator is Fredholm if, and only if, it is invertible modulo the compacts) and by the definition of strictly spectral families of representations in the non-unital case, one can see that condition (ii) in the definition of Fredhom groupoid is equivalent to the fact that $\mathcal{R}(\mathcal{G}_F)$ defines a strictly spectral family of representations of the quotient $C_r^*(\mathcal{G})/\mathcal{K} \simeq C_r^*(\mathcal{G})/C_r^*(\mathcal{G}_U)$. Since $\mathcal{G}$ is second countable, its associated C^*-algebras are separable, hence the set $\{\pi_x | \, x \in F\}$ is strictly spectral if, and only if, it is exhaustive, by Theorem 3.3. The first part is then a consequence of Proposition 3.6.

The second part follows again from the definition of strictly spectral families of representations in the non-unital case, from $C_r^*(\mathcal{G}_U) \simeq \mathcal{K}(L^2(U))$, and from Atkinson's theorem. □

In the next results, we give an abstract characterization of Lie Fredholm groupoids, using some of the ideas of the previous section. We obtain in particular conditions that will make it simpler to check whether a given Lie groupoid is Fredholm. Recall that, in this paper, all groupoids are Hausdorff.

Theorem 4.5. *Assume* $\mathcal{G}$ *is a Lie groupoid. If* $\mathcal{G}$ *is Fredholm, then the following two conditions are satisfied:*

(i) *The canonical projection* $C_r^*(\mathcal{G}) \to C_r^*(\mathcal{G}_F)$ *induces an isomorphism*
$$C_r^*(\mathcal{G})/C_r^*(\mathcal{G}_U) \simeq C_r^*(\mathcal{G}_F), \quad F := M \smallsetminus U\,.$$
(ii) $\mathcal{G}_F$ *has Exel's property.*

(Recall that condition (ii) means that $\mathcal{R}(\mathcal{G}_F)$ is an exhaustive set of representations of $C_r^*(\mathcal{G}_F)$.)

Proof. Let us assume that $\mathcal{G}$ is a Fredholm groupoid and check the two conditions of the statement.

Let $p\colon C_r^*(\mathcal{G}) \to C_r^*(\mathcal{G}_F)$ be the natural projection induced by restriction. It is known that $C_r^*(\mathcal{G}_U) \subseteq \ker(p)$ (see the proof of [85, Prop.II.4.5 (a)]). To prove (i), we need to show that we have equality $C_r^*(\mathcal{G}_U) = \ker(p)$. Let us proceed by contradiction, that is, let us assume that $C_r^*(\mathcal{G}_U) \neq \ker(p)$. Then we can choose $a = a^* \in \ker(p) \smallsetminus C_r^*(\mathcal{G}_U)$. Note that since $\pi_x = \pi_x \circ p$, $x \in F = M \setminus U$, we have $\ker p \subset \ker \pi_x$, in particular, $1 - \pi_x(a) = 1$ is invertible for all $x \notin U$. On the other hand, since a is self-adjoint and non-zero, in the quotient $\ker(p)/C_r^*(\mathcal{G}_U)$, there is $0 \neq \lambda \in \mathbb{R}$ such that $\lambda - a$ is not invertible in $\ker(p)/C_r^*(\mathcal{G}_U)$. By rescaling, we may assume $\lambda = 1$ and thus $1 - a$ is not invertible in $C_r^*(\mathcal{G})/C_r^*(\mathcal{G}_U) \cong C_r^*(\mathcal{G})/\mathcal{K}$. We conclude that, in this case, $\mathcal{R}(\mathcal{G}_F)$ is not a strictly spectral family of representations of $C_r^*(\mathcal{G})/C_r^*(\mathcal{G}_U)$. It then follows that $1-a$ is not Fredholm, however $1-\pi_x(a) = 1$ is invertible for all $x \notin U$. This is a contradiction.

The first part shows that there is an isomorphism

$$C_r^*(\mathcal{G}_F) \simeq C_r^*(\mathcal{G})/C_r^*(\mathcal{G}_U) \simeq C_r^*(\mathcal{G})/\mathcal{K}(L^2(U)).$$

By Fredholmness of $\mathcal{G}$, we have that $\mathcal{R}(\mathcal{G}_F)$ defines a strictly spectral family of representations of $C_r^*(\mathcal{G})/\mathcal{K}$, so $\mathcal{R}(\mathcal{G}_F)$ is also a strictly spectral family of representations of $C_r^*(\mathcal{G}_F)$. Since $\mathcal{G}$ is second countable, it is exhaustive, by Theorem 3.3, and hence $\mathcal{G}_F$ has Exel's property. This proves (ii) and hence the direct implication. □

We remark that we proved in fact that, if $\mathcal{R}(\mathcal{G}_F)$ defines a strictly spectral family of representations of $C_r^*(\mathcal{G})/C_r^*(\mathcal{G}_U)$, then $C_r^*(\mathcal{G})/C_r^*(\mathcal{G}_U) \simeq C_r^*(\mathcal{G}_F)$ and $\mathcal{G}_F$ has Exel's property.

A stronger form of the converse of Theorem 4.5 is contained in the following theorem. Recall that a faithful representation of an ideal in a C^*-algebra extends *uniquely* to the whole algebra.

Theorem 4.6. *Let $\mathcal{G} \rightrightarrows M$ be a Lie groupoid. Let us assume that $U \subset M$ is a dense $\mathcal{G}$-invariant open subset such that $\mathcal{G}_U \simeq U \times U$. Also, let us assume that the two conditions of Theorem 4.5 are satisfied. Then, for any unital C^*-algebra $\mathbf{\Psi}$ containing $M_n(C_r^*(\mathcal{G}))$ as an essential ideal and for any $a \in \mathbf{\Psi}$, we have that $\pi_0(a)$ if Fredholm if, and only if, the image of a in $\mathbf{\Psi}/M_n(C_r^*(\mathcal{G}))$ is invertible and all $\pi_x(a)$, $x \in M \smallsetminus U$, are invertible.*

Proof. The proof is the same for any n (using also Proposition 3.6, so we let $n = 1$ for notational simplicity). Let us assume that conditions (i) and (ii) of Theorem 4.5 are satisfied and let $\mathbf{\Psi}$ be a C^*-algebra containing $C_r^*(\mathcal{G})$ as an essential ideal. Note that π_0 is injective on $\mathbf{\Psi}$ since $C^*(\mathcal{G})$ is an essential ideal of $\mathbf{\Psi}$. By our assumptions on $\mathcal{G}$, we obtain that the algebra $\Psi/C_r^*(\mathcal{G}_U) \simeq \pi_0(\Psi)/\mathcal{K} =: B$ contains $B_0 := \pi_0(C_r^*(\mathcal{G}))/\mathcal{K}$ as an ideal and $B/B_0 \simeq \mathbf{\Psi}/C_r^*(\mathcal{G})$. Moreover, $B_0 \simeq C_r^*(\mathcal{G}_F)$ by (i).

Let now $a \in \mathbf{\Psi}$ be arbitrary. By Atkinson's Theorem, we know that $\pi_0(a)$ is Fredholm if, and only if, its image in B is invertible. But by (ii) and by Corollary 3.5, $c \in B$ is invertible if, and only if, the image of c in $\mathbf{\Psi}/C_r^*(\mathcal{G})$ and all $\pi_x(c)$ are invertible for all $x \in F$. This completes the proof. □

Corollary 4.7. *Let $\mathcal{G} \rightrightarrows M$ be a Lie groupoid and $\emptyset \neq U \subset M$ be a dense, $\mathcal{G}$-invariant, open subset such that $\mathcal{G}_U \simeq U \times U$, and $F = M \setminus U$. Then $\mathcal{G}$ is Fredholm if, and only if, $C_r^*(\mathcal{G})/C_r^*(\mathcal{G}_U) \simeq C_r^*(\mathcal{G}_F)$ and $\mathcal{R}(\mathcal{G}_F)$ is a strictly spectral set of representations of $C_r^*(\mathcal{G}_F)$ (that is, if $\mathcal{G}_F$ has Exel's property).*

Recalling the exact sequence in Proposition 2.6, it follows that, in particular, if $\mathcal{G}$ is metrically amenable, then $\mathcal{G}$ is Fredholdm if, and only if, $\mathcal{G}_F$ has Exel's property.

Remark 4.8. It is immediate to observe that, if the set $\mathcal{R}(\mathcal{G}_F)$ of regular representations π_x, $x \in F$, forms a strictly spectral set of representations of $C^*(\mathcal{G}_F)$, then $\mathcal{G}_F$ has Exel's strong property, and conditions (ii) and (iii) of the Theorem 4.5 are satisfied (using Proposition 2.6).

This remark and the above results then give the following sufficient conditions for a Lie groupoid to be Fredholm.

Corollary 4.9. *Let $\mathcal{G} \rightrightarrows M$ be a Lie groupoid and $U \subset M$ be a dense $\mathcal{G}$-invariant open subset such that $\mathcal{G}_U \simeq U \times U$, and $F = M \setminus U$. Assume that $\mathcal{R}(\mathcal{G}_F)$ forms a strictly spectral set of representations of $C^*(\mathcal{G}_F)$ (equivalently, that $\mathcal{G}_F$ has Exel's strong property). Then $\mathcal{G}$ is a Fredholm groupoid.*

See also [7]. Together with Theorem 3.13, we obtain an important class of Fredholm groupoids. Recall that we assume that all our groupoids are Hausdorff.

Theorem 4.10. *Let $\mathcal{G} \rightrightarrows M$ be a stratified submersion Lie groupoid with filtration U_i such that U_0 is dense and $\mathcal{G}_{U_0} \simeq U_0 \times U_0$. Assume that all isotropy groups $\mathcal{G}_x^x$ are amenable. Then $\mathcal{G}$ is a Fredholm groupoid.*

Proof. By Remark 3.12 applied to $F := M \smallsetminus U_0$ and by Theorem 3.13, we know that the family $\mathcal{R}(\mathcal{G}_F)$ of regular representations of $\mathcal{G}_F$ is an exhaustive family of representations of $C^*(\mathcal{G}_F)$, that is, $\mathcal{G}_F$ has Exel's strong property. The result follows from the previous corollary. □

We notice that we can replace π_x, $x \in F$, with π_b, $b \in B_S$, for the strata corresponding to the boundary. We can further restrict to the set of b's in a dense set of orbits, since they give rise to a faithful family of representations of $\mathcal{G}_F$, again by [44].

4.2. Pseudodifferential operators

So far we have discussed mostly groupoids and operators in their (unitized) C^*-algebras. The study of differential operators on suitable non-compact spaces can be reduced to the study of groupoid algebras using pseudodifferential operators on Lie groupoids. In this subsection, we will explain this reduction procedure. See [42, 72, 90, 98, 108] for introductions to pseudodifferential operators.

Let $\mathcal{G}$ be a Lie groupoid and consider the algebra $\Psi^*(\mathcal{G})$ of *pseudodifferential operators on the groupoid* $\mathcal{G}$, whose definition we now briefly recall [4, 65, 77]. Then, for $m \in \mathbb{Z} \cup \{\pm\infty\}$, $\Psi^m(\mathcal{G})$ consists of smooth families $(P_x)_{x\in M}$ of classical pseudo-differential operators $P_x \in \Psi^m(\mathcal{G}_x)$ of order m, that are right invariant with respect to the action of $\mathcal{G}$ and have compactly supported distribution kernels. In particular, $\Psi^{-\infty}(\mathcal{G})$ is nothing but the convolution algebra of smooth, compactly supported function on $\mathcal{G}$, that is, $\Psi^{-\infty}(\mathcal{G}) \simeq \mathcal{C}_c^\infty(\mathcal{G})$.

This construction extends right away to operators between smooth sections of some vector bundles $E, F \to M$. There are two methods of doing this. The first method is to consider families $(P_x)_{x\in M}$ of operators $P_x \in \Psi^m(\mathcal{G}_x; r^*(E), r^*(F))$. We denote by $\Psi^m(\mathcal{G}; E, F)$ the resulting set of operators. By embedding $E \oplus F$ into a trivial bundle of dimension N, we can identify $\Psi^m(\mathcal{G}; E, F)$ with a subspace of $M_N(\Psi^m(\mathcal{G}))$, when convenient. Similar considerations yield differential operators generated by $\mathcal{V}$ acting between vector bundles E and F endowed with metric preserving connections.

The second method to deal with operators between two vector bundles $E, F \to M$ is as follows. Let us choose an embedding of $E \oplus F \to \underline{\mathbb{C}}^N$ into a trivial vector bundle and denote by e and f the corresponding orthogonal projections onto the images of e and f. Then we can identify

$$\Psi^m(\mathcal{G}; E, F) := f M_N(\Psi^m(\mathcal{G}))e .$$

This idea applies to other types of similar operators (norm closures of operators in $\Psi^m(\mathcal{G})$, differential operators, ...).

One of the main points of considering pseudodifferential operators on a groupoid is that, if we denote $\mathcal{V} := \text{Lie}(\mathcal{G}) := \varrho(\Gamma(A(\mathcal{G})))$, then

$$\text{Diff}^m(\mathcal{V}; E, F) = \Psi^m(\mathcal{G}; E, F) \cap \text{Diff}(M; E, F),$$

where $\text{Diff}^m(\mathcal{V}; E, F)$ is generated by $\mathcal{C}^\infty(M; \text{End}(E \oplus F))$ and $\mathcal{V}$ (using compatible connections on E and F) and $\text{Diff}(M; E, F)$ denotes *all* differential operators on M acting between sections of E and F. See also 4.14.

Another main point for introducing the algebra $\Psi^*(\mathcal{G})$ is related to parametrices and inverses of differential operators. To discuss this, let us denote by $\overline{\Psi}(\mathcal{G})$ the C^*-algebra obtained as the closure of $\Psi^0(\mathcal{G})$ with respect to all contractive $*$-representations of $\Psi^{-\infty}(\mathcal{G})$, as in [49]. As we will see below, this definition extends to operators acting between vector bundles. Let us denote, as usual, by S^*A the set of unit vectors in the Lie algebroid $A^*(\mathcal{G})$ associated with $\mathcal{G}$ (as in Remark 2.11), with respect to some fixed metric on $A^*(\mathcal{G})$. Then $\overline{\Psi}(\mathcal{G})$ fits into the following exact sequence

$$0 \to C^*(\mathcal{G}) \to \overline{\Psi}(\mathcal{G}) \xrightarrow{\sigma_0} \mathcal{C}_0(S^*A) \to 0,$$

where σ_0 is an extension of the principal symbol map (see for instance [48] and the references therein.) Moreover, if $P \in \Psi^m(\mathcal{G}; E, F)$, $m \geq 0$, is such that $P : H^m(M; E) \to L^2(M; F)$ is invertible, then its inverse will be in $\overline{\Psi}(\mathcal{G}; F, E)$, the norm closure of $\Psi^0(\mathcal{G}; F, E)$. Typically, Fredholm conditions are obtained for Fredholm Lie groupoids by applying our results to the algebra $\mathbf{\Psi} = M_N(\overline{\Psi}(\mathcal{G}))$, for some N. Let us see how this is done.

We first need to recall the definition of Sobolev spaces. Let us fix in this subsection a Lie groupoid $\mathcal{G} \rightrightarrows M$. For the purpose of the next result, let us assume that its space of units M has an open, dense, $\mathcal{G}$-invariant subset $U \subset M_0 := M \smallsetminus \partial M$ such that the restriction $\mathcal{G}_U$ is isomorphic to the pair groupoid $U \times U$. Let us also assume that the space of units M is **compact.** This is needed in order to construct canonical Sobolev spaces on the interior of M. Indeed, there is an essentially unique class of metrics on $A(\mathcal{G})$, which, by restriction, gives rise to a class of metrics on U, that are called *compatible* (with the groupoid $\mathcal{G}$), see also [3]. All these metrics are Lipschitz equivalent and complete [2, 49]. In fact, the Sobolev spaces of all these metrics will coincide. They are given as the domains of the powers of $1 + \Delta$, where Δ is the (geometer' s, i.e., positive) Laplacian. We shall denote by $H^s(U) = H^s(M)$ these Sobolev spaces. The Sobolev spaces $H^s(M)$ are discussed in detail in [2]. See also [74] for a review. The operators in

$\Psi^m(\mathcal{G})$ (and their vector bundle analogues) model differential operators associated to (any) compatible metric (see also Remark 4.14 below). By considering a vector bundle $E \to M$ and Δ_E the associated Laplacian on sections of E, we obtain the spaces $H^s(M; E)$.

Note that the Sobolev spaces $H^s(M)$ are *not* defined using the compact manifold structure on M, but rather using the complete metric on $U \subset M$. Recall that if the set $M \setminus U$ is as in Definition 4.3, then the set U is called a manifold with *amenable ends.* If $P \in \Psi^m(\mathcal{G}; E, F)$, then $P : H^s(M; E) \to H^{s-m}(M; F)$ is continuous.

Let $\Psi^m(\mathcal{G}; E, F)$ denote the pseudodifferential operators acting on the lifts of E and F to $\mathcal{G}$ via d. We let $\Psi^m(\mathcal{G}; E) := \Psi^m(\mathcal{G}; E, E)$. Then if $P : H^s(M; E) \to H^{s-m}(M; F)$ is an order m pseudodifferential operator, we replace the study of its Fredholm properties with those of

$$P_1 := \begin{bmatrix} 0 & P & 0 \\ P^* & 0 & 0 \\ 0 & 0 & Q_m \end{bmatrix},$$

where Q_m is an invertible order m pseudodifferential operator acting on the complement of $E \oplus F$. Let G denote the complement of $E \oplus F$ in $\mathbb{C}^N$. We fix the smooth vector bundles $E, F, G \to M$ in what follows. We denote by Δ_E, Δ_F, and Δ_G the associated Laplacians acting on sections of E, respectively, F and G. For instance, $Q_m \in \Psi^m(\mathcal{G}; G)$ could be $(1 + \Delta_G)^{m/2}$ (or a suitable approximation it, see [49]).

One of the main drawbacks of the algebras $\Psi^m(\mathcal{G})$ is that they are too small to contain resolvents. This issue is easily fixed by considering completions with respect to suitable norms. Let us consider the norm $\| \cdot \|_{m,s}$ defined by.

$$\|P\|_{m,s} := \|(1 + \Delta_F)^{(s-m)/2} P (1 + \Delta_E)^{-s/2}\|_{L^2 \to L^2} \tag{2}$$

We then let

$$L^m_s(\mathcal{G}; E, F) := \| \cdot \|_{m,s}\text{-closure of } \Psi^m(\mathcal{G}; E, F) . \tag{3}$$

Recall that $H^s(M; E)$ is the domain of $(1 + \Delta_E)^{s/2}$, if $s \geq 0$, whenever M is compact (see [2, 37, 49]. Hence $L^m_s(\mathcal{G}; E, F)$ is the *norm closure* of $\Psi^m(\mathcal{G}; E, F)$ in the norm topology of continuous linear maps $H^s(M; E) \to H^{s-m}(M; F)$. We note that, if $P \in \Psi^m(\mathcal{G}; E, F)$, then

$$(1 + \Delta_F)^{(s-m)/2} P (1 + \Delta_E)^{-s/2} \in \overline{\Psi}(\mathcal{G}; E, F) =: L^0_0(\mathcal{G}; E, F)$$

by [49] (see also [5]). Moreover, let

$$\mathcal{W}^m(\mathcal{G}) := \Psi^m(\mathcal{G}) + \cap_s L^{-\infty}_s(\mathcal{G}) .$$

Then $\mathcal{W}^m(\mathcal{G}) \subset L^m_s(\mathcal{G})$ and $\mathcal{W}^\infty(\mathcal{G})$ is an algebra of pseudodifferential operators that contains the inverses of its L^2-invertible operators.

Recall that we denote by $\pi_0 : C^*(\mathcal{G}) \to \mathcal{L}(L^2(M))$ the vector representation and that M is compact. It is unitarily equivalent to the regular representations

π_x, $x \in U$. Moreover, since $\mathcal{G}$ is Hausdorff, π_0 is injective. We have the following result from [49].

Proposition 4.11. *Let $\mathcal{G} \rightrightarrows M$ be a Lie groupoid with $\mathcal{G}_U = U \times U$ and M compact. Then $M_N(\overline{\Psi}(\mathcal{G})) = M_N(L^0_0(\mathcal{G}))$ contains $M_N(C^*(\mathcal{G}))$ as an essential ideal. Let $s \in \mathbb{R}$ and $P \in L^m_s(\mathcal{G}; E, F) \supset \Psi^m(\mathcal{G}; E, F)$. Then*

$$a := (1 + \Delta_F)^{(s-m)/2} P (1 + \Delta_E)^{-s/2} \in \overline{\Psi}(\mathcal{G}; E, F),.$$

We have that $P : H^s(M; E) \to H^{s-m}(M; F)$ is Fredholm if, and only if, $a : L^2(M; E) \to L^2(M; F)$ is Fredholm.

This proposition applies, in particular, if $\mathcal{G}$ is a Fredholm Lie groupoid with M compact.

Proof. We can assume $N = 1$. The fact that $\overline{\Psi}(\mathcal{G})$ contains $C^*(\mathcal{G})$ as an essential ideal is a general fact – true for any Lie groupoid. This general fact is true because it is true for any non-compact manifold, in particular, for each of the manifolds $\mathcal{G}_x$. We have, by the definitions of Sobolev spaces and of Fredholm operators, that P is Fredholm if, and only if, a is Fredholm. The fact that $a := (1 + \Delta_F)^{(s-m)/2} P (1 + \Delta_E)^{-s/2} \in \overline{\Psi}(\mathcal{G}; E, F)$ follows from the definition of $L^m_s(\mathcal{G}; E, F)$, Equation (3) and the results in [49], as noticed already above. □

This proposition then gives right away the following result (for technical reasons, we may have to replace P with a in certain proofs). Recall that an operator $P \in \Psi^m(\mathcal{G}; E, F)$ consists of a right invariant family $P = (P_x)$, $x \in M$, with P_x acting between the sections of $r^*(E)$ and $r^*(F)$ on $\mathcal{G}_x$. In particular, P_x is invariant with respect to the free, proper action of $\mathcal{G}^x_x$ on $\mathcal{G}_x$ by right multiplication. For simplicity, we shall drop r^* from the notation below, since there is no danger of confusion. We have $P_x = \pi_x(P)$. The operators P_x, $x \in M \setminus U$, will be called *limit operators* (of P). They go back to [30].

Theorem 4.12. *Let $\mathcal{G} \rightrightarrows M$ be a Fredholm Lie groupoid with M compact and $\emptyset \neq U \subset M$, open such that $\mathcal{G}_U = U \times U$. Let $s \in \mathbb{R}$ and $P \in L^m_s(\mathcal{G}; E, F) \supset \Psi^m(\mathcal{G}; E, F)$. We have*

$$\begin{aligned} P : H^s(M; E) \to H^{s-m}(M; F) \text{ is Fredholm } \quad &\Leftrightarrow \quad P \text{ is elliptic and each} \\ P_x : H^s(\mathcal{G}_x; E) \to H^{s-m}(\mathcal{G}_x; F)\,,\ x \in M \smallsetminus U\,, &\quad \text{is invertible}\,. \end{aligned}$$

Proof. Let us use the notation of Proposition 4.11. By replacing P with a and using Proposition 4.4, we may assume that we work with $N \times N$ matrices of operators. The proof is the same for all N, so we let $N = 1$. Since $\mathcal{G}$ is Fredholm, Theorem 4.6, applied to $\mathbf{\Psi} := \overline{\Psi}(\mathcal{G})$, gives that $a \in \overline{\Psi}(\mathcal{G})$ is Fredholm if, and only if, its image in $\overline{\Psi}(\mathcal{G})/C^*(\mathcal{G})$ is invertible and all the operators $\pi_x(a)$ are invertible. We then notice that $\pi_x(a) = (1 + \Delta_x)^{(s-m)/2} P_x (1 + \Delta_x)^{-s/2}$ since the extension of π_x to operators affiliated to $\overline{\Psi}(\mathcal{G})$ is given by $\pi_x(P) = P_x$ since this is true for $P \in \Psi^0(\mathcal{G})$ and $\pi_x(\Delta) = \Delta_x$, the Laplacian on $\mathcal{G}_x$ by [49]. □

Remark 4.13. To obtain the result as formulated in Theorem 1.2, we let α be the set of orbits of $\mathcal{G}$ acting on $F := M \smallsetminus U$, $M_0 := U$, $M_\alpha := \mathcal{G}_x$, for some x in the orbit α, $G_\alpha := \mathcal{G}_x^x$, and $P_\alpha := \pi_x(P)$. It may be useful to notice here that often the bundle $M_\alpha \to M_\alpha/G_\alpha$ is trivial. This is the case, for example, for stratified submersion groupoids.

Remark 4.14. Let $\mathcal{V} := r_*(\Gamma(A(\mathcal{G}))$ be the vector fields on M coming from the infinitesimal action of $\mathcal{G} \rightrightarrows M$ on M. We denote by $\mathrm{Diff}^m(\mathcal{V})$ the set of *order m* differential operators on M generated by $\mathcal{V}$ and multiplication with functions in $\mathcal{C}^\infty(M)$ and $\mathrm{Diff}(\mathcal{V}) = \cup_m \mathrm{Diff}^m(\mathcal{V})$. We denote by $\mathrm{Diff}^m(\mathcal{V}; E, F)$ the analogous differential operators acting between sections of vector bundles $E, F \to M$. We have that all geometric operators (Laplace, Dirac, Hodge, ...) associated to a compatible metric on M (one that comes by restriction from $A(\mathcal{G})$) belong to $\mathrm{Diff}^m(\mathcal{V}; E, F)$ for suitable $E, F \to M$; moreover, $\pi_x(D)$ for a geometric operator is of the same type as D. [3].

Remark 4.15. If $\mathcal{G}$ is d-connected (in the sense that the fibers of $d : \mathcal{G} \to M$ are connected), then the orbits of the vector fields $\mathcal{V}$ are the same as the orbits of $\mathcal{G}$. If $x \in M$ and $\alpha \subset M$ is its orbit, then this orbit (with its intrinsic manifold topology) is diffeomorphic to the set M_α/G_α. This makes more explicit the data in Remark 4.13.

Although we shall not use this in the present paper, let us record the consequence for essential spectra. Notice that we do not need the closure of the unions for the essential spectra. We denote by $\sigma(Q)$ the spectrum of an operator Q, defined as the set of $\lambda \in \mathbb{C}$ such that $Q - \lambda$ is *not* invertible, and by $\sigma_{\mathrm{ess}}(Q)$ its essential spectrum, defined as the set of $\lambda \in \mathbb{C}$ such that $Q - \lambda$ is *not* Fredholm. Here Q may be unbounded. If $P \in \Psi^m(\mathcal{G}; E)$, $m > 0$, we consider it to be defined on H^m.

Corollary 4.16. *Consider the framework of Theorem* 4.12 *and* $P \in \Psi^m(\mathcal{G}; E)$, $m \geq 0$, *elliptic if* $m > 0$. *We have*

$$\begin{aligned} \sigma_{\mathrm{ess}}(P) &= \cup_{x \in \partial M} \sigma(P_x) \cup \cup_{\xi \in S^*A} \sigma(\sigma_0(P)(\xi)), \quad \text{if } m = 0, \\ \sigma_{\mathrm{ess}}(P) &= \cup_{x \in \partial M} \sigma(P_x), \quad \text{if } m > 0. \end{aligned}$$

See also the discussion in [76] for an explanation of the differences between the cases $m = 0$ and $m > 0$ in the above corollary. Note that, if P is not elliptic, the statement in the case $m > 0$ is not true anymore, and, in fact, the essential spectrum depends on more than just the principal symbol, in general. Thus, in the scalar, order zero case, we have

$$\sigma_{\mathrm{ess}}(P) = \cup_{x \in \partial M} \sigma(P_x) \cup Im(\sigma_0(P)).$$

Combining Theorems 4.12 and 4.10, we obtain the following result (recall that all our groupoids are Hausdorff and second countable).

Theorem 4.17. *Let M be compact and $\mathcal{G} \rightrightarrows M$ be a stratified submersion Lie groupoid with filtration U_i such that U_0 is dense in M and $\mathcal{G}_{U_0} \simeq U_0 \times U_0$. Assume* *that all isotropy groups $\mathcal{G}_x^x$ are amenable. Then, for any smooth vector bundles $E, F \to M$ and any $P \in L_s^m(\mathcal{G}; E, F) \supset \Psi^m(\mathcal{G}; E, F)$, we have*

$$P : H^s(M; E) \to H^{s-m}(M; F) \text{ is Fredholm} \quad \Leftrightarrow \quad P \text{ is elliptic and all}$$
$$P_x : H^s(\mathcal{G}_x; E) \to H^{s-m}(\mathcal{G}_x; F)\,, \ x \in M \smallsetminus U_0\,, \quad \text{are invertible}.$$

We actually have $P_x : H^s(\mathcal{G}_x; r^*(E)) \to H^{s-m}(\mathcal{G}_x; r^*(F))$, but, as we have explained above, we drop r^* from the notation, for the sake of simplicity, since there is no danger of confusion. We will continue to do that in all related statements below.

Remark 4.18. If $\mathcal{G}$ be as in the above theorem (a stratified submersion groupoid with filtration U_i), and assume that it is also d-connected, as in Remark 4.13. Then, when $P = D$ is a differential operator, we can identify the limit operators D_α (and hence also M_α and G_α) in a direct way. Indeed, let α be the $\mathcal{V}$-orbit $=$ the $\mathcal{G}$-orbit through some $x \in M \smallsetminus U_0$. We have $G_\alpha = \mathcal{G}_x^x$, as discussed in Remark 4.13. Then $M_\alpha = \alpha \times G_\alpha$. Let $D \in \mathrm{Diff}(\mathcal{V})$. By linearity, we may assume $D = aX_{i_1}X_{i_2}\cdots X_{i_m}$, with $X_i \in \mathcal{V}$ a *local basis* of $\mathcal{V}$ near x and $a \in \mathcal{C}^\infty(M)$. (We may also assume that i_j form a non decreasing sequence, but that is not necessary.) Since α is an orbit, we have that this basis (X_j) consists of vectors tangent to α. We can choose this basis such that $X_1, \ldots, X_k$ form a local basis of $T_x\alpha$ and X_j, $j > k$, are zero on the orbit α (where they are defined). Consequently, the vector fields X_j, $j > k$, come from a basis $\tilde{X}_j$, $j > k$, of $\mathrm{Lie}(G_\alpha)$. We then have

$$D_\alpha = a|_\alpha Y_1 Y_2 \cdots Y_m\,, \quad \text{near } \{x\} \times G_\alpha \subset \alpha \times G_\alpha =: M_\alpha\,, \tag{4}$$

where $Y_j = X_{i_j}|_\alpha$, if $i_j \leq k$, and $Y_j = \widetilde{X}_{i_j}$ if $i_j > k$. We shall say that the variables Y_j (or X_j) with $j > k$ are *ghost derivatives* at the considered orbit α (see Remark 4.18).

See also [8, 9, 100, 106]. It would be interesting to extend the results of this section to L^p-spaces in view of [1, 82].

5. Examples

In this section, we use the results of the previous section to obtain Fredholm conditions for operators on some standard non-compact manifolds: manifolds with (poly)cylindrical ends, asymptotically Euclidean manifolds, asymptotically hyperbolic manifolds, boundary fibration structures, and others. We tried to write this section in such a way that it can to a large extent be read independently of the rest of the paper. The reader interested only in applications can start reading the paper with this section.

The general setting of this section is that of a (non-compact) smooth manifold M_0 whose geometry is determined by a compactification M to a manifold with corners and a Lie algebra of vector fields $\mathcal{V}$ on M. (Although we shall not use this, let us mention for people familiar with the concept, that $(M, \mathcal{V})$ will be a Lie manifold with some additional properties.) The differential (and pseudodifferential) operators considered and for which we obtain Fredholm conditions are the ones generated by $\mathcal{V}$ and $\mathcal{C}^\infty(M)$, that is, the ones in Diff($\mathcal{V}$) and its variants for vector bundles. (Recall that Diff($\mathcal{V}$) was introduced in Remark 4.14.) The proof of the Fredholm conditions in this section are obtained as particular cases of Theorem 4.12 by showing that appropriate groupoids are Hausdorff stratified submersion Lie groupoids with amenable isotropy groups and hence that they are Fredholm groupoids, in view of Theorem 3.13. Many of the results below can also be obtained from the results in [48, 49], but the approach followed here aims to be more convenient for non-specialists.

In this section, we continue to denote by $\mathcal{G}$ a Lie groupoid with units M, a manifold of dimension n. In the theorem yielding Fredholm conditions, M will be assumed **compact** and endowed with a *smooth* metric h. We assume that h is defined *everywhere* on M, in particular, that it extends to a smooth manifold containing M as a submanifold. Also, usually it will be no loss of generality to assume M connected. Our results are formulated for operators in $L^m_s(\mathcal{G}; E, F)$, which is a suitable completion of $\Psi^m(\mathcal{G}; E, F)$, see Equations (2) and (3). This allows us to greatly enlarge the scope of our results since the completion procedure defining the spaces L^m_s leads to algebras that are closed under taking the inverses of L^2-invertible elements.

5.1. Examples related to group actions

We include first some examples that are closely related to group actions.

5.1.1. The action of a group on a space. Let us assume that Lie group G acts smoothly on a manifold M. This yields the *transformation* (or *group action*) Lie groupoid $\mathcal{G} := M \rtimes G$, which, as a set, consists of $M \times G$ and has units M. We have

$$d(x, g) := x\,,\ r(x, g) := gx\,,\ \text{ and }\ (hx, g)(x, h) := (x, gh)\,.$$

If $M = G$ with G acting by translations, then $M \rtimes G \simeq G \times G$, the product groupoid. Therefore, if $G \subset M$ as a dense, G-invariant open set, and G acts on itself by left translation, we are in the setting in which we can ask if the resulting groupoid is Fredholm Lie groupoid. This groupoid was used in [32, 57, 67]. The groupoids used in [32, 67] turn out to be stratified submersion groupoids. As discussed in those papers, this recovers the classical HVZ-theorem [21, 84, 102] as a particular case of Theorem 4.12 (with $U = G$). We note that $M \times G$ is always Hausdorff (since M and G are Hausdorff). If $V \subset M$ is an open subset, then the reduction groupoid $(M \rtimes G)_V^V$ will be called a *local transformation* (*or action*) groupoid, and will also be Hausdorff. Many related results (including the non Lie case), were also obtained by Böttcher, Chandler-Wilde, Karlovich, Lindner, Rabinovich, Roch, Rozenblyum, Silberman, and many others. See [6, 10, 11, 16, 59, 81–83, 89] and the references therein.

This groupoid models pseudodifferential operators compatible with any G invariant metric on G. Let $\mathfrak{g}$ be the Lie algebra of G. Then $A(\mathcal{G}) = M \times \mathfrak{g}$, with $\mathfrak{g}$

acting on M via the infinitesimal action of G. The associated Lie algebra of vector fields is $\mathrm{Lie}(\mathcal{G}) := \varrho(\Gamma(A(\mathcal{G}))) = \mathcal{V} := \mathcal{C}^\infty(M)\mathfrak{g} \subset \Gamma(T\Omega)$. We have that $\mathrm{Diff}(\mathcal{V})$ is generated by $\mathcal{C}^\infty(M)$ and $\mathfrak{g}$.

5.1.2. The *b*-groupoid. Let M be a manifold with corners. Then the b-groupoid of [49, 61, 65, 77] is defined as a set as the disjoint union

$$\mathcal{G}_b := \sqcup_F (F \times F) \times (\mathbb{R}_+^*)^{k_F} \simeq \sqcup_F (F \times F) \times \mathbb{R}^{k_F} \rightrightarrows \sqcup_F F = M\,,$$

where F ranges through the open, *connected* faces of M, k_F is the codimension of the face F, $F \times F$ is the pair groupoid and $(\mathbb{R}_+^*)^{k_F}$ is a group for componentwise multiplication (and hence also a Lie groupoid).

To obtain the smooth structure on this groupoid, we notice that, locally, it is a transformation groupoid $[0,\infty)^k \rtimes (\mathbb{R}_+^*)^k$. More precisely, let us choose $g \in \mathcal{G}_b$. Then $g = (x, y, v) \in (F \times F) \times \mathbb{R}^{k_F}$ for some *connected* open face $F \subset M$. We can choose a coordinate system $V \subset M$ such that $x, y \in V$, $\overline{V}$ is compact in F, and we have a tubular neighborhood $V \times [0,\epsilon)^{k_F} \subset M$. Then, a neighborhood of g in $\mathcal{G}_b$ is diffeomorphic to the local transformation groupoid obtained by reducing to $V \times [0,\epsilon)^{k_F}$ the action groupoid

$$(\mathbb{R}^{d_F} \times [0,\infty)^{k_F}) \rtimes (\mathbb{R}^{d_F} \times (0,\infty)^{k_F})\,,$$

where $d_F = n - k_F$ is the dimension of F. Here $\mathbb{R}^{d_F}$ acts by translations on itself and $\mathbb{R}_+^* = (0,\infty)$ acts by multiplication on $[0,\infty)$. In applications, it will be, in fact, more convenient to notice that $\mathbb{R}^{d_F} \rtimes \mathbb{R}^{d_F}$ is the pair groupoid, and hence to identify a neighborhood of g with a reduction of

$$(V \times V) \times [0,\infty)^{k_F} \rtimes (0,\infty)^{k_F}\,,$$

the product of the pair groupoid $V \times V$ and the transformation groupoid $[0,\infty)^{k_F} \rtimes (0,\infty)^{k_F}$.

If M has embedded faces, that is, if each hyperface H has a defining function r_H, then we can identify $\mathcal{G}_b$ with an open subset of Monthubert's realization of the b-groupoid [66, Proposition 4.5]

$$\mathcal{M} := \{\, (x, y, t) \in M \times M \times [-1,1]^{\mathcal{H}} \,|\, (1 - t_K)x_K(x) = (1 + t_H)x_H(y) \,\}\,,$$

where $\mathcal{H}$ denotes the set of *hyperfaces* of M and $H, K \in \mathcal{H}$. See also [13, 14, 51, 52, 104].

We see that the b-groupoid is a stratified submersion groupoid as follows. The set U_k, $k \leq n$, is defined as the union of the open faces of codimension k of M. Then $(\mathcal{G}_b)_{U_k \setminus U_{k-1}}$ is isomorphic to the topological disjoint union of the groupoids $(F \times F) \times \mathbb{R}^k$, where F ranges through the set of open faces of codimension k. In particular, $(\mathcal{G}_b)_{U_k \setminus U_{k-1}}$ is isomorphic to the fibered pull-back of a bundle of Lie groups, by Example 2.20. In particular, $(\mathcal{G}_b)_x^x \cong \mathbb{R}^{n-k}$ is amenable, for all $x \in M$.

5.1.3. Manifolds with poly-cylindrical ends. The Lie algebroid of $\mathcal{G}_b$ is identified by

$$\Gamma(A(\mathcal{G}_b)) \simeq \mathcal{V}_b := \{X \in \Gamma(M;TM) |\, X \text{ tangent to all faces of } M\}. \quad (5)$$

Let h be an ordinary metric on M. The general form of a compatible metric on M is then

$$g_b := h + \sum_{H \in \mathcal{H}} \left(\frac{dx_H}{x_H}\right)^2. \quad (6)$$

Manifolds with metrics of this form are called *manifolds with poly-cylindrical ends*, following [63]. The Lie algebroid $A(\mathcal{G}_b)$ is often denoted T^bM. The groupoid $\mathcal{G}_b$ models pseudodifferential operators compatible with the metric g_b. Thus, by definition, the metric g_b comes from restriction from the Lie algebroid $A(\mathcal{G}_b)$, since $A(\mathcal{G}_b)|_{M_0} = TM_0$. Since the base M is compact, all metrics on $A(\mathcal{G}_b)$ will be equivalent, so the constructions will not depend significantly on the choice of the metric. In particular, all geometric operators associated to the metric g_b (Laplace, Dirac, Hodge, ...) will belong to $\Psi^m(\mathcal{G}_b; E, F)$ (which is independent of the metric), for suitable E and F. Moreover, it turns out that Theorem 4.17 applies to $\mathcal{G}_b$ and $\Psi^m(\mathcal{G}_b; E, F)$, provided that M *is compact.* (See also Remark 4.14. Further details can be found in [3].)

The statement of Theorem 4.5 can be (slightly) simplified in this case by noticing the following. The representations π_x and π_y are unitarily equivalent if x and y are in the same open face F, in which case, they will act on $F \times \mathbb{R}^{k_F} \simeq \mathcal{G}_x$. Let π_F be the associated representation. For Fredholm conditions, it is enough to consider the invertibility of the operators $P_H := \pi_x(P)$, $x \in H$, for the faces of maximal dimension (that is, for hyperfaces) in order to obtain Fredholm conditions, since π_x is weakly contained in π_y if x is contained in the closure of the face containing y. Let $\mathcal{H}$ denote the set of hyperfaces of M, as before. Recall that $L^m_s(\mathcal{G}; E, F)$ is a suitable completion of $\Psi^m(\mathcal{G}; E, F)$, see Equations (2) and (3). We obtain the following result [62].

Corollary 5.1. *Let $P \in L^m_s(\mathcal{G}_b; E, F) \supset \Psi^m(\mathcal{G}_b; E, F)$, M compact. We have*

$$P : H^s(M; E) \to H^{s-m}(M; F) \text{ is Fredholm} \quad \Leftrightarrow \quad P \text{ is elliptic and all}$$
$$P_H : H^s(H \times \mathbb{R}; E) \to H^{s-m}(H \times \mathbb{R}; F),\ H \in \mathcal{H}, \quad \text{are invertible}.$$

The operators P_H are invariant with respect to translations by $t \in \mathbb{R}$ acting on the $\mathbb{R}$ component. All geometric differential operators associated to the metric g_b belong to $\mathrm{Diff}^m(\mathcal{V}_b; E, F) \subset \Psi^m(\mathcal{G}_b; E, F)$, for suitable vector bundles $E, F \to M$.

If $P \in \mathrm{Diff}^m(\mathcal{V}_b; E, F)$, the limit operators P_x, $x \in \partial M$, are obtained as explained in Remark 4.18. They are invariant with respect to the action of the isotropy group $\mathcal{G}_x^x$ (this is always the case for operators of the form $\pi_x(a)$). In this example, only derivatives of the form $x_H \partial_{x_H}$ are ghost derivatives (see Remark 4.18).

Remark 5.2. Carvalho and Qiao have constructed in [15] a similar groupoid to the b-groupoid in order to study layer potentials. Their groupoid, however, was not d-connected, in general. Nevertheless, the above corollary generalizes to their setting, after some obvious modifications.

5.1.4. Asymptotically Euclidean spaces. Let us assume that M has a smooth boundary ∂M with defining function $x = x_{\partial M}$ and let $\mathcal{V}_{sc} := x\mathcal{V}_b$. The resulting differential operators $\mathrm{Diff}(\mathcal{V}_{sc})$ and the associated pseudodifferential operators are the SG-operators of [20, 27, 78, 91–93] (called "scattering operators" in [63]). They can be obtained by considering the groupoid

$$\mathcal{G}_{sc} := TM|_{\partial M} \sqcup (M_0 \times M_0) \rightrightarrows \partial M \sqcup M_0 = M .$$

To obtain a manifold structure on $\mathcal{G}_{sc}$, let us consider first $G = \mathbb{R}^n$ and $M =$ the radial compactification of G with the induced action of G. Then $\mathcal{G}_{sc} = M \rtimes G$ is Hausdorff and a stratified submersion groupoid. In general, $\mathcal{G}_{sc}$ is locally of this form (and can be obtained by gluing reductions of such groupoids), and hence it is Hausdorff. It satisfies $\mathrm{Lie}(\mathcal{G}_{sc}) := \Gamma(A(\mathcal{G}_{sc})) = \mathcal{V}_{sc}$. Thus, if g_b denotes a b-metric on M (or rather, on M_0, then the natural metric associated to $\mathcal{V}_{sc}$ is

$$g_{sc} := x^{-2} g_b := x^{-2}\big(h + x^{-2} dx^2\big) , \tag{7}$$

where g_b is as in Equation (6).

We have that the orbits of $\mathcal{G}$ on $F := M \smallsetminus U_0 = \partial M$ are reduced to points and that each stabilizer $\mathcal{G}_x^x = T_x M \simeq \mathbb{R}^n$, for $x \in F$. In particular, $\mathcal{G}_x = \mathcal{G}_x^x$, for $x \in \partial M$ and all derivatives at the boundary are ghost derivatives. If $P \in \mathrm{Diff}^m(\mathcal{V}_{sc}; E, F)$, the limit operators P_x are obtained as explained in Remark 4.18. Theorem 4.17 becomes in our case:

Corollary 5.3. *Let $P \in L^m_s(\mathcal{G}_{sc}; E, F) \supset \Psi^m(\mathcal{G}_{sc}; E, F)$, M compact. We have*

$$P : H^s(M; E) \to H^{s-m}(M; F) \text{ is Fredholm} \quad \Leftrightarrow \quad P \text{ is elliptic and all}$$
$$P_x : H^s(T_x M; E) \to H^{s-m}(T_x M; F) , \ x \in \partial M , \quad \text{are invertible}.$$

The operators P_x are constant coefficient (i.e., P_x is invariant with respect to the action of $T_x M$ on itself by translations). All geometric differential operators associated to the metric g_{sc} belong to $\mathrm{Diff}^m(\mathcal{V}_{sc}; E, F) \subset \Psi^m(\mathcal{G}_{sc}; E, F)$, for suitable vector bundles $E, F \to M$.

We have $P_x = \pi_x(P)$ and P_x is translation invariant (i.e., constant coefficient, in this case), and hence can be studied using the Fourier transform. In this case, all the vector fields yield ghost derivatives.

5.1.5. Asymptotically hyperbolic manifolds. We continue to assume that M is a manifold with smooth boundary ∂M. The groupoid $\mathcal{G}_{ah}$ modeling asymptotically hyperbolic spaces is chosen such that

$$\mathrm{Lie}(\mathcal{G}_{ah}) \simeq \Gamma(A(\mathcal{G}_{ah})) = \mathcal{V}_0 := \{X \in \Gamma(M; TM) |\, X|_{\partial M} = 0\} .$$

Let $\mathcal{L}_x := T_x(\partial M) \rtimes (0, \infty)$ be the semi-direct product as in the previous subsection, with $(0, \infty)$ acting by dilations on $T_x(\partial M)$. Let $\mathcal{L} \to \partial M$ be the bundle of

Lie groups with fiber $\mathcal{L}_x$. Then

$$\mathcal{G}_{ah} := \mathcal{L} \sqcup M_0 \times M_0 \rightrightarrows \partial M \sqcup M_0 .$$

The topology is again locally given by a transformation groupoid. This can be seen in the case of $M = \mathbb{R}^{n-1} \times [0, \infty)$ with the natural action of the semi-direct product (cross-product) $G_n := \mathbb{R}^{n-1} \rtimes (0, \infty)$ obtained by recalling that, as smooth manifolds, we have $\mathbb{R}^{n-1} \rtimes (0, \infty) = \mathbb{R}^{n-1} \times (0, \infty)$. The action of $(0, \infty)$ on $\mathbb{R}^{n-1}$ is by dilations. This groupoid is a particular case of the *edge groupoid*, following next, so the reader can consult he next section for more details. We again have that $\mathcal{G}_x = \mathcal{G}_x^x = \mathcal{L}_x$, if $x \in \partial M$, and that all derivatives are ghost derivatives at the boundary. The metric is

$$g_{ah} := x^{-2} h ,$$

where x is the distance to the boundary (close to the boundary) and h is an everywhere smooth metric on M, as before.

Corollary 5.4. *Let $s \in \mathbb{R}$ and $P \in L_s^m(\mathcal{G}_{ah}; E, F) \supset \Psi^m(\mathcal{G}_{ah}; E, F)$, with M compact. Denote $M_x := T_x\partial M \rtimes \mathbb{R} = (\mathcal{G}_{ah})_x$. We have*

$$P : H^s(M; E) \to H^{s-m}(M; F) \text{ is Fredholm } \Leftrightarrow \text{ } P \text{ is elliptic and all}$$
$$P_x : H^s(M_x; E) \to H^{s-m}(M_x; F) , \ x \in \partial M , \quad \text{are invertible}.$$

The operator P_x is invariant with respect to the action of M_x on itself by right translations, for each $x \in \partial M$. All geometric differential operators associated to the metric g_{ah} belong to $\mathrm{Diff}^m(\mathcal{V}_{ah}; E, F) \subset \Psi^m(\mathcal{G}_{ah}; E, F)$, for suitable vector bundles $E, F \to M$.

We thus need to study the invertibility of certain (right) invariant operators on the group $G_{n-1} := \mathbb{R}^{n-1} \rtimes (0, \infty)$, for which standard methods of representation theory can be used.

5.2. The edge groupoid

This example is motivated by the results in [36, 38, 58, 75, 95], where the original Fredholm results on the edge calculus can also be found. It is a particular case of the next example, that of a desingularization groupoid, but we nevertheless treat it separately, for the benefit of the reader. See also [23, 47, 107]. We consider the following framework.

First, M is a manifold with smooth boundary ∂M. We assume that we are given a smooth fibration $\pi : \partial M \to B$, where B is a smooth manifold (thus without boundary). We fix a tubular neighborhood U of ∂M in M: $U \simeq \partial M \times [0, 1)$. Let $\mathcal{H} := B \times B$ be the pair groupoid. We now construct the so-called *edge groupoid* $\mathcal{G}_e$ that will turn out to be a groupoid with Lie algebroid given by the set $\mathcal{V}_e$ of vector fields on M that are tangent to the fibers of $\pi : \partial M \to B$ (in particular, these vector fields are tangent to the boundary). If $B = \partial M$, then we recover the groupoid that models asymptotically hyperbolic spaces. We denote by $\mathrm{Diff}(\mathcal{V}_e)$ the algebra of differential operators generated by $\mathcal{V}_e$ and by multiplication with functions in $\mathcal{C}^\infty(M)$.

Let $\mathcal{L} := TB \rtimes \mathbb{R}_+^* \to B$ be the bundle of Lie groups obtained from $TB \to B$ (regarded as a bundle of commutative Lie groups) by taking the semi-direct product with $\mathbb{R}_+^* := (0,\infty)$ acting by dilations on the fibers of $TB \to B$. Its pullback $\pi^{\downarrow\downarrow}(\mathcal{L})$ via $\pi : \partial M \to B$ is hence a Lie groupoid with units ∂M. Let $M_0 := M \smallsetminus \partial M$ be the interior of M. Then, as a set, the *edge groupoid* $\mathcal{G}_e$ is the disjoint union

$$\mathcal{G}_e := \pi^{\downarrow\downarrow}(\mathcal{L}) \sqcup (M_0 \times M_0) \rightrightarrows \partial M \sqcup M_0 .$$

To define the smooth structure on this groupoid, we could use either the results in [73] or proceed directly in four steps as follows.

Step 1. We first consider the adiabatic groupoid $\mathcal{H}_{ad}$ of $\mathcal{H} := B \times B$ (this is the *tangent groupoid* of [17]; see [75] for more details and references). The adiabatic groupoid $\mathcal{H}_{ad}$ is a Lie groupoid with units $B \times [0,\infty)$ and Lie algebroid $A(\mathcal{H}_{ad}) = TB \times [0,\infty) \to B \times [0,\infty)$, which, as a vector bundle, is the fibered pull-back of $A(\mathcal{H}) = TB \to B$ to $B \times [0,\infty)$ via the projection $B \times [0,\infty) \to B$. The Lie algebroid structure on the sections of $A(\mathcal{H}_{ad})$ is not that of a fibered pull-back Lie algebroid, but is given by $[X,Y](t) = t[X(t),Y(t)]$. As a set, $\mathcal{H}_{ad}$ is the disjoint union

$$\mathcal{H}_{ad} := A(\mathcal{H}) \times \{0\} \sqcup \mathcal{H} \times (0,\infty) .$$

The groupoid structure of $\mathcal{H}_{ad}$ is such that $A(\mathcal{H}) \times \{0\}$ has the Lie groupoid structure of a bundle of Lie groups and $\mathcal{H} \times (0,\infty)$ has the product Lie groupoid structure with $(0,\infty)$ the groupoid associated to a space (that is $(0,\infty)$ has only units, and all orbits are reduced to a single point). The smooth structure is obtained using the exponential map. See also [105].

Step 2. Let $\pi : \partial M \to B$ be the given fibration map. We denote also by π the resulting map $\partial M \times [0,\infty) \to B \times [0,\infty)$. Then we consider the pullback Lie groupoid $\pi^{\downarrow\downarrow}(\mathcal{H}_{ad})$.

Step 3. Let $\mathbb{R}_+^* = (0,\infty)$ act by dilations on the $[0,\infty)$ variable on $\pi^{\downarrow\downarrow}(\mathcal{H}_{ad})$ and consider the semi-direct product $\pi^{\downarrow\downarrow}(\mathcal{H}_{ad}) \rtimes \mathbb{R}_+^*$ [26]. As a set, it is the disjoint union of $\pi^{\downarrow\downarrow}(\mathcal{L})$ and of the pair groupoid of $\partial M \times (0,\infty)$.

Step 4. Let us identify the tubular neighborhood $U \subset M$ of ∂M with $\partial M \times [0,1)$. Then we can consider the reduction $\mathcal{H}' := \left(\pi^{\downarrow\downarrow}(\mathcal{H}_{ad}) \rtimes \mathbb{R}_+^*\right)_U^U$. By the previous step, this reduction $\mathcal{H}'$ is the disjoint union of $\pi^{\downarrow\downarrow}(\mathcal{L})$ and of the pair groupoid of $\partial M \times (0,1)$, which we can view as a subset of $M_0 \times M_0$, the pair groupoid of M_0. We then glue the reduction $\mathcal{H}' := \left(\pi^{\downarrow\downarrow}(\mathcal{H}_{ad}) \rtimes \mathbb{R}_+^*\right)_U^U$ with $M_0 \times M_0$ by identifying the reduction of $\mathcal{H}'$ to $U \smallsetminus \partial M$ (which is the pair groupoid of $\partial M \times (0,1)$, as we have seen), with its image in $M_0 \times M_0$. This gluing construction is, of course, nothing but a particular case of the gluing construction in [39] and [75].

In any case, we obtain right away from the definition that the edge groupoid $\mathcal{G}_e$ is a Hausdorff stratified submersion groupoid. The set of units of $\mathcal{G}_e$ is M and the representations π_x, $x \in \partial M$, are equivalent precisely when they map to the same point in B and they act on $\pi^{-1}(b) \times \mathcal{L}_b \simeq \pi^{-1}(b) \times T_bB \times \mathbb{R}$.

If $P \in \mathrm{Diff}(\mathcal{V}_e)$ and $b \in B$, we can obtain the limit operators $P_b := \pi_b(P)$ as follows. The restriction of P to an infinitesimal neighborhood of $\pi^{-1}(b)$ in M will have some hidden (ghost) derivatives coming from the Lie algebra of the group $\mathcal{L}_b := T_bB \rtimes \mathbb{R}_+^*$. Let us choose local coordinates (y, z) on ∂M that are compatible with $\partial M \to B$, in the sense that z comes from a coordinate system on B. Let x denote the defining function of ∂M. Then locally, $\mathcal{V}_e$ is generated by $x\partial_x$, ∂_{y_j}, $x\partial_{z_k}$, with j and k (always) ranging through a suitable index set:

$$\mathcal{V}_e = \mathcal{C}^\infty(M)x\partial_x + \sum_j \mathcal{C}^\infty(M)\partial_{y_j} + \sum_k \mathcal{C}^\infty(M)x\partial_{z_k}.$$

With this notation, the Lie algebra $\mathrm{Lie}(\mathcal{L}_b)$ is generated by the vector fields $x\partial_x$ and $x\partial_{z_k}$. These come from non-zero vector fields in $\mathcal{V}_e$ that restrict to 0 on $\pi^{-1}(b)$, they are the ghost derivatives. We let the ghost derivatives act on $\mathcal{L}_b$, and thus we obtain a differential operator P_b on $\pi^{-1}(b) \times \mathcal{L}_b$. It can be obtained by patching together metrics for which $x\partial_x$, ∂_{y_j}, $x\partial_{z_k}$ are an orthonormal set of vectors.

Theorem 4.12 then becomes.

Corollary 5.5. *Let $P \in L_s^m(\mathcal{G}_e; E, F) \supset \Psi^m(\mathcal{G}_e; E, F)$, M compact. We have*

$$P : H^s(M; E) \to H^{s-m}(M; F) \text{ is Fredholm} \Leftrightarrow P \text{ is elliptic and all}$$
$$P_b : H^s(M_b; E) \to H^{s-m}(M_b; F)\,,\ b \in B\,, \text{ are invertible},$$

where $M_b := \pi^{-1}(b) \times T_bB \rtimes \mathbb{R}$. The operator P_b is invariant with respect to the action of $T_bB \rtimes \mathbb{R}$ by right translations on itself. All geometric differential operators associated to g_e belong to $\mathrm{Diff}^m(\mathcal{V}_e; E, F) \subset \Psi^m(\mathcal{G}_e; E, F)$, for suitable $E, F \to M$.

If B is reduced to a point and ∂M is connected, the groupoid $\mathcal{G}_e$ constructed in the last subsection recovers the groupoid of the b-calculus: $\mathcal{G}_e = \mathcal{G}_b$. It models in this case manifolds with cylindrical ends. If $B = \partial M$, the corresponding groupoid models "asymptotically hyperbolic" spaces.

5.3. Desingularization groupoids

One of the nice features of the class of stratified submersion groupoids is that it is invariant with respect to desingularization along suitable submanifolds. In this subsection, we will give an *ad hoc* argument for this statement. Recall the fibered pull-back $\pi^{\downarrow\downarrow}(B)$ of Example 2.19. We proceed in a slightly greater generality than in [75], to which we refer for more details and for the unexplained arguments.

Let us assume that M is a manifold with corners, that H is a hyperface of M and that we are given a *tame* submersion of manifolds with corners $\pi : H \to B$. Let $U \simeq H \times [0, 1)$ be a tubular neighborhood of H in M. The hyperface H will play the role played by the boundary in the previous examples. Typically, M will be the result of a desingularization procedure, such as a blow-up (see Remark 5.6 and the following section).

We assume that we are given a Lie groupoid $\mathcal{G}$ on $M \smallsetminus H$ and a Lie groupoid $\mathcal{H} \rightrightarrows B$ such that the following is satisfied.

Local fibered pull-back structure assumption: *The reduction of $\mathcal{G}$ to $H \times (0,1)$ is isomorphic to the pullback $p^{\downarrow\downarrow}(\mathcal{H})$ via the map $p := \pi \circ p_1 : H \times (0,1) \to B$.*

We fix $\mathcal{H}$ as in the above assumption throughout this subsection. We define then a groupoid $\mathcal{K} \rightrightarrows M$ such that, as a set, it is the disjoint union

$$\mathcal{K} := \pi^{\downarrow\downarrow}(\mathcal{A}(\mathcal{H}) \rtimes \mathbb{R}_+^*) \sqcup \mathcal{G} \rightrightarrows H \sqcup M \smallsetminus H.$$

The groupoid $\mathcal{H}$ replaces the pair groupoid $B \times B$ of the example of the edge groupoid. We proceed as in that example to consider the adiabatic groupoid $\mathcal{H}_{ad}$ of $\mathcal{H}$, which has units $B \times [0,\infty)$. We pull back this groupoid to a groupoid $\pi^{\downarrow\downarrow}(\mathcal{H}_{ad})$ using the map $\pi_1 : H \times [0,\infty) \to B \times [0,\infty)$, and then we consider the semi-direct product $\pi_1^{\downarrow\downarrow}(\mathcal{H}_{ad}) \rtimes \mathbb{R}_+^* = \pi_1^{\downarrow\downarrow}(\mathcal{H}_{ad} \rtimes \mathbb{R}_+^*)$ with $\mathbb{R}_+^*$ acting by dilations on $[0,\infty)$.

As in the previous example of the edge groupoid, we view U as an open subset of $H \times [0,\infty)$ and we consider the groupoid $\mathcal{H}'$ defined as the reduction to U of $\pi^{\downarrow\downarrow}(\mathcal{H}_{ad}) \rtimes \mathbb{R}_+^*$, with $\mathcal{H}$ as in the local fibered pull-back assumption. The construction is completed as in the fourth step of the edge groupoid. That is, we use the invariant subset $H \times \{0\}$ of the units of $\mathcal{H}'$ to write the groupoid as a disjoint union using Remark 2.7. The reduction of $\mathcal{H}'$ to the complement of $H \times \{0\}$, that is, to $H \times (0,1)$, is (isomorphic to) the fibered pull-back $p^{\downarrow\downarrow}(\mathcal{H})$ of $\mathcal{H}$ to $H \times (0,1)$, by the local fibered pull-back structure assumption. We can view this fibered pull-back $p^{\downarrow\downarrow}(\mathcal{H})$ as a subset of $\mathcal{G}$. This gives that we can glue $\mathcal{H}'$ and $\mathcal{G}$ along the common open groupoid $p^{\downarrow\downarrow}(\mathcal{H})$ to obtain a groupoid $\mathcal{K}$.

It can be proved that if $\mathcal{G}$ is Fredholm, then $\mathcal{K}$ is also Fredholm. Moreover, if $\mathcal{G}$ is a stratified submersion groupoid, then $\mathcal{K}$ will also be a stratified submersion groupoid.

Remark 5.6. Typically, we start with a Lie groupoid $\mathcal{G}' \rightrightarrows M'$ and $L \subset M'$ is a submanifold with corners. Then $M := [M' : L]$, the blow-up of M' with respect to L and H is the hyperface corresponding to L in this blow-up, with $\pi : H \to L$ the blow-down map. Finally, $\mathcal{G}$ is the reduction of $\mathcal{G}'$ to the complement of L. Then we denote $[[\mathcal{G}' : L]] := \mathcal{K}$. This is the *desingularization* construction from [75].

5.4. Desingularization and singular spaces

Let us show how to use the desingularization construction in some typical examples. By M_k and L_k we will denote manifolds with corners of depths k. Thus M_0 and L_0 will have, in fact, no corners or boundary (hence they will be "smooth").

5.4.1. The desingularization of a smooth submanifold. The simplest example of a desingularization groupoid is the desingularization of the pair groupoid $\mathcal{G}_0 := M_0 \times M_0$ with respect to a smooth submanifold $L_0 \subset M_0$. Recall that M_0 is also smooth. Thus neither M_0 nor L_0 have corners. The example of this subsection is a particular case of the edge groupoid, and hence it is related to the edge calculus. [38, 58, 70, 95].

Let N be the normal bundle of L_0 in M_0 and denote by $S \subset N$ the set of unit vectors in N, that is, S is the unit sphere bundle of the normal bundle of L_0 in M_0.

We let $\pi : S \to L_0$ be the natural projection. Then the blow-up $M_1 := [M_0 : L_0]$ of M_0 with respect to L_0 is the disjoint union

$$M_1 := [M_0 : L_0] := (M_0 \smallsetminus L_0) \sqcup S\,,$$

with the structure of a manifold with smooth boundary S. We let $\mathcal{G}_1$ to be the associated edge groupoid introduced in Subsection 5.2. It is a Lie groupoid with base $M_1 := [M_0 : L_0]$. We are moreover in the framework of Remark 5.6, so $\mathcal{G}_1 = [[\mathcal{G}_0 : L_0]]$. The filtration of M_1 has two sets, namely, $U_1 = M_1$ and $U_0 := M_0 \smallsetminus L_0 \subset M_1$, both of which are open and invariant for $\mathcal{G}_1$ (but U_0 is not invariant for $\mathcal{G}_0$, in general).

It turns out that $\mathcal{G}_1$ is a stratified submersion Lie groupoid if we consider the two strata U_0 and S. Also, $(\mathcal{G}_0)_{U_0}^{U_0} = U_0 \times U_0$ and $\mathcal{G}_S$ is the fibered pull-back of a bundle of Lie groups $\mathcal{L} \to S$, as described in Subsection 5.2.

5.4.2. The desingularization of a submanifold with boundary. We now generalize the last construction to manifolds with boundary. Let M_1 be a compact manifold with smooth boundary. We denote by $F := \partial M_1$ its boundary and by $G := M_1 \smallsetminus F$ its interior. On M_1 we consider the Lie algebra of vector fields $\mathcal{V}_b$ *tangent* to ∂M_1, as before. It is the space of sections of $T^b M_1$, the "b-tangent bundle" [61, 63] of M_1. Let $\mathcal{G}_1 := \mathcal{G}_b$, as defined in the previous section.

Let $L_1 \subset M_1$ be an embedded smooth submanifold assumed to be such that its boundary is $\partial L_1 = L_1 \cap \partial M_1$ and such that L_1 intersects ∂M_1 *transversely.* Then L_1 has a tubular neighborhood U in M_1, and hence we can consider the blow-up $M_2 := [M_1 : L_1]$, which will be a manifold with corners of rank 2. Moreover, the reduction of $\mathcal{G}_1 = \mathcal{G}_b$ to $U \smallsetminus L_1$ satisfies the local fibered pull-back structure assumption, since it is the b-groupoid of $U \smallsetminus L_1$. In view of Remark 5.6, we can then define $\mathcal{G}_2 := [[\mathcal{G}_1 : L_1]]$.

Again it turns out that $\mathcal{G}_2$ is a stratified submersion Lie manifold; indeed, this is seen by choosing the following filtration of M_2 with three sets:

$$U_0 := M_1 \smallsetminus (L_1 \cup \partial M_1) \subset U_1 := M_1 \smallsetminus L_1 \subset U_2 := M_2\,.$$

The sets U_j are open and $\mathcal{G}_2$ invariant (but not $\mathcal{G}_1$ invariant). The sets U_0 and U_1 are $\mathcal{G}_2$-invariant by construction. Assume they are connected, for simplicity. The restriction $(\mathcal{G}_2)_{U_1}$ coincides with the reduction $(\mathcal{G}_1)_{U_1}^{U_1}$ by the definition of the desingularization groupoid (Remark 5.6). In particular, $(\mathcal{G}_2)_{U_0} = (\mathcal{G}_2)_{U_0}^{U_0} = (\mathcal{G}_1)_{U_0}^{U_0} = U_0^2$ and

$$(\mathcal{G}_2)_{U_1 \smallsetminus U_0} := (\mathcal{G}_1)_{U_1 \smallsetminus U_0}^{U_1 \smallsetminus U_0} = (U_1 \smallsetminus U_0)^2 \times \mathbb{R}_+^*\,,$$

where U_0^2 and $(U_1 \smallsetminus U_0)^2$ are pair groupoids. In the general case, if $U_1 \smallsetminus U_0$ is not connected, we write $U_1 \smallsetminus U_0 = \sqcup V_j$ as the disjoint union decomposition of its connected components, then we have $(\mathcal{G}_2)_{U_1 \smallsetminus U_0} := \sqcup_j V_j^2 \times \mathbb{R}_+^*$. Let us denote as in the boundaryless case by S the unit sphere bundle of the normal bundle to L_1 in M_1. (The set S is the last stratum $U_2 \smallsetminus U_1$ in M_2.) Then the restriction of $\mathcal{G}_1$ to $S := M_2 \smallsetminus U_1$ is the following fibered pull-back groupoid. Let $\pi : S \to L$ be the natural projection, as before. Let again $G_S := T^bL \rtimes \mathbb{R}_+^* \to L$ be the group bundle over L obtained by taking the direct product of the b-tangent bundle to L with the action of $\mathbb{R}_+^*$ by dilation on the fibers of $TL \to L$. Then

$$\mathcal{G}_S := \pi^{\downarrow\downarrow}(G_S).$$

5.4.3. Desingularization of a stratified subset of dimension one. We now deal with a slightly more complicated example by combining the two previous examples. We thus introduce the groupoid that is obtained from the desingularization of a stratified subset of dimension one. Full details as well as applications will be included in a forthcoming paper with Mihai Putinar.

Let M_0 be a smooth, compact manifold (so no corners). Let

$$L_0 := \{P_1, P_2, \dots, P_k\} \subset M_0$$

and let us assume that we are given a subset $\mathcal{S} \subset M_0$ such that

$$\mathcal{S} = L_0 \cup \cup_{j=1}^{l} \gamma_j,$$

where each γ_j is the image of a smooth map $c_j : [0,1] \to M_0$, with the following properties:

(i) $c_j'(t) \neq 0$,
(ii) $c_j(0), c_j(1) \in L_0 := \{P_1, P_2, \dots, P_k\}$,
(iii) $c_j((0,1))$ are disjoint and do not intersect L_0 and
(iv) the vectors $c_j'(0)$ and $c_j'(1)$, $j = 1, \dots, l$, are all distinct.

We now introduce the desingularization of $\mathcal{G}_0 := M_0 \times M_0$ (the pair groupoid) with respect to $\mathcal{S}$. The set $L_0 := \{P_1, P_2, \dots, P_k\} \subset M_0$ satisfies the assumptions of 5.4.1. We can first define $\mathcal{G}_1$ to be the desingularization of $\mathcal{G}_0$ with respect to L_0 as in that example:

$$\mathcal{G}_1 := [[\mathcal{G}_0 : L_0]],$$

which is a groupoid with units $M_1 := [M_0 : L_0]$. The smooth maps c_j then lift to smooth maps

$$\tilde{c}_j : [0,1] \to [M_0 : L_0].$$

The assumption that the vectors $c_j'(0)$ and $c_j'(1)$, $j = 1, \dots, l$, are all distinct then gives that the sets $\tilde{\gamma}_j := \tilde{c}_j([0,1])$ are all disjoint and intersect the boundary of M_1 transversely. Let L_1 be the disjoint union of the embedded curves $\tilde{\gamma}_j$. Then we can perform a further desingularization along L_1, as in 5.4.2, thus obtaining

$$\mathcal{G}_2 := [[\mathcal{G}_1 : L_1]],$$

which is a boundary fibration Lie groupoid. The Lie groupoid $\mathcal{G}_2$ is the *desingularization* groupoid of M_0 with respect to $\mathcal{S}$. Its structure is given as in the previous subsection.

This example can be extended to higher-dimensional cases by using clean intersecting families.

Acknowledgment. We thank Vladimir Georgescu, Marius Mantoiu, Jean Renault, and Wolfgang Schulze for useful discussions.

References

[1] B. Ammann and N. Große, *L^p-spectrum of the Dirac operator on products with hyperbolic spaces.* Calc. Var. Partial Differ. Equ. **55** (2016), 127–163.

[2] B. Ammann, A.D. Ionescu, and V. Nistor, *Sobolev spaces on Lie manifolds and regularity for polyhedral domains.* Doc. Math. **11** (2006), 161–206.

[3] B. Ammann, R. Lauter, and V. Nistor, *On the geometry of Riemannian manifolds with a Lie structure at infinity.* Int. J. Math. Math. Sci. **2004** (2004), 161–193.

[4] B. Ammann, R. Lauter, and V. Nistor, *Pseudodifferential operators on manifolds with a Lie structure at infinity.* Ann. of Math. (2) **165** (2007), 717–747.

[5] B. Ammann, R. Lauter, V. Nistor, and A. Vasy, *Complex powers and non-compact manifolds.* Comm. Partial Differ. Equ. **29** (2004), 671–705.

[6] M.A. Bastos, C. Fernandes, and Yu.I. Karlovich. *A C^*-algebra of singular integral operators with shifts admitting distinct fixed points.* J. Math. Anal. Appl. **413** (2014), 502–524.

[7] I. Beltiţă, D. Beltiţă, and M. Măntoiu, *Symbol calculus of square-integrable operator-valued maps.* Rocky Mountain J. Math. **46** (2016), 1795–1851.

[8] K. Bohlen, *Boutet de Monvel operators on singular manifolds.* C. R. Math. Acad. Sci. Paris **354** (2016), 239–243.

[9] K. Bohlen, *Boutet de Monvel operators on Lie manifolds with boundary.* Adv. Math. **312** (2017), 234–285.

[10] A. Böttcher, Yu.I. Karlovich, and I.M. Spitkovsky, *The C^*-algebra of singular integral operators with semi-almost periodic coefficients.* J. Funct. Anal. **204** (2003), 445–484.

[11] A. Böttcher and B. Silbermann, *Analysis of Toeplitz Operators.* 2nd ed. Springer, Berlin, 2006.

[12] M. Buneci, *Groupoid C^*-algebras.* Surv. Math. Appl. **1** (2006), 71–98.

[13] P. Carrillo Rouse and J.-M. Lescure, *Geometric obstructions for Fredholm boundary conditions for manifolds with corners.* arXiv:1703.05612 [math.DG], 2017.

[14] P. Carrillo Rouse, J.M. Lescure, and B. Monthubert, *A cohomological formula for the Atiyah–Patodi–Singer index on manifolds with boundary.* J. Topol. Anal. **6** (2014), 27–74.

[15] C. Carvalho and Yu Qiao, *Layer potentials C^*-algebras of domains with conical points.* Cent. Eur. J. Math. **11** (2013), 27–54.

[16] S. Chandler-Wilde and M. Lindner, *Limit operators, collective compactness, and the spectral theory of infinite matrices.* Mem. Amer. Math. Soc. **989** (2011).

[17] A. Connes, *Non commutative differential geometry.* Publ. Math. IHES **62** (1985), 41–144.

[18] H. Cordes, *Spectral theory of linear differential operators and comparison algebras.* Cambridge University Press, Cambridge, 1987.

[19] H.O. Cordes and E.A. Herman, *Gelfand theory of pseudo differential operators.* Amer. J. Math. **90** (1968), 681–717.

[20] S. Coriasco and L. Maniccia, *On the spectral asymptotics of operators on manifolds with ends.* Abstr. Appl. Anal. (2013), Art. ID 909782, 21 pages.

[21] H.L. Cycon, R.G. Froese, W. Kirsch, and B. Simon, *Schrödinger Operators, with Application to Quantum Mechanics and Global Geometry.* Springer, Berlin, 1987.

[22] M. Damak and V. Georgescu, *Self-adjoint operators affiliated to C^*-algebras.* Rev. Math. Phys. **16** (2004), 257–280.

[23] A. Dasgupta and M.W. Wong, *Spectral theory of SG pseudo-differential operators on $L^p(\mathbb{R}^n)$.* Studia Math. **187** (2008), 185–197.

[24] M. Dauge, *Elliptic Boundary Value Problems on Corner Domains. Smoothness and Asymptotics of Solutions.* Springer, Berlin, 1988.

[25] C. Debord, J.-M. Lescure, and F. Rochon, *Pseudodifferential operators on manifolds with fibred corners.* Ann. Inst. Fourier **65** (2015), 1799–1880.

[26] C. Debord and G. Skandalis, *Adiabatic groupoid, crossed product by $\mathbb{R}_+^*$ and pseudodifferential calculus.* Adv. Math. **257** (2014), 66–91.

[27] A. Degeratu and M. Stern, *Witten spinors on nonspin manifolds.* Comm. Math. Phys. **324** (2013), 301–350.

[28] J. Dixmier, *Les C^*-Algèbres et Leurs Représentations.* Éditions Jacques Gabay, Paris, 1996.

[29] R. Exel, *Invertibility in groupoid C^*-algebras.* In: "Operator Theory, Operator Algebras and Applications". Oper. Theory Adv. Appl. **242** (2014), 173–183.

[30] J. Favard, *Sur les équations différentielles linéaires à coefficients presque-périodiques.* Acta Math. **51** (1928), 31–81.

[31] V. Georgescu and A. Iftimovici, *Crossed products of C^*-algebras and spectral analysis of quantum Hamiltonians.* Comm. Math. Phys. **228** (2002), 519–560.

[32] V. Georgescu and V. Nistor, *On the essential spectrum of N-body Hamiltonians with asymptotically homogeneous interactions.* J. Operator Theory **77** (2017), 333–376.

[33] C. Gérard, O. Oulghazi, and M. Wrochna, *Hadamard states for the Klein-Gordon equation on Lorentzian manifolds of bounded geometry.* Comm. Math. Phys. **352** (2017), 519–583.

[34] C. Gérard and M. Wrochna, *Hadamard states for the linearized Yang–Mills equation on curved spacetime.* Comm. Math. Phys. **337** (2015), 253–320.

[35] J. Gracia-Bondía, J. Várilly, and H. Figueroa, *Elements of Noncommutative Geometry.* Birkhäuser Boston, Inc., Boston, MA, 2001.

[36] D. Grieser and E. Hunsicker, *Pseudodifferential operator calculus for generalized $\mathbb{Q}$-rank 1 locally symmetric spaces. I.* J. Funct. Anal. **257** (2009), 3748–3801.

[37] N. Große and C. Schneider, *Sobolev spaces on Riemannian manifolds with bounded geometry: general coordinates and traces.* Math. Nachr. **286** (2013), 1586–1613.

[38] V.V. Grušin, *A certain class of elliptic pseudodifferential operators that are degenerate on a submanifold.* Mat. Sb. (N.S.) **84** (1971), 163–195.

[39] M. Gualtieri and Songhao Li, *Symplectic groupoids of log symplectic manifolds.* Int. Math. Res. Not. IMRN **2014** (2014), 3022–3074.

[40] Ph. Higgins and K. Mackenzie, *Algebraic constructions in the category of Lie algebroids.* J. Algebra **129** (1990), 194–230.

[41] Ph. Higgins and K. Mackenzie, *Fibrations and quotients of differentiable groupoids.* J. London Math. Soc. (2) **42** (1990), 101–110.

[42] L. Hörmander, *The Analysis of Linear Partial Differential Operators. III Pseudo-Differential Operators.* Springer, Berlin, 2007.

[43] M. Ionescu and D. Williams, *The generalized Effros–Hahn conjecture for groupoids.* Indiana Univ. Math. J. **58** (2009), 2489–2508.

[44] M. Khoshkam and G. Skandalis, *Regular representation of groupoid C^*-algebras and applications to inverse semigroups.* J. Reine Angew. Math. **546** (2002), 47–72.

[45] V.A. Kondrat′ev, *Boundary value problems for elliptic equations in domains with conical or angular points.* Transl. Moscow Math. Soc. **16** (1967), 227–313.

[46] V. Kozlov, V. Maz′ya, and J. Rossmann, *Spectral Problems Associated with Corner Singularities of Solutions to Elliptic Equations.* American Mathematical Society, Providence, RI, 2001.

[47] T. Krainer, *A calculus of abstract edge pseudodifferential operators of type ρ, δ.* In: "Elliptic and Parabolic Equations. Selected papers based on the presentations at the workshop, Hannover, Germany, September 10–12, 2013". Springer Proceedings in Mathematics & Statistics **119** (2015), 179–207.

[48] R. Lauter, B. Monthubert, and V. Nistor, *Pseudodifferential analysis on continuous family groupoids.* Doc. Math. **5** (2000), 625–655.

[49] R. Lauter and V. Nistor, *Analysis of geometric operators on open manifolds: a groupoid approach.* In: "Quantization of Singular Symplectic Quotients". Progr. Math., vol. 198, pp. 181–229. Birkhäuser, Basel, 2001.

[50] M. Lein, M. Măntoiu, and S. Richard, *Magnetic pseudodifferential operators with coefficients in C^*-algebras.* Publ. Res. Inst. Math. Sci. **46** (2010), 755–788.

[51] M. Lesch, *Operators of Fuchs Type, Conical Singularities, and Asymptotic Methods.* Teubner Verlagsgesellschaft mbH, Stuttgart, 1997.

[52] M. Lesch and B. Vertman, *Regularizing infinite sums of zeta-determinants.* Math. Ann. **361** (2015), 835–862.

[53] K. Mackenzie, *Lie Groupoids and Lie Algebroids in Differential Geometry.* Cambridge University Press, Cambridge, 1987.

[54] K. Mackenzie, *General theory of Lie groupoids and Lie algebroids.* Cambridge University Press, Cambridge, 2005.

[55] M. Măntoiu, *C^*-algebras, dynamical systems at infinity and the essential spectrum of generalized Schrödinger operators.* J. Reine Angew. Math. **550** (2002), 211–229.

[56] M. Măntoiu, R. Purice, and S. Richard, *Spectral and propagation results for magnetic Schrödinger operators; a C^*-algebraic framework.* J. Funct. Anal. **250** (2007), 42–67.

[57] M. Măntoiu, *Essential spectrum and Fredholm properties for operators on locally compact groups.* J. Oper. Theory **77** (2017), 481–501.

[58] R. Mazzeo, *Elliptic theory of differential edge operators. I.* Commun. Partial Differ. Equations **16** (1991), 1615–1664.

[59] M. Melgaard and G. Rozenblum, *Spectral estimates for magnetic operators.* Math. Scand. **79** (1996), 237–254.

[60] S.T. Melo, R. Nest, and E. Schrohe, *C^*-structure and K-theory of Boutet de Monvel's algebra.* J. Reine Angew. Math. **561** (2003), 145–175.

[61] R. Melrose, *The Atiyah–Patodi–Singer Index Theorem.* A.K. Peters, Wellesley, MA., 1993.

[62] R. Melrose and P. Piazza, *Analytic K-theory on manifolds with corners.* Adv. Math. **92** (1992), 1–26.

[63] R.B. Melrose, *Geometric Scattering Theory.* Cambridge University Press, Cambridge, 1995.

[64] I. Moerdijk and J. Mrčun, *Introduction to Foliations and Lie Groupoids.* Cambridge University Press, Cambridge, 2003.

[65] B. Monthubert, *Pseudodifferential calculus on manifolds with corners and groupoids.* Proc. Amer. Math. Soc. **127** (1999), 2871–2881.

[66] B. Monthubert, *Groupoids and pseudodifferential calculus on manifolds with corners.* J. Funct. Anal. **199** (2003), 243–286.

[67] J. Mougel, V. Nistor, and N. Prudhon, *A refined HVZ-theorem for asymptotically homogeneous interactions and finitely many collision planes.* Revue Romaine de Mathématiques Pures et Appliqués **62** (2017), 287–308.

[68] P. Muhly, J. Renault, and D. Williams, *Equivalence and isomorphism for groupoid C^*-algebras.* J. Operator Theory **17** (1987), 3–22.

[69] P.S. Muhly, J. Renault, and D. Williams, *Continuous-trace groupoid C^*-algebras. III.* Trans. Amer. Math. Soc. **348** (1996), 3621–3641.

[70] V.E. Nazaĭkinskiĭ, A.Yu. Savin, B.Yu. Sternin, and B.-W. Shulze, *On the index of elliptic operators on manifolds with edges.* Mat. Sb. **196** (2005), 23–58.

[71] S. Nazarov and B. Plamenevsky, *Elliptic Problems in Domains with Piecewise Smooth Boundaries.* Walter de Gruyter & Co., Berlin, 1994.

[72] F. Nicola and L. Rodino, *Global Pseudo-Differential Calculus on Euclidean Spaces.* Birkhäuser Verlag, Basel, 2010.

[73] V. Nistor, *Groupoids and the integration of Lie algebroids.* J. Math. Soc. Japan **52** (2000), 847–868.

[74] V. Nistor, *Analysis on singular spaces: Lie manifolds and operator algebras.* J. Geom. Phys. **105** (2016), 75–101.

[75] V. Nistor, *Desingularization of Lie groupoids and pseudodifferential operators on singular spaces.* arXiv:1512.08613 [math.DG], 2015. Communications in Analysis and Geometry, to appear.

[76] V. Nistor and N. Prudhon, *Exhausting families of representations and spectra of pseudodifferential operators.* J. Oper. Theory, **78** (2017), 247–279.

[77] V. Nistor, A. Weinstein, and Ping Xu, *Pseudodifferential operators on differential groupoids.* Pacific J. Math. **189** (1999), 117–152.

[78] C. Parenti. *Operatori pseudodifferentiali in $\mathbb{R}^n$ e applicazioni.* Annali Mat. Pura ed App. **93** (1972), 391–406.

[79] A. Paterson, *Groupoids, Inverse Semigroups, and Their Operator Algebras.* Birkhäuser Boston, Boston, MA, 1999.

[80] B.A. Plamenevskiĭ, *Algebras of Pseudodifferential Operators.* Kluwer Academic Publishers Group, Dordrecht, 1989.

[81] V. Rabinovich and S. Roch, *Essential spectrum and exponential decay estimates of solutions of elliptic systems of partial differential equations. Applications to Schrödinger and Dirac operators.* Georgian Math. J. **15** (2008), 333–351.

[82] V. Rabinovich, S. Roch, and B. Silbermann, *Limit Operators and Their Applications in Operator Theory.* Birkhäuser, Basel, 2004.

[83] V. Rabinovich, B.-W. Schulze, and N. Tarkhanov, *C^*-algebras of singular integral operators in domains with oscillating conical singularities.* Manuscripta Math. **108** (2002), 69–90.

[84] M. Reed and B. Simon, *Methods of Modern Mathematical Physics. IV. Analysis of Operators.* Academic Press, New York, 1978.

[85] J. Renault, *A Groupoid Approach to C^*-Algebras.* Springer, Berlin, 1980.

[86] J. Renault, *The ideal structure of groupoid crossed product C^*-algebras.* J. Oper. Theory **25** (1991), 3–36.

[87] M. Rieffel, *Induced representations of C^*-algebras.* Adv. Math. **13** (1974), 176–257.

[88] S. Roch, *Algebras of approximation sequences: structure of fractal algebras.* In "Singular Integral Operators, Factorization and Applications". Oper. Theory Adv. Appl. **142** (2014), 287–310.

[89] S. Roch, P. Santos, and B. Silbermann, *Non-Commutative Gelfand Theories.* Springer-Verlag London, London, 2011.

[90] M. Ruzhansky and V. Turunen, *Pseudo-Differential Operators and Symmetries. Background Analysis and Advanced Topics.* Birkhäuser, Basel, 2010.

[91] E. Schrohe, *The symbols of an algebra of pseudodifferential operators.* Pacific J. Math. **125** (1986), 211–224.

[92] E. Schrohe, *A Ψ^* algebra of pseudodifferential operators on noncompact manifolds.* Arch. Math. (Basel) **51** (1988), 81–86.

[93] E. Schrohe, *Fréchet algebra techniques for boundary value problems on noncompact manifolds: Fredholm criteria and functional calculus via spectral invariance.* Math. Nachr. **199** (1999), 145–185.

[94] E. Schrohe and B.-W. Schulze, *Boundary value problems in Boutet de Monvel's algebra for manifolds with conical singularities. I.* In "Pseudo-Differential Calculus and Mathematical Physics". Math. Top., vol. 5, pp. 97–209. Akademie Verlag, Berlin, 1994.

[95] B.-W. Schulze, *Pseudo-Differential Operators on Manifolds with Singularities.* North-Holland, Amsterdam, 1991.

[96] R.T. Seeley, *Singular integrals on compact manifolds.* Amer. J. Math. **81** (1959), 658–690.

[97] R.T. Seeley, *The index of elliptic systems of singular integral operators.* J. Math. Anal. Appl. **7** (1963), 289–309.

[98] S.R. Simanca. *Pseudo-Differential Operators.* Longman Scientific & Technical, Harlow, Essex, 1990.

[99] A. Sims and D. Williams, *Renault's equivalence theorem for reduced groupoid C^*-algebras.* J. Oper. Theory **68** (2012), 223–239.

[100] B.K. So, *On the full calculus of pseudo-differential operators on boundary groupoids with polynomial growth.* Adv. Math. **237** (2013), 1–32.

[101] M. Taylor, *Gelfand theory of pseudo differential operators and hypoelliptic operators.* Trans. Amer. Math. Soc. **153** (1971), 495–510.

[102] G. Teschl, *Mathematical Methods in Quantum Mechanics. With Applications to Schrödinger Operators.* 2nd ed. American Mathematical Society, Providence, RI, 2014.

[103] J. Tu, *Non-Hausdorff groupoids, proper actions and K-theory.* Doc. Math. **9** (2004), 565–597.

[104] E. Van Erp and R. Yuncken, *A groupoid approach to pseudodifferential operators.* arXiv:1511.01041 [math.DG], 2015.

[105] E. Van Erp and R. Yuncken, On the tangent groupoid of a filtered manifold. arXiv:1611.01081 [math.DG], 2016. Bull. London Mas. Soc., to appear.

[106] S. Vassout, *Unbounded pseudodifferential calculus on Lie groupoids.* J. Funct. Anal. **236** (2006), 161–200.

[107] B. Vertman, *Heat-trace asymptotics for edge Laplacians with algebraic boundary conditions.* J. Anal. Math. **125** (2015), 285–318.

[108] M.I. Višik and V.V. Grušin, *Degenerate elliptic differential and pseudodifferential operators.* Russian Math. Surv. **25** (1970), 21–50.

[109] D. Williams, *Crossed Products of C^*-Algebras.* American Mathematical Society, Providence, RI, 2007.

Catarina Carvalho
Departamento de Matemática
Instituto Superior Técnico
University of Lisbon
Av. Rovisco Pais
1049-001 Lisbon, Portugal
e-mail: `catarina.carvalho@math.tecnico.ulisboa.pt`

Victor Nistor
Université de Lorraine
UFR MIM, Ile du Saulcy
CS 50128
F-57045 Metz, France
and
Inst. Math. Romanian Acad.
PO BOX 1-764
014700 Bucharest, Romania
e-mail: `victor.nistor@univ-lorraine.fr`

Yu Qiao
School of Mathematics and Information Science
Shaanxi Normal University
Xi'an, 710119, China
e-mail: `yqiao@snnu.edu.cn`

Operator Theory:
Advances and Applications, Vol. 267, 123–130

Statistical e-Convergence of Bögel-Type Continuous Functions

Kamil Demirci and Sevda Orhan

Abstract. In the present work, using the concept of statistical e-convergence instead of Pringsheim's sense for double sequences, we obtain a Korovkin type approximation theorem for double sequences of positive linear operators defined on the space of all real-valued B-continuous functions on a compact subset of the real two-dimensional space. Then, displaying an example, it is shown that our new result is stronger than its classical version.

Mathematics Subject Classification (2010). 40B05; 41A36.

Keywords. Double sequences, statistical e-convergence, Korovkin type approximation theorem, B-continuity, positive linear operator.

1. Introduction and Preliminaries

Let $\mathbb{N}$ denote the set of all natural numbers. A double sequence $x = \{x_{m,n}\}$ is convergent in Pringsheim's sense if, for every $\varepsilon > 0$, there exists $N = N(\varepsilon) \in \mathbb{N}$ such that $|x_{m,n} - L| < \varepsilon$ whenever $m, n > N$. In this case L is called the Pringsheim limit of x and is denoted by $P-\lim x = L$ (see [14]). A double sequence is called bounded if there exists a positive number M such that $|x_{m,n}| \leq M$ for all $(m, n) \in \mathbb{N}^2 = \mathbb{N} \times \mathbb{N}$. Note that in contrast to the case for single sequences, a convergent double sequence need not to be bounded. In addition to the Pringsheim convergence, Boos et al. [5, 6] introduced and investigated the following notion of e-convergence of double sequences, which is stronger method than Pringsheim's:

A double sequence $x = \{x_{m,n}\}$ is said to be e-convergent to a number L if

$$\forall\varepsilon > 0 \; \exists n_0 \in \mathbb{N} \; \forall n \geq n_0 \; \exists m_n \in \mathbb{N} \; \forall m \geq m_n : |x_{m,n} - L| < \varepsilon.$$

In this case, we write $e - \lim\limits_{m,n} x_{m,n} = L$.

For related topics about these convergence methods, see [15, 18, 19]. Recently, the statistical e-convergence has been introduced in [16]. Before this convergence method let us first remind of the concept of statistical convergence, which is closely related to the concept of natural density. The notion of statistical convergence of sequences of real numbers was given by Steinhaus [17] and Fast [11], independently.

Let K be a subset of $\mathbb{N}$. Then the natural density of K is given by

$$\delta(K) := \lim_{n} \frac{1}{n} |\{k \leq n : k \in K\}|$$

provided that the limit on the right-hand side exists, where $|A|$ denotes the cardinality of the set A. Then a sequence $x = \{x_n\}$ is said to be statistically convergent to a number L if, for every $\varepsilon > 0$,

$$\delta(\{n \in \mathbb{N} : |x_n - L| \geq \varepsilon\}) = 0.$$

In this case, we write $st - \lim_{n} x_n = L$. It is easy to see that every convergent sequence is statistically convergent but not conversely.

Recently, this concept has been extended to the double sequences by Moricz [13]. If $E \subset \mathbb{N}^2$ is a two-dimensional subset of positive integers, then $E_{j,k} := \{(m,n) \in E : m \leq j,\ n \leq k\}$ and $|E_{j,k}|$ denotes the cardinality of $E_{j,k}$. The double natural density of E is given by

$$\delta_2(E) := P - \lim_{j,k} \frac{1}{jk} |E_{j,k}|,$$

if it exists. The number sequence $x = \{x_{m,n}\}$ is statistically convergent to L provided that for every $\varepsilon > 0$, the set

$$E := E_{j,k}(\varepsilon) := \{m \leq j, n \leq k : \ |x_{m,n} - L| \geq \varepsilon\}$$

has natural density zero; in that case we write $st_2 - \lim_{m,n} x_{m,n} = L$.

Clearly, a P-convergent double sequence is statistically convergent in Pringsheim sense to the same value but its converse is not always true. Also, note that a statistically convergent double sequence need not to be bounded. For example, consider the double sequence $x = \{x_{m,n}\}$ given by

$$x_{m,n} = \begin{cases} mn, & \text{if } m \text{ and } n \text{ are squares,} \\ 1, & \text{otherwise.} \end{cases}$$

Then, clearly $st_2 - \lim_{m,n} x_{m,n} = 1$. Nevertheless, x is neither convergent in Pringsheim sense nor bounded.

Now we recall the statistical e-convergence method.

Definition 1.1 ([16]). A double sequence $x = \{x_{m,n}\}$ is said to be statistically e-convergent to the number L if for every $\varepsilon > 0$,

$$\delta(\{n : \delta(\{m : |x_{m,n} - L| \geq \varepsilon\}) = 0\}) = 1.$$

In this case, we write $st_e - \lim_{m,n} x_{m,n} = L$. If a double sequence $x = \{x_{m,n}\}$ is e-convergent then it is statistically e-convergent. But, the converse of this implication may not be true. Also, a double sequence which is statistically e-convergent need not be statistical convergent (for more properties and details, see also [16]).

Example. Let define $s = \{s_{m,n}\}$ by

$$s_{m,n} := \begin{cases} mn, & m \leq n, \\ m^2 + 1, & m > n \text{ and } m \text{ is square}, \\ 1, & m > n \text{ and } m \text{ is not square}. \end{cases}$$

Then, it is easy to see that

$$st_e - \lim_{m,n} s_{m,n} = 1, \tag{1.1}$$

however, $e - \lim_{m,n} s_{m,n}$, $P - \lim_{m,n} s_{m,n}$ and $st_2 - \lim_{m,n} s_{m,n}$ do not exist.

In this work, using the concept of statistical e-convergence for double sequences, we obtain a Korovkin type approximation theorem for double sequences of positive linear operators defined on the space of all real valued B-continuous functions on a compact subset of the real two-dimensional space. Then, displaying an example, it is shown that our new result is stronger than its classical version.

2. A Korovkin-type approximation theorem

The Korovkin theory has been widely studied in the literature. This theory is connected with the approximation to continuous functions by means of positive linear operators (see, for instance, [1, 12]). In order to improve the classical Korovkin theory, the space of Bögel-type continuous (or, simply, B-continuous) functions instead of the ordinary continuity has been used in [2–4].

The definition of B-continuity was introduced by Bögel [7–9] as follows:

Let X and Y be compact subsets of the real numbers, and let $D = X \times Y$. Then, a function $f : D \to \mathbb{R}$ is called a B-continuous at a point $(x, y) \in D$ if, for every $\varepsilon > 0$, there exists a positive number $\delta = \delta(\varepsilon)$ such that

$$|\Delta_{x,y} [f(u, v)]| < \varepsilon,$$

for any $(u, v) \in D$ with $|u - x| < \delta$ and $|v - y| < \delta$, where the symbol $\Delta_{x,y} [f(u, v)]$ denotes the mixed difference of f defined by

$$\Delta_{x,y} [f(u, v)] = f(u, v) - f(u, y) - f(x, v) + f(x, y).$$

By $C_b(D)$ we denote the space of all B-continuous functions on D. Recall that $C(D)$ and $B(D)$ denote the space of all continuous (in the usual sense) functions and the space of all bounded function on D, respectively and the usual supremum norm on the spaces $B(D)$ is given by

$$\|f\| := \sup_{(x,y)\in D} |f(x, y)| \quad \text{for } f \in B(D).$$

Then, notice that $C(D) \subset C_b(D)$. Moreover, one can find an unbounded B-continuous function, which follows from the fact that, for any function of the type $f(u, v) = g(u) + h(v)$, we have

$$\Delta_{x,y} [f(u, v)] = 0 \quad \text{for all} \quad (x, y), (u, v) \in D.$$

We recall that the following lemma for B-continuous functions was proved by Badea et al. [2].

Lemma 2.1 ([2]). *If $f \in C_b(D)$, then, for every $\varepsilon > 0$, there are two positive numbers $A(\varepsilon) = A(\varepsilon, f)$ and $B(\varepsilon) = B(\varepsilon, f)$ such that*

$$|\Delta_{x,y}[f(u,v)]| \leq \frac{\varepsilon}{3} + A(\varepsilon)(u-x)^2 + B(\varepsilon)(v-y)^2$$

holds for all $(x,y),(u,v) \in D$.

Let T be a linear operator from $C_b(D)$ into $B(D)$. Then, as usual, we say that T is positive linear operator provided that $f \geq 0$ implies $Tf \geq 0$. Also, we denote the value of $T(f)$ at a point $(x,y) \in D$ by $T(f(u,v);x,y)$ or, briefly, $T(f;x,y)$.

Throughout the paper, for fixed $(x,y) \in D$ and $f \in C_b(D)$, we use the function $F_{x,y}$ defined as follows:

$$F_{x,y}(u,v) = f(u,y) + f(x,v) - f(u,v) \qquad \text{for } (u,v) \in D. \tag{2.1}$$

Since

$$\Delta_{x,y}[F_{x,y}(u,v)] = -\Delta_{x,y}[f(u,v)]$$

holds for all (x,y), $(u,v) \in D$, the B-continuity of f implies the B-continuity of $F_{x,y}$ for every fixed $(x,y) \in D$. We also use the following test functions

$$r_0(x,y) = 1,\ r_1(x,y) = x,\ r_2(x,y) = y \text{ and } r_3(x,y) = x^2 + y^2.$$

The following version of the classical Korovkin type approximation theorem was given by [2].

Theorem 2.2 ([2]). *Let $\{T_{m,n}\}$ be a sequence of positive linear operators acting from $C_b(D)$ into $B(D)$. Assume that the following conditions hold, for all $(x,y) \in D$ and $(m,n) \in \mathbb{N}^2$,*

(i) $T_{m,n}(r_0;x,y) = 1$,
(ii) $T_{m,n}(r_1;x,y) = r_1(x,y) + u_{m,n}(x,y)$,
(iii) $T_{m,n}(r_2;x,y) = r_2(x,y) + v_{m,n}(x,y)$,
(iv) $T_{m,n}(r_3;x,y) = r_3(x,y) + w_{m,n}(x,y)$,

where $\{u_{m,n}(x,y)\}$, $\{v_{m,n}(x,y)\}$ and $\{w_{m,n}(x,y)\}$ converge to zero uniformly on D as $m,n \to \infty$ (in any manner). Then the sequence

$$\{T_{m,n}(F_{x,y};x,y)\}$$

converges uniformly to $f(x,y)$ with respect to $(x,y) \in D$, where $F_{x,y}$ is given by (2.1).

Now we have the following main result.

Theorem 2.3. *Let $\{T_{m,n}\}$ be a sequence of positive linear operators acting from $C_b(D)$ into $B(D)$. Assume that the following conditions hold:*

$$\delta(\{n : \delta(\{m : T_{m,n}(r_0;x,y) = 1 \ \text{for all } (x,y) \in D\}) = 1\}) = 1 \tag{2.2}$$

and

$$st_e - \lim_{m,n} \|T_{m,n}(r_i) - r_i\| = 0 \quad \textit{for } i = 1, 2, 3. \tag{2.3}$$

Then, for all $f \in C_b(D)$, we have

$$st_e - \lim_{m,n} \|T_{m,n}(F_{x,y}) - f\| = 0, \tag{2.4}$$

where $F_{x,y}$ is given by (2.1).

Proof. Let $(x, y) \in D$ and $f \in C_b(D)$ be fixed. Putting

$$S := \{m : T_{m,n}(r_0; x, y) = 1 \ \text{ for all } (x, y) \in D\}, \tag{2.5}$$

and

$$T := \{n : \delta(\{m : T_{m,n}(r_0; x, y) = 1 \ \text{ for all } (x, y) \in D\}) = 1\}, \tag{2.6}$$

we get from (2.2) that

$$\delta(\mathbb{N} \backslash S) = 0 \text{ and } \delta(T) = 1. \tag{2.7}$$

Using the B-continuity of the function $F_{x,y}$ given by (2.1), Lemma 2.1 implies that, for every $\varepsilon > 0$, there exist two positive numbers $A(\varepsilon)$ and $B(\varepsilon)$ such that

$$|\Delta_{x,y}[F_{x,y}(u,v)]| \leq \frac{\varepsilon}{3} + A(\varepsilon)(u-x)^2 + B(\varepsilon)(v-y)^2 \tag{2.8}$$

holds for every $(u, v) \in D$. Also, by (2.2), observe that

$$T_{m,n}(F_{x,y}; x, y) - f(x, y) = T_{m,n}(\Delta_{x,y}[F_{x,y}(u,v)]; x, y) \tag{2.9}$$

holds for all $m \in S$ and $n \in T$. Since, $T_{m,n}$ linear and positive, for all $m \in S$ and $n \in T$, it follows from (2.8) and (2.9) that

$$\begin{aligned}
|T_{m,n}(F_{x,y}; x, y) - f(x, y)| &= |T_{m,n}(\Delta_{x,y}[F_{x,y}(u,v)]; x, y)| \\
&\leq T_{m,n}(|\Delta_{x,y}[F_{x,y}(u,v)]|; x, y) \\
&\leq \frac{\varepsilon}{3} + A(\varepsilon) T_{m,n}((u-x)^2; x, y) \\
&\quad + B(\varepsilon) T_{m,n}((v-y)^2; x, y) \\
&\leq \frac{\varepsilon}{3} + C(\varepsilon)\{x^2 + y^2 + T_{m,n}(r_3; x, y) \\
&\quad - 2x T_{m,n}(r_1; x, y) - 2y T_{m,n}(r_2; x, y)\},
\end{aligned}$$

where $C(\varepsilon) = \max\{A(\varepsilon), B(\varepsilon)\}$ and hence,

$$|T_{m,n}(F_{x,y}; x, y) - f(x, y)| \leq \frac{\varepsilon}{3} + C(\varepsilon) \sum_{i=1}^{3} |T_{m,n}(r_i; x, y) - r_i(x, y)| \tag{2.10}$$

holds for all $m \in S$ and $n \in T$.

Now, taking supremum over $(x, y) \in D$ on both sides of inequality (2.10) we have, for all $m \in S$ and $n \in T$,

$$\|T_{m,n}(F_{x,y}) - f\| \leq \frac{\varepsilon}{3} + C(\varepsilon) \sum_{i=1}^{3} \|T_{m,n}(r_i) - r_i\|.$$

For a given $\varepsilon' > 0$, choose $\varepsilon > 0$ such that $\varepsilon < 3\varepsilon'$ and define

$$K := \{n : \delta(\{m : \|T_{m,n}(F_{x,y}) - f\| \geq \varepsilon'\}) = 0\},$$

$$K_i := \left\{n : \delta\left(\left\{m : \|T_{m,n}(r_i) - r_i\| \geq \frac{3\varepsilon' - \varepsilon}{9C(\varepsilon)}\right\}\right) = 0\right\}, \quad i = 1, 2, 3.$$

Then by hypothesis (2.3) we get $\delta(K_i) = 1$, $i = 1, 2, 3$. If we take $K_4 = \bigcap_{i=1}^{3} K_i$, we have $\delta(K_4) = 1$. Hence, $\delta(K_4 \cap T) = 1$.

For each $n \in K_4 \cap T$ we define

$$S_i^n = \left\{m : \|T_{m,n}(r_i) - r_i\| \geq \frac{3\varepsilon' - \varepsilon}{9C(\varepsilon)}\right\}, \quad i = 1, 2, 3.$$

From the inequality (2.10) for each $n \in K_4 \cap T$ we get

$$\{m : \|T_{m,n}(F_{x,y}) - f\| \geq \varepsilon'\} \cap S \subseteq \bigcup_{i=1}^{3} (S_i^n \cap S),$$

since

$$\delta(\{m : \|T_{m,n}(F_{x,y}) - f\| \geq \varepsilon'\} \cap S) \leq \sum_{i=1}^{3} \delta(S_i^n \cap S) \leq \sum_{i=1}^{3} \delta(S_i^n)$$

and $\delta(S_i^n) = 0$, then we obtain

$$\delta(\{m : \|T_{m,n}(F_{x,y}) - f\| \geq \varepsilon'\} \cap S) = 0. \tag{2.11}$$

Also,

$$\begin{aligned}
&\delta(\{m : \|T_{m,n}(F_{x,y}) - f\| \geq \varepsilon'\}) \\
&\quad = \delta(\{m : \|T_{m,n}(F_{x,y}) - f\| \geq \varepsilon'\} \cap S) \\
&\qquad + \delta(\{m : \|T_{m,n}(F_{x,y}) - f\| \geq \varepsilon'\} \cap (\mathbb{N} \backslash S)) \\
&\quad \leq \delta(\{m : \|T_{m,n}(F_{x,y}) - f\| \geq \varepsilon'\} \cap S) + \delta(\mathbb{N} \backslash S)
\end{aligned}$$

and it follows from (2.7) and (2.11) that

$$\delta(\{m : \|T_{m,n}(F_{x,y}) - f\| \geq \varepsilon'\}) = 0.$$

This implies that $K_4 \cap T \subseteq K$. So, $\delta(K) = 1$. That is (2.4) holds, which completes the proof. □

Remark 2.4. We know that, for some $f \in C_b(D)$, the function f may be unbounded on the compact set D. However, we can say from the conditions (2.2), (2.8) and (2.9) that the number

$$\sup_{(x,y) \in D} |T_{m,n}(F_{x,y}; x, y) - f(x, y)|$$

in Theorem 2.3 is finite for each $m \in S$ and $n \in T$, where S and T are given by (2.5) and (2.6).

Remark 2.5. We now show that our result Theorem 2.3 is stronger than its classical version Theorem 2.2. To see this first consider the following Bernstein-type operators introduced by Badea and Cottin [3]:

$$H_{m,n}(f;x,y)=\sum_{s=0}^{m}\sum_{t=0}^{n}F_{x,y}\left(\frac{s}{m},\frac{t}{n}\right)\binom{m}{s}\binom{n}{t}x^sy^t(1-x)^{m-s}(1-y)^{n-t},$$

where $F_{x,y}$ is given by (2.1), and $(x,y)\in I^2=[0,1]\times[0,1]$; $f\in C_b(I^2)$. Then, by Theorem 2.2, we know that, for any $f\in C_b(I^2)$

$$P-\lim_{m,n}\|H_{m,n}(f)-f\|=0. \tag{2.12}$$

Now, we define the following positive linear operators on $C_b(I^2)$ as follows:

$$T_{m,n}(f;x,y)=s_{m,n}H_{m,n}(f;x,y). \tag{2.13}$$

where $\{s_{m,n}\}$ given by (1.1). Then, observe that the sequence of positive linear operators $\{T_{m,n}\}$ satisfy all hypotheses of Theorem 2.3 defined by (2.13). So, by (2.12) and (1.1), we have

$$st_e-\lim_{m,n}\|T_{m,n}(f)-f\|=0.$$

However, since $\{s_{m,n}\}$ is not P-convergent, the sequence $\{T_{m,n}(f;x,y)\}$ given by (2.13) does not converge uniformly to the function $f\in C_b(I^2)$. So, we conclude that Theorem 2.2 does not work for the operators $T_{m,n}$ in (2.13) while our Theorem 2.3 still works. Also, since $\{s_{m,n}\}$ is not statistically convergent, the sequence $\{T_{m,n}(f;x,y)\}$ given by (2.13) does not converge statistically uniformly. Hence, the statistical Korovkin type approximation theorem given by Dirik, Duman and Demirci [10] does not work.

References

[1] F. Altomare and M. Campiti, *Korovkin-type approximation theory and its applications.* de Gruyter Stud. Math. **17**, Walter de Gruyter, Berlin, 1994.

[2] C. Badea, I. Badea, and H.H. Gonska, *A test function and approximation by pseudopolynomials.* Bull. Austral. Math. Soc. **34** (1986), 53–64.

[3] C. Badea and C. Cottin, *Korovkin-type theorems for generalized Boolean sum operators.* Approximation theory (Kecskemét, 1990), 51–68, Colloq. Math. Soc. János Bolyai, **58**, North-Holland, Amsterdam, 1991.

[4] I. Badea, *Modulus of continuity in Bögel sense and some applications for approximation by a Bernstein-type operator.* Studia Univ. Babeş-Bolyai Ser. Math.-Mech. **18** (1973), 69–78 (in Romanian).

[5] J. Boos, *Classical and modern methods in summability.* Oxford University Press Inc., New York, 2000.

[6] J. Boos, T. Leiger, and K. Zeller, *Consistency theory for SM-methods.* Acta. Math. Hungar. **76** (1997), 83–116.

[7] K. Bögel, *Mehrdimensionale Differentiation von Funktionen mehrerer Veränderlicher.* J. Reine Angew. Math. **170** (1934), 197–217.

[8] K. Bögel, *Über mehrdimensionale Differentiation, Integration und beschränkte Variation.* J. Reine Angew. Math. **173** (1935), 5–29.

[9] K. Bögel, *Über die mehrdimensionale Differentiation.* Jahresber. Deutsch. Mat.-Verein. **65** (1962), 45–71.

[10] F. Dirik, O. Duman, and K. Demirci, *Approximation in statistical sense to B-continuous functions by positive linear operators.* Studia Sci. Math. Hungarica **47** (3) (2010), 289–298.

[11] H. Fast, *Sur la convergence statistique.* Colloq. Math. **2** (1951), 241–244.

[12] P.P. Korovkin, *Linear operators and approximation theory.* Hindustan Publ. Co., Delhi, 1960.

[13] F. Moricz, *Statistical convergence of multiple sequences.* Arch. Math. (Basel) **81** (2004), 82–89.

[14] A. Pringsheim, *Zur Theorie der zweifach unendlichen Zahlenfolgen.* Math. Ann. **53** (1900), 289–321.

[15] Y. Sever and Ö. Talo, *e-Core of double sequences.* Acta Math. Hungar. **144** (2014), 236–246.

[16] Y. Sever and Ö. Talo, *Statistical e-convergence of double sequences and its application to Korovkin type approximation theorem for functions of two variables.* Iranian Journal of Science and Technology, Transactions A: Science (2016). DOI 10.1007/s40995-016-0018-7.

[17] H. Steinhaus, *Sur la convergence ordinaire et la convergence asymtotique.* Colloq. Math. **2** (1951), 73–74.

[18] M. Zeltser, *Investigation of double sequence spaces by soft and hard analitical methods.* Dissertationes Mathematicae Universitatis Tartuensis 25, Tartu University Press, Univ. of Tartu, Faculty of Mathematics and Computer Science, (Tartu, 2001).

[19] M. Zeltser, *On conservative and coercive SM-methods.* Proc. Est. Acad. Sci. Phys. Math., **50** (2001), 76–85.

Kamil Demirci and Sevda Orhan
Sinop University
Faculty of Arts and Sciences
Department of Mathematics
TR-57000, Sinop, Turkey
e-mail: kamild@sinop.edu.tr
sevdaorhan@sinop.edu.tr

Operator Theory:
Advances and Applications, Vol. 267, 131–139

Weighted Statistical Relative Approximation by Positive Linear Operators

Kamil Demirci, Sevda Orhan and Burçak Kolay

Abstract. In this paper we define weighted statistical relative uniform convergence by using weighted density and give a Korovkin-type approximation theorem. Then, we construct an example such that our new approximation result works but its weighted statistical and statistical (and classical) cases do not work. We also compute the rates of weighted statistical relative uniform convergence of sequences of positive linear operators.

Mathematics Subject Classification (2010). 40A35, 41A25, 41A36.

Keywords. Statistical convergence, weighted statistical convergence, weighted statistical relative uniform convergence, Korovkin theorem.

1. Introduction and preliminaries

The Korovkin theorem provides a criterion for whether a given sequence $\{L_k\}$ of positive linear operators converges to the identity operator [11]. More recently, general versions of the Korovkin theorem were studied, in which a more general notion of convergence is used. In particular, the use of statistical convergence had a great impulse in recent years. Statistical convergence was first introduced independently by Fast [7] and Steinhaus [15]. The concept of weighted statistical convergence was introduced and studied by Karakaya and Chishti [10] and this convergence was modified by Mursaleen et al. in [13]. But, in [9] Ghosal showed that both definitions of weighted statistical convergence are not well defined in general and modified the definition of weighted statistical convergence (see, [2–4,12]). Recently, Demirci and Orhan [5] define a new type of statistical convergence by using the notions of the natural density and the relative uniform convergence. The main purpose of the present paper is using this thought for defining weighted statistical relative uniform convergence and using this convergence to prove a Korovkin type theorem and also, giving weighted statistical relative rates.

Let K be a subset of $\mathbb{N}$, the set of natural numbers, then *the natural density of* K, denoted by $\delta(K)$, is given by

$$\delta(K) := \lim_n \frac{1}{n} |\{k \leq n : k \in K\}|$$

whenever the limit exists, where $|B|$ denotes the cardinality of the set B [14]. The sequence $x = \{x_k\}$ is said to be *statistically convergent* to L if for every $\varepsilon > 0$,

$$\delta(\{k \in \mathbb{N} : |x_k - L| \geq \varepsilon\}) = 0,$$

i.e., for every $\varepsilon > 0$,

$$\lim_n \frac{1}{n} |\{k \leq n : |x_k - L| \geq \varepsilon\}| = 0.$$

We note that every convergent sequence is statistically convergent, but not conversely.

Let $p = \{p_k\}$ be a sequence of nonnegative real numbers such that $p_1 > 0$, $\liminf_{n\to\infty} p_n > 0$ and $P_n = \sum_{k=1}^{n} p_k \to \infty$ as $n \to \infty$. The *weighted density* of $K \subset \mathbb{N}$ denoted by $\delta_{\overline{N}}(K)$, is given by

$$\delta_{\overline{N}}(K) := \lim_n \frac{1}{P_n} |\{k \leq P_n : k \in K\}|.$$

The sequence $x = \{x_k\}$ is said to be *weighted statistically convergent* (or $S_{\overline{N}}$-*convergent*) to L if for every $\varepsilon > 0$,

$$\lim_n \frac{1}{P_n} |\{k \leq P_n : p_k |x_k - L| \geq \varepsilon\}| = 0.$$

We write $st_{\overline{N}} - \lim x = L$ [9]. If $p_k = 1$ for all k, then weighted statistical convergence is reduced to statistical convergence. In [9] Ghosal showed that statistical convergence and weighted statistical convergence do not imply each other in general.

Let f and f_k belong to $C(X)$, which is the space of all continuous real-valued functions on a compact subset X of the real numbers.

Definition 1.1 ([5]). A sequence $\{f_k\}$ is said to be *statistically relatively uniform convergent* to f on X if there exists a function $\sigma(x)$ satisfying $|\sigma(x)| \neq 0$, called a scale function such that for every $\varepsilon > 0$,

$$\lim_n \frac{1}{n} \left| \left\{ k \leq n : \sup_{x \in X} \left| \frac{f_k(x) - f(x)}{\sigma(x)} \right| \geq \varepsilon \right\} \right| = 0.$$

We can easily get the following definition like the definition of weighted statistical convergence for the number sequences.

Definition 1.2. A sequence $\{f_k\}$ is said to be *weighted statistically uniform convergent* to f on X if for every $\varepsilon > 0$,

$$\lim_n \frac{1}{P_n} \left| \left\{ k \leq P_n : p_k \sup_{x \in X} |f_k(x) - f(x)| \geq \varepsilon \right\} \right| = 0.$$

This limit is denoted by $(w - st)\ f_k \rightrightarrows f$.

Now, we can give the following definition.

Definition 1.3. A sequence $\{f_k\}$ is said to be *weighted statistically relatively uniform convergent* to f on X if there exists a function $\sigma(x)$ satisfying $|\sigma(x)| \neq 0$, called a scale function such that for every $\varepsilon > 0$,

$$\lim_n \frac{1}{P_n} \left| \left\{ k \leq P_n \, : \, p_k \sup_{x \in X} \left| \frac{f_k(x) - f(x)}{\sigma(x)} \right| \geq \varepsilon \right\} \right| = 0.$$

This limit is denoted by $(w - st)\ f_k \rightrightarrows f\ (X; \sigma)$. If $p_k = 1$ for all k, then weighted statistical relative uniform convergence is reduced to statistical relative uniform convergence.

Using the above definitions, we get following result immediately.

Lemma 1.4. *$(w - st)\ f_k \rightrightarrows f$ on X implies $(w - st)\ f_k \rightrightarrows f\ (X; \sigma)$.*

However, one can construct an example which guarantees that the converse of Lemma 1.4 is not always true. Such an example is given as follows.

Example 1.5. For each $k \in \mathbb{N}$, define $g_k : [0,1] \to \mathbb{R}$ by

$$g_k(x) = \frac{k^3 x}{2 + k^4 x^3} \tag{1.1}$$

Then observe that

$$\sup_{x \in [0,1]} \left| \frac{g_k(x) - g(x)}{\sigma(x)} \right| = \frac{k^3}{2 + k^4}$$

and with the choice of $p_k = \sqrt{k}$, we have that $(w - st)\ g_k \rightrightarrows g = 0\ ([0,1]; \sigma)$,

$$\sigma(x) = \begin{cases} \frac{1}{x^2}, & 0 < x \leq 1, \\ 2, & x = 0. \end{cases}$$

However $\{g_k\}$ is not weighted statistically uniform convergent, i.e.,

$$\sup_{x \in [0,1]} |g_k(x) - g(x)| = \frac{\sqrt[3]{k^5}}{3}$$

then

$$\lim_n \frac{1}{P_n} \left| \left\{ k \leq P_n \, : \, \frac{\sqrt[6]{k^{13}}}{3} \geq \varepsilon \right\} \right| \neq 0.$$

Also, $\{g_k\}$ is neither ordinary uniform nor statistically uniform convergent to the function $g = 0$ on the interval $[0,1]$.

2. Weighted statistical relative Korovkin theorem

In this section we use the notion of weighted statistical relative uniform convergence to prove a Korovkin type approximation theorem. Let L be a linear operator from $C(X)$ into itself. Then, as usual, we say that L is a positive linear operator provided that $f \geq 0$ implies $L(f) \geq 0$. Also, we denote the value of $L(f)$ at a point $x \in X$ by $L(f(u); x)$ or, briefly, $L(f; x)$. Throughout the paper we use the test functions $f_i(x) = x^i$ $(i = 0, 1, 2)$ and $\|f\|_{C(X)}$ denotes the usual supremum norm of f in $C(X)$.

Theorem 2.1. *Let $\{L_k\}$ be a sequence of positive linear operators acting from $C(X)$ into itself. Then, for all $f \in C(X)$,*

$$(w - st)\ L_k(f) \rightrightarrows f\ (X; \sigma) \tag{2.1}$$

if and only if

$$(w - st)\ L_k(f_i) \rightrightarrows f_i\ (X; \sigma_i), \quad i = 0, 1, 2, \tag{2.2}$$

where $\sigma(x) = \max\{|\sigma_i(x)| : i = 0, 1, 2\}$, $|\sigma_i(x)| > 0$ and $\sigma_i(x)$ is unbounded, $i = 0, 1, 2$.

Proof. Reasoning exactly as in the proof of Theorem 2 in [5] and because of ε is arbitrary, we arrive at

$$\sup_{x \in X} \left| \frac{L_k(f; x) - f(x)}{\sigma(x)} \right| \leq M \left\{ \sup_{x \in X} \left| \frac{L_k(f_0; x) - f_0(x)}{\sigma_0(x)} \right| + \sup_{x \in X} \left| \frac{L_k(f_1; x) - f_1(x)}{\sigma_1(x)} \right| + \sup_{x \in X} \left| \frac{L_k(f_2; x) - f_2(x)}{\sigma_2(x)} \right| \right\} \tag{2.3}$$

where

$$M = \varepsilon + K + \frac{2K}{\delta^2} \left(\|f_2\|_{C(X)} + 2\|f_1\|_{C(X)} + f_0 \right)$$

and $\sigma(x) = \max\{|\sigma_i(x)| : i = 0, 1, 2\}$. Hence

$$p_k \sup_{x \in X} \left| \frac{L_k(f; x) - f(x)}{\sigma(x)} \right| \leq M \left\{ p_k \sup_{x \in X} \left| \frac{L_k(f_0; x) - f_0(x)}{\sigma_0(x)} \right| + p_k \sup_{x \in X} \left| \frac{L_k(f_1; x) - f_1(x)}{\sigma_1(x)} \right| + p_k \sup_{x \in X} \left| \frac{L_k(f_2; x) - f_2(x)}{\sigma_2(x)} \right| \right\}.$$

Now, for a given $r > 0$, choose $\varepsilon > 0$ such that $\varepsilon < r$. Then,

$$S_n(r) := \left\{ k \leq P_n : p_k \sup_{x \in X} \left| \frac{L_k(f; x) - f(x)}{\sigma(x)} \right| \geq r \right\}$$

and

$$S_{i,n}(r) := \left\{ k \leq P_n \ : \ p_k \sup_{x \in X} \left| \frac{L_k(f_i; x) - f_i(x)}{\sigma_i(x)} \right| \geq \frac{r}{3M} \right\}, \quad i = 0, 1, 2.$$

It follows from (2.3) that $S_n(r) \subset \cup_{i=0}^{2} S_{i,n}(r)$ and

$$\lim_n \frac{|S_n(r)|}{P_n} \leq \sum_{i=0}^{2} \lim_n \frac{|S_{i,n}(r)|}{P_n}.$$

Then using the hypothesis (2.2), we get

$$(w - st) \ L_k(f) \rightrightarrows f \ (X; \sigma)$$

where $\sigma(x) = \max\{|\sigma_i(x)| : i = 0, 1, 2\}$.

This completes the proof of the theorem. □

In the following we construct an example of a sequence of positive linear operators satisfying the conditions of Theorem 2.1 but does not satisfy the conditions of the weighted statistical and statistical (and classical) cases of the Korovkin results introduced in [12] and [8] (and [11]), respectively.

Example 2.2. Let $X = [0, 1]$ and consider the following Bernstein–Kantorovich polynomials

$$U_k(f; x) = (k+1) \sum_{n=0}^{k} \binom{k}{n} x^n (1-x)^{k-n} \int_{n/(k+1)}^{(n+1)/(k+1)} f(t) \, dt$$

where $x \in [0, 1]$, $f \in C[0, 1]$ and $k \in \mathbb{N}$. Using these polynomials, we introduce the following positive linear operators on $C[0, 1]$:

$$T_k(f; x) = (1 + g_k(x)) U_k(f; x), \quad x \in [0, 1] \text{ and } f \in C[0, 1], \tag{2.4}$$

where $g_k(x)$ is given by (1.1). Then, observe that

$$T_k(f_0; x) = (1 + g_k(x)) f_0(x),$$

$$T_k(f_1; x) = (1 + g_k(x)) \left(f_1(x) \frac{k}{k+1} + \frac{1}{2(k+1)} \right),$$

$$T_k(f_2; x) = (1 + g_k(x)) \left(f_2(x) \frac{k(k-1)}{(k+1)^2} + \frac{2kx}{(k+1)^2} + \frac{1}{3(k+1)^2} \right).$$

Let $p_k = \sqrt{k}$ $(k = 1, 2, \ldots)$. Since $(w - st)$ $g_k \rightrightarrows g = 0$ $([0, 1]; \sigma)$, where

$$\sigma(x) = \begin{cases} \frac{1}{x^2}, & 0 < x \leq 1, \\ 2, & x = 0, \end{cases}$$

we conclude that

$$(w - st) \ T_k(f_i) \rightrightarrows f_i \ ([0, 1]; \sigma) \quad \text{for each} \quad i = 0, 1, 2.$$

So, by Theorem 2.1, we immediately see that

$$(w-st)\ T_k(f) \rightrightarrows f\ ([0,1];\sigma) \quad \text{for all} \quad f \in C[0,1].$$

However, since $\{g_k\}$ is not weighted statistically uniform and statistically uniform convergent to the function $g=0$ on the interval $[0,1]$, we can say that weighted statistical and statistical Korovkin theorem does not work for our operators defined by (2.4). Furthermore, since $\{g_k\}$ and hence $\{T_k\}$ is not uniformly convergent (in the ordinary sense) to the function $g=0$ on $[0,1]$, the classical Korovkin theorem does not work either.

3. Weighted statistical relative rates

In this section first we define the rate of weighted statistically relatively uniform convergent sequence and recall that the modulus of continuity of a function $f \in C(X)$. Then, we study the rates of positive linear operators defined on $C(X)$ with the help of this definition and modulus of continuity.

We now present the following definition.

Definition 3.1. A sequence $\{f_n\}$ is said to converge weighted statistical *uniform relative to the scale function* $\sigma(x)$, $|\sigma(x)|>0$, to f on X with the rate of $\beta \in (0,1)$ if for every $\varepsilon>0$,

$$\lim_n \frac{\left|\left\{k \leq P_n : p_k \sup_{x\in X}\left|\frac{f_k(x)-f(x)}{\sigma(x)}\right| \geq \varepsilon\right\}\right|}{n^{1-\beta}P_n}=0.$$

In this case, it is denoted by

$$(w-st)-(f_n-f)=o(n^{-\beta})\ (X;\sigma).$$

Then we first need the following lemma.

Lemma 3.2. *Let $\{f_n\}$ and $\{g_n\}$ be function sequences belonging to $C(X)$. Assume that*

$$(w-st)-(f_n-f)=o(n^{-\beta_0})\ (X;\sigma_0)$$

and

$$(w-st)-(g_n-g)=o(n^{-\beta_1})\ (X;\sigma_1),$$

$|\sigma_i(x)|>0$, $i=0,1$. *Let $\beta=\min\{\beta_0,\beta_1\}$. Then the following statements hold:*

(i) $(w-st)-(f_n+g_n)-(f+g)=o(n^{-\beta})\ (X;\max\{|\sigma_i| : i=0,1\})$,
(ii) $(w-st)-(f_n-f)(g_n-g)=o(n^{-\beta})\ (X;\sigma_0\sigma_1)$,
(iii) $(w-st)-(\lambda(f_n-f))=o(n^{-\beta_0})\ (X;\sigma_0)$ *for any real number* λ,
(iv) $(w-st)-\sqrt{|f_n-f|}=o(n^{-\beta_0})\ (X;\sqrt{|\sigma_0|})$.

Proof. (i) Assume that $(w-st)-(f_n-f)=o(n^{-\beta_0})\ (X;\sigma_0)$ and that $(w-st)-(g_n-g)=o(n^{-\beta_1})\ (X;\sigma_1)$. Also, for every $\varepsilon>0$ define

$$\varphi_n(\varepsilon):=\left|\left\{k \leq P_n : p_k \sup_{x\in X}\left|\frac{(f_k+g_k)(x)-(f+g)(x)}{\sigma(x)}\right| \geq \varepsilon\right\}\right|,$$

$$\varphi_{0,n}(\varepsilon) := \left| \left\{ k \le P_n \; : \; p_k \sup_{x \in X} \left| \frac{f_k(x) - f(x)}{\sigma_0(x)} \right| \ge \frac{\varepsilon}{2} \right\} \right|,$$

$$\varphi_{1,n}(\varepsilon) := \left| \left\{ k \le P_n \; : \; p_k \sup_{x \in X} \left| \frac{g_k(x) - g(x)}{\sigma_1(x)} \right| \ge \frac{\varepsilon}{2} \right\} \right|,$$

where $\sigma(x) = \max\{|\sigma_i(x)| : i = 0, 1\}$. Then observe that

$$\frac{\varphi_n(\varepsilon)}{n^{1-\beta} P_n} \le \frac{\varphi_{0,n}(\varepsilon) + \varphi_{1,n}(\varepsilon)}{n^{1-\beta} P_n}. \tag{3.1}$$

Since $\beta = \min\{\beta_0, \beta_1\}$, by (3.1), we get

$$\frac{\varphi_n(\varepsilon)}{n^{1-\beta} P_n} \le \frac{\varphi_{0,n}(\varepsilon)}{n^{1-\beta_0} P_n} + \frac{\varphi_{1,n}(\varepsilon)}{n^{1-\beta_1} P_n}. \tag{3.2}$$

Now by taking limit as $n \to \infty$ in (3.2) and using the hypotheses, we conclude that

$$\lim_n \frac{\varphi_n(\varepsilon)}{n^{1-\beta} P_n} = 0,$$

which completes the proof of (i). Since the proofs of (ii), (iii) and (iv) are similar, we omit them. □

On the other hand, we recall that the modulus of continuity of a function $f \in C(X)$ is defined by

$$\omega(f, \delta) = \sup_{|y-x| \le \delta, \; x,y \in X} |f(y) - f(x)| \quad (\delta > 0).$$

Then we have the following result.

Theorem 3.3. *Let X be a compact subset of the real numbers, and let $\{L_n\}$ be a sequence of positive linear operators acting from $C(X)$ into itself. Assume that the following conditions hold:*

(a) $(w - st) - (L_n(f_0) - f_0) = o(n^{-\beta_0})\ (X; \sigma_0)$,

(b) $(w - st) - \omega(f, \delta_n) = o(n^{-\beta_1})\ (X; \sigma_1)$, *where* $\delta_n(x) = \sqrt{L_n(\psi^2; x)}$ *with* $\psi(y) = y - x$.

Then we have, for all $f \in C(X)$,

$$(w - st) - (L_n(f) - f) = o(n^{-\beta})\ (X; \sigma),$$

where $\beta = \min\{\beta_0, \beta_1\}$ and

$$\sigma(x) = \max\{|\sigma_0(x)|, |\sigma_1(x)|, |\sigma_0(x)\sigma_1(x)|\},$$

$|\sigma_i(x)| > 0$ *and $\sigma_i(x)$ is unbounded, $i = 0, 1$.*

Proof. Let $f \in C(X)$ and $x \in X$. Then it is well known that

$$\begin{aligned} |L_n(f; x) - f(x)| \le & M |L_n(f_0; x) - f_0(x)| \\ & + \left(L_n(f_0; x) + \sqrt{L_n(f_0; x)} \right) \omega(f, \delta_n) \end{aligned}$$

where $M = \|f\|_{C(X)}$ (see, for instance, [1,6]). This yields that

$$\sup_{x\in X}\left|\frac{L_n(f;x)-f(x)}{\sigma(x)}\right| \le M \sup_{x\in X}\left|\frac{L_n(f_0;x)-f_0(x)}{\sigma_0(x)}\right| + 2\sup_{x\in X}\frac{\omega(f,\delta_n)}{|\sigma_1(x)|}$$
$$+\sup_{x\in X}\frac{\omega(f,\delta_n)}{|\sigma_1(x)|}\sup_{x\in X}\left|\frac{L_n(f_0;x)-f_0(x)}{\sigma_0(x)}\right|$$
$$+\sup_{x\in X}\frac{\omega(f,\delta_n)}{|\sigma_1(x)|}\sup_{x\in X}\sqrt{\left|\frac{L_n(f_0;x)-f_0(x)}{\sigma_0(x)}\right|}.$$

Now considering the above inequality, the hypotheses (a), (b), and Lemma 3.2, the proof is completed at once. □

Remark 3.4. If we replace the conditions (a), (b) in Theorem 3.3 by the following condition:

$$(w-st)-(L_n(f_i)-f_i)=o(n^{-\beta_i})\ (X;\sigma_i), \quad i=0,1,2, \tag{3.3}$$

Then, since

$$L_n(\psi^2;x) \le N\{|L_n(f_0;x)-f_0(x)|$$
$$+|L_n(f_1;x)-f_1(x)|+|L_n(f_2;x)-f_2(x)|\} \tag{3.4}$$

where $N = \|f_2\|_{C(X)} + 2\|f_1\|_{C(X)} + f_0$. We get

$$\sup_{x\in X}\frac{L_n(\psi^2;x)}{|\sigma(x)|} \le N\left\{\sup_{x\in X}\left|\frac{L_n(f_0;x)-f_0(x)}{\sigma_0(x)}\right|\right.$$
$$\left.+\sup_{x\in X}\left|\frac{L_n(f_1;x)-f_1(x)}{\sigma_1(x)}\right|+\sup_{x\in X}\left|\frac{L_n(f_2;x)-f_2(x)}{\sigma_2(x)}\right|\right\}$$

where $\sigma(x)=\max\{|\sigma_i(x)| : i=0,1,2\}$. It follows that (3.3), (3.4) and Lemma 3.2 that

$$(w-st)-\delta_n=\sqrt{L_n(\psi^2)}=o(n^{-\beta})\ (X;\sigma)$$

where $\beta=\min\{\beta_0,\beta_1,\beta_2\}$. So, by Theorem 3.3 we get, for all $f \in C(X)$,

$$(w-st)-(L_n(f)-f)=o(n^{-\beta})\ (X;\sigma).$$

Hence, if we replace the conditions (a), (b) with the condition (3.3) in Theorem 3.3, then we get the rates of weighted statistically relatively uniform convergence of sequence of positive linear operators in Theorem 2.1.

References

[1] F. Altomare and M. Campiti, *Korovkin-Type Approximation Theory and Its Applications.* de Gruyter Stud. Math. 17, Walter de Gruyter, Berlin, 1994.

[2] S. Akdağ, *Weighted equi-statistical convergence of the Korovkin type approximation theorems.* Results. Math., **72** (2017), 1073–1085.

[3] C. Belen and S.A. Mohiuddine, *Generalized weighted statistical convergence and application.* Appl. Math and Comp. **219** (2013), 9821–9826.

[4] N.L. Braha, V. Loku, and H.M. Srivastava, Λ^2*-weighted statistical convergence and Korovkin and Voronovskaya type theorems.* Appl. Math and Comp. **266** (2015), 675–686.

[5] K. Demirci and S. Orhan, *Statistically relatively uniform convergence of positive linear operators.* Results. Math. **69** (2016), 359–367.

[6] R.A. DeVore, *The Approximation of Continuous Functions by Positive Linear Operators.* Lecture Notes in Math., vol. 293, Springer-Verlag, Berlin, 1972.

[7] H. Fast, *Sur la convergence statistique.* Colloq. Math. **2** (1951), 241–244.

[8] A.D. Gadjiev and C. Orhan, *Some approximation theorems via statistical convergence.* Rocky Mountain J. Math. **32** (2002), 129–138.

[9] S. Ghosal, *Weighted statistical convergence of order* α *and its applications.* Journal of the Egyptian Mathematical Society (2016) 24, 60–67.

[10] V. Karakaya and T.A. Chishti, *Weighted statistical convergence.* Iran. J. Sci. Technol. Trans. A Sci. **33** (2009), 219–223.

[11] P.P. Korovkin, *Linear Operators and Approximation Theory.* Hindustan Publ. Co., Delhi, 1960.

[12] S.A. Mohiuddine, A. Alotaibi, and B. Hazarika, *Weighted A-statistical convergence for sequences of positive linear operators.* The Scientific World J. Volume 2014, Article ID 437863, 8 pages.

[13] M. Mursaleen, V. Karakaya, M. Ertürk, and F. Gürsoy, *Weighted statistical convergence and its application to Korovkin type approximation theorem.* Appl. Math. Comput. **218** (2012), 9132–9137.

[14] I. Niven and H.S. Zuckerman, *An Introduction to the Theory of Numbers.* John Wiley & Sons, Fourth Ed., New York, 1980.

[15] H. Steinhaus, *Sur la convergence ordinaire et la convergence asymptotique.* Colloq. Math. **2** (1951), 73–74.

Kamil Demirci, Sevda Orhan and Burçak Kolay
Sinop University
Faculty of Sciences and Arts
Department of Mathematics
57000 Sinop, Turkey
e-mail: `kamild@sinop.edu.tr`
`sevdaorhan@sinop.edu.tr`
`burcakyilmaz@sinop.edu.tr`

Operator Theory:
Advances and Applications, Vol. 267, 141–166

Descriptor Systems Under Feedback and Output Injection

Marija Dodig

Abstract. In this paper we study simultaneous feedback and output injection on descriptor linear system described by a quadruple of matrices (E, A, B, C). We describe the possible Kronecker invariants of the resulting pencil $\lambda E-(A+BF+KC)$, when F and K vary, in the case when the pencil corresponding to the system (E, A, B, C) has no infinite elementary divisors of the second, third and fourth type. The solution is constructive and explicit, and is given over algebraically closed fields.

Mathematics Subject Classification (2010). Primary 15A83; Secondary 93C05, 15A22.

Keywords. Descriptor systems, matrix pencils completion, state feedback, output injection, Carlson problem.

1. Introduction

Let $\mathbb{F}$ be a field. Let $E, A \in \mathbb{F}^{\tau\times\pi}$, $B \in \mathbb{F}^{\tau\times\eta}$ and $C \in \mathbb{F}^{\epsilon\times\pi}$ be matrices. We consider a time-invariant linear system S with outputs

$$E\,\dot{x}(t) = A\,x(t) + B\,u(t),$$
$$y(t) = C\,x(t).$$

Such a system, determined by the quadruple (E, A, B, C), is called a descriptor system (see, e.g., [22]). The corresponding matrix pencil of the system S is

$$\begin{bmatrix} \lambda E - A & -B \\ -C & 0 \end{bmatrix}. \tag{1.1}$$

We emphasize that throughout the paper matrix E does not need to be invertible, and not even a square matrix.

This work was done within the activities of CEAFEL and was partially supported by Fundação para a Ciência e a Tecnologia (FCT), project ISFL-1-1431. The author was supported by the FCT Exploratory Grant "Matrix Completions", number IF/01232/2014. This work was also partially supported by the Ministry of Education, Science, and Technological Development of the Republic of Serbia, project No. 174020.

Our aim is to study the properties of descriptor systems obtained by simultaneous proportional feedback and proportional output injection, i.e., we study the possible Kronecker invariants of a pencil $\lambda E - (A + BF + KC)$, when F and K vary.

We note that without loss of generality we shall assume that matrices B and C are of the full column and row rank, respectively.

Due to generality of the problem, only some partial solutions are known, like in the case $C = 0$ [8, 10] or in the regular case [26]. In addition, only some of the possible invariants of the pencil $\lambda E - (A + BF + KC)$ are determined, like the eigenvalues, i.e., poles and zeros of the corresponding system, see, e.g., [1, 17, 18, 21, 28].

The main difficulty of the problem lies in the fact that simultaneous feedback and output injection on the system (E, A, B, C) imply dealing with special canonical form of the pencil (1.1) (for all details see Section 2, and references therein), which has nine different types of invariants: invariant factors, column minimal indices of the first and of the second type, row minimal indices of the first and of the second type, and infinite elementary divisors of the first, second, third and fourth type.

In this paper we consider a system (E, A, B, C) with restriction only on the type of infinite elementary divisors involved – we study the case when the corresponding pencil has no infinite elementary divisors of the second, third and fourth type.

Apart of theoretical Matrix theory importance, this problem has applications in Control theory. If $E = I$ it appears naturally in the action of the Morse group, see [24, 25]. Also, the pencil $\lambda I - (A + BF + KC)$ arise in combining Kalman filter and Linear-quadratic Regulator (LQR) control in Linear-quadratic Gaussian (LQG) control, for details see, e.g., [5, 14, 29, 30].

The LQG control problem studies linear systems with incomplete state information (i.e., not all the state variables are measured and available for feedback) and undergoing control subject to quadratic costs. Finally the LQG controller is also fundamental to the optimal control of perturbed non-linear systems. For all details and possible applications see, e.g., [4, 5, 14, 29, 30].

As the main result of the paper, we solve the following problem over an algebraically closed field.

Problem 1. Describe the possible Kronecker invariants of

$$\lambda E - (A + BF + KC),$$

when F and K vary, in the case when the pencil (1.1) has no infinite elementary divisors of the second, third and fourth type.

The obtained solution is constructive and explicit.

If $E = I$, as a corollary of our main result we obtain a full description for the pencil $\lambda I - (A + BF + KC)$ to be regular (see also [2, 26, 27]). Note that this result has possible applications in studying transfer functions of LQG, for details see, e.g., [29, 30].

We solve Problem 1 by presenting it as a matrix pencil completion problem (see Section 3). To that end we use a proportional feedback and proportional output injection equivalence on (1.1), that preserves the strict equivalence class of the pencil $\lambda E-(A+BF+KC)$, and the corresponding Kronecker-like canonical form given in [22] (see also [6, 19, 21, 23]) – see Section 2. In the solution we use recent combinatorial results in matrix pencils completion that also involve a solution to the Carlson problem, Young diagrams and LR sequences, obtained in [9, 11] and given in Sections 4 and 5. In order to apply the results from [9, 11] we need the field $\mathbb{F}$ to be algebraically closed. Thus, from now on we shall assume that $\mathbb{F}$ is an algebraically closed field.

2. Feedback equivalence and a canonical form

Let (E,A,B,C) and (E',A',B',C') be two quadruples of matrices, both from $\mathbb{F}^{\tau\times\pi}\times\mathbb{F}^{\tau\times\pi}\times\mathbb{F}^{\tau\times\eta}\times\mathbb{F}^{\epsilon\times\pi}$.

Let

$$\operatorname{rank}\begin{bmatrix}\lambda E-A & -B\\ -C & 0\end{bmatrix}=w.$$

The feedback equivalence relation obtained by proportional feedback and proportional output injection is given in [22] (see also [19]), by the following definition.

Definition 2.1. Two descriptor systems given by quadruples $(E,\ A,\ B,\ C)$ and $(E',\ A',\ B',\ C')$ are feedback equivalent if there exist invertible matrices P, Q, R and T, and matrices X and Y of appropriate sizes, such that

$$\begin{aligned}E'&=PEQ,\\ A'&=P(A+BX+YC)Q,\\ B'&=PBT,\\ C'&=RCQ.\end{aligned}$$

Equivalently:

$$\begin{bmatrix}\lambda E'-A' & -B'\\ -C' & 0\end{bmatrix}=\begin{bmatrix}P & PY\\ 0 & R\end{bmatrix}\begin{bmatrix}\lambda E-A & -B\\ -C & 0\end{bmatrix}\begin{bmatrix}Q & 0\\ XQ & T\end{bmatrix}. \tag{2.1}$$

Lemma 2.2. *Let*

$$(E,A,B,C)\in\mathbb{F}^{\tau\times\pi}\times\mathbb{F}^{\tau\times\pi}\times\mathbb{F}^{\tau\times\eta}\times\mathbb{F}^{\epsilon\times\pi}$$

and

$$(E',A',B',C')\in\mathbb{F}^{\tau\times\pi}\times\mathbb{F}^{\tau\times\pi}\times\mathbb{F}^{\tau\times\eta}\times\mathbb{F}^{\epsilon\times\pi}$$

be two feedback equivalent descriptor systems. Let $M(\lambda)\in\mathbb{F}[\lambda]^{\tau\times\pi}$ be a matrix pencil. There exist matrices F and K such that $\lambda E-(A+BF+KC)$ is strictly equivalent to $M(\lambda)$, if and only if, there exist matrices F' and K' such that $\lambda E'-(A'+B'F'+K'C')$ is strictly equivalent to $M(\lambda)$.

Proof. Since (E, A, B, C) and (E', A', B', C') are feedback equivalent, there exist invertible matrices P, Q, R and T, and matrices X and Y, such that (2.1) holds.

Let $F \in \mathbb{F}^{\eta\times\pi}$ and $K \in \mathbb{F}^{\tau\times\epsilon}$ be matrices. Then we define

$$F' = T^{-1}(F - X)Q, \quad K' = P(K - Y)R^{-1}.$$

It is straightforward to see that

$$\lambda E' - (A' + B'F' + K'C') = P(\lambda E - (A + BF + KC))Q,$$

as wanted. □

The canonical form and invariants for the equivalence given in Definition 2.1 are obtained in [22]. For details and motivations see also [6, 19, 21, 23]. These invariants are:

1. Invariant factors: $\bar{\alpha}_1 \mid \cdots \mid \bar{\alpha}_w$,
2. Column minimal indices of the 1st type:
$$c_1^1 + 1 \geq \cdots \geq c_{m_1}^1 + 1 > 0,$$
3. Column minimal indices of the 2nd type:
$$c_1^2 \geq \cdots \geq c_\rho^2 > c_{\rho+1}^2 = \cdots = c_{m_2}^2 = 0,$$
4. Row minimal indices of the 1st type:
$$r_1^1 + 1 \geq \cdots \geq r_{p_1}^1 + 1 > 0,$$
5. Row minimal indices of the 2nd type:
$$r_1^2 \geq \cdots \geq r_\theta^2 > r_{\theta+1}^2 = \cdots = r_{p_2}^2 = 0,$$
6. Infinite elementary divisors of the 1st type:
$$b_1^1 \geq \cdots \geq b_\nu^1 > b_{\nu+1}^1 = \cdots = b_{b_1}^1 = 1,$$
7. Infinite elementary divisors of the 2nd type:
$$b_1^2 + 1 \geq \cdots \geq b_{b_2}^2 + 1 \geq 1,$$
8. Infinite elementary divisors of the 3rd type:
$$b_1^3 + 1 \geq \cdots \geq b_{b_3}^3 + 1 \geq 1,$$
9. Infinite elementary divisors of the 4th type:
$$b_1^4 + 2 \geq \cdots \geq b_{b_4}^4 + 2 \geq 2.$$

Here we have

$$w = \sum_{i=1}^{w} d(\bar{\alpha}_i) + \sum_{i=1}^{m_1}(c_i^1 + 1) + \sum_{i=1}^{m_2} c_i^2 + \sum_{i=1}^{p_1}(r_i^1 + 1) + \sum_{i=1}^{p_2} r_i^2$$
$$+ \sum_{i=1}^{b_1} b_i^1 + \sum_{i=1}^{b_2}(b_i^2 + 1) + \sum_{i=1}^{b_3}(b_i^3 + 1) + \sum_{i=1}^{b_4}(b_i^4 + 2)$$

and

$$\begin{aligned}\tau &= w + p_2 - b_3 - b_4,\\ \pi &= w + m_2 - b_2 - b_4,\\ \eta &= m_1 + b_2 + b_4,\\ \epsilon &= p_1 + b_3 + b_4.\end{aligned}$$

Let

$$n = w - m_1 - p_1 - b_2 - b_3 - 2b_4.$$

Thus we can define polynomials $\alpha_1 | \cdots | \alpha_n$ by

$$\alpha_i = \bar{\alpha}_{i+m_1+p_1+b_2+b_3+2b_4}, \quad i = 1, \dots, n,$$

and we also have

$$n = \sum_{i=1}^{n} d(\alpha_i) + \sum_{i=1}^{m_1} c_i^1 + \sum_{i=1}^{m_2} c_i^2 + \sum_{i=1}^{p_1} r_i^1 + \sum_{i=1}^{p_2} r_i^2 + \sum_{i=1}^{b_1} b_i^1 + \sum_{i=1}^{b_2} b_i^2 + \sum_{i=1}^{b_3} b_i^3 + \sum_{i=1}^{b_4} b_i^4.$$

In this paper we are interested in pencils when $b_2 = b_3 = b_4 = 0$. Thus, from now on we have

$$\begin{aligned}\tau &= n + p_1 + m_1 + p_2,\\ \pi &= n + p_1 + m_1 + m_2,\\ \eta &= m_1,\\ \epsilon &= p_1.\end{aligned}$$

Throughout the paper we shall consider only monic polynomials. For any polynomial f, $d(f)$ denotes its degree. If

$$f = \lambda^n - a_{n-1}\lambda^{n-1} - \cdots - a_1\lambda - a_0,$$

then the matrix

$$\begin{bmatrix} e_2^n & \cdots & e_n^n & a \end{bmatrix}^T$$

is called the companion matrix of the polynomial f. Here, e_i^n is the ith column of the identity matrix I_n and $a = \begin{bmatrix} a_0 & \cdots & a_{n-1} \end{bmatrix}^T$. Also, we denote

$$C(f) := \lambda I - \begin{bmatrix} e_2^n & \cdots & e_n^n & a \end{bmatrix}^T.$$

Further on in the paper, we shall use the following notation

$$\begin{aligned}
N &= \operatorname{diag}(C(\alpha_1), \dots, C(\alpha_n)),\\
B_1 &= \operatorname{diag}\left(\begin{bmatrix} C(\lambda^{b_1^1}) & e_{b_1^1} \\ (e_1^{b_1^1})^T & 0 \end{bmatrix}, \dots, \begin{bmatrix} C(\lambda^{b_\nu^1}) & e_{b_\nu^1} \\ (e_1^{b_\nu^1})^T & \end{bmatrix}, I_{b_1-\nu} \right),\\
C^1 &= \operatorname{diag}(C(\lambda^{c_1^1+1}), \dots, C(\lambda^{c_{m_1}^1+1})),\\
R^1 &= \operatorname{diag}(C(\lambda^{r_1^1+1}), \dots, C(\lambda^{r_{p_1}^1+1})),\\
C_2 &= \begin{pmatrix} \operatorname{diag}\left(\begin{bmatrix} C(\lambda^{c_1^2}) & e_{c_1^2} \end{bmatrix}, \dots, \begin{bmatrix} C(\lambda^{c_\rho^2}) & e_{c_\rho^2} \end{bmatrix} \right) & 0 \end{pmatrix},
\end{aligned}$$

where the number of zero columns in C_2 is $m_2 - \rho$. Also,

$$R_2 = \left(\begin{array}{c} \operatorname{diag}\left(\begin{bmatrix} C(\lambda^{r_1^2}) \\ (e_1^{r_1^2})^T \end{bmatrix}, \dots, \begin{bmatrix} C(\lambda^{r_\theta^2}) \\ (e_1^{r_\theta^2})^T \end{bmatrix} \right) \\ 0 \end{array} \right),$$

where the number of zero rows in R_2 is $p_2 - \theta$. Also,

$$G_1 = \operatorname{diag}((e_1^{r_1^1+1})^T, \dots, (e_1^{r_{p_1}^1+1})^T), \quad E_1 = \operatorname{diag}(e_{c_1^1+1}, \dots, e_{c_{m_1}^1+1}),$$

where by e_i we have denoted the ith column of the unit matrix $I_i \in \mathbb{F}^{i \times i}$.

The canonical form (Kronecker canonical form) for the pencil

$$\begin{bmatrix} \lambda E - A & -B \\ -C & 0 \end{bmatrix} \tag{2.2}$$

without infinite elementary divisors of the second, third and fourth type, for feedback equivalence given in Definition 2.1, is the following (for details see [19, 22]):

$$\left[\begin{array}{c|c|c|c|c|c||c} N & & & & & & \\ \hline & B_1 & & & & & \\ \hline & & C^1 & & & & E_1 \\ \hline & & & C_2 & & & \\ \hline & & & & R^1 & & \\ \hline & & & & & R_2 & \\ \hline\hline & & & & G_1 & & \end{array} \right]. \tag{2.3}$$

By Lemma 2.2, without loss of generality, from now on we can assume that the system (E, A, B, C), i.e., the corresponding pencil (2.2), is in the canonical form (2.3). This means that from now on

$$\lambda E - A = \left[\begin{array}{c|c|c|c|c|c} N & & & & & \\ \hline & B_1 & & & & \\ \hline & & C^1 & & & \\ \hline & & & C_2 & & \\ \hline & & & & R^1 & \\ \hline & & & & & R_2 \end{array} \right],$$

$$-B = \begin{bmatrix} 0 \\ 0 \\ E_1 \\ 0 \\ 0 \\ 0 \end{bmatrix} \quad \text{and} \quad -C = \begin{bmatrix} 0 & 0 & 0 & 0 & G_1 & 0 \end{bmatrix}.$$

3. Re-writing the main problem as a matrix pencil completion problem

Denote by $A(\lambda) \in \mathbb{F}[\lambda]^{(n+p_1+p_2)\times(n+m_1+m_2)}$ the subpencil of (2.3) obtained from (2.3) after erasing rows and columns corresponding to the units in E_1 and G_1, i.e.,

$$A(\lambda) = \left[\begin{array}{c|c|c|c|c|c} N & & & & & \\ \hline & B_1 & & & & \\ \hline & & C_1 & & & \\ \hline & & & C_2 & & \\ \hline & & & & R_1 & \\ \hline & & & & & R_2 \end{array}\right] \tag{3.1}$$

with

$$C_1 = \left(\operatorname{diag}\left(\left[\begin{array}{cc} C(\lambda^{c_1^1}) & e_{c_1^1} \end{array}\right], \ldots, \left[\begin{array}{cc} C(\lambda^{c_{\rho'}^1}) & e_{c_{\rho'}^1} \end{array}\right]\right) \quad 0 \right),$$

where $\rho' = \max\{i | c_i^1 > 0\}$, and the number of zero columns is $m_1 - \rho'$, and

$$R_1 = \left(\begin{array}{c} \operatorname{diag}\left(\left[\begin{array}{c} (e_1^{r_1^1})^T \\ C(\lambda^{r_1^1}) \end{array}\right], \ldots, \left[\begin{array}{c} (e_1^{r_{\theta'}^1})^T \\ C(\lambda^{r_{\theta'}^1}) \end{array}\right]\right) \\ 0 \end{array} \right),$$

where $\theta' = \max\{i | r_i^1 > 0\}$, and the number of zero rows in R_1 is $p_1 - \theta'$.

The pencil $A(\lambda)$ has the following invariants for the equivalence relation given by (2.1):

1. $\operatorname{rank} A(\lambda) = n$,
2. invariant factors: $\alpha_1 \mid \cdots \mid \alpha_n$,
3. column minimal indices of the 1st type:
$$c_1^1 \geq \cdots \geq c_{m_1}^1 \geq 0,$$
4. column minimal indices of the 2nd type:
$$c_1^2 \geq \cdots \geq c_\rho^2 > c_{\rho+1}^2 = \cdots = c_{m_2}^2 = 0,$$
5. row minimal indices of the 1st type:
$$r_1^1 \geq \cdots \geq r_{p_1}^1 \geq 0,$$
6. row minimal indices of the 2nd type:
$$r_1^2 \geq \cdots \geq r_\theta^2 > r_{\theta+1}^2 = \cdots = r_{p_2}^2 = 0,$$
7. infinite elementary divisors of the 1st type:
$$b_1^1 \geq \cdots \geq b_{b_1}^1 \geq 1.$$

By the form of the pencil (2.3), it is straightforward to see that the problem of describing the possible Kronecker invariants of the pencil $\lambda E-(A+BF+KC)$ when F and K vary, is equivalent to the following completion problem: complete $A(\lambda)$ by m_1 rows and p_1 columns with the fixed positions of λ's (each added row and column has a unique λ corresponding to each column and row minimal index of the second kind, respectively), up to a pencil strictly equivalent to $\lambda E-(A+BF+KC)$. More precisely, let

$$E_2=\operatorname{diag}(e_1^{r_1^1+1},\dots,e_1^{r_{p_1}^1+1})$$

and

$$E_3=\operatorname{diag}((e_{c_1^1+1})^T,\dots,(e_{c_{m_1}^1+1})^T).$$

Let

$$\sum_{j=1}^{n} d(\alpha_j)=t,\quad \sum_{i=1}^{b_1} b_i^1=b,\quad \sum_{i=1}^{m_1} c_i^1=u_1,$$

$$\sum_{i=1}^{m_2} c_i^2=u_2,\quad \sum_{i=1}^{p_1} r_i^1=v_1,\quad \sum_{i=1}^{p_2} r_i^2=v_2.$$

Then Problem 1 is equivalent to the following problem.

Problem 2 (Problem 1 written as a Completion Problem). Find necessary and sufficient conditions for the existence of matrices

$$\begin{array}{lll}
X_1\in\mathbb{F}^{t\times p_1}, & X_2\in\mathbb{F}^{b\times p_1}, & X_3\in\mathbb{F}^{u_1\times p_1},\\
X_4\in\mathbb{F}^{u_2\times p_1}, & X_5\in\mathbb{F}^{(v_1+p_1)\times p_1}, & X_6\in\mathbb{F}^{(v_2+p_2)\times p_1},\\
Y_1\in\mathbb{F}^{m_1\times t}, & Y_2\in\mathbb{F}^{m_1\times b}, & Y_3\in\mathbb{F}^{m_1\times(u_1+m_1)},\\
Y_4\in\mathbb{F}^{m_1\times(u_2+m_2)}, & Y_5\in\mathbb{F}^{m_1\times v_1}, & Y_6\in\mathbb{F}^{m_1\times v_2},
\end{array}$$

and $Y_7\in\mathbb{F}^{m_1\times p_1}$, such that the pencil

$$\left[\begin{array}{c|c|c|c|c|c||c}
N & & & & & & X_1\\ \hline
 & B_1 & & & & & X_2\\ \hline
 & & C_1 & & & & X_3\\ \hline
 & & & C_2 & & & X_4\\ \hline
 & & & & R_1 & & \lambda E_2+X_5\\ \hline
 & & & & & R_2 & X_6\\ \hline\hline
Y_1 & Y_2 & \lambda E_3+Y_3 & Y_4 & Y_5 & Y_6 & Y_7
\end{array}\right] \tag{3.2}$$

has a prescribed set of Kronecker invariants.

By a partition we mean an ordered nonincreasing sequence of nonnegative integers. Also, if $(a_1,\dots,a_s)$ is a partition, then we assume $a_i=+\infty$, for $i\le 0$, and $a_i=-\infty$, for $i\ge s+1$.

In the rest of the paper we shall use the following notation.

Let $\bar{\mathbf{f}}=(\bar f_1,\dots,\bar f_{p_1+p_2})$, where $\bar f_1\ge\dots\ge\bar f_{p_1+p_2}$ is the nonincreasing ordering of

$$(r_1^1+1,\dots,r_{p_1}^1+1)\cup(r_1^2,\dots,r_{p_2}^2),$$

and let $\mathbf{f}=(f_1,\dots,f_{m_1+m_2})$, where $f_1\ge\dots\ge f_{m_1+m_2}$ is the nonincreasing ordering of

$$(c_1^1+1,\dots,c_{m_1}^1+1)\cup(c_1^2,\dots,c_{m_2}^2).$$

Now we can state our main result.

Theorem 3.1. *Let (E, A, B, C) be a descriptor system without infinite elementary divisors of the second, third and fourth type. Let $\bar{\alpha}_1 | \cdots | \bar{\alpha}_{n+m_1+p_1}$ be its invariant factors; let*

$$\alpha_i = \bar{\alpha}_{i+m_1+p_1}, \quad i = 1, \dots, n;$$

let

$$c_1^1 + 1 \geq \cdots \geq c_{m_1}^1 + 1 > 0 \quad \text{and} \quad c_1^2 \geq \cdots \geq c_\rho^2 > c_{\rho+1}^2 = \cdots = c_{m_2}^2 = 0$$

be its column minimal indices of the first and of the second type, respectively; let

$$r_1^1 + 1 \geq \cdots \geq r_{p_1}^1 + 1 > 0 \quad \text{and} \quad r_1^2 \geq \cdots \geq r_\theta^2 > r_{\theta+1}^2 = \cdots = r_{p_2}^2 = 0$$

be its row minimal indices of the first and second type, respectively; and let $b_1^1 \geq \cdots \geq b_{b_1}^1$ be its infinite elementary divisors of the first type. Suppose that $\gamma_1 | \cdots | \gamma_{n+m_1+p_1}$ are polynomials and

$$\begin{aligned} &\bar{b}_1 \geq \cdots \geq \bar{b}_{b_1}, \\ &d_1 \geq \cdots \geq d_{\bar{\rho}} > d_{\bar{\rho}+1} = \cdots = d_{m_2} = 0, \\ &\bar{r}_1 \geq \cdots \geq \bar{r}_{\bar{\theta}} > \bar{r}_{\bar{\theta}+1} = \cdots = \bar{r}_{p_2} = 0 \end{aligned}$$

are nonnegative integers.

There exist matrices F and K such that pencil $\lambda E - (A + BF + KC)$ has $\gamma_1 | \cdots | \gamma_{n+m_1+p_1}$, $\bar{b}_1 \geq \cdots \geq \bar{b}_{b_1}$, $d_1 \geq \cdots \geq d_{\bar{\rho}} > d_{\bar{\rho}+1} = \cdots = d_{m_2} = 0$ and $\bar{r}_1 \geq \cdots \geq \bar{r}_{\bar{\theta}} > \bar{r}_{\bar{\theta}+1} = \cdots = \bar{r}_{p_2} = 0$ as Kronecker invariants (invariant factors, infinite elementary divisors, column and row minimal indices, respectively), if and only if the following conditions hold:

(i) $\rho \leq \bar{\rho} \leq \rho + m_1$ *and* $\theta \leq \bar{\theta} \leq \theta + p_1$,

(ii) $\bar{b}_i = b_i^1$ *for all* $i = 1, \dots, b_1$,

(iii) $\gamma_i \mid \alpha_i \mid \gamma_{i+m_1+p_1}$ *for all* $i = 1, \dots, n$,

(iv) $\bar{r}_i \geq t_i$ *for all* $i = 1, \dots, \bar{\theta}$,

(v) $d_i \geq s_i$ *for all* $i = 1, \dots, \bar{\rho}$,

(vi) $$\sum_{i=1}^{n} d(\alpha_i) + \sum_{i=1}^{m_1} (c_i^1 + 1) + \sum_{i=1}^{m_2} c_i^2 + \sum_{i=1}^{p_1} (r_i^1 + 1) + \sum_{i=1}^{p_2} r_i^2 = \sum_{i=1}^{n+m_1+p_1} d(\gamma_i) + \sum_{i=1}^{m_2} d_i + \sum_{i=1}^{p_2} \bar{r}_i,$$

(vii) $$x_j + y_k \leq \sum_{i=1}^{n+m_1+p_1} d(\gamma_i) - \sum_{i=1}^{n+m_1+p_1-k-j} d(\operatorname{lcm}(\alpha_{i-m_1-p_1+k+j}, \gamma_i))$$
for all $j = 0, \dots, m_1$ *and* $k = 0, \dots, p_1$,

where

$$x_j = \sum_{i=1}^{h_j} f_i - \sum_{i=1}^{h_j - j} d_i, \quad j = 0, \ldots, m_1, \quad h_j = \min\{i | d_{i-j+1} < f_i\}, \quad h_0 := 0,$$

$$y_k = \sum_{i=1}^{\bar{h}_k} \bar{f}_i - \sum_{i=1}^{\bar{h}_k - k} \bar{r}_i, \quad k = 0, \ldots, p_1, \quad \bar{h}_k = \min\{i | \bar{r}_{i-k+1} < \bar{f}_i\}, \quad \bar{h}_0 := 0,$$

and $s_1 \geq \cdots \geq s_{\bar{\rho}}$*, and* $t_1 \geq \cdots \geq t_{\bar{\theta}}$ *are nonincreasing orderings of*

$$(c^2_1, \ldots, c^2_\rho) \cup (c^1_{m_1} + 1, \ldots, c^1_{m_1+\rho+1-\bar{\rho}} + 1)$$

and

$$(r^2_1, \ldots, r^2_\theta) \cup (r^1_{p_1} + 1, \ldots, r^1_{p_1+\theta+1-\bar{\theta}} + 1),$$

respectively.

Before passing to the proof, we give some auxiliary results in Sections 4 and 5 which will be used in the course of the proof.

4. Auxiliary results

If $\psi_1 | \cdots | \psi_n$ is a polynomial chain (e.g., the list of invariant factors of a rank n pencil), then we make the convention that $\psi_i = 1$, for all $i \leq 0$, and $\psi_i = 0$, for all $i \geq n + 1$.

In the paper, we shall use a particular case of [9, Theorem 3]. If in [9, Theorem 3], we additionally have the assumption

$$\sum_{i=1}^{p} (r_i + 1) = \sum_{i=1}^{u+y} d(\tilde{\gamma}_i) - x_l,$$

then condition (41) in [9, Theorem 3] is equal to condition (42) from the same theorem, for $k = 0$. So under this assumption, instead of having two, we only have one condition.

So, if we introduce new notation and put

$$y_k = \sum_{i=1}^{u+y} d(\gamma_i) - x_l - \sum_{i=k+1}^{p} (r_i + 1),$$

then the additional assumption becomes $y_0 = 0$, and we can write this particular case of [9, Theorem 3] in the following way, by using the notation appropriate to this paper.

Theorem 4.1. *Let* n, p_1 *and* m_1 *be nonnegative integers. Let* $x_0, x_1, \ldots, x_{m_1}$ *and* $y_0, y_1, \ldots, y_{p_1}$ *be integers with* $x_0 = y_0 = 0$*, and let* $\gamma_1 | \cdots | \gamma_{n+m_1+p_1}$ *be polynomials from* $\mathbb{F}[\lambda]$*, such that* $\gamma_n = 1$.

There exist homogeneous polynomials $\beta_1|\cdots|\beta_{n+p_1}$ *which satisfy* $\beta_n = 1$ *and*

$$y_k \leq \sum_{i=n+p_1-k+1}^{n+p_1} d(\beta_i), \quad k = 1,\ldots,p_1,$$

$$\sum_{i\geq 1} d(\mathrm{lcm}(\beta_{n+p_1+1-i}, \gamma_{n+m_1+p_1-j+1-i})) \leq \sum_{i=1}^{n+m_1+p_1} d(\gamma_i) - x_j, \; j = 0,\ldots,m_1,$$

$$\sum_{i=1}^{n+p_1} d(\beta_i) = \sum_{i=1}^{n+m_1+p_1} d(\gamma_i) - x_{m_1},$$

$$\gamma_i \mid \beta_i \mid \gamma_{i+m_1}, \quad i = 1,\ldots,n+p_1,$$

if and only if

$$x_j + y_k \leq \sum_{i=n+m_1+p_1-j-k+1}^{n+m_1+p_1} d(\gamma_i), \quad j = 0,\ldots,m_1, \quad k = 0,\ldots,p_1.$$

Definition 4.2. Let $\alpha : \alpha_1|\cdots|\alpha_n$ and $\gamma : \gamma_1|\cdots|\gamma_{n+m}$ be two polynomial chains. By the *minimal path* between α and γ we mean the polynomial chain $\sigma(\alpha,\gamma)$: $\sigma_1(\alpha,\gamma)|\cdots|\sigma_m(\alpha,\gamma)$, where

$$\sigma_i(\alpha,\gamma) = \frac{\pi_i}{\pi_{i-1}}, \quad i = 1,\ldots,m, \quad \pi_i = \prod_{j=1}^{i+n} \mathrm{lcm}(\alpha_{j-i},\gamma_j), \quad i = 0,\ldots,m.$$

For more details see, e.g., [31].

Lemma 4.3 ([11, Lemma 3]). *Let* α, β, $\gamma \in \mathbb{F}[\lambda]$. *Then*

$$\mathrm{lcm}(\alpha,\gamma) \mid \frac{\mathrm{lcm}(\alpha,\beta)\,\mathrm{lcm}(\gamma,\beta)}{\beta}.$$

In [12] we have introduced the following definition of majorization that deals with three different partitions of integers.

Definition 4.4. Consider partitions $\mathbf{a} = (a_1,\ldots,a_s)$, $\mathbf{d} = (d_1,\ldots,d_m)$ and $\mathbf{g} = (g_1,\ldots,g_{m+s})$. If

$$d_i \geq g_{i+s}, \qquad i = 1,\ldots,m, \tag{4.1}$$

$$\sum_{i=1}^{h_j} g_i - \sum_{i=1}^{h_j-j} d_i \leq \sum_{i=1}^{j} a_i, \quad h_j := \min\{i | d_{i-j+1} < g_i\}, \quad j = 1,\ldots,s, \tag{4.2}$$

$$\sum_{i=1}^{m+s} g_i = \sum_{i=1}^{m} d_i + \sum_{i=1}^{s} a_i, \tag{4.3}$$

then we say that $\mathbf{g}$ is *majorized* by the pair $(\mathbf{d},\mathbf{a})$. We call this type of majorization *the generalized majorization*, and we write

$$\mathbf{g} \prec' (\mathbf{d},\mathbf{a}).$$

Lemma 4.5. *Let* $\mathbf{a} = (a_1, \ldots, a_s)$ *(with* $a_1 \geq \cdots \geq a_s \geq 0$*),* $\mathbf{d} = (d_1, \ldots, d_m)$ *(with* $d_1 \geq \cdots \geq d_{\bar{\rho}} > d_{\bar{\rho}+1} = \cdots = d_m = 0$*) and* $\mathbf{g} = (g_1, \ldots, g_{m+s})$ *(with* $g_1 \geq \cdots \geq g_g > g_{g+1} = \cdots = g_{m+s} = 0$*) be partitions such that*

$$\mathbf{g} \prec' (\mathbf{d}, \mathbf{a}). \tag{4.4}$$

Then

$$\bar{\rho} \leq g \leq \bar{\rho} + s. \tag{4.5}$$

Moreover, we have

$$g = \bar{\rho} + \chi, \tag{4.6}$$

where $\chi = \sharp\{i \in \{1, \ldots, s\} | g_{h_i} > 0\}$.

Convention: $h_0 := 0$, $h_{s+1} := m + s + 1$.

Proof. By definition of the generalized majorization, (4.4) gives (4.1), (4.2) and (4.3). We directly obtain that (4.1) implies

$$g - s \leq \bar{\rho}.$$

Now, let us suppose that $\chi > 0$. Then (4.2) and (4.3) give

$$\sum_{i=h_\chi+1}^{m+s} g_i \geq \sum_{i=h_\chi-\chi+1}^{m} d_i + \sum_{i=\chi+1}^{s} a_i. \tag{4.7}$$

Also, we have

$$\sum_{i=h_\chi-\chi+1}^{h_{\chi+1}-\chi-1} d_i \geq \sum_{i=h_\chi+1}^{h_{\chi+1}-1} g_i. \tag{4.8}$$

Indeed, if $\chi = s$, then (4.8) follows from (4.1), and if $\chi < s$, then (4.8) follows from the definition of $h_{\chi+1}$.

Since $g_{h_{\chi+1}} = \cdots = g_{m+s} = 0$ (from the definition of χ), equations (4.7) and (4.8) give

$$\sum_{i=h_\chi-\chi+1}^{h_{\chi+1}-\chi-1} d_i \geq \sum_{i=h_\chi+1}^{h_{\chi+1}-1} g_i \geq \sum_{i=h_\chi-\chi+1}^{m} d_i + \sum_{i=\chi+1}^{s} a_i \geq \sum_{i=h_\chi-\chi+1}^{m} d_i. \tag{4.9}$$

Thus we must have equalities in (4.9), and so we have that

$$d_{h_{\chi+1}-\chi} = \cdots = d_m = 0, \quad a_{\chi+1} = \cdots = a_s = 0,$$

as well as

$$d_{i-\chi} = g_i, \quad i = h_\chi + 1, \ldots, h_{\chi+1} - 1.$$

Thus, we obtain

$$\bar{\rho} = g - \chi \leq g,$$

as wanted.

Now, suppose that $\chi = 0$. Then, $g_{h_1} = 0$, and so we have that $h_1 = m+1$, as well as $d_i \geq g_i$, $i = 1, \ldots, m$. Since (4.3) is valid, we have that in this case $g_i = d_i$ for all $i = 1, \ldots, m$, as well as $a_1 = \cdots = a_s = 0$. Hence, we directly obtain $\bar{\rho} = g$.

Altogether, we have proven (4.5) and (4.6), as wanted. □

By using the concept of the generalized majorization and the minimal path, together with the notation introduced in Section 2, by [13, Remark 4] and by Lemma 4.5, we obtain that [10, Theorem 2] is equivalent to the following.

Theorem 4.6. *Consider the pencil* $A(\lambda) = \operatorname{diag}(B, N, C_1, C_2, R_1, R_2)$ *as in* (3.1). *Let* $\beta_1|\cdots|\beta_{n+m_1}$ *be polynomials, and let*

$$\tilde{b}_1 \geq \cdots \geq \tilde{b}_{b_1}, \quad \tilde{r}_1 \geq \cdots \geq \tilde{r}_{p_1+p_2},$$

and

$$g_1 \geq \cdots \geq g_{\bar{\rho}} > g_{\bar{\rho}+1} = \cdots = g_{m_2} = 0$$

be nonnegative integers. There exist matrices $Y_1, \ldots, Y_6$ *such that the pencil*

$$\left[\begin{array}{c|c|c|c|c|c}
B & & & & & \\ \hline
 & N & & & & \\ \hline
 & & C_1 & & & \\ \hline
 & & & C_2 & & \\ \hline
 & & & & R_1 & \\ \hline
 & & & & & R_2 \\ \hline\hline
Y_1 & Y_2 & \lambda E_3 + Y_3 & Y_4 & Y_5 & Y_6
\end{array}\right] \in \mathbb{F}[\lambda]^{(n+p_1+p_2+m_1)\times(n+m_1+m_2)}$$

has $\beta_1|\cdots|\beta_{n+m_1}$ *as invariant factors,* $\tilde{b}_1 \geq \cdots \geq \tilde{b}_{b_1}$ *as degrees of infinite elementary divisors,* $\tilde{r}_1 \geq \cdots \geq \tilde{r}_{p_1+p_2}$ *as row minimal indices, and*

$$g_1 \geq \cdots \geq g_{\bar{\rho}} > g_{\bar{\rho}+1} = \cdots = g_{m_2} = 0$$

as column minimal indices, if and only if the following conditions are valid:

(i) $\tilde{b}_i = b_i^1$ *for all* $i = 1, \ldots, b_1$,
(ii) $(\tilde{r}_1, \ldots, \tilde{r}_{p_1+p_2}) = (r_1^1, \ldots, r_{p_1}^1) \cup (r_1^2, \ldots, r_{p_2}^2)$,
(iii) $g_i \geq s_i$ *for all* $i = 1, \ldots, \bar{\rho}$,
(iv) $\beta_i|\alpha_i|\beta_{i+m_1}$ *for all* $i = 1, \ldots, n$,
(v) $\mathbf{f} \prec' (\mathbf{g}, \sigma(\alpha, \beta))$,

where $\mathbf{f} = (f_1, \ldots, f_{m_1+m_2})$ *with*

$$(f_1, \ldots, f_{m_1+m_2}) = (c_1^1 + 1, \ldots, c_{m_1}^1 + 1) \cup (c_1^2, \ldots, c_{m_2}^2),$$
$$(s_1, \ldots, s_{\bar{\rho}}) = (c_1^2, \ldots, c_{\rho}^2) \cup (c_{m_1}^1 + 1, \ldots, c_{m_1+\rho+1-\bar{\rho}}^1 + 1),$$

and $\mathbf{g} = (g_1, \ldots, g_{m_2})$.

5. Carlson problem and combinatorial results

In the following sections, we shall use the concept of Young diagrams. A Young diagram is a finite collection of boxes, whose rows are aligned on the left, while the row lengths (numbers of boxes in a row) are weakly decreasing. In particular, a weakly decreasing sequence of nonnegative integers $\mathbf{a} = (a_1, a_2, \ldots, a_n)$ bijectively corresponds to a Young diagram, denoted $Y_{\mathbf{a}}$, which has a_i boxes in the ith row.

Definition 5.1 ([11]). Let $\mathbf{a} = (a_1, a_2, \dots, a_n)$ and $\mathbf{c} = (c_1, c_2, \dots, c_{n+m})$ be partitions such that $c_i \geq a_i \geq c_{i+m}$, $i = 1, \dots, n$. By *the minimal path* between $\mathbf{a}$ and $\mathbf{c}$ we mean the partition $\mathbf{s} = (s_1, s_2, \dots, s_m)$, given by

$$s_i := c_{n+i} + \sum_{j=1}^{n} \max(a_j, c_{j+i-1}) - \sum_{j=1}^{n} \max(a_j, c_{j+i}), \quad i = 1, \dots, m,$$

and we denote it by $s(\mathbf{a}, \mathbf{c})$.

5.1. Littlewood–Richardson coefficients and Carlson problem

We shall recall the definition of the Littlewood–Richardson coefficients, for details see, e.g., [15, 20].

Let $\mathbf{a} = (a_1, a_2, \dots, a_n)$, $\mathbf{b} = (b_1, b_2, \dots, b_m)$, $\mathbf{c} = (c_1, c_2, \dots, c_{n+m})$ be weakly decreasing sequences of nonnegative integers such that

$$\sum_{i=1}^{n+m} c_i = \sum_{i=1}^{n} a_i + \sum_{i=1}^{m} b_i,$$

and the Young diagram of $\mathbf{a}$ is contained in that for $\mathbf{c}$, i.e., $a_i \leq c_i$. Thus, the skew diagram $\mathbf{c} \setminus \mathbf{a}$, consists of $\sum_{i=1}^{m} b_i$ boxes.

Order the boxes of $\mathbf{c} \setminus \mathbf{a}$ by first enumerating the boxes in the top row (from right to left), then the ones in the second row (from right to left), and so on down the array. Then *the Littlewood–Richardson coefficient* $LR^{\mathbf{c}}_{\mathbf{a},\mathbf{b}}$ is the number of ways to fill the boxes of $\mathbf{c} \setminus \mathbf{a}$ with the integers from $\{1, \dots, m\}$, so that the following conditions are satisfied:

(a) entries in any row are weakly increasing from left to right;
(b) entries in the each column are strictly increasing from top to bottom;
(c) the integer i occurs exactly b_i times, for $i \in \{1, \dots, m\}$;
(d) for any integer p, with $1 \leq p < \sum_{i=1}^{m} b_j$, and any positive integer i, the number of times i occurs in the first p boxes of the ordering is, at least, as large as the number of times that $i+1$ occurs in these first p boxes.

Note that the Littlewood–Richardson coefficient $LR^{\mathbf{c}}_{\mathbf{a},\mathbf{b}}$ is positive if and only if there exists at least one filling of the boxes of $\mathbf{c} \setminus \mathbf{a}$ that satisfies the conditions (a)–(d).

Throughout the paper we will deal with (chains of) polynomials from $\mathbb{F}[\lambda]$.

Definition 5.2. Let $\alpha : \alpha_1 | \cdots | \alpha_n$ be a polynomial chain, and let $\psi \in \mathbb{F}[\lambda]$ be an irreducible polynomial. Let a_i denote the degree of ψ in α_{n+1-i}, $i = 1, \dots, n$, i.e.,

$$\alpha_{n+1-i} = \psi^{a_i} q_i, \qquad i = 1, \dots, n,$$

where $q_i \in \mathbb{F}[\lambda]$ are not divisible by ψ, for any $i = 1, \dots, n$. Then, $\mathbf{a} = (a_1, \dots, a_n)$ is called the partition of the ψ-elementary divisors of α.

The following proposition is straightforward.

Proposition 5.3. *Let $\alpha : \alpha_1 | \cdots | \alpha_n$ and $\gamma : \gamma_1 | \cdots | \gamma_{n+m}$ be two polynomial chains. Let ψ be an irreducible polynomial. Let $\sigma = \sigma(\alpha, \gamma)$ be the minimal path between*

α and γ. Let $\mathbf{a} = (a_1, \ldots, a_n)$, $\mathbf{c} = (c_1, \ldots, c_{n+m})$ and $\mathbf{s} = (s_1, \ldots, s_m)$ be the partitions of the ψ-elementary divisors of α, γ and σ, respectively. Then we have

$$\mathbf{s} = s(\mathbf{a}, \mathbf{c}).$$

Lemma 5.4 ([11, Lemma 4]). *Let $\mathbf{a} = (a_1, \ldots, a_n)$ and $\mathbf{c} = (c_1, \ldots, c_{n+m+p})$ be partitions such that $c_i \geq a_i \geq c_{i+m+p}$, $i = 1, \ldots, n$. Let $\mathbf{s} = s(\mathbf{a}, \mathbf{c}) = (s_1, s_2, \ldots, s_{m+p})$ be the minimal path between them. Let $\mathbf{b} = (b_1, \ldots, b_m)$ be a partition, such that $s_i \geq b_i \geq s_{i+p}$, $i = 1, \ldots, m$. Then there exists a partition $\mathbf{d} = (d_1, d_2, \ldots, d_{n+m})$, such that*

$$c_i \geq d_i \geq a_i, \quad i = 1, \ldots, n+m,$$

and the following conditions are valid:

$$s(\mathbf{d}, \mathbf{c}) = s(\mathbf{b}, \mathbf{s}), \text{ and so } d_i \geq c_{i+p}, \quad i = 1, \ldots, n+m,$$
$$LR^{\mathbf{d}}_{\mathbf{a},\mathbf{b}} > 0.$$

Carlson problem. In [15], the Littlewood–Richardson coefficient is related to matrix completion problems. In particular, it appears in the necessary and sufficient condition for a solution of the classical completion problem – the Carlson problem, see [3]. Carlson problem consists of determining the possible invariant polynomials of the square matrix

$$C = \begin{bmatrix} A & 0 \\ X & B \end{bmatrix} \in \mathbb{F}^{(n+m)\times(n+m)}$$

where $A \in \mathbb{F}^{n\times n}$ and $B \in \mathbb{F}^{m\times m}$ are fixed matrices, and $X \in \mathbb{F}^{m\times n}$ varies.

In [11] we have given the following result.

Theorem 5.5 ([11, Theorem 7]). *Let $A(\lambda) \in \mathbb{F}[\lambda]^{n\times n}$, $B(\lambda) \in \mathbb{F}[\lambda]^{m\times m}$ be regular matrix pencils, with $\alpha : \alpha_1 | \cdots | \alpha_n$ and $\beta : \beta_1 | \cdots | \beta_m$, as homogeneous invariant factors, respectively. Let $\gamma : \gamma_1 | \cdots | \gamma_{n+m}$ be homogeneous polynomials. Let $\psi_0, \psi_1, \ldots, \psi_k$ be the irreducible factors of $\alpha_n \beta_m$. For every $i = 0, \ldots, k$, denote by a^i, b^i and c^i the partitions corresponding to the ψ_i-elementary divisors of α, β and γ, respectively.*

If for every $i = 0, \ldots, k$, the Littlewood–Richardson coefficient $LR^{c^i}_{a^i,b^i}$ is positive, then there exists a matrix pencil $X(\lambda) \in \mathbb{F}[\lambda]^{n\times m}$, such that the homogeneous invariant polynomials of the regular pencil

$$C(\lambda) = \begin{bmatrix} A(\lambda) & X(\lambda) \\ 0 & B(\lambda) \end{bmatrix}$$

are $\gamma_1 | \cdots | \gamma_{n+m}$.

Now we can give a proof of Theorem 3.1. In Subsection 6.1 we prove the necessity of conditions (i)–(vii). The proof of sufficiency of the conditions is given in Subsection 6.2. First, in Subsection 6.2.1 we prove the sufficiency of conditions (i)–(vii) in the case when $\alpha_n = 1$, and finally, in Subsection 6.2.2 we prove the sufficiency of conditions (i)–(vii) without restrictions.

6. Proof of Theorem 3.1

Recall that in Section 3, we have proved that Problem 1 is equivalent to a matrix pencil completion problem (Problem 2). Thus, in order to prove Theorem 3.1, it is enough to prove that conditions (i)–(vii) are necessary and sufficient for the existence of matrices $X_1,\dots,X_6,Y_1,\dots,Y_7$ such that the pencil (3.2) has $\gamma_1|\cdots|\gamma_{n+m_1+p_1}$ as its invariant factors, $\bar{b}_1,\dots,\bar{b}_{b_1}$ as degrees of its infinite elementary divisors, $d_1,\dots,d_{m_2}$ as its column minimal indices, and $\bar{r}_1,\dots,\bar{r}_{p_2}$ as its row minimal indices.

First consider the column completion of the pencil $\operatorname{diag}(N,B_1,C_1,C_2,R_1,R_2)$ given by

$$\left[\begin{array}{c|c|c|c|c|c||c} N & & & & & & X_1 \\ \hline & B_1 & & & & & X_2 \\ \hline & & C_1 & & & & X_3 \\ \hline & & & C_2 & & & X_4 \\ \hline & & & & R_1 & & \lambda E_2 + X_5 \\ \hline & & & & & R_2 & X_6 \end{array}\right] \tag{6.1}$$

Denote the invariant factors of (6.1) by $\beta_1|\cdots|\beta_{n+p_1}$, row minimal indices of (6.1) by $\tilde{r}_1 \geq \cdots \geq \tilde{r}_{\tilde{\theta}} > \tilde{r}_{\tilde{\theta}+1} = \cdots = \tilde{r}_{p_2} = 0$, column minimal indices by $g_1 \geq \cdots \geq g_{m_1+m_2}$, and degrees of infinite elementary divisors by $\tilde{b}_1 \geq \cdots \geq \tilde{b}_{b_1}$. By applying transposed version of Theorem 4.6, we obtain that this completion exists if and only if $\tilde{b}_i = b_i^1$, $i=1,\dots,b_1$, $g_1 \geq \cdots \geq g_{m_1+m_2}$ is nonincreasing ordering of $(c_1^1,\dots,c_{m_1}^1)\cup(c_1^2,\dots,c_{m_2}^2)$, and the following conditions are valid

$$\tilde{r}_i \geq t_i, \quad i=1,\dots,\tilde{\theta}, \tag{6.2}$$

$$\beta_i|\alpha_i|\beta_{i+p_1}, \quad i=1,\dots,n, \tag{6.3}$$

$$\bar{\mathbf{f}} \prec' (\tilde{\mathbf{r}}, \sigma(\alpha,\beta)), \tag{6.4}$$

with $\bar{\mathbf{f}} = (\bar{f}_1,\dots,\bar{f}_{p_1+p_2})$, $\bar{f}_1 \geq \cdots \geq \bar{f}_{p_1+p_2}$ is nonincreasing ordering of $(r_1^1+1,\dots,r_{p_1}^1+1)\cup(r_1^2,\dots,r_{p_2}^2)$, $t_1 \geq \cdots \geq t_{\tilde{\theta}}$ is nonincreasing ordering of $(r_1^2,\dots,r_{\theta}^2)\cup(r^1_{p_1+\theta+1-\tilde{\theta}}+1,\dots,r^1_{p_1}+1)$, and $\tilde{\mathbf{r}} = (\tilde{r}_1,\dots,\tilde{r}_{p_2})$.

Now we consider completion by rows of (6.1) to (3.2). By performing strictly equivalent operations on (3.2) we can put (6.1) in the Kronecker canonical form. Thus, this row completion, becomes equivalent to finding necessary and sufficient conditions for existence of matrices $\bar{Y}_1,\dots,\bar{Y}_5$ such that the pencil

$$\left[\begin{array}{c|c|c|c|c} \bar{N} & & & & \\ \hline & B_1 & & & \\ \hline & & C_1 & & \\ \hline & & & C_2 & \\ \hline & & & & R \\ \hline\hline \bar{Y}_1 & \bar{Y}_2 & \lambda E_3 + \bar{Y}_3 & \bar{Y}_4 & \bar{Y}_5 \end{array}\right]$$

has $\gamma_1|\cdots|\gamma_{n+m_1+p_1}$ as its invariant factors, $\bar{b}_1,\ldots,\bar{b}_{b_1}$ as degrees of its infinite elementary divisors, $d_1,\ldots,d_{m_2}$ as its column minimal indices, and $\bar{r}_1,\ldots,\bar{r}_{p_2}$ as its row minimal indices. Here $\bar{N}=\operatorname{diag}(C(\beta_1),\ldots,C(\beta_{n+p_1}))$,

$$\bar{R}=\begin{pmatrix} \operatorname{diag}\left(\begin{bmatrix} C(\lambda^{\tilde{r}_1}) \\ (e_1^{\tilde{r}_1})^T \end{bmatrix},\ldots,\begin{bmatrix} C(\lambda^{\tilde{r}_{\tilde{\theta}}}) \\ (e_1^{\tilde{r}_{\tilde{\theta}}})^T \end{bmatrix}\right) \\ 0 \end{pmatrix},$$

where the number of zero rows in $\bar{R}$ is $p_2-\tilde{\theta}$.

As before, by Theorem 4.6 we obtain that this completion exists if and only if the following conditions are valid:

$$\bar{b}_i=b_i^1, \qquad i=1,\ldots,b_1, \tag{6.5}$$

$$\bar{r}_i=\tilde{r}_i, \qquad i=1,\ldots,p_2, \tag{6.6}$$

$$d_i\geq s_i, \qquad i=1,\ldots,\bar{\rho}, \tag{6.7}$$

$$\gamma_i|\beta_i|\gamma_{i+m_1}, \qquad i=1,\ldots,n+p_1, \tag{6.8}$$

$$\mathbf{f}\prec'(\mathbf{d},\sigma(\beta,\gamma)). \tag{6.9}$$

Since we have (6.6), by the definition of $\sigma(\alpha,\beta)$, y_k, and by the definition of the generalized majorization, we have that (6.4) is equivalent to:

$$\bar{r}_i\geq\bar{f}_{i+p_1}, \quad i=1,\ldots,p_2, \tag{6.10}$$

$$y_k\leq\sum_{i=1}^{n+p_1}d(\beta_i)-\sum_{i=1}^{n+p_1-k}d(\operatorname{lcm}(\alpha_{i-p_1+k},\beta_i)), \quad k=1,\ldots,p_1, \tag{6.11}$$

$$\sum_{i=1}^{n}d(\alpha_i)+\sum_{i=1}^{p_2}r_i^2+\sum_{i=1}^{p_1}(r_i^1+1)=\sum_{i=1}^{n+p_1}d(\beta_i)+\sum_{i=1}^{p_2}\bar{r}_i. \tag{6.12}$$

Moreover, by Lemma 4.5 we also have

$$\theta\leq\bar{\theta}\leq\theta+p_1. \tag{6.13}$$

Analogously, by the definition of $\sigma(\beta,\gamma)$, x_j, and by the definition of the generalized majorization, we have that (6.9) is equivalent to:

$$d_i\geq f_{i+m_1}, \quad i=1,\ldots,m_2, \tag{6.14}$$

$$x_j\leq\sum_{i=1}^{n+m_1+p_1}d(\gamma_i)-\sum_{i=1}^{n+p_1+m_1-j}d(\operatorname{lcm}(\beta_{i-m_1+j},\gamma_i),j=1,\ldots,m_1, \tag{6.15}$$

$$\sum_{i=1}^{n+p_1}d(\beta_i)+\sum_{i=1}^{m_2}c_i^2+\sum_{i=1}^{m_1}(c_i^1+1)=\sum_{i=1}^{n+m_1+p_1}d(\gamma_i)+\sum_{i=1}^{m_2}d_i. \tag{6.16}$$

We note that conditions (6.2) and (6.7) imply conditions (6.10) and (6.14), respectively. By Lemma 4.5, we also have

$$\rho\leq\bar{\rho}\leq\rho+m_1. \tag{6.17}$$

By unifying these results, we obtain that the existence of matrices $X_1, \ldots, X_6$, $Y_1, \ldots, Y_7$, such that (3.2) has $\gamma_1|\cdots|\gamma_{n+m_1+p_1}$ as its invariant factors, $\bar{b}_1, \ldots, \bar{b}_{b_1}$ as degrees of its infinite elementary divisors, $d_1, \ldots, d_{m_2}$ as its column minimal indices, and $\bar{r}_1, \ldots, \bar{r}_{p_2}$ as its row minimal indices, is equivalent to the existence of polynomials $\beta_1|\cdots|\beta_{n+p_1}$ which satisfy the following conditions:

$$\bar{b}_i = b_i^1, \qquad i = 1, \ldots, b_1, \tag{6.18}$$

$$\bar{r}_i \geq t_i, \qquad i = 1, \ldots, \bar{\theta}, \tag{6.19}$$

$$\beta_i|\alpha_i|\beta_{i+p_1}, \qquad i = 1, \ldots, n, \tag{6.20}$$

$$\bar{\mathbf{f}} \prec' (\bar{\mathbf{r}}, \sigma(\alpha, \beta)), \tag{6.21}$$

$$d_i \geq s_i, \qquad i = 1, \ldots, \bar{\rho}, \tag{6.22}$$

$$\gamma_i|\beta_i|\gamma_{i+m_1}, \qquad i = 1, \ldots, n + p_1, \tag{6.23}$$

$$\mathbf{f} \prec' (\mathbf{d}, \sigma(\beta, \gamma)). \tag{6.24}$$

Hence, we are left with proving that conditions (i)–(vii) are necessary and sufficient for the existence of such polynomials $\beta_1|\cdots|\beta_{n+p_1}$.

Now we can give a proof of Theorem 3.1. We start with proving the necessity of conditions (i)–(vii).

6.1. Necessity of conditions (i)–(vii)

Let us suppose that there exist polynomials $\beta_1|\cdots|\beta_{n+p_1}$ which satisfy conditions (6.18)–(6.24).

Condition (6.18) is equal to (ii). As we have showed in Lemma 4.5, conditions (6.21) and (6.24) imply (6.13) and (6.17), respectively, thus we have (i). Conditions (6.20) and (6.23) together give (iii). Conditions (6.19) and (6.22) are equal to (iv) and (v), respectively. Since (6.21) is equivalent to (6.10), (6.11) and (6.12), while (6.24) is equivalent to (6.14), (6.15) and (6.16), we have that (6.12) and (6.16) give (vi). So we are left with obtaining the condition (vii).

By summing (6.11) and (6.15) we have that for every $j = 0, \ldots, m_1$, and every $k = 0, \ldots, p_1$, the following is valid:

$$\begin{aligned} x_j + y_k \leq & \sum_{i=1}^{n+m_1+p_1} d(\gamma_i) - \sum_{i=1}^{n+m_1+p_1-k-j} d(\operatorname{lcm}(\beta_{i-m_1+j}, \gamma_i)) \\ & - \sum_{i=n+m_1+p_1-k-j+1}^{n+m_1+p_1-k} d(\operatorname{lcm}(\beta_{i-m_1+j}, \gamma_i)) \\ & + \sum_{i=1}^{n+p_1} d(\beta_i) - \sum_{i=1}^{n+p_1-k} d(\operatorname{lcm}(\alpha_{i-p_1+k}, \beta_i)). \end{aligned}$$

By applying Lemma 4.3, on every summand of the second sum on the right-hand side, we obtain the following inequality

$$\begin{aligned} x_j + y_k \leq & \sum_{i=1}^{n+m_1+p_1} d(\gamma_i) - \sum_{i=1}^{n+m_1+p_1-k-j} d(\operatorname{lcm}(\alpha_{i-m_1-p_1+k+j}, \gamma_i)) \\ & - \sum_{i=1}^{n+m_1+p_1-k-j} d(\beta_{i-m_1+j}) \\ & + \sum_{i=1}^{n+m_1+p_1-k-j} d(\operatorname{lcm}(\alpha_{i-m_1-p_1+k+j}, \beta_{i-m_1+j})) \\ & - \sum_{i=n+m_1+p_1-k-j+1}^{n+m_1+p_1-j} d(\operatorname{lcm}(\beta_{i-m_1+j}, \gamma_i)) \\ & + \sum_{i=1}^{n+p_1} d(\beta_i) - \sum_{i=1}^{n+p_1-k} d(\operatorname{lcm}(\alpha_{i-p_1+k}, \beta_i)). \end{aligned}$$

Finally, since

$$\sum_{i=n+p_1-k+1}^{n+p_1} d(\beta_i) \leq \sum_{i=n+p_1-k+1}^{n+p_1} d(\operatorname{lcm}(\beta_i, \gamma_{i+m_1-j})),$$

we obtain (vii), as wanted.

6.2. Sufficiency of conditions (i)–(vii)

The proof of the sufficiency of conditions (i)–(vii) will be split in two cases. First in Subsection 6.2.1, we consider the case of Theorem 3.1 when $\alpha_n = 1$, and in Section 6.2.2 we shall consider the general case, i.e., when $\alpha_n \neq 1$.

6.2.1. Case $\alpha_n = 1$. Let $\alpha_1 = \cdots = \alpha_n = 1$. Let conditions (i)–(vii) be valid. We are left with proving the existence of polynomials $\beta_1 | \cdots | \beta_{n+p_1}$ which satisfy conditions (6.18)–(6.24).

By (iii) we have $\gamma_i = 1$, $i = 1, \ldots, n$. Conditions (iv) and (v) are equal to (6.19) and (6.22), respectively (thus implying (6.10) and (6.14), as well). Also, condition (ii) is equal to (6.18). By Theorem 4.1, condition (vii) implies the existence of polynomials $\beta_1 | \cdots | \beta_{n+p_1}$ which satisfy $\beta_n = 1$ and

$$y_k \leq \sum_{i=n+p_1-k+1}^{n+p_1} d(\beta_i), \quad k = 1, \ldots, p_2,$$

$$\sum_{i\geq 1} d(\operatorname{lcm}(\beta_{n+p_1+1-i}, \gamma_{n+p_1+m_1-j+1-i}) \leq \sum_{i=1}^{n+m_1+p_1} d(\gamma_i) - x_j,\ j = 0, \ldots, m_1,$$

$$\sum_{i=1}^{n+p_1} d(\beta_i) = \sum_{i=1}^{n+m_1+p_1} d(\gamma_i) - x_{m_1}, \tag{6.25}$$

$$\gamma_i \mid \beta_i \mid \gamma_{i+m_1}, \quad i = 1, \ldots, n+p_1.$$

Since we are considering the case $\alpha_n = 1$, in this way we have obtained polynomials $\beta_1 | \cdots | \beta_{n+p_1}$ which satisfy conditions (6.20), (6.23), (6.11), and (6.15). Thus by the definition of generalized majorization, in order to prove (6.21) and (6.24), we are left with proving (6.16) and (6.12). Since (vi) is valid, it is enough to prove the validity of one of them, e.g., (6.16).

Conditions (vi), and (vii) for $j = m_1$ and $k = p_1$, give

$$0 \leq \sum_{i=h_{m_1}+1}^{m_1+m_2} f_i - \sum_{i=h_{m_1}-m_1+1}^{m_2} d_i + \sum_{i=\bar{h}_{p_1}+1}^{p_1+p_2} \bar{f}_i - \sum_{i=\bar{h}_{p_1}-p_1+1}^{p_2} \bar{r}_i. \tag{6.26}$$

Since (6.10) and (6.14) are valid, we have that

$$0 \geq \sum_{i=h_{m_1}+1}^{m_1+m_2} f_i - \sum_{i=h_{m_1}-m_1+1}^{m_2} d_i + \sum_{i=\bar{h}_{p_1}+1}^{p_1+p_2} \bar{f}_i - \sum_{i=\bar{h}_{p_1}-p_1+1}^{p_2} \bar{r}_i. \tag{6.27}$$

Therefore, one must have the equality in both (6.26) and (6.27), as well as in (6.10) for $i = h_{m_1} - m_1 + 1, \ldots, m_2$, and in (6.14) for $i = \bar{h}_{p_1} - p_1 + 1, \ldots, p_2$. Hence

$$x_{m_1} = \sum_{i=1}^{h_{m_1}} f_i - \sum_{i=1}^{h_{m_1}-m_1} d_i = \sum_{i=1}^{m_1+m_2} f_i - \sum_{i=1}^{m_2} d_i,$$

and

$$y_{p_1} = \sum_{i=1}^{\bar{h}_{p_1}} \bar{f}_i - \sum_{i=1}^{\bar{h}_{p_1}-p_1} d_i = \sum_{i=1}^{p_1+p_2} \bar{f}_i - \sum_{i=1}^{p_2} \bar{r}_i,$$

which, taking into account (6.25), finishes the proof.

Now we can pass to the case $\alpha_n \neq 1$.

6.2.2. Sufficiency of conditions (i)–(vii) if $\alpha_n \neq 1$. Consider the pencil $A(\lambda)$ in its canonical form (3.1), i.e., $\operatorname{diag}(N, B_1, C_1, C_2, R_1, R_2)$. By applying the result of the previous section, when $\alpha_n = 1$, we have that there exists matrices $X_2, \ldots, X_6, Y_2, \ldots, Y_7$ such that the pencil

$$\left[\begin{array}{c|c|c|c|c||c} B_1 & & & & & X_2 \\ \hline & C_1 & & & & X_3 \\ \hline & & C_2 & & & X_4 \\ \hline & & & R_1 & & \lambda E_2 + X_5 \\ \hline & & & & R_2 & X_6 \\ \hline\hline Y_2 & \lambda E_3 + Y_3 & Y_4 & Y_5 & Y_6 & Y_7 \end{array}\right]$$
$$\in \mathbb{F}[\lambda]^{(n-t+p_1+p_2+m_1)\times(n-t+m_1+m_2+p_1)}, \tag{6.28}$$

has rank equal to $n - t + m_1 + p_1$, and has $d_1, \ldots, d_{m_2}$, as its column minimal indices, $\bar{r}_1, \ldots, \bar{r}_{p_2}$ as its row minimal indices, $\bar{b}_1, \ldots, \bar{b}_{b_1}$ as degrees of its infinite elementary divisors, and

$$\underbrace{1|\cdots|1}_{n-t}|\sigma_1(\alpha,\gamma)|\cdots|\sigma_{m_1+p_1}(\alpha,\gamma)$$

as its invariant factors.

Indeed, the pencil $\mathrm{diag}(B_1, C_1, C_2, R_1, R_2)$ has the same column, row minimal indices and degrees of infinite elementary divisors as $A(\lambda)$ and has only trivial invariant factors – exactly $n-t$ of them. Denote those invariant factors by $\alpha'_1|\cdots|\alpha'_{n-t} = 1$. Since (i)–(vii) are valid for α_i's, γ_i's, b_i's, c^1_i's, c^2_i's, r^1_i's, r^2_i's, d_i's, $\bar{b}_i$'s and $\bar{r}_i$'s, from the form of the conditions, it is straightforward to see that they are also satisfied for α'_i's and

$$\underbrace{1|\cdots|1}_{n-t}|\sigma_1(\alpha,\gamma)|\cdots|\sigma_{m_1+p_1}(\alpha,\gamma)$$

instead of α_i's and γ_i's, respectively, while keeping all other invariants the same. Therefore by Subsection 6.2.1, there exists a completion (6.28).

Next, consider the subpencil

$$\left[\begin{array}{c|c|c|c|c||c} B_1 & & & & & X_2 \\ \hline & C_1 & & & & X_3 \\ \hline & & C_2 & & & X_4 \\ \hline & & & R_1 & & \lambda E_2 + X_5 \\ \hline & & & & R_2 & X_6 \end{array}\right]$$
$$\in \mathbb{F}[\lambda]^{(n-t+p_1+p_2)\times(n-t+m_1+m_2+p_1)}. \tag{6.29}$$

Completely analogously as in the beginning of Section 6, by applying Theorem 4.6 twice – once for the completion from $\mathrm{diag}(B, C_1, C_2, R_1, R_2)$ up to (6.29), and next for the completion from (6.29) to (6.28), we obtain that (6.29) has $(c^1_1, \ldots, c^1_{m_1}) \cup (c^2_1, \ldots, c^2_{m_2})$ as its column minimal indices, $\bar{r}_1, \ldots, \bar{r}_{p_2}$ as its row minimal indices, and $\bar{b}_1, \ldots, \bar{b}_{b_1}$ as degrees of its infinite elementary divisors. Also, since (6.29) has rank equal to $n-t+p_1$ and it has at least $n-t$ trivial invariant factors, we denote its invariant factors by

$$\underbrace{1|\cdots|1}_{n-t}|\bar{\beta}_1|\cdots|\bar{\beta}_{p_1}.$$

Now, it is straightforward to see that for the completion from (6.29) up to (6.28), we can consider (6.29) in the canonical form: $\mathrm{diag}(B, C_1, C_2, N(\bar{\beta}), \bar{R})$, where $N(\bar{\beta}) = \mathrm{diag}(C(\bar{\beta}_1), \ldots, C(\bar{\beta}_{p_1}))$, and

$$\bar{R} = \mathrm{diag}\left(\begin{array}{c} \left[\begin{array}{c} C(\lambda^{\bar{r}_1}) \\ (e_{\bar{r}_1})^T \end{array}\right], \ldots, \left[\begin{array}{c} C(\lambda^{\bar{r}_{\bar{\theta}}}) \\ (e_{\bar{r}_{\bar{\theta}}})^T \end{array}\right] \\ 0 \end{array} \right),$$

where the number of the zero rows equals $p_2 - \bar{\theta}$.

Thus, we have that there exist $\bar{Y}_2, \ldots, \bar{Y}_6$, such that

$$\left[\begin{array}{c|c|c|c|c} B_1 & & & & \\ \hline & C_1 & & & \\ \hline & & C_2 & & \\ \hline & & & N(\beta) & \\ \hline & & & & \bar{R} \\ \hline\hline \bar{Y}_2 & \lambda E_3 + \bar{Y}_3 & \bar{Y}_4 & \bar{Y}_5 & \bar{Y}_6 \end{array}\right],$$

is strictly equivalent to (6.28). Thus, by applying Theorem 4.6, we obtain that the following conditions hold:

$$d_i \geq s_i, \quad i = 1, \ldots, \bar{\rho}, \tag{6.30}$$

$$\sigma_i(\alpha, \gamma) | \bar{\beta}_i | \sigma_{i+m_1}(\alpha, \gamma), \quad i = 1, \ldots, p_1, \tag{6.31}$$

$$\mathbf{f} \prec' (\mathbf{d}, \sigma(\bar{\beta}, \sigma(\alpha, \gamma))).$$

Before proceeding, we note that by applying [10, Lemmas 3–4] and [7, Lemma 7], without loss of generality, we can consider $X_2 = 0$, $X_3 = 0$ and $X_4 = 0$ in the pencil (6.29).

Let $\psi_1, \ldots, \psi_k$ be all distinct irreducible factors of $\gamma_{n+m_1+p_1}$. For every $i = 1, \ldots, k$, let

$$\begin{aligned} a^i &= (a_1^i, a_2^i, \ldots, a_n^i), \\ b^i &= (b_1^i, b_2^i, \ldots, b_{p_1}^i), \\ c^i &= (c_1^i, c_2^i, \ldots, c_{n+m_1+p_1}^i), \\ s^i &= (s_1^i, s_2^i, \ldots, s_{m_1+p_1}^i) \end{aligned}$$

be the weakly decreasing partitions corresponding to the ψ_i-elementary divisor of the polynomial chains α, $\bar{\beta}$, γ and σ, respectively. More precisely:

$$\begin{aligned} \alpha_{n+1-j} &:= \psi_1^{a_j^1} \psi_2^{a_j^2} \cdots \psi_k^{a_j^k}, \quad j = 1, \ldots, n, \\ \bar{\beta}_{p_1+1-j} &:= \psi_1^{b_j^1} \psi_2^{b_j^2} \cdots \psi_k^{b_j^k}, \quad j = 1, \ldots, p_1, \\ \gamma_{n+m_1+p_1+1-j} &:= \psi_1^{c_j^1} \psi_2^{c_j^2} \cdots \psi_k^{c_j^k}, \quad j = 1, \ldots, n + m_1 + p_1, \\ \sigma_{m_1+p_1+1-j} &:= \psi_1^{s_j^1} \psi_2^{s_j^2} \cdots \psi_k^{s_j^k}, \quad j = 1, \ldots, m_1 + p_1. \end{aligned}$$

Note that by Proposition 5.3, we have that $s^i = s(a^i, c^i)$, for all i. Also for all i, by (i) we have $c_j^i \geq a_j^i \geq c_{j+m_1+p_1}^i$, $j = 1, \ldots, n$, and by (6.31) we have $s_j^i \geq b_j^i \geq s_{j+m_1}^i$, $j = 1, \ldots, p_1$.

Now, by Lemma 5.4, for every $i = 1, \ldots, k$, there exists a nonincreasing partition $d^i = (d_1^i, d_2^i, \ldots, d_{n+p_1}^i)$ such that

$$c_j^i \geq d_j^i \geq a_j^i, \quad j = 1, \ldots, n + p_1, \tag{6.32}$$

and

$$s(d^i, c^i) = s(b^i, s^i), \tag{6.33}$$

$$d^i_j \geq c^i_{j+m_1}, \quad j = 1, \dots, p_1, \tag{6.34}$$

$$LR^{d^i}_{a^i, b^i} > 0. \tag{6.35}$$

Finally, we define the homogeneous polynomials $\phi : \phi_1 | \cdots | \phi_{n+p_1}$ by

$$\phi_{n+p_1+1-j} := \psi_1^{d^1_j} \psi_2^{d^2_j} \cdots \psi_k^{d^k_j}, \quad j = 1, \dots, n + p_1.$$

From the definition of the polynomials $\phi_1 | \cdots | \phi_{n+p_1}$ and from (6.35), by Theorem 5.5 (Carlson problem), there exists a pencil Y such that

$$\left[\begin{array}{c|c} N & Y \\ 0 & N(\bar{\beta}) \end{array}\right]$$

is strictly equivalent to $\mathrm{diag}(C(\phi_1), \dots, C(\phi_{n+p_1}))$. Thus, since (6.29) is strictly equivalent to $\mathrm{diag}(B_1, C_1, C_2, N(\bar{\beta}), \bar{R})$, we have that there exist pencils W_1 and W_2, and X_1 such that the pencil

$$\left[\begin{array}{c||c|c|c|c|c||c} N & 0 & 0 & 0 & W_1 & W_2 & X_1 \\ \hline\hline & B & & & & & 0 \\ \hline & & C_1 & & & & 0 \\ \hline & & & C_2 & & & 0 \\ \hline & & & & R_1 & & \lambda E_2 + X_5 \\ \hline & & & & & R_2 & X_6 \end{array}\right] \tag{6.36}$$

has $\phi_1 | \cdots | \phi_{n+p_1}$ as its invariant factors, $\bar{b}_1, \dots, \bar{b}_{b_1}$ as degrees of its infinite elementary divisors, $\bar{r}_1, \dots, \bar{r}_{p_2}$ as its row minimal indices, and $(c^1_1, \dots, c^1_{m_1}) \cup (c^2_1, \dots, c^2_{m_2})$ as its column minimal indices.

By [7, Lemma 7], without loss of generality, we can consider $W_1 = 0$ and $W_2 = 0$.

From the definition of ϕ_i, by (6.32) and (6.34), we have that

$$\gamma_i \mid \phi_i \mid \gamma_{i+m_1}, \quad i = 1, \dots, n + p_1, \tag{6.37}$$

and (6.33) gives

$$\sigma(\phi, \gamma) = \sigma(\bar{\beta}, \sigma(\alpha, \gamma)).$$

Hence (6.24) becomes

$$\mathbf{f} \prec' (\mathbf{d}, \sigma(\phi, \gamma)). \tag{6.38}$$

Finally, since the subpencil $\mathrm{diag}(B_1, C_1, C_2)$ of (6.36) has zeros in corresponding rows and columns, the pencil (6.36) is strictly equivalent to

$$\mathrm{diag}(B_1, C_1, C_2, N(\phi), \bar{R}),$$

with

$$N(\phi) = \mathrm{diag}(C(\phi_1), \dots, C(\phi_{n+p_1})),$$

in a way that the block $\mathrm{diag}(B_1, C_1, C_2)$ remains unchanged.

Hence, by applying Theorem 4.6, we have that conditions (6.30), (6.37) and (6.38) imply the existence of matrices $Y_1,\dots,Y_7$ with m_1 rows, such that the pencil

$$\left[\begin{array}{c|c|c|c|c|c||c}
N & & & & & & X_1 \\ \hline
 & B_1 & & & & & X_2 \\ \hline
 & & C_1 & & & & X_3 \\ \hline
 & & & C_2 & & & X_4 \\ \hline
 & & & & R_1 & & \lambda E_2 + X_5 \\ \hline
 & & & & & R_2 & X_6 \\ \hline\hline
Y_1 & Y_2 & \lambda E_3 + Y_3 & Y_4 & Y_5 & Y_6 & Y_7
\end{array}\right]$$

has $d_1,\dots,d_{m_2}$, as its column minimal indices, $\bar{r}_1,\dots,\bar{r}_{p_2}$ as its row minimal indices, and $\gamma_1|\cdots|\gamma_{n+m_1+p_1}$ as invariant factors, and $\bar{b}_1,\dots,\bar{b}_{b_1}$ as degrees of its infinite elementary divisors. This finishes our proof.

7. Corollary

If $E = I$ then we have that the pencil $\lambda I - (A + BF + KC)$ is regular, and its (finite) invariant factors form a complete set of its Kronecker invariants. Moreover, in this case we have $m_2 = p_2 = 0$. So, we directly obtain the following result.

Corollary 7.1. *For a given system (I, A, B, C), there exist matrices F and K such that $\lambda I - (A + BF + KC)$ is regular with $\gamma_1|\cdots|\gamma_{n+m_1+p_1}$ as invariant factors, if and only if the following conditions are valid:*

(i) $\gamma_i \mid \alpha_i \mid \gamma_{i+m_1+p_1}, \quad i = 1,\dots,n,$

(ii)
$$\sum_{i=1}^{n} d(\alpha_i) + \sum_{i=1}^{m_1}(c_i^1 + 1) + \sum_{i=1}^{p_1}(r_i^1 + 1) = \sum_{i=1}^{n+m_1+p_1} d(\gamma_i),$$

(iii)
$$\sum_{i=1}^{j}(c_i^1 + 1) + \sum_{i=1}^{k}(r_i^1 + 1) \leq \sum_{i=1}^{n+m_1+p_1} d(\gamma_i) - \sum_{i=1}^{n+m_1+p_1-k-j} d(\operatorname{lcm}(\alpha_{i-m_1-p_1+k+j}, \gamma_i)),$$
$$j = 0,\dots,m_1, \quad k = 0,\dots,p_1.$$

Remark 7.2. By using the classical majorization [16], conditions (ii) and (iii) can be unified as

$$(c_1 + 1,\dots,c_{m_1} + 1) \cup (r_1 + 1,\dots,r_{p_1} + 1) \prec (d(\sigma_{m_1+p_1}(\alpha,\gamma)),\dots,d(\sigma_1(\alpha,\gamma))).$$

We note that Corollary 7.1 can also be obtained as a corollary of [11, Theorem 1] (see also [2, 26, 27]).

Acknowledgment

We would like to thank the referee for hers/his valuable comments and suggestions, that have significantly improved the presentation of the paper.

References

[1] G. Basile and G. Marro, *Controlled and Conditional Invariants in Linear Systems Theory.* Prentice-Hall, Englewood Cliffs, NJ, 1992.

[2] I. Cabral and F.C. Silva, *On feedback equivalence and completion problems.* Linear Algebra Appl. **265** (1997) 231–245.

[3] D. Carlson, *Inequalities relating the degrees of elementary divisors within a matrix.* Simon Stevin **44** (1970), 3–10.

[4] B.M. Chen, *Robust and H_∞ Control.* Springer, New York, 2000.

[5] B.M. Chen, Z. Lin, and Y. Shamash, *Linear Systems Theory: A Structural Decomposition Approach.* Birkhäuser, Boston, 2004.

[6] B.W. Dickinson, *On the fundamental theorem of linear state variable feedback.* IEEE Trans. Automat. Control AC-19 (1974) 577–579.

[7] M. Dodig, *Feedback invariants of matrices with prescribed rows.* Linear Algebra Appl. **405** (2005) 121–154.

[8] M. Dodig, *Pole placement problem for singular systems.* Numerical Linear Algebra Appl. **18** (2011), 283–297.

[9] M. Dodig, *Completion of quasi-regular matrix pencils.* Linear Algebra Appl. **501** (2016), 198–241.

[10] M. Dodig and M. Stošić, *Singular systems, state feedback problem.* Linear Algebra Appl. **431** (2009), 1267–1292.

[11] M. Dodig and M. Stošić, *Similarity class of a matrix with prescribed submatrix.* Linear and Multilinear Algebra **57** (2009), 217–245.

[12] M. Dodig and M. Stošić, *Combinatorics of column minimal indices and matrix pencil completion problems.* SIAM J. Matrix Anal. & Appl. **31** (2010), 2318–2346.

[13] M. Dodig and M. Stošić, *The rank distance problem for pairs of matrices and a completion of quasi-regular matrix pencils.* Linear Algebra Appl. **457** (2014), 313–347.

[14] J.C. Doyle, *Guaranteed margins for LQG regulators.* IEEE Trans. on AC **23** (1978), 756–757.

[15] W. Fulton, *Eigenvalues, invariant factors, highest weights, and Schubert calculus.* Bull. Amer. Math. Soc. **37** (2000), 209–249.

[16] G. Hardy, J.E. Littlewood, and G. Pólya, *Inequalities.* Cambridge University Press, Cambridge, 1991.

[17] V. Kučera, *Analysis and Design of Discrete Linear Control Systems.* Academia, Prague, 1991.

[18] V. Kučera and P. Zagalak, Fundamental theorem of state feedback for singular systems, Automatica **24** (1988), 653–658.

[19] G. Lebrete and J.J. Loiseau, *Proportional and proportional-derivative canonical forms for descriptor systems with outputs.* Automatica **30** (1994), 847–864.

[20] D.E. Littlewood and A.R. Richardson, *Group characters and algebra.* Phil. Trans. Roy. Soc. London A **223** (1934), 99–141.

[21] J.J. Loiseau, *Some geometric considerations about the Kronecker normal form.* Int. J. Control **42** (1985), 1411–1431.

[22] J.J. Loiseau and G. Lebrete, *A new canonical form for descriptor systems with outputs*. Analysis and Optimization of Systems, vol. 144, Springer, New York, 1990.

[23] M. Malabre, *A structural approach for linear singular systems*. International Mini Symposium on Singular Systems, Atlanta, 1987.

[24] S. Mondié, P. Zagalak, and V. Kučera, *State feedback in linear control theory*. Linear Algebra Appl. **317** (2000), 177–192.

[25] A.S. Morse, *Structural invariants of linear multivariable systems*. SIAM J. Control **11** (1973), 446–465.

[26] A. Roca and I. Zaballa, *Invariant factor assignment under state feedback and output injection*. Linear Algebra Appl. **332–334** (2001), 401–436.

[27] A. Roca and I. Zaballa, *A polynomial approach to the invariant factors assignment problem via state feedback and output injection*. Linear Algebra Appl. **401** (2005), 135–158.

[28] H.H. Rosenbrock, *State-Space and Multivariable Theory*. Thomas Nelson and Sons, London, 1970.

[29] A. Saberi, B.M. Chen, and P. Sannuti, *Loop Transfer Recovery: Analysis and Design*. Springer, London, 1993.

[30] G. Stein and M. Athans, *The LQG/LTR procedure for multivariable control design*. IEEE Trans. on AC **32** (1987), 105–114.

[31] I. Zaballa, *Matrices with prescribed rows and invariant factors*. Linear Algebra Appl. **87** (1987), 113–146.

Marija Dodig
CEAFEL, Departamento de Matématica
Universidade de Lisboa
Edificio C6, Campo Grande
1749-016 Lisbon, Portugal

and

Mathematical Institute SANU
Knez Mihajlova 36
11000 Beograd, Serbia
e-mail: `msdodig@fc.ul.pt`

Operator Theory:
Advances and Applications, Vol. 267, 167–183

Hermitian Geometry on Resolvent Set

Ronald G. Douglas and Rongwei Yang

Abstract. For a tuple $A = (A_1, A_2, \dots, A_n)$ of elements in a unital Banach algebra $\mathcal{B}$, its *projective joint spectrum* $P(A)$ is the collection of $z \in \mathbb{C}^n$ such that $A(z) = z_1A_1 + z_2A_2 + \cdots + z_nA_n$ is not invertible. It is known that the $\mathcal{B}$-valued 1-form $\omega_A(z) = A^{-1}(z)dA(z)$ contains much topological information about the joint resolvent set $P^c(A)$. This paper studies geometric properties of $P^c(A)$ with respect to Hermitian metrics defined through the $\mathcal{B}$-valued *fundamental form* $\Omega_A = -\omega_A^* \wedge \omega_A$ and its coupling with *faithful* states ϕ on $\mathcal{B}$, i.e., $\phi(\Omega_A)$. The connection between the tuple A and the metric is the main subject of this paper. In particular, it shows that the Kählerness of the metric is tied with the commutativity of the tuple, and its completeness is related to the Fuglede–Kadison determinant.

Mathematics Subject Classification (2010). Primary 47A13; Secondary 53A35.

Keywords. Maurer–Cartan form, projective joint spectrum, Hermitian metric, Ricci tensor, Fuglede–Kadison determinant.

0. Introduction

In [6], Cowen and the first-named author introduced geometric concepts such as holomorphic bundle and curvature into Operator Theory. This gave rise to complete and computable invariants for certain operators with rich spectral structure, i.e., the Cowen–Douglas operators. The idea was extended to multivariable cases in the study of Hilbert modules in analytic function spaces, where curvature invariant is defined for some canonical tuples of commuting operators. We refer readers to [8–10, 24] and the references therein for more information. This paper aims to study general (possibly non-commuting) tuples using geometric ideas. The approach is based on the newly emerged concept of projective joint spectrum $P(A)$ introduced by the second-named author (cf. [29]). Some preliminary findings are reported here. They shall help to build a foundation for subsequent and more in-depth studies.

Let $\mathcal{B}$ be a complex Banach algebra with unit I and

$$A = (A_1, A_2, \dots, A_n)$$

be a tuple of linearly independent elements in $\mathcal{B}$. The multiparameter pencil

$$A(z) := z_1A_1 + z_2A_2 + \cdots + z_nA_n$$

is an important subject of study in various fields, for example in algebraic geometry [28], mathematical physics [1, 21], PDE [2], group theory [13], etc., and more recently in the settlement of the Kadison–Singer Conjecture ([22]). In these studies, the primary interest was in the case when A is a tuple of self-adjoint operators. For general tuples, the notion of projective joint spectrum is defined in [29] and its property is investigated in a series of papers (cf. [3–5, 25]).

Definition. For a tuple $A = (A_1, A_2, \dots, A_n)$ of elements in a unital Banach algebra $\mathcal{B}$, its projective joint spectrum is defined as

$$P(A) = \{z \in \mathbb{C}^n | A(z) \text{ is not invertible}\}.$$

The *projective resolvent set* refers to the complement $P^c(A) = \mathbb{C}^n \setminus P(A)$.

Various notions of joint spectra for (commuting) tuples have been defined in the past, for instance the Harte spectrum ([16, 17]) and the Taylor spectrum ([11, 26, 27]), and they are key ingredients in multivariable operator theory. The projective joint spectrum, however, has one notable distinction: it is "base free" in the sense that, instead of using I as a base point and looking at the invertibility of $(A_1 - z_1I, A_2 - z_2I, \dots, A_n - z_nI)$ in various constructions, it uses the much simpler pencil $A(z)$. This feature makes the projective joint spectrum computable in many interesting noncommuting examples, for example the tuple of general compact operators ([25]), the generating tuple for the free group von Neumann algebra ([3]), the Cuntz tuple of isometries $(S_1,\ S_2, \dots,\ S_n)$ ([7, 19]) satisfying

$$S_i^*S_j = \delta_{ij}I \quad \text{for } 1 \leq i, j \leq n, \quad \sum_{i=1}^{n} S_iS_i^* = I,$$

where I is the identity. For the generating tuple $(1,\ a,\ t)$ of the infinite dihedral group

$$D_\infty =< a,\ t \mid a^2 = t^2 = 1 >,$$

the projective spectrum has found convincing applications to the study of group of intermediate growth (finitely generated group whose growth is faster than polynomial growth but slower than exponential growth (cf. [14])).

For a general tuple A, it can happen that $P(A) = \mathbb{C}^n$ and hence $P^c(A)$ is empty. But in this paper we shall always assume that $P^c(A)$ is non-empty. It is known that every path-connected component of $P^c(A)$ is a domain of holomorphy. This fact is a consequence of a result in [30], and it is also proved independently in [20]. From this point of view, $P^c(A)$ has good analytic properties. On $P^c(A)$, the holomorphic Maurer–Cartan type $\mathcal{B}$-valued $(1,0)$-form

$$\omega_A(z) = A^{-1}(z)dA(z) = \sum_{j=1}^{n} A^{-1}(z)A_j dz_j$$

played an important role in [29], where it is shown to contain much information about the topology of $P^c(A)$. For example, the de Rham cohomology $H^*(P^c(A), \mathbb{C})$ can be computed by coupling ω_A with invariant multilinear functionals or cyclic cocycles ([4, 29]). Two simple facts about ω_A are useful here.

1) By differentiating the equation $A(z)A(z)^{-1} = I$, one easily verifies that

$$dA^{-1}(z) = -A^{-1}(z)(dA(z))A^{-1}(z),$$

and consequently,

$$d\omega_A = dA^{-1}(z) \wedge dA(z) = -\omega_A \wedge \omega_A. \tag{0.1}$$

Here d is complex differentiation (see Section 1) and $\wedge$ is the wedge product.

2) ω_A is invariant under left multiplication on the tuple A by invertible elements. To be precise, we let $GL(\mathcal{B})$ denote the set of invertible elements in $\mathcal{B}$. Then for any $L \in GL(\mathcal{B})$ and $B = LA = (LA_1, LA_2, \ldots, LA_n)$, we have for the multiparameter pencils $B(z) = LA(z)$, and hence $P(B) = P(A)$. Moreover,

$$\omega_B = B^{-1}(z)dB(z) = A^{-1}(z)L^{-1}LdA(z) = \omega_A. \tag{0.2}$$

This shows that ω_A is invariant under the left action of $GL(\mathcal{B})$ on the tuple A.

For simplicity, we shall use Einstein convention for summation in many places in this paper. For example, we shall write $\omega_A(z) = A^{-1}(z)A_j dz_j$. When $\mathcal{B}$ is a C^*-algebra, the adjoint of $\omega_A(z)$ is the $\mathcal{B}$-valued $(0,1)$-form

$$\omega_A^*(z) = (A^{-1}(z)A_j)^* d\bar{z}_j.$$

Now we define the *fundamental form* for the tuple A as

$$\Omega_A = -\omega_A^* \wedge \omega_A = (A^{-1}(z)A_j)^* A^{-1}(z)A_k dz_k \wedge d\bar{z}_j. \tag{0.3}$$

Here the negative sign exists in the first equality because it is conventional to write a $(1,1)$-form as linear combinations of $dz_j \wedge d\bar{z}_k$. Often, the factor $\frac{i}{2}$ is used to make $\frac{i}{2}\Omega_A$ a *self-adjoint* form (cf. (1.2)), but it is not important in this paper.

For a suitable choice of linear functional ϕ, such as a faithful state on $\mathcal{B}$, the evaluation

$$\phi(\Omega_A) = \phi\left((A^{-1}(z)A_j)^* A^{-1}(z)A_k\right) dz_k \wedge d\bar{z}_j$$

induces a positive definite bilinear form on the holomorphic tangent bundle of $P^c(A)$, thus giving a Hermitian metric on $P^c(A)$. The connection between the metric and the tuple A shall be the primary concern of this paper. In particular, it shows that the Kählerness of the metric is tied with the commutativity of the tuple (cf. Theorem 2.4). A notable feature of this metric is that it has singularities at the joint spectrum $P(A)$. So completeness of the metric is an important issue.

1. $\mathcal{B}$-valued differential forms

Let M be a complex manifold of dimension n. If $z = (z_1, z_2, \dots, z_n)$ is the coordinate in a local chart, then ∂_i stands for $\frac{\partial}{\partial z_i}$, and $\bar{\partial}_i$ stands for $\frac{\partial}{\partial \bar{z}_i}$. As a convention, we let $\partial = \sum_i \partial_i$, $\bar{\partial} = \sum_i \bar{\partial}_i$, and $d = \partial + \bar{\partial}$. Consider a globally defined smooth $(1,1)$-form $\Phi(z) = g_{jk}(z) dz_j \wedge d\bar{z}_k$ expressed in the local chart with the Einstein summation convention. Φ induces a bilinear form $\hat{\Phi}(z) = g_{jk}(z) dz_j \otimes d\bar{z}_k$ on the holomorphic tangent bundle $T(M)$ over the local chart such that

$$\hat{\Phi}(z)(\partial_j,\ \partial_k) = g_{jk}(z).$$

We say that Φ defines a Hermitian metric on M if the $n \times n$ matrix function $g(z) = (g_{jk}(z))$ is positive definite for each $z \in M$. In this case $g(z)$ is called the associated metric matrix, and $\frac{i}{2}\Phi(z)$ is called the fundamental form of the metric. Here, the constant $\frac{i}{2}$ is to normalize the fundamental form so that

(a) $\frac{i}{2}\Phi(z)$ is real, i.e., $\frac{i}{2}\Phi(z) = \overline{\frac{i}{2}\Phi(z)}$;
(b) in the one variable case $z = x + iy$, we have $\frac{i}{2} dz \wedge d\bar{z} = dx \wedge dy$.

Some geometric concepts are relevant to the study here.

1) A Hermitian metric induced by $\Phi(z)$ is said to be Kähler if $d\Phi = 0$, i.e., $\Phi(z)$ is closed.
2) The Ricci curvature tensor is the $n \times n$ matrix $R(z) = (R_{jk}(z))$, where $R_{jk} = -\partial_j \bar{\partial}_k \log \det g(z)$, and the Ricci form is
$$\mathrm{Ric}_\Phi(z) = R_{jk}(z) dz_j \wedge d\bar{z}_k.$$
3) The metric is said to be Ricci flat if $R(z) = 0$ for each $z \in M$.
4) The metric is said to be Einstein if $\mathrm{Ric}_\Phi = \lambda\Phi$ for some constant λ, and it is Calabi–Yau if it is Kähler and Ricci flat (i.e., $\lambda = 0$).

One checks that a Hermitian $(1,1)$-form $\Phi(z) = g_{ij}(z) dz_i \wedge d\bar{z}_j$ induces a Kähler metric if and only if

$$\partial_k g_{ij}(z) = \partial_i g_{kj}(z), \quad z \in M, \tag{1.1}$$

for all $1 \le i, j, k \le n$. Further, Φ is Ricci flat if and only if $\log \det g(z)$ is pluriharmonic (cf. [23]). Hence in this case $\log \det g(z) = f(z) + \overline{f(z)}$ holds locally for some holomorphic function f, i.e.,

$$\det g(z) = e^{f(z) + \overline{f(z)}} = |e^{f(z)}|^2.$$

Now we consider a smooth $\mathcal{B}$-valued (p,q)-form $\omega^{p,q}(z)$ defined on a complex domain $R \subset \mathbb{C}^n$, where $0 \le p, q \le n$. The set of such forms is denoted by $\Lambda^{p,q}(R, \mathcal{B})$. Then $\omega^{p,q}(z)$ is said to be ∂-closed or $\overline{\partial}$-closed if $\partial\omega^{p,q}(z) = 0$, or respectively $\overline{\partial}\omega^{p,q}(z) = 0$ on R. If $\omega^{p,q}$ is both ∂-closed and $\overline{\partial}$ closed, then $d\omega^{p,q} = 0$, and in this case we simply say $\omega^{p,q}$ is closed.

For the rest of the paper, unless stated otherwise, we shall assume $\mathcal{B}$ is a C^*-algebra. Recall that for a tuple A, the fundamental form Ω_A is given by (0.3).

Now we check that $\frac{i}{2}\Omega_A$ is *self-adjoint* in the sense that

$$\left(\frac{i}{2}\Omega_A\right)^* = \frac{i}{2}\Omega_A. \tag{1.2}$$

Indeed, one checks by (0.3) that

$$\begin{aligned}\left(\frac{i}{2}\Omega_A\right)^* &= \frac{-i}{2}(-\omega_A^* \wedge \omega_A)^* \\ &= \frac{-i}{2}\left((A^{-1}(z)A_j)^* A^{-1}(z)A_k dz_k \wedge d\overline{z}_j\right)^* \\ &= \frac{-i}{2}\left(A^{-1}(z)A_k\right)^* A^{-1}(z)A_j d\overline{z}_k \wedge dz_j) \\ &= \frac{-i}{2}(\omega_A^* \wedge \omega_A) = \frac{i}{2}\Omega_A.\end{aligned}$$

Further, in the case $n = 1$ and A_1 is invertible, one has $A(z) = zA_1$, $P^c(A) = \mathbb{C} \setminus \{0\}$, and $\omega_A(z) = \frac{dz}{z}$. Writing $z = x + yi$ we have

$$\frac{i}{2}\Omega_A = \frac{i(dz \wedge d\overline{z})}{2|z|^2} = \frac{dx \wedge dy}{x^2 + y^2},$$

which is real. For a linear functional $\phi \in \mathcal{B}^*$,

$$\phi\left(\frac{i}{2}\Omega_A\right) = \frac{i}{2}\phi\left(A^{-1}(z)A_j)^* A^{-1}(z)A_k\right) dz_k \wedge d\overline{z}_j,$$

which is a regular $(1,1)$-form. Moreover, if ϕ is positive then by (1.2)

$$\overline{\phi\left(\frac{i}{2}\Omega_A\right)} = \phi\left(\frac{-i}{2}\Omega_A^*\right) = \phi\left(\frac{i}{2}\Omega_A\right),$$

and hence $\phi\left(\frac{i}{2}\Omega_A\right)$ is a regular real $(1,1)$-form. The following definitions are $\mathcal{B}$-valued versions of Hermitian metric and Kähler metric.

Definition. A smooth $\mathcal{B}$-valued $(1,1)$-form $\Phi(z) = \Phi_{jk}(z)dz_j \wedge d\overline{z}_k$ on a complex domain R is said to be Hermitian if for every $z \in R$ and any set of n complex numbers $v_1, v_2, \ldots, v_n$ the double sum $v_i\Phi_{jk}(z)\overline{v_k}$ is a positive element in $\mathcal{B}$.

Definition. A smooth $\mathcal{B}$-valued Hermitian $(1,1)$-form $\Phi(z)$ is said to be Kähler if it is closed, i.e., $d\Phi(z) = 0$.

Two nuances are worth mentioning. First, there is a difference between a self-adjoint $\mathcal{B}$-valued $(1,1)$-form as defined in (1.2) and a Hermitian $\mathcal{B}$-valued $(1,1)$-form defined above – the latter requires positivity of $\mathcal{B}$-valued matrix $(\Phi_{ij}(z))$. Second, in the scalar-valued case, for a Hermitian metric defined by $\Omega(z) = g_{jk}(z)dz_j \wedge d\bar{z}_k$, the complex metric matrix $(g_{jk}(z))$ needs to be positive definite, but in the $\mathcal{B}$-valued case, it is only required that $v_j\Phi_{jk}(z)\overline{v_k} \geq 0$ in $\mathcal{B}$. For the fundamental form Ω_A and any set of n complex numbers $v_1, v_2, \dots, v_n$, one checks by (0.3) that the double sum

$$\begin{aligned} v_j A_k^*(A^{-1}(z))^* A^{-1}(z) A_j \bar{v_k} &= \bar{v_k} A_k^*(A^{-1}(z))^* A^{-1}(z) v_j A_j) \\ &= (A^{-1}(z)A(v))^*(A^{-1}(z)A(v) \geq 0. \end{aligned} \tag{1.3}$$

So Ω_A is Hermitian.

In this paper, the extended tuple $\hat{A} = (I, A_1, A_2, \dots, A_n)$ is used in some places for normalization purposes. In this case we can regard I as A_0 and $P(\hat{A})$ is then clearly a subset in $\mathbb{C}^{n+1}$. This extension guarantees two convenient facts:

1) obviously, $(1, 0, 0, \dots, 0) \in P^c(\hat{A})$;
2) the identity I is in the range of $\hat{A}(z)$ as z varies in $\mathbb{C}^{n+1}$.

Now we are in position to state the first result.

Proposition 1.1. *For the tuple $A = (A_1, A_2, \dots, A_n)$, the following statements are equivalent.*

(i) *A is commuting.*
(ii) *$\omega_A(z)$ is closed.*
(iii) *$\Omega_{\hat{A}}(z)$ is Kähler.*

Proof. Using (0.1) and the fact that $dz_j \wedge dz_k = -dz_k \wedge dz_j$, one checks that

$$\begin{aligned} d\omega_A &= -\omega_A \wedge \omega_A \\ &= A^{-1}(z)A_j \; A^{-1}(z)A_k dz_j \wedge dz_k \\ &= \sum_{j<k} [A^{-1}(z)A_j, \; A^{-1}(z)A_k] dz_j \wedge dz_k, \end{aligned} \tag{1.4}$$

where $[a, b] = ab - ba$. So when A is a commuting tuple we have

$$[A^{-1}(z)A_j, \; A^{-1}(z)A_k] = 0, \quad z \in P^c(A),$$

and consequently $d\omega_A = \partial\omega_A = 0$. This shows (i) implies (ii).

Recall that $\Omega_A(z) = -\omega_A^* \wedge \omega_A$. Since $\omega_A(z)$ is holomorphic

$$\begin{aligned} d\Omega_A &= -(\partial + \bar{\partial})(\omega_A^* \wedge \omega_A) \\ &= -(\partial\omega_A^* + \bar{\partial}\omega_A^*) \wedge \omega_A + \omega_A^* \wedge (\partial\omega_A + \bar{\partial}\omega_A) \\ &= -\big((\partial\omega_A)^* \wedge \omega_A\big) + (\omega_A^* \wedge \partial\omega_A). \end{aligned} \tag{1.5}$$

So if $\omega_A(z)$ is closed then $d\Omega_A = -(0 \wedge \omega_A) + (\omega_A^* \wedge 0) = 0$, i.e., Ω_A is Kähler. Since $\hat{A}$ is commuting, the forgoing arguments in particular imply that $\Omega_{\hat{A}}$ is Kähler as well. This shows that (ii) implies (iii). Now if $\Omega_{\hat{A}}$ is Kähler then by (1.5)

$$0 = d\Omega_{\hat{A}} = -(\bar{\partial}\omega_{\hat{A}}^*) \wedge \omega_{\hat{A}} + \omega_{\hat{A}}^* \wedge \partial\omega_{\hat{A}}. \tag{1.6}$$

Since the first term in (1.6) is a $(1,2)$-form and the second is a $(2,1)$-form, the sum is 0 only if both terms are 0. Furthermore, by (1.4) and (1.6) we have

$$0 = \omega_{\hat{A}}^* \wedge d\omega_{\hat{A}} = (\hat{A}^{-1}(z)A_i)^* \sum_{0 \leq j < k} [\hat{A}^{-1}(z)A_j, \; \hat{A}^{-1}(z)A_k] d\bar{z}_i \wedge dz_j \wedge dz_k,$$

which implies that on $P^c(\hat{A})$,

$$(\hat{A}^{-1}(z)A_i)^*[\hat{A}^{-1}(z)A_j,\ \hat{A}^{-1}(z)A_k] = 0, \quad \forall\ 0 \le i,\ j,\ k \le n.$$

Setting $i = 0$ and using the fact $\hat{A}(z) = A_0 = I$ at $z = (1, 0, 0, \dots, 0) \in \mathbb{C}^{n+1}$, we have

$$A_k A_j = A_j A_k,\ \forall 1 \le j,\ k \le n,$$

i.e., the tuple A is commuting. So (iii) implies (i). □

To prove that (iii) implies (i), we used the fact that the tuple $\hat{A}$ includes I. In fact, as the proof indicates that if $A(z) = I$ for some $z \in \mathbb{C}^n$ then there is no need to use $\hat{A}$. But this kind of requirement is indispensable.

Example 1.2. Consider a pair $A = (A_1, A_2)$, where A_1 is invertible. Then writing $A(z) = A_1(z_1 + z_2 A_1^{-1} A_2)$ and using (0.2), we have

$$\omega_A = (z_1 I + z_2 A_1^{-1} A_2)^{-1} d(z_1 I + z_2 A_1^{-1} A_2).$$

Since I commutes with $A_1^{-1}A_2$, the fundamental form Ω_A is Kähler by Proposition 1.1. But clearly A_1 may not commute with A_2.

For every $\phi \in \mathcal{B}^*$, we have $d\phi(\omega_A) = \phi(d\omega_A)$. We let $H^q(P^c(A), \mathbb{C})$ denote the qth de Rham cohomology of $P^c(A)$. The following is then an easy consequence of Proposition 1.1.

Corollary 1.3. *Let A be a commuting tuple. Then $\phi(\omega_A) \in H^1(P^c(A), \mathbb{C})$ for every $\phi \in \mathcal{B}^*$.*

Proposition 1.1 raises a natural question: is ω_A exact, i.e., is there a $\mathcal{B}$-valued smooth function $f(z)$ on $P^c(A)$ such that $\omega_A = df$? We take time to address this issue now. For every $\phi \in \mathcal{B}^*$ and a $\mathcal{B}$-valued smooth (p, q)-form $\omega^{p,q}(z)$ on a complex domain R, the coupling $\phi(\omega^{p,q}(z))$ is a scalar-valued smooth (p, q)-form on R. For example by (0.3),

$$\phi(\Omega_A) = \phi\left(A^{-1}(z)A_j)^* A^{-1}(z)A_k\right) dz_k \wedge d\overline{z}_j.$$

Moreover, $\omega^{p,q}(z) = 0$ if and only if $\phi(\omega^{p,q}(z)) = 0$ for every $\phi \in \mathcal{B}^*$. We let the set of $\mathcal{B}$-valued (p, q)-forms on R be denoted by $\omega^{p,q}(R, \mathcal{B})$. The fact

$$\partial\phi(\omega^{p,q}(z)) = \phi(\partial\omega^{p,q}(z))$$

is nicely expressed in the following commuting diagram

$$\begin{array}{ccc} \Lambda^{p,q}(R, \mathcal{B}) & \xrightarrow{\partial_p} & \Lambda^{p+1,q}(R, \mathcal{B}) \\ \downarrow \phi & & \downarrow \phi \\ \Lambda^{p,q}(R, \mathbb{C}) & \xrightarrow{\partial_p} & \Lambda^{p+1,q}(R, \mathbb{C}), \end{array} \tag{1.7}$$

and the parallel diagram regarding $\overline{\partial}$. Here the subscript p in ∂_p is to indicate the space on which ∂ acts. Since $d = \partial + \overline{\partial}$, we see that $\omega^{p,q}(z)$ is closed, i.e., $d\omega^{p,q} = 0$, if and only if $\phi(\omega^{p,q}(z))$ is closed for every $\phi \in \mathcal{B}^*$. Since by [20, 30] the joint resolvent set $P^c(A)$ is a union of domains of holomorphy, we shall only consider the case when R is a domain of holomorphy, in which case its de Rham cohomology $H^p(R,\mathbb{C})$ (or $H^p(R,\mathcal{B})$) can be computed through holomorphic forms. We denote the set of $\mathcal{B}$-valued holomorphic p-forms by $\Lambda^p(R,\mathcal{B})$. Note again that in this case $d=\partial$. We refer the readers to [23] for details on domains of holomorphy and holomorphic forms.

Since $\partial_p\partial_{p-1}=0$, one naturally has $\mathcal{B}$-valued cohomology groups

$$H^p(R,\mathcal{B}) = \ker \partial_p / im\partial_{p-1}, \quad p \geq 0,$$

where ∂_{-1} is the trivial inclusion map $\{0\} \to \Lambda^0(R,\mathcal{B})$. It then follows from the following commuting diagram for holomorphic forms

$$\begin{array}{ccc} \Lambda^p(R,\mathcal{B}) & \xrightarrow{\partial_p} & \Lambda^{p+1}(R,\mathcal{B}) \\ \downarrow \phi & & \downarrow \phi \\ \Lambda^p(R,\mathbb{C}) & \xrightarrow{\partial_p} & \Lambda^{p+1}(R,\mathbb{C}), \end{array}$$

that a form ω is in $H^p(R,\mathcal{B})$ if and only if $\phi(\omega)$ is in $\Lambda^p(R,\mathbb{C})$ for every $\phi \in \mathcal{B}^*$. It was shown in [29] that if $\mathcal{B}$ is a Banach algebra with a trace ϕ, then for every tuple A such that $P^c(A)$ is nonempty, the 1-form $\phi(\omega_A)$ is a non-trivial element in $H^1(P^c(A),\mathbb{C})$. The above observations thus lead to the following.

Corollary 1.4. *Let $\mathcal{B}$ be a unital Banach algebra with a trace and A be a commuting tuple of elements in $\mathcal{B}$. Then ω_A is a non-trivial element in $H^1(P^c(A),\mathcal{B})$.*

Proof. By Proposition 1.1, we see that ω_A is closed when A is a commuting tuple. If it were exact, then there exists $\mathcal{B}$-valued smooth function $f(z)$ on $P^c(A)$ such that $\omega_A = df$. If ϕ is the trace, then it follows that

$$\phi(\omega_A) = \phi(df) = d\phi(f).$$

Since $\phi(f)$ is globally defined on $P^c(A)$, the 1-form $\phi(\omega_A)$ is a trivial element in $H^1(P^c(A),\mathbb{C})$ contradicting with the above mentioned fact in [29] that $\phi(\omega_A)$ is a non-trivial element in $H^1(P^c(A),\mathbb{C})$. □

Proposition 1.1 indicates that the commutativity of tuples A has a natural homological interpretation.

2. Hermitian metric on $P^c(A)$

Recall that for a C^*-algebra $\mathcal{B}$, a smooth $\mathcal{B}$-valued (1,1)-form

$$\Phi(z) = \Phi_{jk}(z)dz_j \wedge d\bar{z}_k$$

on a domain R is said to be Hermitian if for any set of n complex numbers $v_1, v_2, \ldots, v_n$ the double sum $v_j\Phi_{jk}(z)\overline{v_k}$ is a positive element in $\mathcal{B}$ for every $z \in R$. It follows that if $\phi \in \mathcal{B}^*$ is a positive linear functional, then the matrix $\left(\phi(\Phi_{jk}(z))\right)$ is positive semi-definite for every $z \in R$. By (1.3), we see that the fundamental

(1,1)-form $\Omega_A = -\omega_A^* \wedge \omega_A$ is Hermitian. This section defines some natural Hermitian metrics on $P^c(A)$ through Ω_A and its pairing with certain states on $\mathcal{B}$.

A bounded linear functional ϕ on a C^*-algebra $\mathcal{B}$ is called a state if it is positive and $\phi(I) = 1$. Consider a tuple $A = (A_1, A_2, \ldots, A_n)$ of elements in $\mathcal{B}$. Given a state ϕ, then on $P^c(A)$ we compute that

$$\begin{aligned}\phi(\Omega_A(z)) &= -\phi(\omega_A^* \wedge \omega_A)\\ &= \phi(A_k^*(A^{-1}(z))^* A^{-1}(z) A_j) dz_j \wedge d\bar{z}_k\\ &:= g_{jk}(z) dz_j \wedge d\bar{z}_k. \end{aligned} \tag{2.1}$$

As observed in the previous paragraph, the $n \times n$ matrix $g(z) = (g_{jk}(z))$ is positive semi-definite for every $z \in P^c(A)$. So it becomes a natural question for which ϕ the matrix $g(z)$ is positive definite on $P^c(A)$. In other words, when does $\phi(\Omega_A)$ define a Hermitian metric on $P^c(A)$? To answer this question, we consider the operator space $\mathcal{H}_A = \text{span}\{A_1, A_2, \ldots, A_n\}$. A state ϕ on $\mathcal{B}$ is said to be *faithful on* $\mathcal{H}_A$ if for every nonzero element $h \in \mathcal{H}_A$ we have $\phi(h^* h) > 0$.

Proposition 2.1. *Let ϕ be a state on a C^*-algebra $\mathcal{B}$. Then for any tuple A the (1,1)-form $\phi(\Omega_A)$ induces a Hermitian metric on $P^c(A)$ if and only if ϕ is faithful on $\mathcal{H}_A$.*

Proof. For every fixed $z \in P^c(A)$, since $A(z) = z_1 A_1 + z_2 A_2 + \cdots + z_n A_n$ is invertible, there are numbers $\alpha > \beta > 0$ such that

$$\beta I \leq (A^{-1}(z))^* A^{-1}(z) \leq \alpha I.$$

Note that α and β depend on z, but it is not important here. For every nonzero vector $v = (v_1, v_2, \ldots, v_n) \in \mathbb{C}^n$, one checks that

$$\begin{aligned} v_j g_{jk}(z) \bar{v_k} &= v_j \phi(A_k^*(A^{-1}(z))^* A^{-1}(z) A_j) \bar{v_k}\\ &= \phi(\bar{v_k} A_k^*(A^{-1}(z))^* A^{-1}(z) v_j A_j)\\ &= \phi\left((A^{-1}(z)A(v))^*(A^{-1}(z)A(v))\right)\\ &= \phi\left(A^*(v)(A^{-1}(z))^* A^{-1}(z) A(v)\right). \end{aligned}$$

Since

$$\beta A^*(v)A(v) \leq A^*(v)(A^{-1}(z))^* A^{-1}(z) A(v) \leq \alpha A^*(v) A(v)$$

and ϕ is positive, it follows that

$$\beta \phi(A^*(v)A(v)) \leq v_j g_{jk}(z) \bar{v_k} \leq \alpha \phi(A^*(v)A(v)).$$

Hence $(g_{jk}(z))$ is positive definite for every $z \in P^c(A)$ if and only if ϕ is faithful on $\mathcal{H}_A$. □

Example 2.2. A state ϕ on a C^*-algebra $\mathcal{B}$ is called a faithful tracial state if $\phi(ab) = \phi(ba)$ for any $a, b \in \mathcal{B}$ and $\phi(a^* a) > 0$ when $a \neq 0$. If ϕ is a faithful tracial state, then it is clearly a faithful state on $\mathcal{H}_A$. In this case, we shall write the metric matrix $(g_{ij}(z))$ as $g_A(z)$.

Example 2.3. Assume $\mathcal{B}$ is a C^*-algebra of operators acting on a Hilbert space $\mathcal{H}$. Let ϕ_x be a vector state defined on $\mathcal{B}$ by $\phi_x(a) = \langle ax, x\rangle$, where $a \in \mathcal{B}$ and x is fixed with $\|x\| = 1$. If $\{A_1x, A_2x, \ldots, A_nx\}$ is linearly independent, then for any nonzero vector $(c_1, c_2, \ldots, c_n) \in \mathbb{C}^n$, we have

$$\phi_x((c_iA_i)^*(c_iA_i)) = \|c_iA_ix\|^2 > 0,$$

i.e., ϕ is faithful on $\mathcal{H}_A$.

It follows from Propositions 1.1, 2.1 and the commuting diagram (1.7) that for a commuting tuple $A = (A_1, A_2, \ldots, A_n)$, and every state ϕ faithful on $\mathcal{H}_{\hat{A}}$,

$$d\phi(\Omega_{\hat{A}}) = \phi(d\Omega_{\hat{A}}) = 0.$$

Hence $\phi(\Omega_{\hat{A}})$ defines a Kähler metric on $P^c(\hat{A})$. Interestingly, the converse is also true if ϕ is faithful on the entire $\mathcal{B}$.

Theorem 2.4. *Let ϕ be a faithful state on $\mathcal{B}$. Then $\hat{A} = (I, A_1, A_2, \ldots, A_n)$ is commuting if and only if $\phi(\Omega_{\hat{A}})$ is Kähler.*

Proof. It only remains to check sufficiency. If $\phi(\Omega_{\hat{A}})$ is Kähler, then we have $0 = d\phi(\Omega_{\hat{A}}) = \phi(d\Omega_{\hat{A}})$. Writting I as A_0, then by (1.6), we have

$$\begin{aligned} 0 &= \phi(\omega_{\hat{A}}^* \wedge \partial\omega_{\hat{A}}) \\ &= \sum_i \sum_{0\le j<k} \phi\left((\hat{A}^{-1}(z)A_i)^*[\hat{A}^{-1}(z)A_j,\ \hat{A}^{-1}(z)A_k]\right) d\bar{z}_i \wedge dz_j \wedge dz_k, \end{aligned}$$

which implies

$$\phi\left((\hat{A}^{-1}(z)A_i)^*[\hat{A}^{-1}(z)A_j,\ \hat{A}^{-1}(z)A_k]\right) = 0, \quad \forall 0 \le i, j, k \le n. \tag{2.2}$$

Applying $\bar{\partial}_m$ to both sides of (2.2) and using the fact $\bar{\partial}_m\hat{A}^{-1}(z) = 0$ and

$$\bar{\partial}_m(\hat{A}^{-1}(z))^* = (\partial_m\hat{A}^{-1}(z))^* = -(\hat{A}^{-1}(z)A_m\hat{A}^{-1}(z))^*$$

we have

$$\phi\left((\hat{A}^{-1}(z)A_m\hat{A}^{-1}(z)A_i)^*[\hat{A}^{-1}(z)A_j,\ \hat{A}^{-1}(z)A_k]\right) = 0,\ \forall 0 \le i, j, k, m \le n. \tag{2.3}$$

In (2.3), setting (m, i) to (j, k), then to (k, j) and taking the difference, we have

$$\phi\left([\hat{A}^{-1}(z)A_j,\ \hat{A}^{-1}(z)A_k]^*[\hat{A}^{-1}(z)A_j,\ \hat{A}^{-1}(z)A_k]\right) = 0, \quad \forall j, k.$$

Since ϕ is faithful, $[\hat{A}^{-1}(z)A_j, \hat{A}^{-1}(z)A_k]{=}0$ for all j, k. Setting z to $(1, 0, 0, \ldots, 0)$, we have $[A_j, A_k] = 0$ for all j, k. □

Again, as indicated by Example 1.2 and the remarks before it, the requirement that I be in the tuple A (or in $\mathcal{H}_A$) is indispensable in Theorem 2.4. We end this section by a question motivated by Theorem 2.4.

Problem 1. Find conditions on commuting tuple A and state ϕ such that $\phi(\Omega_A)$ defines a Ricci-flat metric (i.e., Calabi–Yau metric) on $P^c(A)$.

3. Completeness

Once a metric is defined on a set, an immediate question is whether the set is complete under the metric. In this section we will study this problem for $P^c(A)$ with respect to the metric defined by g_A as in Example 2.2. Since $\omega_A(z) = A^{-1}(z)dA(z)$ resembles the derivative of logarithmic function, it is natural to expect that log shall play an important role here.

In [12], Fuglede and Kadison defined the following notion of determinant

$$\det x = \exp(\phi(\log |x|))$$

for invertible elements x in a finite von Neumann algebra $\mathcal{B}$ with a normalized trace ϕ. Here $|x| = \sqrt{x^*x}$ and $\log |x|$ is defined by the functional calculus, i.e.,

$$\log |x| = \int_{\sigma(|x|)} \log \lambda dE(\lambda),$$

where $E(\lambda)$ is the associated projection-valued spectral measure. Then

$$\phi(\log |x|) = \int_{\sigma(|x|)} \log \lambda d\phi(E(\lambda)). \tag{3.1}$$

Since x is invertible, the spectrum $\sigma(|x|)$ is bounded away from 0. Hence the integral in (3.1) is greater than $-\infty$, which means $\det x \neq 0$. However, there are non-invertible (singular) elements x for which the improper integral in (3.1) is convergent (hence $\det x \neq 0$). This is essentially due to the absolute convergence of

$$\int_0^1 \log t \, dt.$$

An example of such element is given in [12]. For elements x such that the integral in (3.1) is equal to $-\infty$, $\det x$ is naturally defined to be 0. This extends det to all elements in $\mathcal{B}$, and by [12] it is continuous at non-singular elements and upper semi-continuous at singular elements with respect to the norm topology of $\mathcal{B}$.

FK-determinant has been well studied in many papers, and we refer readers to [15] for a survey. In particular, it was generalized to C^*-algebras in [18] as follows. Assume $GL(\mathcal{B})$ is path-connected, $x \in GL(\mathcal{B})$ and $x(t)$ is a piece-wise smooth path in $GL(\mathcal{B})$ such that $x(0) = I$ and $x(1) = x$, then when the quantity

$$\det_* x = \exp\left(\int_0^1 \phi(x^{-1}(t)dx(t))\right)$$

is independent of path $x(t)$, it defines a notion of determinant for invertible elements x. One sees that $\det_*$ may take on complex values, and it is shown in [15] that

$$|\det_* x| = \exp\left(\mathrm{Re}\left(\int_0^1 \phi(x^{-1}(t)dx(t))\right)\right) = \det x.$$

Clearly, $\det I = 1$, and for a fixed $0 \leq s \leq 1$ we have

$$\det_* x(s) = \exp\left(\int_0^s \phi\left(x^{-1}(t)dx(t)\right)\right),$$

so it follows that

$$\phi(x^{-1}(t)dx(t)) = d\log \det_* x(t). \tag{3.2}$$

Here d is the differential with respect to t. The following definition is needed for our study.

Definition. Consider a C^*-algebra $\mathcal{B}$ with a faithful tracial state ϕ. Then an element x will be called ϕ-singular if its FK-determinant $\det x = 0$. Furthermore, for a tuple A of elements in $\mathcal{B}$, a point $p \in P(A)$ is said to be ϕ-singular if $A(p)$ is ϕ-singular.

Now we begin to study the completeness of the metric defined by g_A on $P^c(A)$ in Example 2.2. Let $[P^c(A)]$ denote the completion of $P^c(A)$ with respect to the metric g_A. First, consider two points p and q that lie in the same connected component of $P^c(A)$, and let $\gamma = \{z(t) : \ 0 \le t \le 1\}$ be a piecewise smooth path such that $z(0) = p$ and $z(1) = q$. The length of γ with respect to the metric $g_A = (g_{ij})$ is

$$L(\gamma) = \int_0^1 \sqrt{z_i'(t) g_{ij}(z(t)) \overline{z_j'(t)}}dt.$$

The distance from p to q with respect to the metric g_A is the infimum of $L(\gamma)$ over all such paths, i.e.,

$$\operatorname{dist}(p,q) = \inf_\gamma L(\gamma).$$

Lemma 3.1. *Let $\mathcal{B}$ be a C^*-algebra with a faithful tracial state ϕ. If p and q are in the same connected component of $P^c(A)$ then*

$$\operatorname{dist}(p,q) \ge |\phi(\log|A(p)|) - \phi(\log|A(q)|)|.$$

Proof. First of all, by (2.1) we have $g_{ij}(z) = \phi\big((A^{-1}(z)A_j)^* A^{-1}(z)A_i\big)$. Let p and q be two points in the connected component U of $P^c(A)$, and let $\gamma = \{z(t) | \ 0 \le t \le 1\}$ be a piecewise smooth path such that $z(0) = p$ and $z(1) = q$. Then on γ we have $A'(z(t)) = z_j'(t)A_j$, and the length of γ can be computed as

$$\begin{aligned}
L(\gamma) &= \int_0^1 \sqrt{z_i'(t) g_{ij}(z(t)) \overline{z_j'(t)}}dt \\
&= \int_0^1 \sqrt{z_i'(t)\phi\big((A^{-1}(z)A_j)^* A^{-1}(z)A_i\big)\overline{z_j'(t)}}dt \\
&= \int_0^1 \sqrt{\phi\big((A^{-1}(z(t))z_j'(t)A_j)^* A^{-1}(z(t))z_i'(t)A_i\big)}dt \\
&= \int_0^1 \sqrt{\phi\big(\,(A^{-1}(z(t))A'(z(t))^*\, A^{-1}(z(t))A'(z(t))\big)}dt.
\end{aligned} \tag{3.3}$$

Since ϕ is a state, for every $a, b \in \mathcal{B}$ one has $|\phi(ab)|^2 \leq \phi(a^*a)\phi(b^*b)$. In particular, $|\phi(a)|^2 \leq \phi(a^*a)$. Hence

$$L(\gamma) \geq \int_0^1 |\phi(A^{-1}(z(t))A'(z(t))|dt = \int_0^1 |\phi[A^{-1}(z(t))dA(z(t))]|$$
$$= \int_0^1 |\phi(\omega_A(z(t))| \geq \left|\int_0^1 \phi(\omega_A(z(t))\right|.$$

By (3.2) we have

$$\begin{aligned} L(\gamma) &\geq \left|\int_0^1 d\log \det_* A(z(t))\right| \\ &= \left|\log \det_* A(p) - \log \det_* A(q)\right| \\ &= \Big|\big(\log|\det_* A(p)| - \log|\det_* A(q)|\big) \\ &\quad + i\big(\operatorname{Arg}\det_* A(p) - \operatorname{Arg}\det_* A(q)\big)\Big| \\ &\geq \left|\log|\det_* A(p)| - \log|\det_* A(q)|\right| \\ &= |\log \det A(p) - \log \det A(q)|\,. \end{aligned}$$

Since γ is arbitrary, we have

$$\begin{aligned} \operatorname{dist}(p,q) = \inf_\gamma L(\gamma) &\geq |\log \det A(p) - \log \det A(q)| \\ &= |\phi(\log|A(p)|) - \phi(\log|A(q)|)|\,. \end{aligned}$$

□

The proof of Lemma 3.1 can be modified a little to accommodate the case in which q is a boundary point of a path-connected component of $P^c(A)$. First, we observe that in this case $q \in P(A)$. Suppose there exists a piece-wise smooth path $z(t)$ such that $z(t) \in P^c(A)$, $0 \leq t < 1$ but $q = z(1) \in P(A)$. Then by the proof above we have

$$\begin{aligned} L(\gamma) &\geq \lim_{s\to 1^-} \int_0^s |\phi(\omega_A(z(t))| \\ &\geq \limsup_{s\to 1^-} |\log \det A(p) - \log \det A(z(s))|\,. \end{aligned} \tag{3.4}$$

If q is a ϕ-singular point, we have $\det A(z(1)) = \det A(q) = 0$. Note that 0 is the minimum possible value for det. Hence the upper semi-continuity of $\det A(z(s))$ implies

$$\lim_{s\to 1} \det A(z(s)) = 0.$$

Therefore by (3.4) we have $L(\gamma) = \infty$ for every such path γ in $P^c(A)$ connecting $p \in P^c(A)$ to $q \in P(A)$, and consequently

$$\operatorname{dist}(p,q) = \inf_\gamma L(\gamma) = \infty.$$

Since q has infinite distance to every $p \in P^c(A)$, we have the following.

Theorem 3.2. *For a tuple A in a C^*-algebra $\mathcal{B}$ with a faithful tracial state ϕ, if $q \in \partial P^c(A)$ is ϕ-singular, then $q \notin [P^c(A)]$.*

If $\mathcal{B}$ is a matrix subalgebra in $M_k(\mathbb{C})$, we let Tr and det stand for the ordinary trace and respectively determinant of $k \times k$ matrices. Then for any n-tuple A, the joint spectrum $P(A)$ is the hypersurface $\{\det A(z) = 0\}$ in $\mathbb{C}^n$. Clearly, $\partial P^c(A) = P(A)$. Let $\phi = \frac{1}{k}\operatorname{Tr}$. Then ϕ is a tracial state on $M_k(\mathbb{C})$, and in this case the FK-determinant and the usual determinant satisfies the relation $\det x = |\det x|^{1/k}$ (cf. [12]). So in this case every point in $P(A)$ is ϕ-singular. Since the factor $1/k$ is not important, we have the following.

Corollary 3.3. *For every tuple A of $k \times k$ matrices, the metric on $P^c(A)$ defined by $\operatorname{Tr}(\Omega_A)$ is complete.*

The general linear group $GL(k, \mathbb{C})$ is an open subset in $M_k(\mathbb{C})$, hence it is clearly not complete with respect to the usual Euclidean metric. Corollary 3.3 implies that we can endow $GL(k)$ with a complete metric.

Corollary 3.4. *Let $A_1, A_2, \dots, A_n$ be a basis for the vector space $M_k(\mathbb{C})$, where $n = k^2$. Then $\operatorname{Tr}(\Omega_A)$ defines a complete, left invariant, Ricci flat, but non-Kähler metric on the complex general linear group $GL(k)$.*

Proof. First, since $A_1, A_2, \dots, A_n$ is a basis for the matrix algebra $M_k(\mathbb{C})$, every matrix is of the form $A(z)$ for some z. Hence $z \in P^c(A)$ if and only if $A(z) \in GL(k)$. And this provides an identification of $P^c(A)$ with $GL(k)$. So in particular $\operatorname{Tr}(\Omega_A)$ defines a metric on $GL(k)$. Recall that

$$\operatorname{Tr}(\Omega_A) = -\operatorname{Tr}(\omega_A^* \wedge \omega_A) = \operatorname{Tr}\left((A^{-1}(z)A_j)^* A^{-1}(z)A_i\right) dz_i \wedge d\bar{z}_j.$$

For $1 \le i \le n$, we write $A_i = (a_1^i, a_2^i, \dots, a_k^i)$, where a_m^i, $1 \le m \le k$, is the mth column of A_i. Then

$$\begin{aligned}(A^{-1}(z)A_j)^* A^{-1}(z)A_i =& \left(A^{-1}(z)a_1^j, A^{-1}(z)a_2^j, \dots, A^{-1}(z)a_k^j\right)^* \\ & \times \left(A^{-1}(z)a_1^i, A^{-1}(z)a_2^i, \dots, A^{-1}(z)a_k^i\right),\end{aligned}$$

and hence

$$\operatorname{Tr}\left((A^{-1}(z)A_j)^* A^{-1}(z)A_i\right) = \sum_m \langle A^{-1}(z)a_m^i,\ A^{-1}(z)a_m^j\rangle_{\mathbb{C}^k}.$$

Let α be the $n \times n$-matrix $\left(a_m^i\right)_{1 \le m \le k, 1 \le i \le n}$, whose ith column (in $\mathbb{C}^n$) is denoted by α_i, and let I_k be the $k \times k$ identity matrix. Then $A^{-1}(z) \otimes I_k$ is the $n \times n$ block matrix with $A^{-1}(z)$ on the diagonal. So we can write the sum in (3.3) as

$$\langle (A^{-1}(z) \otimes I_k)\alpha_i,\ (A^{-1}(z) \otimes I_k)\alpha_j\rangle_{\mathbb{C}^n}.$$

Since $A_1, A_2, \dots, A_n$ are linearly independent, α is invertible, and one checks by direct computation that the metric matrix

$$\begin{aligned} g(z) &= \big(\mathrm{Tr}((A^{-1}(z)A_j)^* A^{-1}(z)A_i\big)_{n\times n} \\ &= \big(\langle (A^{-1}(z)\otimes I_k)\alpha_i,\ (A^{-1}(z)\otimes I_k)\alpha_j\rangle\big)_{n\times n} \\ &= \alpha^*(A^{-1}(z)\otimes I_k)^*(A^{-1}(z)\otimes I_k)\alpha. \end{aligned}$$

It follows that $\det g(z) = |\det\alpha|^2 |\det A^{-k}(z)|^2$. Since $\det A^{-k}(z)$ is holomorphic on $P^c(A)$, the metric defined by $\mathrm{Tr}\,\Omega_A$ is Ricci flat by the remark after (1.1).

Moreover, because $A_1, A_2, \dots, A_n$ is a basis for $M_k(\mathbb{C})$, for every fixed $x \in GL(k)$, $xA(z) = A(w)$, where $w \in P^c(A)$ and is uniquely determined by z. This gives rise to a natural action L of $GL(k)$ on $P^c(A)$ given by $L_x(z) = w$. Since by (0.2), ω_A is invariant under left action of $GL(k)$, the metric defined by $\mathrm{Tr}\,\Omega_A$ is invariant under the action of L. The completeness of the metric follows from Corollary 3.3. Since I is of the form $A(z)$ for some $z \in \mathbb{C}^n$ in this case, the non-Kählerness follows from Theorem 2.4. □

We end the paper by posing the following.

Problem 2. Is the converse of Theorem 3.2 true?

Acknowledgment

The second-named author would like to thank Guoliang Yu and the Department of Mathematics at Texas A&M University for their support and hospitality during his visit.

References

[1] M. Andersson and J. Söstrand, *Functional calculus for non-commuting operators with real spectra via an iterated Cauchy formula.* J. Funct. Anal. **210** (2004), 341–375.

[2] F.V. Atkinson, *Multiparameter Eigenvalue Problems.* Academic Press, New York and London, 1972.

[3] P. Cade, J. Bannon, and R. Yang, *On the spectrum of operator-valued entire functions.* Illinois J. Math. **55** (2011), 1455–1465.

[4] P. Cade and R. Yang, *Projective spectrum and cyclic cohomology.* J. Funct. Anal. **265** (2013), 1916–1933.

[5] I. Chagouel, M. Stessin, and K. Zhu, *Geometric spectral theory for compact operators.* Trans. Amer. Soc. **368** (2016), 1559–1582.

[6] M. Cowen and R.G. Douglas, *Complex geometry and operator theory.* Acta Math. **141** (1978), 187–261.

[7] J. Cuntz, *Simple C^*-algebras generated by isometries.* Comm. Math. Phys. **57** (1977), 173–185.

[8] R.G. Douglas, *Invariants for Hilbert modules.* In: "Operator Theory: Operator Algebras and Applications", Part 1 (Durham, NH, 1988), Proc. Sympos. Pure Math., 51, Part 1, Amer. Math. Soc., Providence, RI, 1990, pp. 179–196.

[9] R.G. Douglas and G. Misra, *Geometric invariants for resolutions of Hilbert modules.* In: "Nonselfadjoint Operator Algebras, Operator Theory, and Related Topics". Oper. Theory Adv. Appl. **104**, 1998, pp. 83–112.

[10] R.G. Douglas and V. Paulsen, *Hilbert Modules Over Function Algebras.* Pitman Research Notes in Mathematics Series, 217. Longman Scientific & Technical, Harlow; copublished in the United States with John Wiley & Sons, Inc., New York, 1989.

[11] A. Fainshtein, *Taylor joint spectrum for family of operators generating nil-potent Lie algebras.* J. Oper. Theory **29** (1993), 3–27.

[12] B. Fuglede and R. Kadison, *Determinant theory in finite factors*, Annals Math. **55** (1952), 520–530.

[13] R. Grigorchuk and Z. Šunić, *Schreier spectrum of the Hanoi Towers group on three pegs.* Proc. of Symposia in Pure Math. Vol. 77, 2008, pp. 183–198.

[14] R. Grigorchuk and R. Yang, *Joint spectrum and infinite dihedral group.* Proc. Steklov Institute Math. **297** (2017), 145–178.

[15] P. de la Harpe, *The Fuglede–Kadison determinant, theme and variations.* Proc. Nat. Academy Sci. USA **110** (2013), 15864–15877.

[16] R. Harte, *Spectral mapping theorems.* Proc. Royal Irish Acad. **72** (1972), 89–107.

[17] R. Harte, *Spectral mapping theorems for quasi-commuting systems.* Proc. Royal Irish Acad. **73** (1973), 7–18.

[18] P. de la Harpe and G. Skandalis, *Déterminant associé à une trace sur une algèbre de Banach.* Ann. Inst. Fourier (Grenoble) **34** (1984), 241–260 (in French).

[19] W. He, X. Wang, and R. Yang, *Projective spectrum and kernel bundle* (II). J. Oper. Theory, to appear.

[20] W. He and R. Yang, *Projective spectrum and kernel bundle.* Sci. China. Math. **57** (2014), 1–10.

[21] B. Jefferies, *Spectral Properties of Noncommuting Operators.* Lecture Notes in Math, 1843, Springer-Verlag, Berlin, 2004.

[22] A. Marcus, D. Spielman, and N. Srivastava, *Interlacing families II: Mixed characteristic polynomials and the Kadison–Singer problem.* Annals Math. **182** (2015), 327–350.

[23] R.M. Range, *Holomorphic Functions and Integral Representations in Several Complex Variables.* Graduate Texts in Mathematics 108, Springer-Verlag, New York, 1986.

[24] J. Sarkar, *An introduction to Hilbert module approach to multivariable operator theory.* In: "Operator Theory". Spinger, Basel, 2015, pp. 969–1033.

[25] M. Stessin, K. Zhu, and R. Yang, *Analyticity of a joint spectrum and a multivariable analytic Fredholm theorem.* New York J. Math. **17A** (2011), 39–44.

[26] J.L. Taylor, *A joint spectrum for several commutative operators.* J. Funct. Anal. **6** (1970), 172–191.

[27] J.L. Taylor, *A general framework for multi-operator functional calculus.* Adv. Math. **9** (1972), 137–182.

[28] V. Vinnikov, *Determinantal representation of algebraic curves.* In: "Linear Algebra in Signals, Systems and Control (Boston, MA, 1986)", SIAM, Philadelphia, PA, 1988, pp. 73–99.

[29] R. Yang, *Projective spectrum in Banach algebras.* J. Topol. Anal. **1** (2009), 289–306.

[30] M. Zaidenberg, S. Krein, P. Kuchment, and A. Pankov, *Banach bundles and linear operators.* Russian Math. Surveys **30** (1975), 115–175.

Ronald G. Douglas
Department of Mathematics
Texas A&M University
College Station, TX 77843, USA
e-mail: `rdouglas@math.tamu.edu`

Rongwei Yang
School of Mathematical Sciences
Tianjin Normal University
Tianjin 300387, P.R. China

and

Department of Mathematics and Statistics
SUNY at Albany
Albany, NY 12222, USA
e-mail: `ryang@albany.edu`

Operator Theory:
Advances and Applications, Vol. 267, 185–198

Spectral Algorithms for MRA Orthonormal Wavelets

F. Gómez-Cubillo and S. Villullas

Abstract. Operator techniques lead to spectral algorithms to compute scaling functions and wavelets associated with multiresolution analyses (MRAs). The spectral algorithms depend on the choice of pairs of suitable orthonormal bases (ONBs). This work presents the spectral algorithms for three different pairs of ONBs: Haar bases, Walsh–Paley bases and trigonometric bases. The Walsh–Paley bases connect wavelet theory and dyadic harmonic analysis. The results for trigonometric bases are the first viable attempt to do a discrete Fourier analysis of the problem.

Mathematics Subject Classification (2010). Primary 42C40; Secondary 47N40.

Keywords. Orthonormal wavelets, multiresolution analysis, spectral algorithms.

1. Introduction

On the Hilbert space $L^2(\mathbb{R})$ of square integrable complex functions consider the translation and dilation operators, T and D, given by

$$[Tf](x) := f(x-1)\,, \quad [Df](x) := 2^{1/2}\, f(2x)\,, \quad (f \in L^2(\mathbb{R}))\,.$$

By an *orthonormal wavelet on* $\mathbb{R}$ we mean here a function $\psi \in L^2(\mathbb{R})$ such that the family of dilated-translated versions of ψ

$$\{\psi_{j,k}(x) := [D^j T^k \psi](x) = 2^{j/2}\psi(2^j x - k)\}_{j,k\in\mathbb{Z}}$$

is an orthonormal basis (shortly ONB) of $L^2(\mathbb{R})$.

Certain orthonormal wavelets ψ, in particular, the compactly supported orthonormal wavelets, are related to *multiresolution analyses* (MRAs), i.e., increasing sequences $\{V_j\}_{j\in\mathbb{Z}}$ of closed subspaces of $L^2(\mathbb{R})$ such that $\cap_{j\in\mathbb{Z}} V_j = \{0\}$, $\overline{\cup_{j\in\mathbb{Z}} V_j} = L^2(\mathbb{R})$, $f \in V_0 \Leftrightarrow f_{j,0} = D^j f \in V_j$, and there exists a function

This work was partially supported by research projects MTM2012-31439 and MTM2014-57129-C2-1-P (Secretaría General de Ciencia, Tecnología e Innovación, Ministerio de Economía y Competitividad, Spain).

$\varphi \in L^2(\mathbb{R})$, the *scaling function*, whose integer translates $\{\varphi_{0,k} = T^k\varphi\}_{k\in\mathbb{Z}}$ form an ONB of V_0. In such case, $\psi_{j,k} \in V_j$ for every $j,k \in \mathbb{Z}$, and a pair of discrete quadrature mirror filters $(h_k)_{k\in\mathbb{Z}}$ and $((-1)^{1-k}h_{1-k})_{k\in\mathbb{Z}}$ of $l^2(\mathbb{Z})$ are involved in the *two-scale relations*

$$\varphi_{j,0} = \sum_{k\in\mathbb{Z}} h_k\, \varphi_{j+1,k}\,, \qquad (j \in \mathbb{Z})\,, \tag{1.1}$$

$$\psi_{j,0} = \sum_{k\in\mathbb{Z}} (-1)^{1-k} h_{1-k}\, \varphi_{j+1,k}\,, \quad (j \in \mathbb{Z})\,. \tag{1.2}$$

The two-scale relations (1.1) and (1.2) are the point of departure to obtain algorithms that compute the graph of MRA scaling functions and wavelets. Examples are the Daubechies [2] cascade algorithm or the Daubechies–Lagarias [5] infinite product matrix representation. In the Fourier domain, under suitable conditions, from relation (1.1) one gets

$$\hat{\varphi}(\theta) = \prod_{a=0}^{\infty} \frac{h(e^{-\pi i\theta/2^a})}{2^{1/2}}\, \hat{\varphi}(0)\,. \tag{1.3}$$

where $h(e^{2\pi i\theta}) := \sum_{k\in\mathbb{Z}} h_k\, e^{2\pi ik\theta}$ is the transfer function of the filter (h_k). Conditions for the infinite product in (1.3) determining the Fourier transform of a convenient scaling function can be found in [11].

Fourier transform is closely related to translation operators. On the basis of Fourier analysis, operator techniques have been widely used in wavelet theory, see, for example, [1, 9, 14]. Beyond Fourier analysis, different choices of ONBs may be related to spectral and shift representations of translation and dilation operators on $L^2(\mathbb{R})$. Such ONBs allow to describe every orthonormal wavelet and MRA in terms of rigid operator-valued functions defined on the corresponding functional spectral spaces. By-products of the picture are "spectral formulas" useful to design computational algorithms for orthonormal wavelets and MRAs.

In particular, in [7] the authors propose to consider arbitrary ONBs $\{L_i^{(0)}\}_{i\in\mathbb{I}}$ of $L^2[0,1)$ and $\{K_{\pm,j}^{(0)}\}_{j\in\mathbb{J}}$ of $L^2[\pm1,\pm2)$, and extend them to a pair of ONBs of $L^2(\mathbb{R})$ by means of translations of $\{L_i^{(0)}\}_{i\in\mathbb{I}}$ and dilations of $\{K_{\pm,j}^{(0)}\}_{j\in\mathbb{J}}$. The two resultant ONBs of $L^2(\mathbb{R})$,

$$\{L_i^{(n)} := T^n L_i^{(0)}\}_{i\in\mathbb{I},n\in\mathbb{Z}}\,, \quad \{K_{s,j}^{(m)} := D^m K_{s,j}^{(0)}\}_{s=\pm,j\in\mathbb{J},m\in\mathbb{Z}}\,, \tag{1.4}$$

lead to shift and spectral representations of the operators T and D, respectively. The computational usefulness of this approach has been investigated in [6, 8] for a particular choice of the ONBs $\{L_i^{(0)}\}$ and $\{K_{\pm,j}^{(0)}\}$, the so-called Haar bases.

This work explores two new choices of the ONBs $\{L_i^{(0)}\}$ and $\{K_{\pm,j}^{(0)}\}$ and the corresponding spectral algorithms for compactly supported MRA orthonormal wavelets. The first new bases considered are related with the Walsh–Paley system and the dyadic harmonic analysis [15]. The second bases are related with the trigonometric system and the usual harmonic analysis. For the sake of completeness we include the generic spectral formulas [7] and some known results [8] for Haar bases, which are of interest here.

2. Spectral formulas for MRA orthonormal wavelets

As we have indicated in the Introduction, suitable ONBs of $L^2(\mathbb{R})$ lead to spectral formulas for MRA scaling functions and wavelets. One begins by considering ONBs $\{L_i^{(n)}\}_{i\in\mathbb{I},n\in\mathbb{Z}}$ and $\{K_{s,j}^{(m)}\}_{s=\pm,j\in\mathbb{J},m\in\mathbb{Z}}$ of $L^2(\mathbb{R})$ of the type given by (1.4). For each $f \in L^2(\mathbb{R})$ one has (in L^2-sense)

$$f = \sum_{i,n} \hat{f}_i^{(n)} L_i^{(n)} \quad \text{and} \quad f = \sum_{s,j,m} \tilde{f}_{s,j}^{(m)} K_{s,j}^{(m)}$$

(shortly, $f = \{\hat{f}_i^{(n)}\} = \{\tilde{f}_{s,j}^{(m)}\}$). The change of representation between both expansions is governed by the matrix $\left(\alpha_{i,n}^{s,j,m}\right)$, where

$$\alpha_{i,n}^{s,j,m} := \int_{\mathbb{R}} L_i^{(n)}(x)\, \overline{K_{s,j}^{(m)}(x)}\, dx\,.$$

For an MRA of $L^2(\mathbb{R})$ with associated filter (h_k), scaling function φ and corresponding wavelet ψ, once the ONBs $\{L_i^{(n)}\}$ and $\{K_{\pm,j}^{(m)}\}$ have been fixed, Theorem 2.1 below replaces the classical formulas (1.1) and (1.2) for φ and ψ by the *spectral formulas* (2.1) and (2.2). These spectral formulas are sets of linear equations for the components $\{\hat{\varphi}_i^{(n)}\}$ and $\{\hat{\psi}_i^{(n)}\}$ with coefficients written in terms of the h_k's and $\alpha_{i,n}^{s,j,m}$'s.

Theorem 2.1. *Let $\varphi = \{\hat{\varphi}_i^{(n)}\}$ be the scaling function for an MRA of $L^2(\mathbb{R})$ with associated filter (h_k) satisfying (1.1). Then*

$$\hat{\varphi}_p^{(q)} = \sum_{s,j,m} \overline{\alpha_{p,q}^{s,j,m}} \sum_{i,n} \alpha_{i,n}^{s,j,m-1} \sum_k h_k\, \hat{\varphi}_i^{(n-k)}\,, \quad (p \in \mathbb{I},\, q \in \mathbb{Z})\,, \tag{2.1}$$

Moreover, the function $\psi = \{\hat{\psi}_i^{(n)}\} \in L^2(\mathbb{R})$ determined by the relations

$$\hat{\psi}_p^{(q)} = \sum_{s,j,m} \overline{\alpha_{p,q}^{s,j,m}} \sum_{i,n} \alpha_{i,n}^{s,j,m-1} \sum_k (-1)^{1-k} h_{1-k}\, \hat{\varphi}_i^{(n-k)}\,, \quad (p \in \mathbb{I},\, q \in \mathbb{Z})\,, \tag{2.2}$$

is an orthonormal wavelet associated with the MRA generated by φ.

For a proof of Theorem 2.1 and further details see Section 5.1 of [7].

3. Spectral algorithms for compactly supported MRA orthonormal wavelets

The spectral formulas (2.1) and (2.2) of Theorem 2.1 lead to algorithms to compute the components $\varphi_i^{(n)}$ and $\psi_i^{(n)}$ of the scaling function φ and orthonormal wavelet ψ associated to an MRA with filter (h_k). Since coefficients of the spectral formulas depend on the h_k's as well as the $\alpha_{i,n}^{s,j,m}$'s, different choices of the ONBs $\{L_i^{(n)}\}$ and $\{K_{\pm,j}^{(m)}\}$ will lead to different algorithms. Here we deal with three different families of ONBs $\{L_i^{(n)}\}$ and $\{K_{\pm,j}^{(m)}\}$: the Haar bases, the Walsh–Paley bases and the trigonometric bases.

For practical purposes the attention is focused on compactly supported scaling functions and wavelets, so that the filter (h_k) has only a finite number of nonzero elements h_k. In what follows we will assume that there exists $N \in \mathbb{N}$ such that supp $\varphi \subseteq [0, N]$. Thus, $\hat{\varphi}_i^{(n)} \neq 0$ only for $n = 0, 1, \dots, N-1$, and for each $i \in \mathbb{I}$ we will consider the vector $\hat{\boldsymbol{\varphi}}_i \in \mathbb{C}^N$ given by

$$\hat{\boldsymbol{\varphi}}_i := \left(\hat{\varphi}_i^{(0)}, \hat{\varphi}_i^{(1)}, \dots, \hat{\varphi}_i^{(N-1)}\right)^T.$$

In such case, $h_k \neq 0$ only if $0 \leq k \leq N$ (see Proposition 7.2 in [11]) and two $N \times N$ matrices, here denoted by H_0 and H_1, play a prominent role in the algorithms deduced below for the three choices of ONBs considered. The matrices H_0 and H_1 are determined by the filter (h_k) as follows:

$$[H_0]_{lk} := h_{2l-k}, \quad [H_1]_{lk} := h_{2l-k+1}, \qquad (0 \leq l, k < N). \tag{3.1}$$

The following additional $N \times N$ matrices, defined from H_0 and H_1, will be used too (I denotes the imaginary unit):

$$\tilde{H}_0 := 2^{-1/2}\big[H_0 + H_1\big], \quad \tilde{H}_1 := 2^{-1/2}\big[H_0 - H_1\big], \tag{3.2}$$

$$\tilde{H}_{k,a} := 2^{-1/2}\big[H_0 + e^{-\pi Ik/2^a} H_1\big], \quad (k \in \mathbb{Z},\ a \in \mathbb{N} \cup \{0\}). \tag{3.3}$$

It is worth remembering that distinct approaches, as for example fixed point iterative approximation methods developed by Micchelli and Prautzsch [12] and Daubechies and Lagarias [4, 5], fall also back on the matrices H_0 and H_1. For a profuse study of such pairs of matrices see Protasov [13] and references therein.

The spectral algorithms given below determine the coordinates $\hat{\varphi}_i^{(n)}$ of the scaling function φ. Once $\varphi = \{\hat{\varphi}_i^{(n)}\}$ is given, the wavelet ψ can be calculated from (1.2) or (2.2). The coordinates $\hat{\varphi}_i^{(n)}$ and $\hat{\psi}_i^{(n)}$ identify the functions φ and ψ in L^2-sense:

$$\varphi(x) = \sum_{i,n} \hat{\varphi}_i^{(n)} L_i^{(n)}(x), \quad \psi(x) = \sum_{i,n} \hat{\psi}_i^{(n)} L_i^{(n)}(x), \quad (\text{in } L^2\text{-sense}). \tag{3.4}$$

Lawton's conditions [10] on the filter (h_k) will be of interest here.

Theorem 3.1 (Lawton). *Let $N \in \mathbb{N}$ and let φ be a function in $L^2(\mathbb{R})$ such that* supp $\varphi \subseteq [0, N]$ *and the two-scale relation* (1.1) *is satisfied for a sequence $(h_k)_{k\in\mathbb{Z}}$ (thus, $h_k = 0$ for $k < 0$ or $k > N$). Then, φ is the scaling function for an MRA of $L^2(\mathbb{R})$ if and only if*

(a) $\sum_{k\in\mathbb{Z}} h_k = 2^{1/2}$;

(b) $\sum_{k\in\mathbb{Z}} \overline{h_k}\, h_{k+2n} = \begin{cases} 1, & \text{if } n = 0, \\ 0, & \text{if } n \in \mathbb{Z}\backslash\{0\}; \end{cases}$

(c) *for the* $[2N-1]\times[2N-1]$*-dimensional Lawton matrix* L *defined by*

$$L_{lj}=\sum_{k=0}^{N} h_k\,\overline{h_{j-2l+k}}\,,\quad -N+1\le l,j\le N-1\,,$$

the eigenvalue 1 *is nondegenerate.*

3.1. Haar bases

Let $\mathfrak{h}_0$ and $\mathfrak{h}_1$ be the well-known *Haar scaling function and wavelet* defined by

$$\mathfrak{h}_0:=\chi_{[0,1)}\,,\quad \mathfrak{h}_1:=\chi_{[0,1/2)}-\chi_{[1/2,1)}\,,$$

where χ denotes the characteristic function:

$$\chi_{[a,b)}(x):=\begin{cases}1, & \text{if } x\in[a,b),\\ 0, & \text{otherwise,}\end{cases}$$

For this choice both sets of indices are $\mathbb{I}=\mathbb{J}=\mathbb{N}\cup\{0\}$ when one considers the *Haar ONBs* $\{[L^{\mathfrak{H}}]_i^{(0)}\}_{i\in\mathbb{N}\cup\{0\}}$ of $L^2[0,1)$ and $\{[K^{\mathfrak{H}}]_{\pm,j}^{(0)}\}_{j\in\mathbb{N}\cup\{0\}}$ of $L^2[\pm1,\pm2)$ given by

$$[L^{\mathfrak{H}}]_0^{(0)}:=[\mathfrak{h}_0]_{0,0}\,,\quad [K^{\mathfrak{H}}]_{+,0}^{(0)}:=[\mathfrak{h}_0]_{0,1}\,,\quad [K^{\mathfrak{H}}]_{-,0}^{(0)}:=[\mathfrak{h}_0]_{0,-2}\,,$$

and, for $p\in\mathbb{N}\cup\{0\}$ and $q=0,1,\ldots,2^p-1$,

$$[L^{\mathfrak{H}}]_{2^p+q}^{(0)}:=[\mathfrak{h}_1]_{p,q}\,,[K^{\mathfrak{H}}]_{+,2^p+q}^{(0)}:=[\mathfrak{h}_1]_{p,2^p+q}\,,[K^{\mathfrak{H}}]_{-,2^p+q}^{(0)}:=[\mathfrak{h}_1]_{p,-2^{p+1}+q}\,.$$

Now, according to (1.4), extend them to ONBs of $L^2(\mathbb{R})$:

$$\{[L^{\mathfrak{H}}]_i^{(n)}(x)\}_{i\in\mathbb{N}\cup\{0\},n\in\mathbb{Z}}\quad\text{and}\quad \{[K^{\mathfrak{H}}]_{s,j}^{(m)}\}_{s=\pm,j\in\mathbb{N}\cup\{0\},m\in\mathbb{Z}}\,.$$

After determining the change of representation matrix $\left(\alpha_{i,n}^{s,j,m}\right)$ for these bases one obtains the following spectral algorithm for the Haar coordinates of a compactly supported scaling function [8, Theorem 3.1]:

Theorem 3.2. *Let* $N\in\mathbb{N}$ *and let* $\varphi\in L^2(\mathbb{R})$ *be a scaling function with* $\operatorname{supp}\varphi\subseteq[0,N]$ *and associated MRA filter* $(h_k)_{k\in\mathbb{Z}}$ *satisfying* (1.1)*. Consider the Haar ONB* $\{[L^{\mathfrak{H}}]_i^{(n)}\}$ *given above and the corresponding expansion* $\varphi=\{[\hat\varphi^{\mathfrak{H}}]_i^{(n)}\}$*, i.e.,*

$$\varphi(x)=\sum_{i,n}[\hat\varphi^{\mathfrak{H}}]_i^{(n)}\,[L^{\mathfrak{H}}]_i^{(n)}(x)\quad (\textit{in } L^2\textit{-sense}). \tag{3.5}$$

Then the coordinates

$$\hat\varphi_i^{\mathfrak{H}}:=\left([\hat\varphi^{\mathfrak{H}}]_i^{(0)},[\hat\varphi^{\mathfrak{H}}]_i^{(1)},\ldots,[\hat\varphi^{\mathfrak{H}}]_i^{(N-1)}\right)^T$$

satisfy the following relations:

$$\hat\varphi_0^{\mathfrak{H}}=\tilde H_0\,\hat\varphi_0^{\mathfrak{H}}\,, \tag{3.6}$$

$$\hat\varphi_1^{\mathfrak{H}}=\tilde H_1\,\hat\varphi_0^{\mathfrak{H}}\,, \tag{3.7}$$

and, for $i>1$ *with* $i=\sum_{k=0}^{r} i_k 2^k$, $i_k=0$ *or* 1*, and* $i_r=1$,

$$\hat\varphi_i^{\mathfrak{H}}=H_{i_{r-1}}\cdots H_{i_2}H_{i_1}H_{i_0}\,\hat\varphi_1^{\mathfrak{H}}\,, \tag{3.8}$$

where H_0, H_1, $\tilde H_0$, $\tilde H_1$ *are the* $N\times N$ *matrices given in* (3.1) *and* (3.2).

Thus, according to (3.6), $\hat{\varphi}_0^{\mathfrak{H}}$ is an eigenvector for the eigenvalue 1 of the matrix $\tilde{H}_0$, which exists by conditions on the filter (h_k). Then, $\hat{\varphi}_1^{\mathfrak{H}}$ is obtained from (3.7), and the rest of $\hat{\varphi}_i^{\mathfrak{H}}$, $i > 1$, using the iterative formula (3.8).

As was to be expected, due to the form of the Haar basis $\{[L^{\mathfrak{H}}]_i^{(n)}\}$, iterative formula (3.8) is closely related to Daubechies–Lagarias [5] matrix product approximations at dyadic points. Note, nevertheless, that here the spectral formulas lead to both the initial conditions (3.6)–(3.7) and the iterative formula (3.8).

The coordinates $[\hat{\varphi}^{\mathfrak{H}}]_i^{(n)}$ determine the scaling function φ in L^2-sense by means of the identities (3.5). By construction, the supports of the functions $[L^{\mathfrak{H}}]_i^{(n)}$ form a family of sparse sets. From such sparsity of supports one can deduce, among other things, the following sufficient conditions on H_0 and H_1 for the uniform convergence on $\mathbb{R}$ of the series in (3.5) (see [8, Proposition 3.1] for details):

Proposition 3.3. *Under the conditions of Theorem* 3.2, *suppose that one of the following two conditions is satisfied:*

(i) *The matrices H_0 and H_1 are diagonalizable and*

$$\rho := \max\left\{|\lambda| : \lambda \textit{ is eigenvalue of } H_0 \textit{ or } H_1\right\} < 1 .$$

(ii) *One has*

$$\rho := \max\left\{\sum_{k=n_1}^{n_2-1} |h_{2l-k}|, \sum_{k=n_1}^{n_2-1} |h_{2l-k+1}| : n_1 \le l < n_2\right\} < 1 .$$

Then the series in (3.5) *converge uniformly on* $\mathbb{R}$.

3.2. Walsh–Paley bases

Let $\mathfrak{r}$ be the function defined by

$$\mathfrak{r}(x) := \begin{cases} 1 & , \text{ if } x \in [0, \frac{1}{2}) \\ -1 & , \text{ if } x \in [\frac{1}{2}, 1), \end{cases}$$

extended to $\mathbb{R}$ by periodicity with period 1. The functions of the *Rademacher system* $\{\mathfrak{r}_n\}_{n\in\mathbb{N}\cup\{0\}}$ are of the form

$$\mathfrak{r}_n(x) := \mathfrak{r}(2^n x), \quad (x \in \mathbb{R},\, n \in \mathbb{N} \cup \{0\}).$$

The *Walsh–Paley basis* $\{\mathfrak{w}_n\}_{n\in\mathbb{N}\cup\{0\}}$ is built from Rademacher functions as follows: if $n \in \mathbb{N} \cup \{0\}$ has binary expansion $n = \sum_{k=0}^{\infty} n_k 2^k$, with $n_k \in \{0, 1\}$, then

$$\mathfrak{w}_n := \prod_{k=0}^{\infty} \mathfrak{r}_k^{n_k} .$$

This product is always finite because $n_k = 0$ for k large enough. Moreover, $\mathfrak{w}_0 = 1$ and $\mathfrak{w}_{2^n} = \mathfrak{r}_n$, for all $n \in \mathbb{N} \cup \{0\}$. In what follows we restrict the domain of the functions $\mathfrak{w}_n$ to the interval $[0, 1)$.

The *Walsh–Paley ONBs*

$$\{[L^{\mathfrak{W}}]_i^{(0)}\}_{i\in\mathbb{I}} \text{ of } L^2[0,1) \quad \text{and} \quad \{[K^{\mathfrak{W}}]_{\pm,j}^{(0)}\}_{j\in\mathbb{J}} \text{ of } L^2[\pm 1, \pm 2)$$

are given by

$$\begin{aligned}
&[L^{\mathfrak{W}}]_i^{(0)}(x) := \mathfrak{w}_i(x)\,, && (i \in \mathbb{N} \cup \{0\})\,,\\
&[K^{\mathfrak{W}}]_{+,j}^{(0)}(x) := T\mathfrak{w}_j(x) = \mathfrak{w}_j(x-1)\,, && \\
&[K^{\mathfrak{W}}]_{-,j}^{(0)}(x) := T^{-2}\mathfrak{w}_j(x) = \mathfrak{w}_j(x+2)\,, && (j \in \mathbb{N} \cup \{0\})\,.
\end{aligned}$$

Thus, also in this case $\mathbb{I} = \mathbb{J} = \mathbb{N} \cup \{0\}$. Finally, according to (1.4), extend them to ONBs $\{[L^{\mathfrak{W}}]_i^{(n)}(x)\}_{i\in\mathbb{N}\cup\{0\},n\in\mathbb{Z}}$ and $\{[K^{\mathfrak{W}}]_{s,j}^{(m)}\}_{s=\pm,j\in\mathbb{N}\cup\{0\},m\in\mathbb{Z}}$ of $L^2(\mathbb{R})$.

The spectral algorithm for the Walsh–Paley coefficients (Theorem 3.5 below) may be obtained by determining explicitly the change of representation matrix $(\alpha_{i,n}^{s,j,m})$ for these bases, or using the fact that the usual Walsh–Paley and Haar bases are related by the Hadamard transform [15]. The first approach can be found in [16, Section 3.3.2]. Here we introduce the second one.

For each $p \in \mathbb{N}$, set

$$a_{ql}^{(p)} := 2^{-\frac{p}{2}}\mathfrak{w}_q\Big(\frac{l}{2^p}\Big), \quad (0 \le l, q < 2^p\,, l, q \in \mathbb{N})\,.$$

By a *Hadamard matrix* we shall mean a matrix of the form

$$A^{(p)} := \Big(a_{ql}^{(p)}\Big)_{l,q=0}^{2^p-1}, \quad (p \in \mathbb{N})\,.$$

Since $\mathfrak{w}_l(q/2^p) = \mathfrak{w}_q(l/2^p)$, each Hadamard matrix is a real, symmetric and orthogonal $2^p \times 2^p$ matrix. An inductive argument establishes the following recursive formula

$$\begin{aligned}
A^{(0)} &:= (1)\,,\\
A^{(p+1)} &= \begin{pmatrix} A^{(p)} & A^{(p)} \\ A^{(p)} & -A^{(p)} \end{pmatrix}, \quad (p \in \mathbb{N} \cup \{0\})\,.
\end{aligned} \tag{3.9}$$

The *Hadamard transform* $\perp$ is defined on the vector space of real sequences as follows. Given a real sequence $\mathbf{b} = (b_n)_{n\in\mathbb{N}}$, define $\mathbf{b}^\perp = (b_n^\perp)_{n\in\mathbb{N}}$ by

$$\begin{aligned}
b_0^\perp &:= b_0\,,\\
b_{2^p+q}^\perp &:= \textstyle\sum_{l=0}^{2^p-1} a_{ql}^{(p)} b_{2^p+l}\,, \quad (0 \le q < 2^p,\, p \in \mathbb{N})\,.
\end{aligned}$$

Since the Hadamard matrices are orthogonal, it is clear that $(\mathbf{b}^\perp)^\perp = \mathbf{b}$.

A proof of the following result can be found in [15, page 22].

Lemma 3.4. *The Walsh–Paley and Haar systems are Hadamard transforms of each other, i.e., for each $n \in \mathbb{Z}$ and $x \in \mathbb{R}$,*

$$\begin{aligned}
[L^{\mathfrak{W}}]_0^{(n)}(x) &= [L^{\mathfrak{H}}]_0^{(n)}(x)\,,\\
[L^{\mathfrak{W}}]_{2^p+q}^{(n)}(x) &= \textstyle\sum_{l=0}^{2^p-1} a_{ql}^{(p)}\, [L^{\mathfrak{H}}]_{2^p+l}^{(n)}(x)\,, \quad (0 \le q < 2^p,\, p \in \mathbb{N})\,.
\end{aligned}$$

Now, the spectral algorithm for the Walsh–Paley coefficients can be deduced from Theorem 3.2.

Theorem 3.5. *Let $N \in \mathbb{N}$ and let $\varphi \in L^2(\mathbb{R})$ be a scaling function with* $\operatorname{supp} \varphi \subseteq [0, N]$ *and associated MRA filter $(h_k)_{k\in\mathbb{Z}}$ satisfying* (1.1). *Consider the Walsh–Paley ONB $\{[L^{\mathfrak{W}}]_i^{(n)}\}$ given above and the corresponding expansion $\varphi = \{[\hat{\varphi}^{\mathfrak{W}}]_i^{(n)}\}$, i.e.,*

$$\varphi(x) = \sum_{i,n} [\hat{\varphi}^{\mathfrak{W}}]_i^{(n)} \, [L^{\mathfrak{W}}]_i^{(n)}(x) \quad (\textit{in } L^2\textit{-sense}). \tag{3.10}$$

Then the coordinates

$$\hat{\varphi}_i^{\mathfrak{W}} := \left([\hat{\varphi}^{\mathfrak{W}}]_i^{(0)}, [\hat{\varphi}^{\mathfrak{W}}]_i^{(1)}, \ldots, [\hat{\varphi}^{\mathfrak{W}}]_i^{(N-1)}\right)^T$$

satisfy the following relations:

$$\hat{\varphi}_0^{\mathfrak{W}} = \tilde{H}_0 \, \hat{\varphi}_0^{\mathfrak{W}} , \tag{3.11}$$

and, for $i > 1$ with $i = \sum_{k=0}^{r} i_k 2^k$, $i_k = 0$ or 1, and $i_r = 1$,

$$\hat{\varphi}_i^{\mathfrak{W}} = \tilde{H}_{i_0} \tilde{H}_{i_1} \tilde{H}_{i_2} \cdots \tilde{H}_{i_r} \, \hat{\varphi}_0^{\mathfrak{W}} , \tag{3.12}$$

where $\tilde{H}_0$ and $\tilde{H}_1$ are the $N \times N$ matrices given in (3.2).

Proof. The definition of the matrices H_0, H_1, $\tilde{H}_0$ and $\tilde{H}_1$ given in (3.1) and (3.2) and the recursive formula (3.9) imply that, for $p \in \mathbb{N}$, $0 \le q, j < 2^p$,

$$\tilde{H}_{(2^p+j)_0} \tilde{H}_{(2^p+j)_1} \cdots \tilde{H}_{(2^p+j)_{p-1}} = \sum_{l=0}^{2^p-1} a_{ql}^{(p)} H_{(2^p+l)_{p-1}} H_{(2^p+l)_{p-1}} \cdots H_{(2^p+l)_0},$$

where $2^p + j = \sum_{k=0}^{p} (2^p + j)_k 2^k$. Use these relations and Lemma 3.4 to deduce (3.11) and (3.12) from (3.6)–(3.8). $\square$

Once more, the coordinates $[\hat{\varphi}^{\mathfrak{W}}]_i^{(n)}$ determine the scaling function φ in L^2-sense by means of the identities (3.10). In this case, conditions (i) or (ii) in Proposition 3.3 also imply the uniform convergence on $\mathbb{R}$ of the series in (3.10). See [16, Proposition 47] for details.

3.3. Trigonometric bases

Let $\{\mathfrak{e}_k(x) := \exp(2\pi I k x)\}_{k\in\mathbb{Z}}$ be the usual trigonometric basis of $L^2[0,1)$. Define the *trigonometric ONBs* $\{[L^{\mathfrak{T}}]_i^{(0)}\}_{i\in\mathbb{I}}$ of $L^2[0,1)$ and $\{[K^{\mathfrak{T}}]_{\pm,j}^{(0)}\}_{j\in\mathbb{J}}$ of $L^2[\pm 1, \pm 2)$ by

$$[L^{\mathfrak{T}}]_i^{(0)}(x) := \mathfrak{e}_i(x), \qquad (i \in \mathbb{Z}),$$

$$\left.\begin{aligned} [K^{\mathfrak{T}}]_{+,j}^{(0)}(x) &:= T\mathfrak{e}_j(x) = \mathfrak{e}_j(x-1), \\ [K^{\mathfrak{T}}]_{-,j}^{(0)}(x) &:= T^{-2}\mathfrak{e}_j(x) = \mathfrak{e}_j(x+2), \end{aligned}\right. \quad (j \in \mathbb{Z}).$$

In this case $\mathbb{I} = \mathbb{J} = \mathbb{Z}$ and one obtains the ONBs

$$\{[L^{\mathfrak{T}}]_i^{(n)}\}_{i\in\mathbb{Z},n\in\mathbb{Z}} \quad \text{and} \quad \{[K^{\mathfrak{T}}]_{s,j}^{(m)}\}_{s=\pm,j\in\mathbb{Z},m\in\mathbb{Z}}$$

of $L^2(\mathbb{R})$ following relations (1.4). In what follows, given $\varphi \in L^2(\mathbb{R})$, we will study the coefficients $[\hat{\varphi}^{\mathfrak{T}}]_i^{(n)}$ corresponding to the expansion of φ in the ONB $\{[L^{\mathfrak{T}}]_i^{(n)}\}_{i\in\mathbb{Z},n\in\mathbb{Z}}$:

$$\varphi(x) = \sum_{i,n} [\hat{\varphi}^{\mathfrak{T}}]_i^{(n)} \, [L^{\mathfrak{T}}]_i^{(n)}(x) \quad (\text{in } L^2\text{-sense}).$$

The operator norm for $N \times N$ matrices H is defined by

$$||H|| := \sup_{||\boldsymbol{v}||_2 \neq 0} \frac{||H\boldsymbol{v}||_2}{||\boldsymbol{v}||_2},$$

where

$$||\boldsymbol{v}||_2 := \left(\sum_{n=1}^{N} |v_n|^2 \right)^{1/2}$$

is the Euclidean norm of $\boldsymbol{v} = (v_1, \ldots, v_N) \in \mathbb{C}^N$.

Theorem 3.6. *Let $N \in \mathbb{N}$ and consider the two-scale equation* (1.1)*, where $h_k = 0$ for $k < 0$ or $k > N$ and*

$$\sum_{n\in\mathbb{Z}} h_{2n} = \sum_{n\in\mathbb{Z}} h_{2n+1} = 2^{-1/2}. \tag{3.13}$$

Let H_0, H_1 and $\tilde{H}_{i,a}$ be the $N \times N$ matrices given in (3.1) *and* (3.3)*. Define E_1 to be the $(N-1)$-dimensional subspace of $\mathbb{C}^N$ orthogonal to $\boldsymbol{e}_1 = (1, \ldots, 1)$, the common left eigenvector of H_0 and H_1 for the eigenvalue $2^{-1/2}$. Assume that there exist $\lambda < 1$ and $C > 0$ such that, for all $m \in \mathbb{N}$,*

$$\max_{\substack{d_j \in \{0,1\} \\ j=1,2,\ldots,m}} ||H_{d_1} H_{d_2} \cdots H_{d_m}|_{E_1}|| \leq C \left(\frac{\lambda}{2^{1/2}} \right)^m. \tag{3.14}$$

Then, there is a solution φ to (1.1) *such that* supp $\varphi \subseteq [0, N]$ *and φ is Hölder continuous,*

$$|\varphi(x) - \varphi(y)| \leq C \, |x - y|^{|\ln(\lambda)|/\ln(2)}. \tag{3.15}$$

Moreover, for each $i \in \mathbb{Z}$, the following limit of matrix products exists

$$\lim_{m\to\infty} \prod_{a=0}^{m} \tilde{H}_{i,a} = \lim_{m\to\infty} \tilde{H}_{i,0}\tilde{H}_{i,1} \cdots \tilde{H}_{i,m} =: \prod_{a=0}^{\infty} \tilde{H}_{i,a} = \tilde{H}_{i,0}\tilde{H}_{i,1}\tilde{H}_{i,2} \cdots$$

and

$$\left(\prod_{a=0}^{\infty} \tilde{H}_{i,a} \right)_{n,l} = [\hat{\varphi}^{\mathfrak{T}}]_i^{(n)}, \quad (0 \leq n, l < N). \tag{3.16}$$

Proof. Under the conditions of the theorem, the existence of a solution φ to the two-scale equation (1.1) such that supp $\varphi \subseteq [0, N]$ and φ satisfies (3.15) is part of Theorem 2.2 in [5], taking into account that the matrices T_0 and T_1 considered by

Daubechies and Lagarias in [5] are in correspondence with our $2^{1/2}H_0$ and $2^{1/2}H_1$, respectively. Now, for $x \in [0,1)$, consider its binary expansion

$$x = \sum_{j=1}^{\infty} d_j(x)\, 2^{-j}\,, \text{where } d_j(x) = 0 \text{ or } 1 \text{ for all } j\,.$$

It is known (see Remark 2 on page 1040 of [5]) that, under the conditions of Theorem,

$$\prod_{j=1}^{\infty} 2^{1/2} H_{d_j(x)} = \begin{pmatrix} \varphi(x) & \varphi(x) & \dots & \varphi(x) \\ \varphi(x+1) & \varphi(x+1) & \dots & \varphi(x+1) \\ \vdots & \vdots & & \vdots \\ \varphi(x+N-1) & \varphi(x+N-1) & \dots & \varphi(x+N-1) \end{pmatrix},$$

where the convergence of the product is uniform on $x \in [0,1)$. Indeed, since $\boldsymbol{e}_1 2^{1/2} H_0 = \boldsymbol{e}_1 2^{1/2} H_1 = \boldsymbol{e}_1$, for all $x \in [0,1)$, $s > r \in \mathbb{N}$ and $\boldsymbol{v} \in \mathbb{C}^N$,

$$\boldsymbol{e}_1 \Big[\prod_{j=r+1}^{s} 2^{1/2} H_{d_j(x)} - Id \Big] \boldsymbol{v}^T = 0,$$

i.e., $\big[\prod_{j=r+1}^{s} 2^{1/2} H_{d_j(x)} - Id\big]\boldsymbol{v}^T \in E_1$, so that, by (3.14),

$$\Big\| \prod_{j=1}^{s} 2^{1/2} H_{d_j(x)} - \prod_{j=1}^{r} 2^{1/2} H_{d_j(x)} \Big\| \le C\lambda^r \overset{r\to\infty}{\longrightarrow} 0\,.$$

Then, given $\epsilon > 0$, there exists $r_1 \in \mathbb{N}$ such that, for all $r \geq r_1$,

$$\left| \Big(\prod_{j=1}^{r} 2^{1/2} H_{d_j(x)} \Big)_{n,l} - \varphi(x+n-1) \right| < \frac{\epsilon}{2}, \quad (1 \le n, l < N,\ x \in [0,1))\,. \tag{3.17}$$

On the other hand, for $p \in \mathbb{Z}$ and $r \in \mathbb{N} \cup \{0\}$,

$$\prod_{a=0}^{r} \tilde{H}_{p,a} = \prod_{a=0}^{r} \Big(\frac{H_0 + e^{-\frac{\pi p I}{2^a}} H_1}{2^{1/2}} \Big) = \frac{1}{2^{(r+1)/2}} \sum_{m=0}^{2^{r+1}-1} e^{-\frac{\pi p I m}{2^r}} \prod_{k=0}^{r} H_{m_{r-k}}\,, \tag{3.18}$$

where $m = \sum_{k=0}^{r} m_k 2^k$. For each such $0 \le m < 2^{r+1}$, put

$$x_r^{(m)} := \frac{m}{2^{r+1}} = \sum_{k=0}^{r} m_k 2^{k-r-1} = \sum_{j=1}^{r+1} m_{r+1-j} 2^{-j}\,,$$

so that (3.18) can be written as

$$\prod_{a=0}^{r} \tilde{H}_{p,a} = \frac{1}{2^{r+1}} \sum_{m=0}^{2^{r+1}-1} e^{-2\pi p I x_r^{(m)}} \prod_{j=1}^{r+1} 2^{1/2} H_{d_j(x_r^{(m)})}\,. \tag{3.19}$$

Since φ is Hölder continuous and supp $\varphi \subseteq [0, N]$, the function $e^{-2\pi p I \cdot}\varphi(\cdot)$ is Riemann integrable and one can approximate its integral on each interval $[n, n+1]$, $(0 \leq n < N)$, by means of Riemann sums:

$$\frac{1}{2^{r+1}} \sum_{m=0}^{2^{r+1}-1} e^{-2\pi p I \frac{m}{2^{r+1}}} \varphi\Big(\frac{m}{2^{r+1}} + n\Big) \overset{r\to\infty}{\longrightarrow} \int_n^{n+1} e^{-2\pi p I x}\varphi(x)\, dx = [\hat{\varphi}^{\mathfrak{T}}]_p^{(n)},$$

that is, given $\epsilon > 0$, there exists $r_2 \in \mathbb{N}$ such that for all $r \geq r_2$,

$$\left| \frac{1}{2^{r+1}} \sum_{m=0}^{2^{r+1}-1} e^{-2\pi p I x_r^{(m)}} \varphi\big(x_r^{(m)} + n\big) - [\hat{\varphi}^{\mathfrak{T}}]_p^{(n)} \right| < \frac{\epsilon}{2}. \tag{3.20}$$

Thus, from (3.17), (3.19) and (3.20), for $r \in \mathbb{N}$ such that $r \geq \max\{r_1, r_2\}$ and $0 \leq n, l < N$,

$$\begin{aligned}
&\left| \Big(\prod_{a=0}^{r} \tilde{H}_{p,a}\Big)_{n,l} - [\hat{\varphi}^{\mathfrak{T}}]_p^{(n)} \right| \\
&= \left| \frac{1}{2^{r+1}} \sum_{m=0}^{2^{r+1}-1} e^{-2\pi p I x_r^{(m)}} \Big(\prod_{j=1}^{r+1} 2^{1/2} H_{d_j(x_r^{(m)})}\Big)_{n,l} - [\hat{\varphi}^{\mathfrak{T}}]_p^{(n)} \right| \\
&\leq \left| \frac{1}{2^{r+1}} \sum_{m=0}^{2^{r+1}-1} e^{-2\pi p I x_r^{(m)}} \left[\Big(\prod_{j=1}^{r+1} 2^{1/2} H_{d_j(x_r^{(m)})}\Big)_{n,l} - \varphi\big(x_r^{(m)} + n\big) \right] \right| \\
&\quad + \left| \frac{1}{2^{r+1}} \sum_{m=0}^{2^{r+1}-1} e^{-2\pi p I x_r^{(m)}} \varphi\big(x_r^{(m)} + n\big) - [\hat{\varphi}^{\mathfrak{T}}]_p^{(n)} \right| \\
&< \frac{1}{2^{r+1}} \sum_{m=0}^{2^{r+1}-1} \left| \Big(\prod_{j=1}^{r+1} 2^{1/2} H_{d_j(x_r^{(m)})}\Big)_{n,l} - \varphi\big(x_r^{(m)} + n\big) \right| + \frac{\epsilon}{2} < \epsilon
\end{aligned}$$

and one gets (3.16). □

Remarks 3.7.

(1) Conditions (a) and (b) in Theorem 3.1 imply (3.13). See Remark 5 on page 144 and the beginning of Section 6.2 in [3].

(2) If $\lambda < 1/2$ in (3.14), then (3.15) implies that $|\varphi(x) - \varphi(y)| \leq C\,|x - y|^{\mu}$ with $\mu > 1$. In such case, φ must be constant and, since supp $\varphi \subseteq [0, N]$, one has $\varphi = 0$. Thus, λ in (3.14) necessarily satisfies $\lambda \in [1/2, 1)$.

(3) The condition (3.14) is equivalent to the following one: $\lambda_m < 2^{-1/2}$ for some $m \in \mathbb{N}$, where

$$\lambda_m := \max_{\substack{d_j \in \{0,1\} \\ j=1,2,\ldots,m}} \|H_{d_1} H_{d_2} \cdots H_{d_m}|_{E_1}\|^{1/m}.$$

This condition can also be interpreted in terms of the joint spectral radius for $H_0|_{E_1}$ and $H_1|_{E_1}$. See Lemmas 2.3 and 2.4 in [5] for details.

(4) With respect to the regularity of the solutions to the two-scale relation (1.1) Daubechies and Lagarias proved the following result [4, Theorem 5.1]: Let $N \in \mathbb{N}$ and consider the two-scale equation (1.1), where $h_k = 0$ for $k < 0$ or $k > N$ and $\sum_{k=0}^{N} h_k = 2^{1/2}$ (condition (a) in Theorem 3.1). Define $M = (M_{lk})$ to be the $[N-1] \times [N-1]$ matrix

$$M_{lk} := h_{2l-k+1}\,, \quad (0 \le l, k < N-1)\,.$$

If there exists a non-trivial L^1-solution φ of (1.1) (thus, with $\operatorname{supp}\varphi \subseteq [0, N]$), which is in $\mathcal{C}^m(\mathbb{R})$, then $2^{-1/2}, 2^{-(1+1/2)}, \dots, 2^{-(m+1/2)}$ belong to the spectrum of M. In particular, $m < N-1$.

Corollary 3.8. *Under the conditions of Theorem* 3.6, *the coordinates*

$$\hat{\varphi}_i^{\mathfrak{T}} := ([\hat{\varphi}^{\mathfrak{T}}]_i^{(0)}, [\hat{\varphi}^{\mathfrak{T}}]_i^{(1)}, \dots, [\hat{\varphi}^{\mathfrak{T}}]_i^{(N-1)})^T$$

satisfy the following relations: for $i \in \mathbb{Z}$,

$$\hat{\varphi}_i^{\mathfrak{T}} = \Big(\prod_{a=0}^{\infty} \tilde{H}_{i,a}\Big)\hat{\varphi}_0^{\mathfrak{T}}\,. \tag{3.21}$$

where $\tilde{H}_{k,a}$ *are the matrices given in* (3.3). *Then, for* $i \in \mathbb{Z}$ *and* $b \in \mathbb{N}$,

$$\hat{\varphi}_{2^b i}^{\mathfrak{T}} = \tilde{H}_0^b\, \hat{\varphi}_i^{\mathfrak{T}}\,, \tag{3.22}$$

where $\tilde{H}_0$ *is the* $N \times N$ *matrix given in* (3.2). *In particular,*

$$\hat{\varphi}_0^{\mathfrak{T}} = \tilde{H}_0\, \hat{\varphi}_0^{\mathfrak{T}}\,. \tag{3.23}$$

Proof. Formula (3.21) is a direct consequence of (3.16) and the usual assumption for scaling functions

$$\sum_n [\hat{\varphi}^{\mathfrak{T}}]_0^{(n)} = \hat{\varphi}(0) = 1\,.$$

(Necessarily, $|\hat{\varphi}(0)| = 1$, so that the above assumption means a choice for the phase of φ, see [3, Section 6.2] for details). According to (3.16) and the definition of the matrices $\tilde{H}_{k,a}$ given in (3.3), for $i \in \mathbb{Z}$ and $b \in \mathbb{N}$,

$$\begin{aligned}
\hat{\varphi}_{2^b i}^{\mathfrak{T}} &= \frac{1}{N}\Big[\prod_{a=0}^{\infty} \tilde{H}_{2^b i,a}\Big] e_1^T = \frac{1}{N}\Big[\prod_{a=0}^{b-1} \tilde{H}_{2^b i,a}\Big]\Big[\prod_{a=b}^{\infty} \tilde{H}_{2^b i,a}\Big] e_1^T \\
&= \frac{1}{N}\Big[\prod_{a=0}^{b-1} \tilde{H}_0\Big]\Big[\prod_{a=0}^{\infty} \tilde{H}_{i,a}\Big] e_1^T = \tilde{H}_0^b\, \hat{\varphi}_i^{\mathfrak{T}}\,,
\end{aligned}$$

that is, (3.22) is satisfied. Now, to get (3.23), it suffices to take $i = 0$ and $b = 1$ in (3.22). □

Taking into account that the usual Fourier series of an MRA scaling function and orthogonal wavelet are unuseful, because all their Fourier coefficients (except for $k = 0$) are null (see [3, Section 6.2]), Corollary 3.8 can be seen as an attempt to achieve a discrete frequencial analysis for these functions. Unfortunately, relation

(3.21) does not disengage us from the infinite products already present in the continuous Fourier formula (1.3).

There are other ways to obtain the results of Theorem 3.6 and Corollary 3.8. One of them is to consider the change of representation matrix $(\alpha_{i,n}^{s,j,m})$ for trigonometric bases. The details can be found in Appendix A.3 and Section 3.3.3 of [16]. A second method uses the additional functions $e^{-2\pi \frac{i}{2^a} I x}$, for $a \geq 0$. Let us put

$$\hat{\varphi}^{\mathfrak{T}}_{\frac{i}{2^a}} := \left([\hat{\varphi}^{\mathfrak{T}}]^{(0)}_{\frac{i}{2^a}}, [\hat{\varphi}^{\mathfrak{T}}]^{(1)}_{\frac{i}{2^a}}, \ldots, [\hat{\varphi}^{\mathfrak{T}}]^{(N-1)}_{\frac{i}{2^a}}\right)^T,$$

where

$$[\hat{\varphi}^{\mathfrak{T}}]^{(n)}_{\frac{i}{2^a}} := \int_n^{n+1} \varphi(x)\, e^{-2\pi \frac{i}{2^a} I x}\, dx\,, \quad (0 \leq n < N)\,.$$

Under the conditions of Theorem 3.6, the relationship between trigonometric and Haar bases leads to

$$\hat{\varphi}^{\mathfrak{T}}_{\frac{i}{2^a}} = \frac{-2^a I}{2\pi i}\left[\left(1 - e^{-\frac{\pi i I}{2^{a-1}}}\right)\hat{\varphi}^{\mathfrak{H}}_0 + \sum_{u=0}^{\infty} 2^u \left(1 - e^{-\frac{\pi i I}{2^{a+u}}}\right)^2 \left(\prod_{v=0}^{u-1} \tilde{H}_{i,a+v}\right)\hat{\varphi}^{\mathfrak{H}}_1\right],$$

so that

$$\hat{\varphi}^{\mathfrak{T}}_{\frac{i}{2^a}} = \tilde{H}_{p,a}\, \hat{\varphi}^{\mathfrak{T}}_{\frac{i}{2^{a+1}}}\,, \quad (a \geq 0)\,.$$

A recursive use of this formula imply, in particular, (3.21). See again [16, Section 3.3.3] for details.

4. Conclusions

Operator techniques developed in [7] lead to spectral formulas for scaling functions and orthonormal wavelets associated with MRAs of $L^2(\mathbb{R})$. The approach takes advantage of good properties of pairs of suitable ONBs $\{L_i^{(n)}\}$ and $\{K_{\pm,j}^{(m)}\}$ relative to translation and dilation operators on $L^2(\mathbb{R})$. The spectral formulas are sets of linear equations for the coordinates of the scaling functions and wavelets in their expansions with respect to the ONB $\{L_i^{(n)}\}$ (see Theorem 2.1 and (3.4)).

Once the ONBs $\{L_i^{(n)}\}$ and $\{K_{\pm,j}^{(m)}\}$ are fixed, a spectral algorithm derives from the spectral formulas to compute scaling functions and wavelets. Here we have presented the spectral algorithms for three different families of ONBs: Haar bases in Section 3.1, Walsh–Paley bases in Section 3.2 and trigonometric bases in Section 3.3.

Particular choices of ONBs focus attention on concrete properties of scaling functions and wavelets. So, for example, Haar bases target at point properties, like regularity, whereas trigonometric bases address a discrete frequencial analysis (recall that Fourier series are unusable for compactly supported scaling functions and wavelets). Other choices will be studied elsewhere.

References

[1] X. Dai and D.R. Larson, *Wandering vectors for unitary systems and orthogonal wavelets.* Mem. Amer. Math. Soc. **134** (1998), pp. viii+68.

[2] I. Daubechies, *Orthonormal bases of compactly supported wavelets.* Comm. Pure Appl. Math. **41** (1988), 909–996.

[3] I. Daubechies, *Ten lectures on wavelets*, vol. 61 of CBMS-NSF Regional Conference Series in Applied Mathematics, Society for Industrial and Applied Mathematics (SIAM), Philadelphia, PA, 1992.

[4] I. Daubechies and J.C. Lagarias, *Two-scale difference equations. I. Existence and global regularity of solutions.* SIAM J. Math. Anal. **22** (1991), 1388–1410.

[5] I. Daubechies and J.C. Lagarias, *Two-scale difference equations. II. Local regularity, infinite products of matrices and fractals.* SIAM J. Math. Anal. **23** (1992), 1031–1079.

[6] F. Gómez-Cubillo and Z. Suchanecki, *Inner functions and local shape of orthonormal wavelets.* Appl. Comput. Harmon. Anal. **30** (2011), 273–287.

[7] F. Gómez-Cubillo and Z. Suchanecki, *Spectral models for orthonormal wavelets and multiresolution analysis of $L^2(\mathbb{R})$.* J. Fourier Anal. Appl. **17** (2011), 191–225.

[8] F. Gómez-Cubillo, Z. Suchanecki, and S. Villullas, *Orthonormal MRA wavelets: spectral formulas and algorithms.* Int. J. Wavelets Multiresolut. Inf. Process. **10** (2012), pp. 1250008, 19.

[9] T.N.T. Goodman, S.L. Lee, and W.S. Tang, *Wavelets in wandering subspaces.* Trans. Amer. Math. Soc. **338** (1993), 639–654.

[10] W.M. Lawton, *Necessary and sufficient conditions for constructing orthonormal wavelet bases.* J. Math. Phys. **32** (1991), 57–61.

[11] S. Mallat, *A Wavelet Tour of Signal Processing.* Academic Press, Inc., San Diego, CA, 1998.

[12] C.A. Micchelli and H. Prautzsch, *Uniform refinement of curves.* Linear Algebra Appl. **114/115** (1989), 841–870.

[13] V. Protasov, *Refinement equations and corresponding linear operators.* Int. J. Wavelets Multiresolut. Inf. Process. **4** (2006), 461–474.

[14] A. Ron and Z. Shen, *Affine systems in $L_2(\mathbb{R}^d)$: the analysis of the analysis operator.* J. Funct. Anal. **148** (1997), 408–447.

[15] F. Schipp, W.R. Wade, and P. Simon, *Walsh Series. An Introduction to Dyadic Harmonic Analysis.* Adam Hilger, Ltd., Bristol, 1990.

[16] S. Villullas, *Fórmulas y algoritmos espectrales para wavelets ortonormales y wavelet frames ajustados. Aplicaciones en el procesamiento digital de imágenes.* PhD thesis, IMUVa-LPI, Universidad de Valladolid, Valladolid, Spain, 2016.

F. Gómez-Cubillo and S. Villullas
Dpto de Análisis Matemático
Universidad de Valladolid, Facultad de Ciencias
E-47011 Valladolid, Spain
e-mail: `fgcubill@am.uva.es`
`svillu@am.uva.es`

Operator Theory:
Advances and Applications, Vol. 267, 199–220

The NIEP

Charles R. Johnson, Carlos Marijuán,
Pietro Paparella and Miriam Pisonero

Abstract. The nonnegative inverse eigenvalue problem (NIEP) asks which lists of n complex numbers (counting multiplicity) occur as the eigenvalues of some n-by-n entry-wise nonnegative matrix. The NIEP has a long history and is a known hard (perhaps the hardest in matrix analysis?) and sought after problem. Thus, there are many subproblems and relevant results in a variety of directions. We survey most work on the problem and its several variants, with an emphasis on recent results, and include 130 references. The survey is divided into: a) the single eigenvalue problems; b) necessary conditions; c) low-dimensional results; d) sufficient conditions; e) appending 0's to achieve realizability; f) the graph NIEP's; g) Perron similarities; and h) the relevance of Jordan structure.

Mathematics Subject Classification (2010). Primary 15A18, 15A29, 15A42.

Keywords. Nonnegative matrices, nonnegative inverse eigenvalue problem (NIEP), symmetric realizability.

1. Introduction

The **N**onnegative **I**nverse **E**igenvalue **P**roblem (NIEP) asks which collections of n complex numbers (counting multiplicities) occur as the eigenvalues of an n-by-n matrix, all of whose entries are nonnegative real numbers. This is a long standing problem that is very difficult and, perhaps, the most prominent problem in matrix analysis. Unlike many other inverse eigenvalue problems, tools are limited, and seemingly small results are very welcome. Nonetheless, the problem is very attractive and has been attacked by many excellent researchers (see references). There have been prior surveys of work on the NIEP [4, 7, 20, 27, 92, 117, 129], but none recently. With a great deal of important recent activity, a new survey, emphasizing this activity, will be welcome to anyone considering inquiry in this area. Our intent here is to be broad and informal.

Partially supported by MTM2015-365764-C-1-P(MINECO/FEDER).

Another intriguing aspect of the NIEP is how many avenues can be taken to gain insight into it. Thought about it spawns endless challenging and worthy specific questions. An example is how many interesting variations there are. The real NIEP (R-NIEP) restricts the question to real spectra, and the symmetric NIEP (S-NIEP) further restricts to real spectra that are realizable by symmetric nonnegative matrices. It is known [48] that the row stochastic NIEP (restriction to nonnegative matrices with row sums 1) is equivalent to the general NIEP, but the doubly stochastic NIEP (DS-NIEP), restriction to matrices with both row and column sums 1, is properly more restrictive, but no less difficult. Of course, there are variations on the DS-NIEP, as well: the real DS-NIEP and the symmetric DS-NIEP. The Jordan NIEP (J-NIEP) asks about possible Jordan canonical forms, when there are repeated eigenvalues, and the diagonalizable NIEP (D-NIEP) is the special case in which the realizing matrix is diagonalizable. Finally, there are graph NIEP's when we consider only nonnegative matrices with a given (directed or undirected) graph G (G-NIEP), or only matrices subordinate to same. Further particularizations may also be imagined, but these are the primary ones, so far. We should also mention that the NIEP and each variation also has a trace 0 version. This considers only spectra, the sum of whose elements is 0, or, equivalently, nonnegative matrices, each of whose diagonal entries is 0. In some cases, this is sufficiently restrictive, so as to make the problem easier.

We denote by NIEP_n the set of all spectra enjoyed by some n-by-n nonnegative matrix. Such a spectrum is called **realizable** and a nonnegative matrix with the spectrum is called a **realization** or **realizing matrix**. The notation and terminology are similar for NIEP variants. A solution to the NIEP (or variants), for a particular n, is an "explicit" description of NIEP_n, viewed as a subset of vectors in $\mathbb{C}^n$. NIEP_n is a closed set that is connected, and even star-shaped from the origin (or from the all 1's vector e). The set is also semi-algebraic, but is not generally convex. If $\Lambda = \{\lambda_1, \dots, \lambda_n\}$ is a proposed spectrum, we denote by $\mathrm{Tr}(\Lambda)$ the sum of its components, counting multiplies. Other power sums $s_k(\Lambda) = \sum_{i=1}^{n} \lambda_i^k$, called ***k*th moments**, are also important, and we usually reserve λ_1 as the spectral radius of Λ.

The Perron–Frobenius theory of nonnegative matrices (*e.g.*, [43]) provides several simple, but important, necessary conditions for the NIEP (and variations). These basic necessary conditions are presented in Section 3, followed by several more subtle necessary conditions. The "JLL" conditions are now essential to work on the problem.

The NIEP has a long history since its proposal by Kolmogorov [62], and there are many substantial results. We begin at the natural starting point of the single eigenvalue problem (Section 2) for both the row stochastic NIEP and the DS-NIEP: which individual complex numbers occur in the spectra of row stochastic and of doubly stochastic matrices. The former is solved, but there have been important advances in the description of the solution and in realizations. The latter is unsolved, beyond $n = 4$, though there have been important recent developments.

Section 4 summarizes low-dimensional complete results for certain NIEP variants. General sufficient conditions for the realizability of spectra are given in Section 5. It has long been known that n-fold spectra that meet simple necessary conditions, but are not realizable, may be made realizable by appending of 0 eigenvalues. Information about this phenomenon is given in Section 6. We turn to what is known about the graph-NIEP's in Section 7. The new idea of Perron similarities – studying the diagonalizable NIEP's via the diagonalizing similarities – is discussed in Section 8, and the role of Jordan structure in the NIEP and R-NIEP in Section 9.

2. The Single Eigenvalue Problems

In [62], Kolmogorov posed the problem, denoted by SISEP (**S**tochastic **I**nverse **S**ingle **E**igenvalue **P**roblem), of characterizing the subset of the complex plane, denoted by Θ_n, consisting of the individual eigenvalues of all n-by-n stochastic matrices.

It can easily be verified that for each $n \geq 2$, the region Θ_n is closed, inscribed in the unit-disc, star-convex (with star-centers at zero and one), and symmetric with respect to the real-axis. Furthermore, it is clear that $\Theta_n \subseteq \Theta_{n+1}$, $\forall n \in \mathbb{N}$. In view of these properties, each region Θ_n is determined by its boundary, which consists of so-called **extremal** numbers, *i.e.*, $\partial\Theta_n = \{\lambda \in \Theta_n : \alpha\lambda \notin \Theta_n, \forall \alpha > 1\}$.

In pursuit of characterizing Θ_n, Dmitriev and Dynkin [24] (and independently Karpelevich [57]) showed that if $\lambda = a + bi \in \Theta_n$ and $b \neq 0$, then

$$a + |b| \tan\left(\frac{\pi}{n}\right) \leq 1.$$

Let A be a nonnegative matrix of order n with spectral radius ρ and associated directed graph G. Let m be the length of the longest simple circuit of G. Kellogg and Stephens [59, Theorem 1] showed that if $m = 2$, all eigenvalues of A are real. If $2 < m \leq n$, and if $\lambda = a + bi$ is an eigenvalue of A, then

$$a + |b| \tan\left(\frac{\pi}{m}\right) \leq \rho.$$

For further results on this topic, see [49].

Karpelevich [57, Theorem B], expanding on the work of Dmitriev and Dynkin in [25], resolved SISEP by showing that the boundary of Θ_n consists of curvilinear arcs (called K-arcs), whose points satisfy a polynomial equation that is determined by the endpoints of the arc (which are consecutive roots of unity). The statement of this theorem is unwieldy, but Ito [46, Theorem 2] provides a useful simplification of the result.

Noticeably absent in the Karpelevich Theorem (and other works) are *realizing-matrices* (*i.e.*, a matrix whose spectrum contains a given eigenvalue) for points on the K-arcs. Dmitriev and Dynkin [25, Basic Theorem] give a schematic description of such matrices for points on the boundary of $\Theta_n \backslash \Theta_{n-1}$ and Swift [129, §2.2.2] provides such matrices for $3 \leq n \leq 5$. Recently, Johnson and Paparella [55] provide, for every n and for each K-arc, a single parametric stochastic matrix that realizes the entire K-arc as the parameter runs from 0 to 1.

For $n \in \mathbb{N}$, denote by Π_n the convex-hull of the nth roots-of-unity, *i.e.*,

$$\Pi_n = \left\{ \sum_{k=0}^{n-1} \alpha_k \exp(2\pi i k/n) \in \mathbb{C} : \alpha_k \geq 0, \sum_{k=0}^{n-1} \alpha_k = 1 \right\}.$$

Denote by Ω_n the subset of the complex-plane containing all single eigenvalues of all n-by-n doubly stochastic matrices. Perfect and Mirsky [102] conjectured that $\Omega_n = \bigcup_{k=1}^{n} \Pi_k$ and proved their conjecture when $1 \leq n \leq 3$. Levick et al. [75] proved the Perfect–Mirsky conjecture when $n = 4$ but a counterexample when $n = 5$ was given by Mashreghi and Rivard [87]. Recently, Levick *et al.* conjectured that $\Omega_n = \Theta_{n-1} \cup \Pi_n$ ([75, Conjecture 1]), but there is computational evidence that suggests that the $n = 5$ case is either a rare exception, or the only exception, to Perfect–Mirsky.

3. Necessary conditions for the NIEP's

For a proposed spectrum $\Lambda = \{\lambda_1, \lambda_2, \ldots, \lambda_n\}$, repeats allowed, to be (NIEP-) realizable, a number of necessary conditions are known. The most basic of these follow from the fact that a nonnegative matrix has real entries and nonnegative trace, and from the Perron–Frobenius theory of nonnegative matrices. (We assume the reader is familiar with the Perron–Frobenius theory, which is recounted, for example in [43].) For simplicity, we label these according to simple titles. Since a nonnegative matrix is real, it must have the eigenvalues of a real matrix, *i.e.*, the characteristic polynomial must have real coefficients, or, as a list

(Reality) $$\bar{\Lambda} = \Lambda.$$

Since the trace of a matrix is the sum of the eigenvalues and the trace of a nonnegative matrix is nonnegative, we also have

(Trace) $$\operatorname{Tr} \Lambda = \sum_{i=1}^{n} \lambda_i \geq 0.$$

Of course, it follows that if Λ is realizable and $\operatorname{Tr}(\Lambda) = 0$, then all diagonal entries of any realizing matrix must be 0. Higher power sums are also nonnegative as they are traces of positive integral powers of nonnegative matrices:

(kth moment) $$s_k(\Lambda) = \lambda_1^k + \cdots + \lambda_n^k \geq 0, \quad k \geq 1.$$

According to the Perron–Frobenius theory, the spectral radius of a nonnegative matrix must, itself, be an eigenvalue, so among the eigenvalues Λ must be a nonnegative one, at least as big in absolute value as any others. Without loss of generality, this one may be taken to be λ_1. So, we have

$$\text{(Perron)} \qquad \lambda_1 \geq |\lambda_i|, \quad i = 2, \ldots, n,$$

and if any of the other eigenvalues are the same as λ_1, any realizing matrix must be reducible. Moreover, if all the inequalities are strict, any irreducible realizing matrix must be primitive.

The above conditions are necessary for all the NIEP's, but there are some necessary conditions for particular NIEP's.

Further necessary conditions are more subtle, but many have been noticed. The most ubiquitous of these was noticed independently in [48] and in [77] and is usually referred to as the J-LL condition. They generalize the trace condition and follow from the fact that every power of a nonnegative matrix is nonnegative and that positive diagonal entries must contribute to positive diagonal entries in powers (*e.g.*, $s_p(\Lambda) > 0$ implies $s_{pq}(\Lambda) > 0$ for integers $p \geq 1$ and $q \geq 1$). The general (quantitative) version is

$$\text{(JLL)} \qquad (s_k(\Lambda))^m \leq n^{m-1} s_{km}(\Lambda), \quad k,\, m = 1, 2, \ldots.$$

Let A be an n-by-n real matrix with spectrum $\Lambda = \{\lambda_1, \ldots, \lambda_n\}$. Denote the principal submatrix of A lying in the rows and columns given by the index set $\alpha \subseteq \{1, \ldots, n\}$ by $A[\alpha]$. Define the kth elementary symmetric function

$$E_k(\Lambda) = \sum_{1 \leq i_1 < \cdots < i_k \leq n} \lambda_{i_1} \cdots \lambda_{i_k}$$

and the kth Newton coefficient

$$c_k(\Lambda) = \frac{E_k(\Lambda)}{\binom{n}{k}}, \quad k = 1, \ldots, n, \text{ with } c_0 \equiv 1.$$

Since $E_k(\Lambda) = \sum_{|\alpha|=k} \det A[\alpha]$, as well, $c_k(\Lambda)$ may be viewed as the average value of the k-by-k principal minors of A. The spectrum Λ is called **Newton** if

$$c_k(\Lambda)^2 \geq c_{k-1}(\Lambda) c_{k+1}(\Lambda), \quad k = 1, \ldots, n-1,$$

and these inequalities are referred to as the **Newton inequalities** [42, 51].

They hold for $\Lambda \geq 0$ [96], $\Lambda \subset \mathbb{R}$ [80], and they are valid for real diagonal matrices, diagonalizable matrices with real spectra (the c_k are invariant under similarity), and matrices with real spectra. In [42] it was proved that the Newton inequalities also hold for M-matrices and, thus, inverse M-matrices and it was observed that if A is a nonnegative matrix with spectral radius $\rho(A)$, then $\rho(A)I - A$ is an M-matrix and, therefore, must satisfy the Newton inequalities. If we denote its spectrum $\{\rho(A) - \lambda_1, \ldots, \rho(A) - \lambda_n\}$ by $\rho(A) - \Lambda$, then we have new necessary conditions:

$$\text{(H)} \qquad c_k(\rho(A) - \Lambda)^2 \geq c_{k-1}(\rho(A) - \Lambda) c_{k+1}(\rho(A) - \Lambda), \quad k = 1, \ldots, n-1.$$

In [130] the authors did not focus the attention directly on the spectrum but on the coefficients of the characteristic polynomial. Thus, the NIEP that they consider is: *given real numbers* $k_1, k_2, \ldots, k_n$*, find necessary and sufficient conditions for the existence of a nonnegative matrix of order* n *with characteristic polynomial* $x^n + k_1x^{n-1} + k_2x^{n-2} + \cdots + k_n$.

The coefficients of the characteristic polynomial are closely related to the cyclic structure of the weighted digraph associated with the matrix A, as established by the Coefficients Theorem [22, Theorem 1.3*]. The authors [130] introduce graphic tools to study the NIEP from the characteristic polynomial and use the following method: if $P(x)$ is a realizable polynomial, in the sense that there exists a nonnegative matrix with characteristic polynomial $P(x)$, we try to maximize each coefficient k_j as a function of the previous coefficients, preserving the realizability for a polynomial of degree n with the same previous coefficients. Note that k_j is a continuous function (sum of determinants) of the entries of a nonnegative matrix A realizing the polynomial $P(x)$ and that, as the previous coefficients are bounded above, then the entries of A involved in the expression of k_j are also bounded above (Coefficients Theorem); therefore, this maximum is attained. In this way, new necessary conditions on the three first coefficients are obtained [130, Theorem 3]:

$$k_1 \leq 0;$$

(TAAMP)
$$k_2 \leq \frac{n-1}{2n} k_1^2;$$

$$k_3 \leq \begin{cases} \frac{n-2}{n}\left(k_1k_2 + \frac{n-1}{3n}\left(\left(k_1^2 - \frac{2nk_2}{n-1}\right)^{3/2} - k_1^3\right)\right) & \text{if } \frac{(n-1)(n-4)}{2(n-2)^2}k_1^2 < k_2, \\ k_1k_2 - \frac{(n-1)(n-3)}{3(n-2)^2}k_1^3 & \text{if } k_2 \leq \frac{(n-1)(n-4)}{2(n-2)^2}k_1^2. \end{cases}$$

Another necessary condition in terms of the kth moments $s_k = \operatorname{Tr}(A^k)$, $k = 1, 2, 3$, is obtained in [21, Theorem 3]:

(CL)
$$\Phi := n^2s_3 - 3ns_1s_2 + 2s_1^3 + \tfrac{n-2}{\sqrt{n-1}}(ns_2 - s_1^2)^{3/2} \geq 0.$$

The necessary conditions previous to (JLL) are not independent. In fact, (kth moment) implies (Perron) [33] and also (Reality) [77]. For $k = 1$, (JLL) is reduced to $(s_1(\Lambda))^m \leq n^{m-1}s_m(\Lambda)$, and so, if $s_1(\Lambda) \geq 0$, then (JLL) implies (kth moment).

On the other hand, if we denote the two bounds given for the coefficient k_3 in (TAAMP) by $k_3^{\max 1}$ and $k_3^{\max 2}$, we can use the Newton identities

$$s_m + k_1s_{m-1} + \cdots + k_{m-1}s_1 + k_mm = 0, \quad k, m = 1, 2, 3, \tag{1}$$

to rewrite the condition (CL) in the form

$$\Phi = 3n^2(k_3^{\max 1} - k_3).$$

In the first case, *i.e.*, if $k_2 > \frac{(n-1)(n-4)}{2(n-2)^2}k_1^2$, we have that $\Phi \geq 0$ implies $k_3^{\max 1} \geq k_3$. In the contrary case, $\Phi = 3n^2(k_3^{\max 1} - k_3) \geq 3n^2(k_3^{\max 2} - k_3) \geq 0$. So, in any case, the condition on the coefficient k_3 is stronger than the condition (CL).

In [42] it was proved that (kth moment), (JLL) and (H) are mutually independent. In [83, Theorem 11] that (JLL) for $k = 1$ and $m = 2$, $s_1(\Lambda)^2 \leq ns_2(\Lambda)$, the first (H), $c_1(\rho-\Lambda)^2 \geq c_2(\rho-\Lambda)$, and (TAAMP) over the second coefficient, $k_2(\Lambda) \leq \frac{n-1}{2n}k_1(\Lambda)^2$, are equivalent. But, in general, at least (JLL) and (TAAMP) are independent of the others, and this tandem implies (kth moment). Some examples: the spectra $\{20, -18, 5\sqrt{2} \pm 5\sqrt{2}i\}$ and $\{1, 1, 1, 0, 0\}$ satisfy (JLL), (H) and (CL), but not (TAAMP); the spectra $\{2, -2, -2, 1 \pm i\}$ and $\{3, 1, 1, 1, 1, 1, -2, -2, -2, -2\}$ satisfy (H) and (TAAMP), but not (JLL). That the necessary conditions (JLL), (H) and (TAAMP) are not sufficient for the NIEP is proved by the non-realizable list $\{3, 3, -\sqrt{3} \pm i\}$.

Conjecture. (JLL) and (TAAMP) imply all known necessary conditions. (It is enough to prove that (JLL) and (TAAMP) imply (H).)

In [68] a necessary condition for trace 0 and n, odd was obtained, given by

(LM) $$(s_2(\Lambda))^2 \leq (n-1)s_4(\Lambda).$$

This condition has been generalized in terms of the coefficients by the following result [130, Lemma 37]: if $x^n + k_p x^{n-p} + \cdots + k_{2p}x^{n-2p} + \cdots + k_n$, $k_p \neq 0$, is the characteristic polynomial of a nonnegative matrix, then

$$k_{2p} \leq \frac{1}{2}\left(1 - \frac{1}{\lfloor n/p \rfloor}\right) k_p^2. \tag{2}$$

We can use the Newton identities (1) to express the inequality (2) in terms of the kth moments: if $s_1 = \cdots = s_{p-1} = 0$, we have the more general condition

$$s_p^2 \leq p\left\lfloor \frac{n}{p} \right\rfloor s_{2p}, \quad p = 1, \ldots, \frac{n}{2}$$

that coincides with (LM) in the particular case $p = 2$ and n odd.

Note also that the (JLL) condition for $m = 2$ is $s_p^2 \leq ns_{2p}$, and that $p\left\lfloor \frac{n}{p} \right\rfloor \leq n$. Then the expression (2) is a restricted refinement of the (JLL) conditions.

4. Low-dimensional results

The NIEP for $n \leq 3$ was solved independently by Oliveira [97, Theorem (6.2)] and Loewy–London [77]. For the non-real case:

$$\{\lambda, z, \bar{z}\} \text{ is NIEP-realizable } \iff z \in \lambda\Pi_3 = \{\lambda z' : z' \in \Pi_3\}$$

Meehan [89] solved the NIEP for $n = 4$ in terms of the kth moments, and Torre–Mayo *et al.* [130] in terms of the coefficients of the characteristic polynomial using the necessary conditions (TAAMP). For $n \geq 5$ it remains unsolved.

The R-NIEP for $n \leq 4$ was solved independently by Perfect [99], Oliveira [97, §9] and Loewy–London [77]; in these cases the necessary conditions (Trace) and (Perron) are also sufficient.

Perfect [101, Theorem 4] solved the R-NIEP for $n = 3$ with fixed diagonal entries: $\{d_1, d_2, d_3\}$ is the diagonal of a 3-by-3 nonnegative matrix with spectra $\{\lambda_1, \lambda_2, \lambda_3\}$, where $\lambda_1 \geq \lambda_2 \geq \lambda_3$, if and only if

$$0 \leq d_i \leq \lambda_1; \ \sum_{i=1}^{3} d_i = \sum_{i=1}^{3} \lambda_i; \ \sum_{i \neq j} d_i d_j \geq \sum_{i \neq j} \lambda_i \lambda_j; \ \max\{d_i\} \geq \lambda_2.$$

The S-NIEP and the R-NIEP are equivalent for $n \leq 4$, and remain unsolved for $n \geq 5$. Fiedler [32, Theorem 4.8] solved the S-NIEP in the case $n = 3$ with fixed diagonal entries: $\{d_1, d_2, d_3\}$, with $d_1 \geq d_2 \geq d_3 \geq 0$, is the diagonal of a 3-by-3 symmetric nonnegative matrix with spectra $\{\lambda_1, \lambda_2, \lambda_3\}$, where $\lambda_1 \geq \lambda_2 \geq \lambda_3$, if and only if $\{\lambda_1, \lambda_2, \lambda_3\}$ majorizes $\{d_1, d_2, d_3\}$ and $d_1 \geq \lambda_2$.

Johnson–Laffey–Loewy [50] showed that the R-NIEP and the S-NIEP are different, and Egleston–Lenker–Narayan [27] proved that they are different for $n \geq 5$.

The S-NIEP for $n = 5$ has been widely studied [27, 78, 88], but not fully resolved. It is common to study it considering the number of positive eigenvalues. When there are 1, 4 or 5 positive eigenvalues the answer for the S-NIEP is straightforward. Recently, Johnson–Marijuán–Pisonero [53] resolved all cases with 2 positive eigenvalues, and they give a method, based upon the eigenvalue interlacing inequalities for symmetric matrices, to rule out many unresolved spectra with 3 positive eigenvalues. In particular, this method shows that the nonnegative realizable spectrum $\{6, 3, 3, -5, -5\}$ is not symmetrically realizable.

Also recently, Loewy–Spector [79, Theorem 4] characterize the case $n = 5$ in a particular case: $\Lambda = \{\lambda_1, \lambda_2, \ldots, \lambda_5\}$, where $\lambda_1 \geq \lambda_2 \geq \cdots \geq \lambda_5$ and $2s_1(\Lambda) \geq \lambda_1$, is (S-NIEP) realizable if and only if (Perron), $\lambda_2 + \lambda_5 \leq \mathrm{Tr}\,(\Lambda)$, $\lambda_3 \leq \mathrm{Tr}\,(\Lambda)$. This last condition is implied by the constraint $2s_1(\Lambda) \geq \lambda_1$.

The trace 0 NIEP has also been extensively studied. Reams [106] solved the case $n = 4$: $\{\lambda_1, \lambda_2, \lambda_3, \lambda_4\}$ is trace 0 (NIEP)-realizable if and only if $s_1 = 0$, $s_2, s_3 \geq 0$ and $s_2^2 \leq 4s_4$. The case $n = 5$ was first studied by Reams [106] and he gave a sufficient condition. The case $n = 5$ was finally solved by Laffey–Meehan [69]: $\{\lambda_1, \ldots, \lambda_5\}$ is trace 0 (NIEP)-realizable if and only if $s_1 = 0$, $s_2, s_3 \geq 0$, $s_2^2 \leq 4s_4$ and $12s_5 + 5s_3\sqrt{4s_4 - s_2^2} \geq 5s_2s_3$.

Torre–Mayo *et al.* [130] generalize these solutions in terms of the coefficients of the characteristic polynomial: the polynomial $x^n + k_p x^{n-p} + \cdots + k_{n-1}x + k_n$, with $2 \leq p \leq n \leq 2p+1$, is (NIEP)-realizable if and only if

$$k_p, \ldots, k_{2p-1} \leq 0; \quad k_{2p} \leq \tfrac{k_p^2}{4};$$

$$k_{2p+1} \leq \begin{cases} k_p k_{p+1} & \text{if } k_{2p} \leq 0, \\ k_{p+1}\left(\frac{k_p}{2} - \sqrt{\frac{k_p^2}{4} - k_{2p}}\right) & \text{if } k_{2p} > 0. \end{cases}$$

Spector [127] characterized trace 0 S-NIEP realizability for $n = 5$ by the conditions $\lambda_2 + \lambda_5 \leq 0$ and $s_3 \geq 0$.

5. Sufficient conditions

The first known sufficient condition for the NIEP, that in fact is for the R-NIEP, was announced by Suleĭmanova [128] in 1949 and proved by Perfect [100] in 1953:

$$\left.\begin{array}{r}\Lambda=\{\lambda_1,\ldots,\lambda_n\} \text{ real}, \lambda_1 \geq |\lambda| \text{ for } \lambda\in\Lambda \\ \text{and } \lambda_1+\sum_{\lambda_i<0}\lambda_i \geq 0\end{array}\right\} \Longrightarrow \Lambda \text{ is realizable.}$$

Several other proofs have been given, *e.g.*, [98]. There are several sufficient conditions for the R-NIEP that are checkable in a straightforward way, that is, one only needs to check a few algebraic inequalities, perhaps after ordering the spectrum. The authors of them are: Ciarlet in 1968 [19], Kellogg in 1971 [58], Salzmann in 1972 [111] and Fiedler in 1974 [32]. Borobia in 1995 [6] extended Kellogg's condition by grouping negative eigenvalues.

Other sufficient conditions for the R-NIEP involve partitions of the spectra considered, such as an immediate piece-wise extension of the Suleĭmanova condition. We will name this condition **Suleĭmanova–Perfect** [100, 128]:

$$\left.\begin{array}{r}\Lambda=\{\lambda_1,\lambda_{11},\ldots,\lambda_{1t_1},\ldots,\lambda_r,\lambda_{r1},\ldots,\lambda_{rt_r}\} \text{ real}, \\ \lambda_1\geq|\lambda| \text{ for } \lambda\in\Lambda \\ \text{and } \lambda_j+\sum\limits_{\lambda_{ji}<0}\lambda_{ji}\geq 0 \text{ for } j=1,\ldots,r\end{array}\right\} \Longrightarrow \Lambda \text{ is realizable.}$$

Other sufficient conditions of this type are more elaborate and we will name them by their authors: **Perfect 1** in 1953 [100], Soto 2 in 2003 [114] or its extension Soto p in 2013 [118].

Some of the sufficient conditions besides partitions involve the knowledge of the diagonal entries of a realization of part of the spectrum. The first condition of this type is due to Perfect in 1955 [101]:

$$\left.\begin{array}{r}\Lambda= \quad \{\lambda_1,\ldots,\lambda_r\} \qquad\qquad\qquad\qquad\\ \cup\{\lambda_{11},\ldots,\lambda_{1t_1}\}\cup\cdots\cup\{\lambda_{r1},\ldots,\lambda_{rt_r}\} \\ \{\lambda_1,\ldots,\lambda_r\} \text{ the spectrum of a nonnegative} \\ \text{matrix with diagonal } d_1,\ldots,d_r, \\ \lambda_{ji}\leq 0 \text{ for } j=1,\ldots,r \text{ and } i=1,\ldots,t_j, \\ \lambda_1\geq|\lambda| \text{ for } \lambda\in\Lambda\,,\ \sum\limits_{\lambda\in\Lambda}\lambda\geq 0\,, \\ \text{and } d_j+\sum\limits_{1\leq i\leq t_j}\lambda_{ji}\geq 0\,,\ j=1,\ldots,r\end{array}\right\} \Rightarrow \Lambda \text{ is realizable.}$$

When $\lambda_j \geq 0$, for $j=1,\ldots,r$, we call this condition **Perfect 2**$^+$ (see [85]). There are two equivalent conditions that extend this condition: Soto–Rojo in 2006 [122] and Soto–Rojo–Manzaneda in 2011 [123].

Other well-known sufficient conditions manipulate certain spectra to get a new realizable spectrum: Guo in 1997 [40, Theorems 2.1 and 3.1] or C-realizability [11] that we name the **game** condition [84].

In order to construct a map of sufficient conditions for the R-NIEP, Marijuán–Pisonero–Soto compared these conditions and established inclusion relations or independence relations between them, [84, 85]:

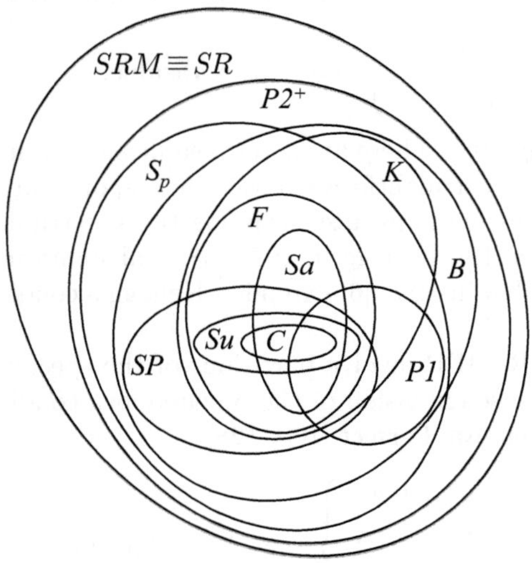

[6] Borobia = B
[19] Ciarlet = C
[32] Fiedler = F
[58] Kellogg = K
[100] Perfect 1 = $P1$
[101] Perfect 2^+ = $P2^+$
[111] Salzmann = Sa
[118] Soto p = S_p
[122] Soto–Rojo = SR
[123] Soto–Rojo–Manzaneda = SRM
[128] Suleĭmanova = Su
[100] Suleĭmanova–Perfect = SP

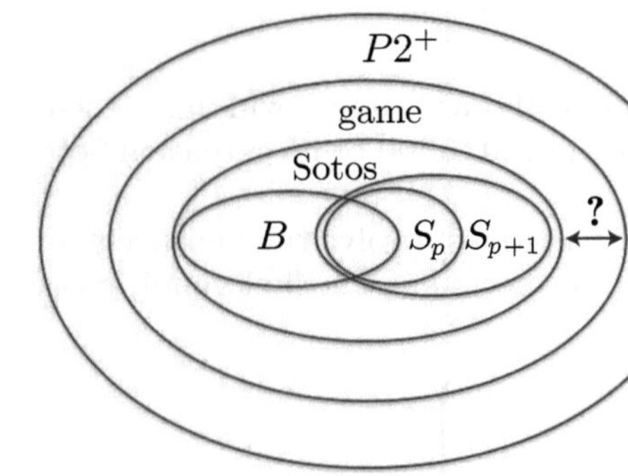

[6] Borobia = B
[101] Perfect 2^+ = $P2^+$
[118] Soto p = S_p
[118] Soto p + 1 = S_{p+1}
[118] Sotos = $\bigcup_{p\geq 2} Soto\ p$

The first known sufficient condition for the S-NIEP is due to Perfect–Mirsky in 1965 [102] for doubly stochastic matrices, and Fiedler in 1974 [32] gave the first one for symmetric nonnegative matrices. Several sufficient conditions which were first obtained for the R-NIEP have later been shown to be valid also for the S-NIEP as well. Fiedler [32], Radwan [105] and Soto [116] showed, respectively, that Kellogg [58], Borobia [6] and Soto 2 [114] are also symmetric sufficient conditions.

Soules in 1983 [126] gave two constructive sufficient conditions for symmetric realization. The inequalities that appear in these conditions are obtained by requiring the diagonal entries of the matrix $R\text{diag}(\lambda_1, \ldots, \lambda_n)R^T$ to be nonnegative, in which R is an orthogonal matrix with a certain pattern. For a particular R, this condition is the Perfect–Mirsky condition.

Soto–Rojo–Moro–Borobia gave in 2007 [124] a symmetric version of the Soto–Rojo condition. Soto p [118] and Soto–Rojo–Manzaneda [123] have also symmetric versions. Laffey–Šmigoc in 2007 [71] gave the symmetric realizability of a spectrum by manipulating two spectra.

Again in order to construct a map of sufficient conditions for the S-NIEP, Marijuán–Pisonero–Soto [86] compared these conditions and established inclusion relations or independence relations between them:

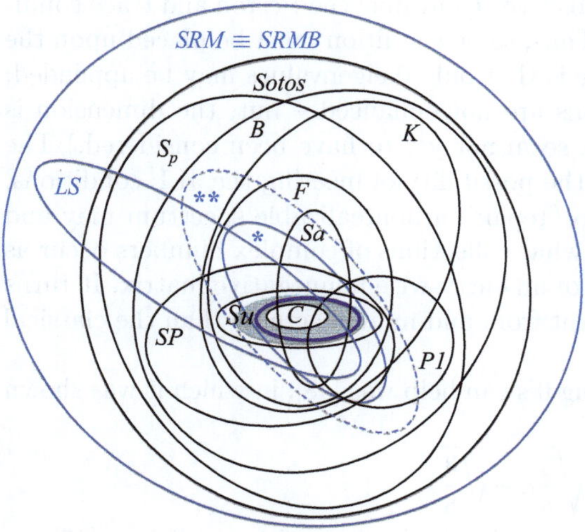

[6] Borobia $= B$
[19] Ciarlet $= C$
[32] Fiedler $= F$
[58] Kellogg $= K$
[71] Laffey–Šmigoc $= LS$
[100] Perfect 1 $= P1$
[102] Perfect–Mirsky = ▬
[111] Salzmann $= Sa$
[118] Soto p $= S_p$
[118] Sotos $= \bigcup_{p\geq 2} Soto\ p$
[123] Soto–Rojo–Manzaneda $= SRM$
[124] Soto–Rojo–Moro–Borobia $= SRMB$
[126] Soules 1 = ▪
[126] Soules 2 $= **$
[126] Soules 2 corollary $=*$
[128] Suleĭmanova $= Su$
[100] Suleĭmanova–Perfect $= SP$

The discontinuous line for Soules 2 in the map means that we only conjecture this position for this sufficient condition.

Recently, Ellard–Šmigoc [29] have modified the Laffey–Šmigoc condition and the Soules 2 condition and they have proved the following equivalence:

$$\text{Soules 2 modified} \iff \text{Laffey–Šmigoc modified} \iff \text{game} \iff \text{Sotos.}$$

This implies the symmetric realizably of the game condition and resolves the question on the RNIEP diagram about whether there exists something between Sotos and game.

There are also some sufficient conditions for the NIEP. Guo in 1997 [40, Theorems 2.1 and 3.1] gave results about changing a realizable spectrum to obtain a realizable spectrum. Šmigoc in 2004 [112, Theorems 10 and 14] gave other such results. Finally, let us mention some generalizations of the Suleĭmanova condition: Laffey–Šmigoc in 2006 [70, Theorems 1 and 3] and Borobia–Moro–Soto in 2008 [11, Theorem 3.3].

6. Embedding spectra, by adding 0's, to achieve realizability

Consider now spectra that meet the most basic necessary conditions for NIEP-realizability: (Perron), (Trace) and (Reality). Even when these conditions are strictly met, the proposed spectrum, σ, need not be realizable (*e.g.*, $\{3,3,-\sqrt{3}\pm i\}$). This raises the natural question of whether the spectrum may be embedded in a larger one, that is realizable, by appending some additional eigenvalues, *e.g.*, $\sigma \hookrightarrow \sigma \cup \tau$. If we are too liberal about what eigenvalues may be appended, this question becomes trivial. For example, realizability may always be achieved by appending a single, sufficiently large, positive eigenvalue; the Perron and trace conditions may be arbitrarily improved. Thus, some condition must be placed upon the appended eigenvalues. A natural one is that only 0 eigenvalues may be appended; now, the Perron and trace conditions are not enhanced – but, the dimension is increased. (Intermediate restrictions seem not yet to have been considered.) The increase in dimension does improve the possibility of meeting the JLL conditions.

The notion of appending 0's to "repair" a nonrealizable spectrum may and has often been viewed another way: what collections of complex numbers occur as the "nonzero part" of the spectrum of an entry-wise nonnegative matrix. It turns out that this question is quite different from and more tractable than the classical NIEP.

The first to show that appending 0's can help was [48], in which it was shown that the spectrum

$$1, \sqrt{\frac{3}{8}}i, -\sqrt{\frac{3}{8}}i$$

is not realizable in dimension 3 (because the single eigenvalue conditions [57] are not met, or the JLL conditions are not met), but the spectrum

$$1, \sqrt{\frac{3}{8}}i, -\sqrt{\frac{3}{8}}i, 0$$

is realizable in dimension 4 by the matrix

$$\begin{pmatrix} 1/4 & 0 & 3/4 & 0 \\ 1/4 & 1/4 & 0 & 1/2 \\ 0 & 3/4 & 1/4 & 0 \\ 7/24 & 0 & 11/24 & 1/4 \end{pmatrix}.$$

Later in [12], it was shown, remarkably, that if a spectrum (with no 0's) meets the basic necessary conditions strictly, then it is the nonzero part of the spectrum of a nonnegative matrix. Sufficiently many 0's may be appended to achieve realizability. Of course, the number of 0's that need be added may be very large (not uniformly bounded in terms of n) because of JLL. Interestingly, the newish method of symbolic dynamics was used in an import way, though many easily proven matricial lemmas were needed as well. Specifically the result is

Theorem 1. *The list of nonzero complex numbers* $\Lambda = \{\lambda_1, \lambda_2, \ldots, \lambda_n\}$ *is the nonzero spectrum of a primitive matrix if and only if*

1. **strict Perron condition**: *Λ contains a positive eigenvalue of multiplicity one that is greater in absolute value than all other λ_i's;*
2. **reality condition**: $p(t) = \prod_{i=1}^{n} (t - \lambda_i)$ *has real coefficients;*

   and
3. **extended trace condition**: $\sum_{i=1}^{n} \lambda_i^{mk} \geq 0$ *for all* k *and* $\sum_{i=1}^{n} \lambda_i^k > 0$ *implies*

$$\sum_{i=1}^{n} \lambda_i^{mk} > 0 \quad \textit{for all } m.$$

The necessity of these conditions is easily verified (the last one via JLL, for example), and the interesting point is that necessary conditions become sufficient when primitivity is the goal and the dimension may be arbitrarily increased. When the Perron condition is not strict it may not be possible to "save" a spectrum meeting obvious necessary conditions. The familiar example $3, 3, -2, -2, -2$ is not only not realizable, but is never the nonzero part of the spectrum of a nonnegative matrix. However, both $3 + \epsilon, 3, -2, -2, -2$ and $3 + \epsilon, 3 - \epsilon, -2, -2, -2$ are both the nonzero parts of the spectra of primitive matrices for arbitrarily small $\epsilon > 0$ (though they are not realizable for ϵ small).

More recently there have been further developments about realizability after appending 0's to a spectrum. In [65] there is matricial proof of key results from [12], which is much more explicit. Though the number of 0's needed to make a spectrum realizable may be very large (and not easy to estimate from [12]), estimates have recently been given under some circumstances by bringing $s_2(\Lambda)$, as well as the trace, into play [70].

7. The graph NIEP

Not surprisingly, realizable spectra that are, in some way extremal, are often realizable by nonnegative matrices with many 0 entries, as are many non-extremal spectra. This raises the question that, if we fix the 0-pattern of a nonnegative matrix, how is the NIEP restricted, *i.e.*, which realizable spectra occur? A natural way to describe a particular 0-pattern is via a graph, which could be directed or undirected. For a particular graph G on n vertices, consider the set of nonnegative n-by-n matrices $N(G)$ for which $A = (a_{ij})$ satisfies $a_{ij} > 0$ if and only if (i, j) is a directed edge, $i \neq j$, of G. (Here, we consider an undirected edge $\{i, j\}$, if G is undirected, to consist of two directed edges (i, j) and (j, i).) For simplicity we consider graphs without loops, and no restriction is placed by the graph upon the diagonal entries, other than nonnegativity. The G-NIEP then just asks which spectra occur among matrices in $N(G)$? (if we wish to emphasize the dimension, which we generally take to be implicit, we may write $N_n(G)$.) Of course the NIEP is just the union of the solutions to the G-NIEP's, over all directed graphs G. The same is true for various variations upon the NIEP. For example, the solution to the S-NIEP is just the solution to the G-NIEP, restricted to symmetric matrices, over all undirected graphs. Thus far, the G-NIEP has been considered only for undirected graphs G, symmetric matrices and real eigenvalues. This seems a fertile area for future work.

A prototype of the G-NIEP, though not presented in graph terms, appeared in [34], in which tridiagonal matrices were considered. Notice that the spectrum of a nonnegative tridiagonal matrix is not only necessarily real but also the spectrum of a symmetric tridiagonal nonnegative matrix. Also, tridiagonal matrices are the case in which G is a path (and an edge only requires nonnegativity of the entry); off-diagonal 0 entries are important. We say that a matrix A is **subordinate** to a graph G if $G(A)$ has the same vertices as G and the edges of $G(A)$ are contained among those of G; equivalently, for $A = (a_{ij})$, $a_{ij} \neq 0$ implies $\{i, j\}$ is an edge of G. The main result of [34] is for the path P on n vertices. Then, for $\lambda_1 \geq \lambda_2 \geq \cdots \geq \lambda_n$, there is an n-by-n nonnegative matrix $A^T = A$ subordinate to P and with eigenvalues $\lambda_1, \lambda_2, \ldots, \lambda_n$ if and only if

$$\lambda_i + \lambda_{n-i+1} \geq 0$$

$i = 1, 2, \ldots, n$. Additional conditions, when the graph is precisely P, are also discussed, but the results are incomplete.

In [48, 74], the observation of [34] is dramatically generalized. Recall that a path is a tree and that all trees are bipartite (but many non-trees are bipartite as well). All bipartite graphs are considered in [74]. The main result is that $\lambda_1 \geq \cdots \geq \lambda_n$ are the eigenvalues of an n-by-n nonnegative symmetric matrix A, subordinate to a given bipartite graph G on n vertices, if and only if

$$\begin{gathered}
\lambda_1 + \lambda_n \geq 0 \\
\lambda_2 + \lambda_{n-1} \geq 0 \\
\vdots \\
\lambda_m + \lambda_{n-m+1} \geq 0 \\
\lambda_{m+1} \geq 0 \\
\vdots \\
\lambda_{n-m} \geq 0
\end{gathered}$$

in which m is the matching number of G. (Note that $\lfloor \frac{n}{2} \rfloor$ is the matching number of a path on n vertices, so that the result of [34] is a special case.) Further observations about the S-NIEP for matrices subordinate to a given graph G are also made.

8. Perron similarities

If S is an invertible matrix and there is a real, diagonal, nonscalar matrix D with $SDS^{-1} \geq 0$, then S is called a *Perron similarity*. Perron similarities were introduced by Johnson and Paparella to study the diagonalizable R-NIEP and the S-NIEP.

Perron similarities were characterized in several ways in [54] and it was shown that $\mathcal{C}(S) := \{x \in \mathbb{R}^n : SD_xS^{-1} \geq 0\}$ is a *polyhedral cone*, *i.e.*, a convex cone in $\mathbb{R}^n$ with finitely-many extremals. This was called the *(Perron) spectracone of* S, and a certain cross-section, a polytope called the *(Perron) spectratope of* S, was also discussed.

These polyhedral sets were used to verify the known necessary and sufficient conditions for the R-NIEP and the S-NIEP for orders up to four. For orders 1, 2, and 4, it is shown that a finite number of Perron similarities are required to cover the realizable region, whereas when $n = 3$, it is shown, via the *relative gain array* (see, *e.g.*, [44]), that an uncountable number of Perron similarities is required to cover the realizable region.

For every $n \geq 1$, the spectracone and spectratope of H_n were characterized, where H_n denotes the *canonical Hadamard matrix of order* $2n$. More specifically, the spectracone of H_n is the conical hull of its rows and the spectratope of H_n is the convex hull of its rows. The spectratope of a general, normalized Hadamard matrix was used to give a constructive proof, for *Hadamard orders*, of a result by Fiedler [32], that every *Suleĭmanova spectrum* is the spectrum of a symmetric nonnegative matrix (in fact, this result was strengthened to show that the constructed realizing matrix is also *doubly stochastic*). It is still an open problem to find a constructive proof that every Suleĭmanova spectrum is SNIEP-realizable. A constructive proof of the Boyle–Handelman theorem for Suleĭmanova spectrum: augmenting any such spectrum with zeros up to a Hadamard order yields a spectrum realizable by a nonnegative matrix that is symmetric and doubly stochastic.

9. Jordan structure and the NIEP's

When there are repeated eigenvalues in a proposed spectrum, a natural question is whether Jordan structure for the repeated eigenvalues can play a role in realizability. The J-NIEP asks which particular Jordan canonical forms occur for n-by-n nonnegative matrices and the D-NIEP is the special case of which spectra occur among diagonalizable n-by-n nonnegative matrices. Of course, the D-NIEP and the NIEP are the same for spectra with distinct eigenvalues. It is also a simple exercise that any spectrum that is D-NIEP realizable by a positive matrix is also J-NIEP realizable for any Jordan canonical form possible for its eigenvalues. However, whether D-NIEP realizability always implies J-NIEP realizability for any possible Jordan form is unclear. This is so for Suleĭmanova spectra [23, 120, 121].

There are spectra that are realizable (in fact, R-NIEP realizable), but are not diagonalizably realizable. The smallest dimension in which this occurs is 5. For $n \leq 4$, any realizable spectrum with (non-real) complex eigenvalues is diagonalizably realizable (the only difficulty could come from a multiple Perron root and that is easily handled by a reducibility argument). For $n \leq 4$, R-NIEP and

S-NIEP realizability are the same, which settles the matter. However, for $n = 5$, the spectrum

$$3+t, 3-t, -2, -2, -2$$

is realizable for $t > (16\sqrt{6})^{1/2} - 39 \approx 0.437\ldots$, [69]. However, it is diagonalizably realizable iff $t \geq 1$ (in which case it is also symmetrically realizable). Thus, for $0.437\ldots < t < 1$, this 5-spectrum is realizable, but not diagonalizable so. This suggests that this phenomenon is fairly common.

References

[1] R.B. Bapat and T.E.S. Raghavan, *Nonnegative Matrices and Applications.* Cambridge University Press, Cambridge, 1997.

[2] L. Benvenuti, *A geometrical representation of the spectra of four dimensional nonnegative matrices.* Linear Algebra Appl. **445** (2014), 162–180.

[3] L. Benvenuti, *A note on eigenvalues location for trace zero doubly stochastic matrices.* Electron. J. Linear Algebra **30** (2015), 599–604.

[4] A. Berman and R.J. Plemmons, *Nonnegative Matrices in the Mathematical Sciences.* SIAM Publications, Philadelphia, 1994.

[5] G. Bharali and O. Holtz, *Functions preserving nonnegativity of matrices.* SIAM J. Matrix Anal. Appl. **30** (2008), 84–101.

[6] A. Borobia, *On the nonnegative eigenvalue problem.* Linear Algebra Appl. **223–224** (1995), 131–140.

[7] A. Borobia, *Inverse eigenvalue problems.* In: "Handbook of Linear Algebra", chapter 20. 20-1–20-14, Chapman & Hall/CRC, USA 2007.

[8] A. Borobia and J. Moro, *On non-negative matrices similar to positive matrices.* Linear Algebra Appl. **266** (1997) 365–379.

[9] A. Borobia and J. Moro, *On the boundary of the set of real spectra of non-negative matrices.* Linear Algebra Appl. **278** (1998), 287–293.

[10] A. Borobia, J. Moro, and R.L. Soto, *Negativity compensation in the nonnegative inverse eigenvalue problem.* Linear Algebra Appl. **393** (2004), 73–89.

[11] A. Borobia, J. Moro, and R.L. Soto, *A unified view on compensation criteria in the real nonnegative inverse eigenvalue problem.* Linear Algebra Appl. **428** (2008), 2574–2584.

[12] M. Boyle and D. Handelman, *The spectra of nonnegative matrices via symbolic dynamics.* Ann. Math. (2) **133** (1991), 249–316.

[13] A. Brauer, *Limits for the characteristic roots of a matrix, IV: Applications to stochastic matrices.* Duke Math. J. **19** (1952), 75–91.

[14] R.A. Brualdi, *Spectra of digraphs.* Linear Algebra Appl. **432** (2010), 2181–2213.

[15] J. Ccapa and R.L. Soto, *On spectra perturbation and elementary divisors of positive matrices.* Electron. J. Linear Algebra **18** (2009), 462–481.

[16] J. Ccapa and R.L. Soto, *On elementary divisors perturbation of nonnegative matrices.* Linear Algebra Appl. **432** (2010), 546–555.

[17] M.T. Chu and K.R. Driessel, *Constructing symmetric nonnegative matrices with prescribed eigenvalues by differential equations.* SIAM J. Math. Anal. **22** (1991), 1372–1387.

[18] M.T. Chu and G.H. Golub, *Inverse Eigenvalue Problems: Theory, Algorithms and Applications.* Oxford University Press, New York, 2005.

[19] P.G. Ciarlet, *Some results in the theory of nonnegative matrices.* Linear Algebra Appl. **1** (1968), 139–152.

[20] A.G. Cronin, *Characterizing the spectra of nonnegative matrices.* PhD thesis, School of Mathematical Sciences, University College Dublin, 2012.

[21] A.G. Cronin and T.J. Laffey, *An inequality for the spectra of nonnegative matrices.* Linear Algebra Appl. **436** (2012), 3225–3238.

[22] D.M. Cvetković, M. Doob, and H. Sachs, *Spectra of Graphs.* Johann Ambrosius Barth, Heidelberg, 1995.

[23] R.C. Díaz and R.L. Soto, *Nonnegative inverse elementary divisors problem in the left half plane.* Linear Multilinear Algebra **64** (2016), 258–268.

[24] N. Dmitriev and E. Dynkin, *On the characteristic numbers of a stochastic matrix.* Dokl. Akad. Nauk SSSR **49** (1945), 159–162 (in Russian).

[25] N. Dmitriev and E. Dynkin, *Eigenvalues of a stochastic matrix.* Izv. Akad. Nauk SSSR Ser. Mat. **10** (1946), 167–184 (in Russian).

[26] D.Ž. Dokovič, *Cyclic polygons, roots of polynomials with decreasing nonnegative coefficients, and eigenvalues of stochastic matrices.* Linear Algebra Appl. **142** (1990), 173–193.

[27] P.D. Egleston, T.D. Lenker, and S.K. Narayan, *The nonnegative inverse eigenvalue problem.* Linear Algebra Appl. **379** (2004), 475–490.

[28] R. Ellard and H. Šmigoc, *Constructing new realisable lists from old in the NIEP.* Linear Algebra Appl. **440** (2014), 218–232.

[29] R. Ellard and H. Šmigoc, *Connecting sufficient conditions for the symmetric nonnegative inverse eigenvalue problem.* Linear Algebra Appl. **498** (2016), 521–552.

[30] L. Elsner, R. Nabben, and M. Neumann, *Orthogonal bases which lead to symmetric nonnegative basis.* Linear Algebra Appl. **271** (1998), 323–343.

[31] G.M. Engel, H. Schneider, and S. Sergeev, *On sets of eigenvalues of matrices with prescribed row sums and prescribed graph.* Linear Algebra Appl. **455** (2014), 187–209.

[32] M. Fiedler, *Eigenvalues of nonnegative symmetric matrices.* Linear Algebra Appl. **9** (1974), 119–142.

[33] S. Friedland, *On an inverse problem for nonnegative and eventually nonnegative matrices.* Israel J. Math. **29** (1978), 43–60.

[34] S. Friedland and A.A. Melkman, *On the eigenvalues of nonnegative Jacobi matrices.* Linear Algebra Appl. **25** (1979), 239–253.

[35] G. Frobenius, *Über Matrizen aus positiven Elementen.* S.-B. K. Preuss. Akad. Wiss. Berlin, 1908 p. 471–476, 1909 p. 514–518.

[36] G. Frobenius, *Über Matrizen aus nicht negativen Elementen.* S.-B. K. Preuss. Akad. Wiss. Berlin, 1912 p. 456–477.

[37] R. Gómez and J.M. Salazar, *Spanning tree invariants, loop systems and doubly stochastic matrices.* Linear Algebra Appl. **432** (2010), 556–565.

[38] S. Guo and W. Guo, *Perturbing non-real eigenvalues of nonnegative matrices.* Linear Algebra Appl. **426** (2007), 199–203.

[39] W. Guo, *An inverse eigenvalue problem for nonnegative matrices.* Linear Algebra Appl. **249** (1996), 67–78.

[40] W. Guo, *Eigenvalues of nonnegative matrices.* Linear Algebra Appl. **266** (1997), 261–270.

[41] D. Hershkowits, *Existence of Matrices Satisfying Prescribed Conditions.* M. S. Thesis, Technion-Israel Inst. of Technology, Haifa, Israel, 1978.

[42] O. Holtz, *M-matrices satisfy Newton's inequalities* Proc. Amer. Math. Soc. **133** (2005), 711–717.

[43] R.A. Horn and C.R. Johnson, *Matrix Analysis.* Cambridge University Press, Cambridge 1990. Corrected reprint of the 1985 original.

[44] R.A. Horn and C.R. Johnson, *Topics in Matrix Analysis.* Cambridge University Press, Cambridge 1991.

[45] S.-G. Hwang and S.-S. Pyo, *The inverse eigenvalue problem for symmetric doubly stochastic matrices.* Linear Algebra Appl. **379** (2004), 77–83.

[46] H. Ito, *A new statement about the theorem determining the region of eigenvalues of stochastic matrices.* Linear Algebra Appl. **267** (1997), 241–246.

[47] C.R. Johnson, *An inclusion for the field of values of a doubly stochastic matrix based on its graph.* Aequationes Mathematicae **17** (1978), 305–310.

[48] C.R. Johnson, *Row stochastic matrices similar to doubly stochastic matrices.* Linear Multilinear Algebra **10** (1981), 113–130.

[49] C.R. Johnson, R.B. Kellogg, and A.B. Stephens, *Complex eigenvalues of a nonnegative matrix with a specified graph. II.* Linear Multilinear Algebra **7** (1979), 129–143.

[50] C.R. Johnson, T.J. Laffey, and R. Loewy, *The real and the symmetric nonnegative inverse eigenvalue problems are different.* Proc. Amer. Math. Soc. **124** (1996), 3647–3651.

[51] C.R. Johnson, C. Marijuán, and M. Pisonero, *Matrices and spectra satisfying the Newton inequalities.* Linear Algebra Appl. **430** (2009), 3030–3046.

[52] C.R. Johnson, C. Marijuán, and M. Pisonero, *Symmetric nonnegative realizability via partitioned majorization.* Linear Multilinear Algebra **65** (2017), 1417–1426.

[53] C.R. Johnson, C. Marijuán, and M. Pisonero, *Ruling out certain 5-spectra for the symmetric nonnegative inverse eigenvalue problem.* Linear Algebra Appl. **512** (2017), 129–135.

[54] C.R. Johnson and P. Paparella, *Perron spectratopes and the real nonnegative inverse eigenvalue problem.* Linear Algebra Appl. **493** (2016), 281–300.

[55] C.R. Johnson and P. Paparella, *A matricial view of the Karpelevič theorem.* Linear Algebra Appl. **520** (2017), 1–15.

[56] A.I. Julio and R.L. Soto, *Persymmetric and bisymmetric nonnegative inverse eigenvalue problem.* Linear Algebra Appl. **469** (2015), 130–152.

[57] F. Karpelevich, *On the eigenvalues of a matrix with nonnegative elements.* Izv. Akad. Nauk SSR Ser. Mat. **15** (1951), 361–383 (in Russian).

[58] R. Kellogg, *Matrices similar to a positive or essentially positive matrix.* Linear Algebra Appl. **4** (1971), 191–204.

[59] R.B. Kellogg and A.B. Stephens, *Complex eigenvalues of a non-negative matrix with a specified graph.* Linear Algebra Appl. **20** (1978), 179–187.

[60] K.H. Kim, N.S. Ormes, and F.W. Roush, *The spectra of nonnegative integer matrices via formal power series.* J. Amer. Math. Soc. **13** (2000), 773–806.

[61] J. Knudsen and J.J. McDonald, *A note on the convexity of the realizable set of eigenvalues for nonnegative symmetric matrices.* Electron. J. Linear Algebra **8** (2001), 110–114.

[62] A.N. Kolmogorov, *Markov chains with a countable number of possible states.* Bull. Moskow Gos. Univ. Mat. Meh. **1** (3)(1937), 1–16.

[63] T.J. Laffey, *Realizing matrices in the nonnegative inverse eigenvalue problem.* Matrices and group representations (Coimbra, 1998), vol. 19 of Textos Mat. Sér. B, 21–32. Univ. Coimbra, Coimbra, 1999.

[64] T.J. Laffey, *Perturbing non-real eigenvalues of nonnegative real matrices.* Electron. J. Linear Algebra **12** (2005), 73–76.

[65] T.J. Laffey, *A constructive version of the Boyle–Handelman theorem on the spectra of nonnegative matrices.* Linear Algebra Appl. **436** (2012), 1701–1709.

[66] T.J. Laffey, R. Loewy, and H. Šmigoc, *Nonnegative matrices that are similar to positive matrices.* SIAM J. Matrix Anal. Appl. **31** (2009), 629–649.

[67] T.J. Laffey, R. Loewy, and H. Šmigoc, *Power series with positive coefficients arising from the characteristic polynomials of positive matrices.* Math. Ann. **364** (2016), 687–707.

[68] T.J. Laffey and E. Meehan, *A refinament of an inequality of Johnson, Loewy and London on nonnegative matrices and some applications.* Electron. J. Linear Algebra **3** (1998), 119–128.

[69] T.J. Laffey and E. Meehan, *A characterization of trace zero nonnegative* 5×5 *matrices.* Linear Algebra Appl. **302/303** (1999), 295–302.

[70] T.J. Laffey and H. Šmigoc, *Nonnegative realization of spectra having negative real parts.* Linear Algebra Appl. **416** (2006), 148–159.

[71] T.J. Laffey and H. Šmigoc, *Construction of nonnegative symmetric matrices with given spectrum.* Linear Algebra Appl. **421** (2007), 97–109.

[72] T.J. Laffey and H. Šmigoc, *On a classic example in the nonnegative inverse eigenvalue problem.* Electron. J. Linear Algebra. **17** (2008), 333–342.

[73] T.J. Laffey and H. Šmigoc, *Nonnegatively realizable spectra with two positive eigenvalues.* Linear Multilinear Algebra **58** (2010), 1053–1069.

[74] A. Leal-Duarte and C.R. Johnson, *Resolution of the symmetric nonnegative inverse eigenvalue problem for matrices subordinate to a bipartite graph.* Positivity **8** (2004), 209–213.

[75] J. Levick, R. Pereira, and D.W. Kribs, *The four-dimensional Perfect–Mirsky conjecture.* Proc. Amer. Math. Soc. **143** (2014), 1951–1956.

[76] Y.J. Ley, W.R. Xu, Y. Lu, and Y.R. Niu, *On the symmetric doubly stochastic inevrese eigenvalue problem.* Linear Algebra Appl. **445** (2014), 181–205.

[77] R. Loewy and D. London, *A note on the inverse problem for nonnegative matrices.* Linear Multilinear Algebra **6** (1978), 83–90.

[78] R. Loewy and J.J. McDonald, *The symmetric nonnegative inverse eigenvalue problem for* 5×5 *matrices.* Linear Algebra Appl. **393** (2004), 275–298.

[79] R. Loewy and O. Spector, *A necessary condition for the spectrum of nonnegative symmetric* 5×5 *matrices.* Linear Algebra Appl. **528** (2017), 206–272.

[80] C. Maclaurin, *A second letter to Martin Folkes, Esq.; concerning the roots of eqnarrays, with the demonstration of other rules in algebra.* Phil. Transactions **36** (1729), 59–96.

[81] M. Marcus and H. Minc, *A Survey of Matrix Theory and Matrix Inequalities.* Allyn and Bacon, Boston, 1964.

[82] C. Marijuán and M. Pisonero, *Relaciones entre condiciones suficientes en el problema espectral real inverso no negativo.* Proceedings V Jornadas de Matemática Discreta y Algorítmica (2006), 335–342.

[83] C. Marijuán and M. Pisonero, *Polinomios característicos de digrafos pesados y desigualdades de Newton.* Proceedings VII Jornadas de Matemática Discreta y Algorítmica (2010), 447–458.

[84] C. Marijuán and M. Pisonero, *On sufficient conditions for the RNIEP and their margins of realizability.* Electronic Notes in Discrete Mathematics **46** (2014), 201–208.

[85] C. Marijuán, M. Pisonero, and R.L. Soto, *A map of sufficient conditions for the real nonnegative inverse eigenvalue problem.* Linear Algebra Appl. **426** (2007), 690–705.

[86] C. Marijuán, M. Pisonero, and R.L. Soto, *A map of sufficient conditions for the symmetric nonnegative inverse eigenvalue problem.* Linear Algebra Appl. **530** (2017), 344–365.

[87] J. Mashreghi and R. Rivard, *On a conjecture about the eigenvalues of doubly stochastic matrices.* Linear Multilinear Algebra **55** (2007), 491–498.

[88] J.J. McDonald and M. Neumann, *The Soules approach to the inverse eigenvalue problem for nonnegative symmetric matrices of order* $n \leq 5$. Contemp. Math. **259** (2000), 387–407.

[89] E. Meehan, *Some Results On Matrix Spectra.* PhD thesis, National University of Ireland, Dublin, 1998.

[90] H. Minc, *Inverse elememtary divisor problem for nonnegative matrices.* Proc. Amer. Math. Soc. **83** (1981), 665–669.

[91] H. Minc, *Inverse elememtary divisor problem for doubly stochastic matrices.* Linear Multilinear Algebra **11** (1982), 121–131.

[92] H. Minc, *Non-Negative Matrices.* John Wiley and Sons, New York, 1988.

[93] L. Mirsky, *Results and problems in the theory of doubly-stochastic matrices.* Z. Wahrscheinlichkeitstheorie 1 (1963), 319–334.

[94] B. Mourad, *An inverse problem for* $n \times n$ *symmetric doubly stochastic matrices.* Inverse Problems **19** (2003), 821–831.

[95] B. Mourad, *A note on the boundary of the set where the decreasingly ordered spectra of symmetric doubly stochastic matrices lie.* Linear Algebra Appl. **416** (2006), 546–558.

[96] I. Newton, *Arithmetica universalis: sive de compositione et resolutione arithmetica liber.* 1707.

[97] G.N. de Oliveira, *Sobre matrizes estocásticas e duplamente estocásticas.* PhD thesis, Coimbra, 1968.

[98] P. Paparella, *Realizing Suleimanova-type spectra via permutative matrices.* Electron. J. Linear Algebra **31** (2016), 306–312.

[99] H. Perfect, *On positive stochastic matrices with real characteristic roots.* Proc. Cambridge Phil. Soc. **48** (1952), 271–276.

[100] H. Perfect, *Methods of constructing certain stochastic matrices.* Duke Math. J. **20** (1953), 395–404.

[101] H. Perfect, *Methods of constructing certain stochastic matrices II.* Duke Math. J. **22** (1955), 305–311.

[102] H. Perfect and L. Mirsky, *Spectral properties of doubly stochastic matrices.* Monatsh. Math. **69** (1965), 35–57.

[103] H. Perfect and L. Mirsky, *The distribution of positive elements in doubly-stochastic matrices.* J. London Math. Soc. **40** (1965), 689–698.

[104] O. Perron, *Zur Theorie der Matrizen.* Math. Ann. **64** (1907), 248–263.

[105] N. Radwan, *An inverse eigenvalue problem for symmetric and normal matrices.* Linear Algebra Appl. **248** (1996), 101–109.

[106] R. Reams, *An inequality for nonnegative matrices and inverse eigenvalue problem.* Linear Multilinear Algebra **41** (1996), 367–375.

[107] R. Reams, *Construction of trace zero symmetric stochastic matrices for the inverse eigenvalue problem.* Electron. J. Linear Algebra **9** (2002), 270–275.

[108] O. Rojo and R.L. Soto, *Existence and construction of nonnegative matrices with complex spectrum.* Linear Algebra Appl. **368** (2003), 53–69.

[109] O. Rojo and R.L. Soto, *Guo perturbations for symmetric nonnegative circulant matrices.* Linear Algebra Appl. **431** (2009), 594–607.

[110] V. Romanovsky, *Recherches sur les chaînes de Markoff* (*Premier Mémoire*). Acta Math. **66** (1936), 147–251.

[111] F. Salzmann, *A note on the eigenvalues of nonnegative matrices.* Linear Algebra Appl. **5** (1972), 329–338.

[112] H. Šmigoc, *The inverse eigenvalue problem for nonnegative matrices.* Linear Algebra Appl. **393** (2004), 365–374.

[113] H. Šmigoc, *Construction of nonnegative matrices and the inverse eigenvalue problem.* Linear Multilinear Algebra **53** (2005), 85–96.

[114] R.L. Soto, *Existence and construction of nonnegative matrices with prescribed spectrum.* Linear Algebra Appl. **369** (2003), 169–184.

[115] R.L. Soto, *Realizability by symmetric nonnegative matrices.* Proyecciones **24** (2005), 65–78.

[116] R.L. Soto, *Realizability criterion for the symmetric nonnegative inverse eigenvalue problem.* Linear Algebra Appl. **416** (2006), 783–794.

[117] R.L. Soto, *Nonnegative Inverse Eigenvalue Problem, Linear Algebra – Theorems and Applications.* Prof. Hassan Yasser (ed.), InTech, DOI: 10.5772/48279.

Available from: http://www.intechopen.com/books/linear-algebra-theorems-and-applications/nonnegative-inverse-eigenvalue-problem, 2012.

[118] R.L. Soto, *A family of realizability criteria for the real and symmetric nonnegative inverse eigenvalue problem.* Numer. Linear Algebra Appl. **20** (2013), 336–348.

[119] R.L. Soto, A. Borobia, and J. Moro, *On the comparison of some realizability criteria for the real nonnegative inverse eigenvalue problem.* Linear Algebra Appl. **396** (2005), 223–241.

[120] R.L. Soto and J. Ccapa, *Nonnegative matrices with prescribed elementary divisors.* Electron. J. Linear Algebra **17** (2008), 287–303.

[121] R.L. Soto, R.C. Díaz, H. Nina, and M. Salas, *Nonnegative matrices with prescribed spectrum and elementary divisors.* Linear Algebra Appl. **439** (2013), 3591–3604.

[122] R.L. Soto and O. Rojo, *Applications of a Brauer theorem in the nonnegative inverse eigenvalue problem.* Linear Algebra Appl. **416** (2006), 844–856.

[123] R.L. Soto, O. Rojo, and C.B. Manzaneda, *On nonnegative realization of partitioned spectra* Electron. J. Linear Algebra **22** (2011), 557–572.

[124] R.L. Soto, O. Rojo, J. Moro, and A. Borobia, *Symmetric nonnegative realization of spectra.* Electron. J. Linear Algebra **16** (2007), 1–18.

[125] R.L. Soto, M. Salas, and C.B. Manzaneda, *On nonnegative realization of complex spectra.* Electron. J. Linear Algebra **20** (2010), 595–609.

[126] G. Soules, *Constructing symmetric nonnegative matrices.* Linear Multilinear Algebra **13** (1983), 241–251.

[127] O. Spector, *A characterization of trace zero symmetric nonnegative 5x5 matrices.* Linear Algebra Appl. **434** (2011), 1000–1017.

[128] H.R. Suleĭmanova, *Stochastic matrices with real characteristic values.* Dokl. Akad. Nauk SSSR. **66** (1949), 343–345 (in Russian).

[129] J. Swift, *The Location of Characteristic Roots of Stochastic Matrices.* M.Sc. thesis, McGill University, Montreal, 1972.

[130] J. Torre–Mayo, M.R. Abril-Raymundo, E. Alarcia-Estévez, C. Marijuán, and M. Pisonero, *The nonnegative inverse eigenvalue problem from the coefficients of the characteristic polynomial. EBL digraphs.* Linear Algebra Appl. **426** (2007), 729–773.

Charles R. Johnson
Department of Mathematics
College of William and Mary
Williamsburg, VA 23187-8795, USA
e-mail: crjohn@wm.edu

Pietro Paparella
Div. of Engineering & Mathematics
University of Washington Bothell
Bothell, WA 98011-8246, USA
e-mail: pietrop@uw.edu

Carlos Marijuán
Dpto. Matemática Aplicada
E.I. Informática
Paseo de Belén 15
E-47011 Valladolid, Spain
e-mail: marijuan@mat.uva.es

Miriam Pisonero
Dpto. Matemática Aplicada
E.T.S. de Arquitectura
Avenida de Salamanca 18
E-47014 Valladolid, Spain
e-mail: mpisoner@maf.uva.es

Operator Theory:
Advances and Applications, Vol. 267, 221–246

Semi-Fredholmness of Weighted Singular Integral Operators with Shifts and Slowly Oscillating Data

Alexei Yu. Karlovich, Yuri I. Karlovich and Amarino B. Lebre

Abstract. Let α, β be orientation-preserving homeomorphisms of $[0,\infty]$ onto itself, which have only two fixed points at 0 and ∞, and whose restrictions to $\mathbb{R}_+ = (0,\infty)$ are diffeomorphisms, and let U_α, U_β be the corresponding isometric shift operators on the space $L^p(\mathbb{R}_+)$ given by $U_\mu f = (\mu')^{1/p}(f \circ \mu)$ for $\mu \in \{\alpha, \beta\}$. We prove sufficient conditions for the right and left Fredholmness on $L^p(\mathbb{R}_+)$ of singular integral operators of the form $A_+ P_\gamma^+ + A_- P_\gamma^-$, where $P_\gamma^\pm = (I \pm S_\gamma)/2$, S_γ is a weighted Cauchy singular integral operator, $A_+ = \sum_{k\in\mathbb{Z}} a_k U_\alpha^k$ and $A_- = \sum_{k\in\mathbb{Z}} b_k U_\beta^k$ are operators in the Wiener algebras of functional operators with shifts. We assume that the coefficients a_k, b_k for $k \in \mathbb{Z}$ and the derivatives of the shifts α', β' are bounded continuous functions on $\mathbb{R}_+$ which may have slowly oscillating discontinuities at 0 and ∞.

Mathematics Subject Classification (2010). 45E05, 47A53, 47G10, 47G30.

Keywords. Right Fredholmness, left Fredholmness, slowly oscillating shift, Wiener algebra of functional operators, weighted singular integral operator, Mellin pseudodifferential operator.

1. Introduction

Let $\mathcal{B}(X)$ denote the Banach algebra of all bounded linear operators acting on a Banach space X. Recall that an operator $A \in \mathcal{B}(X)$ is said to be left invertible (resp. right invertible) if there exists an operator $B \in \mathcal{B}(X)$ such that $BA = I$ (resp. $AB = I$) where $I \in \mathcal{B}(X)$ is the identity operator on X. The operator B is called a left (resp. right) inverse of A. An operator $A \in \mathcal{B}(X)$ is said to be

This work was partially supported by the Fundação para a Ciência e a Tecnologia (Portuguese Foundation for Science and Technology) through the projects UID/MAT/00297/2013 (Centro de Matemática e Aplicações) and UID/MAT/04721/2013 (Centro de Análise Funcional, Estruturas Lineares e Aplicações). The second author was also supported by the SEP-CONACYT Project No. 168104 (México).

invertible if it is left invertible and right invertible simultaneously. We say that A is strictly left (resp. right) invertible if it is left (resp. right) invertible, but not invertible. If the operator A is invertible only from one side, then the corresponding inverse is not uniquely defined. We refer to [10, Section 2.5] for further properties of one-sided invertible operators acting on Banach spaces.

Let $\mathcal{K}(X)$ be the closed two-sided ideal of all compact operators in $\mathcal{B}(X)$, and let $\mathcal{B}^\pi(X) := \mathcal{B}(X)/\mathcal{K}(X)$ be the Calkin algebra of the cosets $A^\pi := A + \mathcal{K}(X)$ where $A \in \mathcal{B}(X)$. Following [4, Chap. XI, Definition 2.3], an operator $A \in \mathcal{B}(X)$ is said to be left Fredholm (resp. right Fredholm) if the coset A^π is left invertible (resp. right invertible) in the Calkin algebra $\mathcal{B}^\pi(X)$. An operator $A \in \mathcal{B}(X)$ is said to be semi-Fredholm if it is left or right Fredholm. We will write $A \simeq B$ if $A - B \in \mathcal{K}(X)$.

Let $C_b(\mathbb{R}_+)$ denote the C^*-algebra of all bounded continuous functions on the positive half-line $\mathbb{R}_+ := (0, +\infty)$. Following Sarason [31, p. 820], a function $f \in C_b(\mathbb{R}_+)$ is called slowly oscillating (at 0 and ∞) if

$$\lim_{r\to s} \sup_{t,\tau\in[r,2r]} |f(t) - f(\tau)| = 0 \quad \text{for} \quad s \in \{0, \infty\}.$$

The set $SO(\mathbb{R}_+)$ of all slowly oscillating functions is a C^*-algebra. This algebra properly contains $C(\overline{\mathbb{R}}_+)$, the C^*-algebra of all continuous functions on the two-point compactification $\overline{\mathbb{R}}_+ := [0, +\infty]$ of $\mathbb{R}_+$.

Suppose α is an orientation-preserving homeomorphism of $\overline{\mathbb{R}}_+$ onto itself, which has only two fixed points 0 and ∞, and whose restriction to $\mathbb{R}_+$ is a diffeomorphism. We say that α is a slowly oscillating shift if $\log \alpha'$ is bounded and $\alpha' \in SO(\mathbb{R}_+)$. The set of all slowly oscillating shifts is denoted by $SOS(\mathbb{R}_+)$. By [15, Lemma 2.2], an orientation-preserving diffeomorphism α of $\mathbb{R}_+$ onto itself belongs to $SOS(\mathbb{R}_+)$ if and only if $\alpha(t) = te^{\omega(t)}$ for $t \in \mathbb{R}_+$ and a real-valued function $\omega \in SO(\mathbb{R}_+) \cap C^1(\mathbb{R}_+)$ is such that the function ψ given by $\psi(t) := t\omega'(t)$ also belongs to $SO(\mathbb{R}_+)$ and $\inf_{t\in\mathbb{R}_+} \big(1 + t\omega'(t)\big) > 0$. The real-valued slowly oscillating function

$$\omega(t) := \log[\alpha(t)/t], \quad t \in \mathbb{R}_+,$$

is called the exponent function of $\alpha \in SOS(\mathbb{R}_+)$.

Through the paper, we will suppose that $1 < p < \infty$ and will use the following notation:

$$\mathcal{B} := \mathcal{B}(L^p(\mathbb{R}_+)), \quad \mathcal{K} := \mathcal{K}(L^p(\mathbb{R}_+)).$$

It is easily seen that if $\alpha \in SOS(\mathbb{R}_+)$, then the weighted shift operator defined by

$$U_\alpha f := (\alpha')^{1/p}(f \circ \alpha)$$

is an isometric isomorphism of the Lebesgue space $L^p(\mathbb{R}_+)$ onto itself. It is clear that $U_\alpha^{-1} = U_{\alpha_{-1}}$, where α_{-1} is the inverse function to α. For $k \in \mathbb{N}$, we denote by U_α^{-k} the operator $(U_\alpha^{-1})^k$. Let $W_{\alpha,p}^{SO}$ denote the collection of all operators of the form

$$A = \sum_{k\in\mathbb{Z}} a_k U_\alpha^k \tag{1.1}$$

where $a_k \in SO(\mathbb{R}_+)$ for all $k \in \mathbb{Z}$ and

$$\|A\|_{W^{SO}_{\alpha,p}} := \sum_{k\in\mathbb{Z}} \|a_k\|_{C_b(\mathbb{R}_+)} < +\infty. \tag{1.2}$$

The set $W^{SO}_{\alpha,p}$ is, actually, a Banach algebra with respect to the usual operations and the norm (1.2). By analogy with the Wiener algebra of absolutely convergent Fourier series, we will call $W^{SO}_{\alpha,p}$ the Wiener algebra.

Let $\Re\gamma$ and $\Im\gamma$ denote the real and imaginary part of $\gamma \in \mathbb{C}$, respectively. If $\gamma \in \mathbb{C}$ satisfies

$$0 < 1/p + \Re\gamma < 1, \tag{1.3}$$

then the operator

$$(S_\gamma f)(t) := \frac{1}{\pi i} \int_{\mathbb{R}_+} \left(\frac{t}{\tau}\right)^\gamma \frac{f(\tau)}{\tau - t} d\tau, \tag{1.4}$$

where the integral is understood in the principal value sense, is bounded on the Lebesgue space $L^p(\mathbb{R}_+)$ (see, e.g., [30, Proposition 4.2.11]). Put

$$P^\pm_\gamma := (I \pm S_\gamma)/2.$$

This paper is a continuation of our recent works [8, 9, 19, 20] (see also references therein). Let α, β belong to $SOS(\mathbb{R}_+)$ and $a_k, b_k \in SO(\mathbb{R}_+)$ for all $k \in \mathbb{Z}$. In [19, 20] we found criteria for the Fredholmness and a formula permitting to calculate the index of the weighted singular integral operator of the form

$$M := (a_0 I + a_1 U_\alpha)P^+_\gamma + (b_0 I + b_1 U_\beta)P^-_\gamma.$$

In this paper we assume that

$$A_+ := \sum_{k\in\mathbb{Z}} a_k U^k_\alpha \in W^{SO}_{\alpha,p}, \quad A_- := \sum_{k\in\mathbb{Z}} b_k U^k_\beta \in W^{SO}_{\beta,p} \tag{1.5}$$

and consider the weighted singular integral operator of the form

$$N := A_+ P^+_\gamma + A_- P^-_\gamma. \tag{1.6}$$

Criteria for the Fredholmness of the operator N in the particular case of $\alpha = \beta$ and $\gamma = 0$ were obtained in [9]. The proof of the sufficiency portion is based on the Allan–Douglas local principle and follows ideas of [15]. In this paper we will show that the localization technique is flexible enough to treat also the case of the left and right Fredholmness for arbitrary shifts α, β and arbitrary γ satisfying (1.3), provided that there are one-sided inverses of A_+ and A_- belonging to the Wiener algebras $W^{SO}_{\alpha,p}$ and $W^{SO}_{\beta,p}$, respectively. We show that the required result on one-sided inverses can be obtained from [8].

By $M(\mathfrak{A})$ we denote the maximal ideal space of a unital commutative Banach algebra $\mathfrak{A}$. Identifying the points $t \in \overline{\mathbb{R}}_+$ with the evaluation functionals $t(f) = f(t)$ for $f \in C(\overline{\mathbb{R}}_+)$, we get $M(C(\overline{\mathbb{R}}_+)) = \overline{\mathbb{R}}_+$. Consider the fibers

$$M_s(SO(\mathbb{R}_+)) := \{\xi \in M(SO(\mathbb{R}_+)) : \xi|_{C(\overline{\mathbb{R}}_+)} = s\}$$

of the maximal ideal space $M(SO(\mathbb{R}_+))$ over the points $s \in \{0, \infty\}$. By [23, Proposition 2.1], the set

$$\Delta := M_0(SO(\mathbb{R}_+)) \cup M_\infty(SO(\mathbb{R}_+))$$

coincides with $\operatorname{clos}_{SO^*}\mathbb{R}_+ \setminus \mathbb{R}_+$, where $\operatorname{clos}_{SO^*}\mathbb{R}_+$ is the weak-star closure of $\mathbb{R}_+$ in the dual space of $SO(\mathbb{R}_+)$. Then $M(SO(\mathbb{R}_+)) = \Delta \cup \mathbb{R}_+$. In what follows we write $a(\xi) := \xi(a)$ for every $a \in SO(\mathbb{R}_+)$ and every $\xi \in \Delta$.

With the operators $A_\pm$ defined by (1.5), we associate the functions $a_\pm$ defined on $\mathbb{R}_+ \times \mathbb{R}$ by

$$a_+(t,x) := \sum_{k\in\mathbb{Z}} a_k(t)e^{ik\omega(t)x}, \quad a_-(t,x) := \sum_{k\in\mathbb{Z}} b_k(t)e^{ik\eta(t)x}, \tag{1.7}$$

where $\omega, \eta \in SO(\mathbb{R}_+)$ are the exponent functions of α, β, respectively. Since the series in (1.5) converge absolutely, we have $a_\pm(\cdot, x) \in SO(\mathbb{R}_+)$ for all $x \in \mathbb{R}$. With the operator N we associate the function n defined on $\mathbb{R}_+ \times \mathbb{R}$ by

$$n(t,x) = a_+(t,x)p_\gamma^+(x) + a_-(t,x)p_\gamma^-(x),$$

where

$$p_\gamma^\pm(x) := (1 \pm s_\gamma(x))/2, \quad s_\gamma(x) := \coth[\pi(x + i/p + i\gamma)], \quad x \in \mathbb{R}. \tag{1.8}$$

Since $a_\pm(\cdot, x), n(\cdot, x) \in SO(\mathbb{R}_+)$ for every $x \in \mathbb{R}$, taking the Gelfand transform of $n(\cdot, x)$, we obtain for $(\xi, x) \in (\Delta \cup \mathbb{R}_+) \times \mathbb{R}$,

$$n(\xi,x) := a_+(\xi,x)p_\gamma^+(x) + a_-(\xi,x)p_\gamma^-(x), \tag{1.9}$$

which gives extensions of the functions $n(\cdot, x)$ to $M(SO(\mathbb{R}_+))$.

Theorem 1.1 (Main result). *Let $1 < p < \infty$ and let $\gamma \in \mathbb{C}$ satisfy* (1.3). *Suppose $a_k, b_k \in SO(\mathbb{R}_+)$ for all $k \in \mathbb{Z}$ and $\alpha, \beta \in SOS(\mathbb{R}_+)$. If*

(i) *the functional operators*

$$A_+ := \sum_{k\in\mathbb{Z}} a_k U_\alpha^k \in W_{\alpha,p}^{SO}, \quad A_- := \sum_{k\in\mathbb{Z}} b_k U_\beta^k \in W_{\beta,p}^{SO}$$

are left (resp. right) invertible on the space $L^p(\mathbb{R}_+)$;

(ii) *for every $\xi \in \Delta$, the function n defined by* (1.7)–(1.9) *satisfies the inequality*

$$\inf_{x\in\mathbb{R}} |n(\xi, x)| > 0; \tag{1.10}$$

then the operator $N = A_+P_\gamma^+ + A_-P_\gamma^-$ is left (resp. right) Fredholm on the space $L^p(\mathbb{R}_+)$.

We conjecture that conditions (i) and (ii) of Theorem 1.1 are also necessary for the one-sided Fredholmness of the operator N.

The paper is organized as follows. Section 2 contains some auxiliary results. In Section 3, on the basis of recent results from [8], we show that if an operator $A \in W_{\alpha,p}^{SO}$ is left (resp. right) invertible, then at least one of its left (resp. right) inverses belongs to the same algebra $W_{\alpha,p}^{SO}$.

Section 4 is devoted to the algebra $\mathcal{A}$ generated by the identity operator I and the operator S_0. We recall that $\mathcal{A}$ is the smallest Banach subalgebra of $\mathcal{B}$

that contains all operators similar to Mellin convolution operators with continuous symbols. In particular, the algebra $\mathcal{A}$ contains the operator S_γ and the operator R_γ with fixed singularities defined by

$$(R_\gamma f)(t) := \frac{1}{\pi i}\int_{\mathbb{R}_+}\left(\frac{t}{\tau}\right)^\gamma \frac{f(\tau)}{\tau+t}d\tau, \tag{1.11}$$

where the integral is understood in the principal value sense. We also recall the description of the maximal ideal space of the algebra $\mathcal{A}^\pi := (\mathcal{A}+\mathcal{K})/\mathcal{K}$.

In Section 5, we recall a version of the Allan–Douglas local principle suitable for the study of one-sided invertibility in subalgebras of the Calkin algebra $\mathcal{B}^\pi = \mathcal{B}/\mathcal{K}$ (see [3, Theorem 1.35(a)]). Following [15, Section 6], we consider the algebra $\mathcal{Z}$ generated by $\mathcal{A}\cup\mathcal{K}$ and the operators of the form cR_0, where $c \in SO(\mathbb{R}_+)$. We recall that the maximal ideal space of $\mathcal{Z}^\pi := \mathcal{Z}/\mathcal{K}$ is homeomorphic to the set $\{-\infty,+\infty\}\cup(\Delta\times\mathbb{R})$. Further, we consider the algebra Λ of operators of local type that consists of all operators $A \in \mathcal{B}$ such that $AC - CA \in \mathcal{K}$ for all $C \in \mathcal{Z}$. Since $\mathcal{Z}^\pi$ is a commutative central subalgebra of $\Lambda^\pi := \Lambda/\mathcal{K}$, we can apply the Allan–Douglas local principle to $\Lambda^\pi \subset \mathcal{B}^\pi$ and its central subalgebra $\mathcal{Z}^\pi$. In particular, an operator $T \in \Lambda$ is left (resp. right) Fredholm if certain cosets $T^\pi + \mathcal{J}^\pi_{\xi,x}$, $T^\pi + \mathcal{J}^\pi_{+\infty}$, and $T^\pi + \mathcal{J}^\pi_{-\infty}$ are left (resp. right) invertible in the corresponding local algebras $\Lambda^\pi_{\xi,x}$, $\Lambda^\pi_{+\infty}$, and $\Lambda^\pi_{-\infty}$. Here (ξ, x) runs through $\Delta\times\mathbb{R}$. This result is applicable to the operator N because it belongs to the algebra $\mathcal{F}_{\alpha,\beta}$ generated by the operators S_0, $U_\alpha^{\pm1}$, $U_\beta^{\pm1}$ and the multiplication operators cI with $c \in SO(\mathbb{R}_+)$. In turn, this algebra is contained in the algebra Λ of operators of local type.

In Section 6, we recall the definition of the algebra $\mathcal{E}(\mathbb{R}_+, V(\mathbb{R}))$ of slowly oscillating functions on $\mathbb{R}_+$ with values in the algebra $V(\mathbb{R})$ of all absolutely continuous functions of finite total variation. This algebra is important for our purposes because Mellin pseudodifferential operators with symbols in $\mathcal{E}(\mathbb{R}_+, V(\mathbb{R}))$ commute modulo compact operators. Moreover, if $\alpha \in SOS(\mathbb{R}_+)$, then $U_\alpha R_\gamma$ is similar to a Mellin pseudodifferential operator with symbol in $\mathcal{E}(\mathbb{R}_+, V(\mathbb{R}))$ up to a compact operator. These results are important ingredients of the proof of two-sided invertibility of the cosets $N^\pi + \mathcal{J}^\pi_{\xi,x}$ in the quotient algebras $\Lambda^\pi_{\xi,x}$ for $(\xi, x) \in \Delta\times\mathbb{R}$ under condition (ii) of Theorem 1.1.

In Section 7, we prove Theorem 1.1. Since, according to Section 3, there are left/right inverses of A_+ (resp. A_-) belonging to $W^{SO}_{\alpha,p} \subset \Lambda$ (resp. $W^{SO}_{\beta,p} \subset \Lambda$), the left/right invertibility of $A_\pm$ implies the left/right invertibility of the coset $A^\pi_\pm + \mathcal{J}^\pi_{\pm\infty} = N^\pi + \mathcal{J}^\pi_{\pm\infty}$ in the local algebra $\Lambda^\pi_{\pm\infty}$. Finally, with the aid of the results of Section 6, we show that condition (ii) of Theorem 1.1 is sufficient for the two-sided invertibility of the cosets $N^\pi + \mathcal{J}^\pi_{\xi,x}$ in the local algebras $\Lambda^\pi_{\xi,x}$ for all $(\xi, x) \in \Delta\times\mathbb{R}$. To complete the proof of Theorem 1.1, it remains to apply the Allan–Douglas local principle (see Section 5).

Finally, in Section 8 we formulate criteria for the two-sided and one-sided invertibility of a binomial functional operator with shift in the form $A = aI - bU_\alpha$, which were obtained in [18]. These results together with Theorem 1.1 imply more effective sufficient conditions for the left, right, and two-sided Fredholmness of the operator $(aI - bU_\alpha)P_\gamma^+ + (cI - dU_\beta)P_\gamma^-$ with $a, b, c, d \in SO(\mathbb{R}_+)$ and $\alpha, \beta \in SOS(\mathbb{R}_+)$.

2. Auxiliary results

2.1. One-sided invertibility of operators on Hilbert spaces

Lemma 2.1. *Let $\mathcal{H}$ be a Hilbert space and $A \in \mathcal{B}(\mathcal{H})$.*

(a) *The operator A is left invertible on the space $\mathcal{H}$ if and only if the operator A^*A is invertible on the space $\mathcal{H}$. In this case, one of the left inverses of A is given by $A^L = (A^*A)^{-1}A^*$.*
(b) *The operator A is right invertible on the space $\mathcal{H}$ if and only if the operator AA^* is invertible on the space $\mathcal{H}$. In this case, one of the right inverses of A is given by $A^R = A^*(AA^*)^{-1}$.*

This statement is known, although we are not able to provide a precise reference. The proof of the sufficiency portion of part (a) is a trivial computation. Now assume that $\langle \cdot, \cdot \rangle$ is the inner product of $\mathcal{H}$ and $A^L \in \mathcal{B}(\mathcal{H})$ is a left inverse of A. Then for every $f \in \mathcal{H}$,

$$\|f\|^2 \le \|A^L\|^2\|Af\|^2 = \|A^L\|^2|\langle Af, Af\rangle| = \|A^L\|^2|\langle A^*Af, f\rangle|.$$

In view of the previous inequality, the invertibility of the operator A^*A follows from the Lax–Milgram theorem (see, e.g., [34, Chap. III, Section 7]). This completes the proof of part (a). The proof of part (b) is reduced to the previous one by passing to adjoint operators.

Another proof of the above lemma can be obtained from general results for C^*-algebras contained in [25, § 23, Corollaries 2–3].

Lemma 2.1 can also be deduced from more general results on the Moore–Penrose invertibility of operators on a Hilbert space (see [13, Example 2.16] or [2, Theorem 4.24]). Notice that the operator A^L (resp. A^R) is the Moore–Penrose inverse of the operator A.

2.2. Fundamental property of slowly oscillating functions

Lemma 2.2 ([23, Proposition 2.2]). *Let $\{a_k\}_{k=1}^\infty$ be a countable subset of $SO(\mathbb{R}_+)$ and $s \in \{0, \infty\}$. For each $\xi \in M_s(SO(\mathbb{R}_+))$ there exists a sequence $\{t_n\}_{n\in\mathbb{N}} \subset \mathbb{R}_+$ such that $t_n \to s$ as $n \to \infty$ and*

$$a_k(\xi) = \lim_{n\to\infty} a_k(t_n) \quad \text{for all} \quad k \in \mathbb{N}. \tag{2.1}$$

Conversely, if $\{t_n\}_{n\in\mathbb{N}} \subset \mathbb{R}_+$ is a sequence such that $t_n \to s$ as $n \to \infty$ and the limits $\lim_{n\to\infty} a_k(t_n)$ exist for all $k \in \mathbb{N}$, then there exists a functional $\xi \in M_s(SO(\mathbb{R}_+))$ such that (2.1) holds.

2.3. Properties of iterations of slowly oscillating shifts

In this subsection we collect some properties of iterations of slowly oscillating shifts. For $t \in \mathbb{R}_+$ and $k \in \mathbb{N}$, let

$$\alpha_0(t) := t, \quad \alpha_k(t) := \alpha[\alpha_{k-1}(t)], \quad \alpha_{-k}(t) := \alpha_{-1}[\alpha_{-k+1}(t)].$$

Lemma 2.3 ([16, Corollary 2.5]). *If $\alpha \in SOS(\mathbb{R}_+)$, then $\alpha_k \in SOS(\mathbb{R}_+)$ for every $k \in \mathbb{Z}$.*

Lemma 2.4 ([15, Lemma 2.3]). *If $c \in SO(\mathbb{R}_+)$ and $\alpha \in SOS(\mathbb{R}_+)$, then $c \circ \alpha$ belongs to $SO(\mathbb{R}_+)$ and*

$$\lim_{t\to s}(c(t) - c[\alpha(t)]) = 0 \quad \textit{for} \quad s \in \{0, \infty\}.$$

Lemma 2.5 ([17, Lemma 2.6]). *Let $\alpha \in SOS(\mathbb{R}_+)$ and α_{-1} be the inverse function to α. If ω and ω_{-1} are the exponent functions of α and α_{-1}, respectively, then $\omega(\xi) = -\omega_{-1}(\xi)$ for all $\xi \in \Delta$.*

Lemma 2.6. *Let $\alpha \in SOS(\mathbb{R}_+)$ and let ω be its exponent function. If $k \in \mathbb{Z}$ and ω_k is the exponent function of α_k, then $\omega_k(\xi) = k\omega(\xi)$ for every $\xi \in \Delta$.*

Proof. For $k = 0, 1$, the statement is trivial. If $k > 1$, then

$$\omega_k(t) = \log \frac{\alpha_k(t)}{t} = \log\left(\prod_{j=0}^{k-1} \frac{\alpha[\alpha_j(t)]}{\alpha_j(t)}\right) = \sum_{j=0}^{k-1} \omega[\alpha_j(t)], \quad t \in \mathbb{R}_+. \tag{2.2}$$

Since $\omega \in SO(\mathbb{R}_+)$, we deduce from Lemmas 2.3–2.4 that for every integer $j \in \{0, \dots, k-1\}$, the function $\omega \circ \alpha_j$ belongs to $SO(\mathbb{R}_+)$ and

$$\lim_{t\to s}(\omega(t) - \omega[\alpha_j(t)]) = 0, \quad s \in \{0, \infty\}. \tag{2.3}$$

Fix $s \in \{0, \infty\}$ and $\xi \in M_s(SO(\mathbb{R}_+))$. By Lemma 2.2, there is a sequence $\{t_n\}_{n\in\mathbb{N}} \subset \mathbb{R}_+$ such that $t_n \to s$ as $n \to \infty$ and

$$\omega(\xi) = \lim_{n\to\infty} \omega(t_n), \quad (\omega \circ \alpha_j)(\xi) = \lim_{n\to\infty} \omega[\alpha_j(t_n)]. \tag{2.4}$$

Equalities (2.3)–(2.4) imply that for $j \in \{0, \dots, k-1\}$,

$$(\omega \circ \alpha_j)(\xi) - \omega(\xi) = \lim_{n\to\infty} (\omega[\alpha_j(t_n)] - \omega(t_n)) = 0.$$

We derive from (2.2) and the above equalities that

$$\omega_k(\xi) = \sum_{j=0}^{k-1} (\omega \circ \alpha_j)(\xi) = k\omega(\xi),$$

which completes the proof for $k > 1$.

If $k < 0$, then we have $\omega_{-k}(\xi) = -k\omega(\xi)$ by the statement just proved. On the other hand, we deduce from Lemma 2.5 that $\omega_k(\xi) = -\omega_{-k}(\xi)$. Thus, $\omega_k(\xi) = -(-k)\omega(\xi) = k\omega(\xi)$ for all $\xi \in \Delta$. □

3. Weak one-sided inverse closedness of the algebra $W^{SO}_{\alpha,p}$

3.1. Inverse closedness of the algebra $W^{SO}_{\alpha,p}$ in the algebra $\mathcal{B}$

Let $\mathfrak{A} \subset \mathfrak{B}$ be two Banach algebras with the same unit element. Recall that the algebra $\mathfrak{A}$ is said to be inverse closed in the algebra $\mathfrak{B}$ if for every element $a \in \mathfrak{A}$ invertible in the algebra $\mathfrak{B}$ its inverse a^{-1} belongs to the algebra $\mathfrak{A}$.

We say that the algebra $\mathfrak{A}$ is weakly left (resp. right) inverse closed in the algebra $\mathfrak{B}$ if for every element $a \in \mathfrak{A}$, which is left (resp. right) invertible in the algebra $\mathfrak{B}$, there exists at least one its left (resp. right) inverse $a^{(-1)}$ that belongs to the algebra $\mathfrak{A}$.

Theorem 3.1 **([8, Theorem 7.4]).** *For every $p \in (1,\infty)$, the algebra $W^{SO}_{p,\alpha}$ is inverse closed in the algebra $\mathcal{B}$.*

3.2. One-sided inverses belonging to the algebra $W^{\mathfrak{S}}_{\alpha,p}$

A function $f \in L^\infty(\mathbb{R}_+)$ is said to be essentially slowly oscillating (at 0 and ∞) if for each (equivalently, for some) $\lambda \in (0,1)$,

$$\lim_{r\to s} \operatorname*{ess\,sup}_{t,\tau\in[\lambda r,r]} |f(t)-f(\tau)| = 0, \quad s \in \{0,\infty\}.$$

Fix $\alpha \in SOS(\mathbb{R}_+)$ and $\tau \in \mathbb{R}_+$. Consider the semi-segment $\gamma \subset \mathbb{R}_+$ with the endpoints τ and $\alpha(\tau)$ such that $\tau \in \gamma$ and $\alpha(\tau) \notin \gamma$. Following [8, Section 3.2], let $\mathfrak{S}$ denote the C^*-subalgebra of $L^\infty(\mathbb{R}_+)$ consisting of all functions on $\mathbb{R}_+$ that are continuous on every semi-segment $\alpha_n(\gamma)$ with $n \in \mathbb{Z}$, have one-sided limits at the points $\alpha_n(\tau)$ for $n \in \mathbb{Z}$, and are essentially slowly oscillating at 0 and ∞. Let $W^{\mathfrak{S}}_{\alpha,p}$ be the unital Banach algebra of operators of the form (1.1) with $a_k \in \mathfrak{S}$ for all $k \in \mathbb{Z}$ and the norm

$$\|A\|_{W^{\mathfrak{S}}_{\alpha,p}} = \sum_{k\in\mathbb{Z}} \|a_k\|_{L^\infty(\mathbb{R}_+)} < \infty.$$

From [8, Theorems 6.3–6.4] we get the following.

Theorem 3.2. *Let $1 < p < \infty$, $\alpha \in SOS(\mathbb{R}_+)$, $a_k \in SO(\mathbb{R}_+)$ for all $k \in \mathbb{Z}$, and*

$$A = \sum_{k\in\mathbb{Z}} a_k U_\alpha^k \in W^{SO}_{\alpha,p}.$$

If A is left (resp. right) invertible on $L^p(\mathbb{R}_+)$, then there exists a left inverse A^L (resp. right inverse A^R) of A such that $A^L \in W^{\mathfrak{S}}_{\alpha,p}$ (resp. $A^R \in W^{\mathfrak{S}}_{\alpha,p}$).

3.3. Weak one-sided inverse closedness of the algebra $W^{SO}_{\alpha,p}$ in $\mathcal{B}$

We will show that the algebra $W^{SO}_{\alpha,p}$ is weakly left and right inverse closed in the algebra $\mathcal{B}$. For every operator $A \in W^{SO}_{\alpha,p}$ of the form (1.1), define its formally adjoint $A^\diamond$ by

$$A^\diamond := \sum_{k\in\mathbb{Z}} (\overline{a_k} \circ \alpha_{-k}) U_\alpha^{-k} \in W^{SO}_{\alpha,p}.$$

Theorem 3.3. *Let $1 < p < \infty$, $\alpha \in SOS(\mathbb{R}_+)$, $a_k \in SO(\mathbb{R}_+)$ for all $k \in \mathbb{Z}$, and*

$$A = \sum_{k\in\mathbb{Z}} a_k U_\alpha^k \in W_{\alpha,p}^{SO}.$$

(a) *If A is left invertible on $L^p(\mathbb{R}_+)$, then the operator $A^\diamond A$ is invertible on the space $L^p(\mathbb{R}_+)$, the operator $A^L := (A^\diamond A)^{-1}A^\diamond$ is a left inverse of A, and $A^L \in W_{\alpha,p}^{SO}$.*

(b) *If A is right invertible on $L^p(\mathbb{R}_+)$, then the operator $AA^\diamond$ is invertible on the space $L^p(\mathbb{R}_+)$, the operator $A^R := A^\diamond(AA^\diamond)^{-1}$ is a right inverse of A, and $A^R \in W_{\alpha,p}^{SO}$.*

Proof. Along with the operator U_α acting on $L^p(\mathbb{R}_+)$, consider the operator $U_{\alpha,2}$ acting on $L^2(\mathbb{R}_+)$ and defined by the same rule $U_{\alpha,2}f = (\alpha')^{1/2}(f\circ\alpha)$. Then we can define the canonical isometric isomorphisms of Banach algebras

$$\Psi_{\mathfrak{S}} : W_{\alpha,p}^{\mathfrak{S}} \to W_{\alpha,2}^{\mathfrak{S}}, \quad \Psi_{SO} : W_{\alpha,p}^{SO} \to W_{\alpha,2}^{SO}$$

by the formulas

$$\Psi_{\mathfrak{S}}\left(\sum_{k\in\mathbb{Z}} c_k U_\alpha^k\right) = \sum_{k\in\mathbb{Z}} c_k U_{\alpha,2}^k, \quad \Psi_{SO}\left(\sum_{k\in\mathbb{Z}} c_k U_\alpha^k\right) = \sum_{k\in\mathbb{Z}} c_k U_{\alpha,2}^k,$$

respectively.

If $A \in W_{\alpha,p}^{SO}$ is left invertible on the space $L^p(\mathbb{R}_+)$, then by Theorem 3.2, there exists an operator $\widetilde{A}^L \in W_{\alpha,p}^{\mathfrak{S}}$ such that $\widetilde{A}^L A = I$ on $L^p(\mathbb{R}_+)$. Hence $\Psi_{\mathfrak{S}}(\widetilde{A}^L)\Psi_{\mathfrak{S}}(A) = I$ on $L^2(\mathbb{R}_+)$. Therefore, the operator

$$A_2 := \Psi_{SO}(A) = \Psi_{\mathfrak{S}}(A) \in W_{\alpha,2}^{SO}$$

is left invertible on $L^2(\mathbb{R}_+)$. Hence, in view of Lemma 2.1, the operator $A_2^*A_2$ is invertible on $L^2(\mathbb{R}_+)$. Observe that $A_2^* \in W_{\alpha,2}^{SO}$ and $A^\diamond = \Phi_{SO}^{-1}(A_2^*) \in W_{\alpha,p}^{SO}$. Since $A_2^*A_2 \in W_{\alpha,2}^{SO}$, we deduce from the inverse closedness of the algebra $W_{\alpha,2}^{SO}$ in the algebra $\mathcal{B}$ (see Theorem 3.1) that $(A_2^*A_2)^{-1} \in W_{\alpha,2}^{SO}$. Then $\Psi_{SO}^{-1}((A_2^*A_2)^{-1}) \in W_{\alpha,p}^{SO}$. Now it is easy to check that

$$(A^\diamond A)^{-1} = \Psi_{SO}^{-1}((A_2^*A_2)^{-1}).$$

Hence $A^L = (A^\diamond A)^{-1}A$ is a left inverse to A. Part (a) is proved.

(b) The proof of part (b) is reduced to the previous one by passing to adjoint operators. ◻

4. Algebra $\mathcal{A}$ of singular integral operators

4.1. Fourier and Mellin convolution operators

Let $\mathcal{F} : L^2(\mathbb{R}) \to L^2(\mathbb{R})$ denote the Fourier transform,

$$(\mathcal{F}f)(x) := \int_{\mathbb{R}} f(y)e^{-ixy}dy, \quad x \in \mathbb{R},$$

and let $\mathcal{F}^{-1} : L^2(\mathbb{R}) \to L^2(\mathbb{R})$ be the inverse of $\mathcal{F}$. A function $a \in L^\infty(\mathbb{R})$ is called a Fourier multiplier on $L^p(\mathbb{R})$ if the mapping $f \mapsto \mathcal{F}^{-1} a \mathcal{F} f$ maps $L^2(\mathbb{R}) \cap L^p(\mathbb{R})$ into itself and extends to a bounded operator on $L^p(\mathbb{R})$. The latter operator is then denoted by $W^0(a)$. We let $\mathcal{M}_p(\mathbb{R})$ stand for the set of all Fourier multipliers on $L^p(\mathbb{R})$. One can show that $\mathcal{M}_p(\mathbb{R})$ is a Banach algebra under the norm

$$\|a\|_{\mathcal{M}_p(\mathbb{R})} := \|W^0(a)\|_{\mathcal{B}(L^p(\mathbb{R}))}.$$

Let $d\mu(t) = dt/t$ be the (normalized) invariant measure on $\mathbb{R}_+$. Consider the Fourier transform on $L^2(\mathbb{R}_+, d\mu)$, which is usually referred to as the Mellin transform and is defined by

$$\mathcal{M} : L^2(\mathbb{R}_+, d\mu) \to L^2(\mathbb{R}), \quad (\mathcal{M} f)(x) := \int_{\mathbb{R}_+} f(t) t^{-ix} \frac{dt}{t}.$$

This operator is invertible, with inverse given by

$$\mathcal{M}^{-1} : L^2(\mathbb{R}) \to L^2(\mathbb{R}_+, d\mu), \quad (\mathcal{M}^{-1} g)(t) = \frac{1}{2\pi} \int_{\mathbb{R}} g(x) t^{ix} \, dx.$$

Let E be the isometric isomorphism

$$E : L^p(\mathbb{R}_+, d\mu) \to L^p(\mathbb{R}), \quad (Ef)(x) := f(e^x), \quad x \in \mathbb{R}. \tag{4.1}$$

Then the map

$$A \mapsto E^{-1} A E \tag{4.2}$$

transforms the Fourier convolution operator $W^0(a) = \mathcal{F}^{-1} a \mathcal{F}$ to the Mellin convolution operator

$$\mathrm{Co}(a) := \mathcal{M}^{-1} a \mathcal{M}$$

with the same symbol a. Hence the class of Fourier multipliers on $L^p(\mathbb{R})$ coincides with the class of Mellin multipliers on $L^p(\mathbb{R}_+, d\mu)$.

4.2. Continuous and piecewise continuous multipliers

We denote by PC the C^*-algebra of all bounded piecewise continuous functions on $\dot{\mathbb{R}} = \mathbb{R} \cup \{\infty\}$. By definition, $a \in PC$ if and only if $a \in L^\infty(\mathbb{R})$ and the one-sided limits

$$a(x_0 - 0) := \lim_{x \to x_0 - 0} a(x), \quad a(x_0 + 0) := \lim_{x \to x_0 + 0} a(x)$$

exist for each $x_0 \in \dot{\mathbb{R}}$. If a function a is given everywhere on $\mathbb{R}$, then its total variation is defined by

$$V(a) := \sup \sum_{k=1}^{n} |a(x_k) - a(x_{k-1})|,$$

where the supremum is taken over all $n \in \mathbb{N}$ and

$$-\infty < x_0 < x_1 < \cdots < x_n < +\infty.$$

If a has a finite total variation, then it has finite one-sided limits $a(x - 0)$ and $a(x + 0)$ for all $x \in \dot{\mathbb{R}}$, that is, $a \in PC$ (see, e.g., [26, Chap. VIII, Sections 3 and 9]). The following theorem gives an important subset of $\mathcal{M}_p(\mathbb{R})$. Its proof can be found, e.g., in [1, Theorem 17.1] or [6, Theorem 2.11].

Theorem 4.1 (Stechkin's inequality). *If $a \in PC$ has finite total variation $V(a)$, then $a \in \mathcal{M}_p(\mathbb{R})$ and*

$$\|a\|_{\mathcal{M}_p(\mathbb{R})} \leq \|S_{\mathbb{R}}\|_{\mathcal{B}(L^p(\mathbb{R}))}\big(\|a\|_{L^\infty(\mathbb{R})} + V(a)\big),$$

where $S_{\mathbb{R}}$ is the Cauchy singular integral operator on $\mathbb{R}$.

According to [6] or [1, p. 325], let PC_p be the closure in $\mathcal{M}_p(\mathbb{R})$ of the set of all functions $a \in PC$ with finite total variation on $\mathbb{R}$. Following [1, p. 331], put

$$C_p(\overline{\mathbb{R}}) := PC_p \cap C(\mathbb{R}),$$

where $\overline{\mathbb{R}} := [-\infty, +\infty]$. It is easy to see that PC_p and $C_p(\overline{\mathbb{R}})$ are Banach algebras.

4.3. Maximal ideal space of the algebras $\mathcal{A}$ and $\mathcal{A}^\pi$

Let $\mathfrak{A}$ be a Banach algebra and $\mathfrak{C}$ be a subset of $\mathfrak{A}$. Following [3, Section 3.45], we denote by $\mathrm{alg}_{\mathfrak{A}}\mathfrak{C}$ the smallest closed subalgebra of $\mathfrak{A}$ containing $\mathfrak{C}$ and by $\mathrm{id}_{\mathfrak{A}}\mathfrak{C}$ the smallest closed two-sided ideal of $\mathfrak{A}$ containing $\mathfrak{C}$.

Put $\mathcal{A} := \mathrm{alg}_{\mathcal{B}}\{I, S_0\}$. Obviously, the algebra $\mathcal{A}$ is commutative. Consider the isometric isomorphism

$$\Phi : L^p(\mathbb{R}_+) \to L^p(\mathbb{R}_+, d\mu), \quad (\Phi f)(t) := t^{1/p} f(t), \quad t \in \mathbb{R}_+.$$

The following result is well known. It is essentially due to Duduchava [6, 7] and Simonenko, Chin Ngok Minh [32]. It can be found with a proof in [5, Section 1.10.2], [12, Section 2.1.2], [30, Sections 4.2.2–4.2.3].

Theorem 4.2.

(a) *The algebra $\mathcal{A}$ is the smallest closed subalgebra of $\mathcal{B}$ that contains the operators $\Phi^{-1}\mathrm{Co}(a)\Phi$ with $a \in C_p(\overline{\mathbb{R}})$.*

(b) *The maximal ideal space of the commutative Banach algebra $\mathcal{A}$ is homeomorphic to $\overline{\mathbb{R}}$. In particular, the operator $\Phi^{-1}\mathrm{Co}(a)\Phi$ with $a \in C_p(\overline{\mathbb{R}})$ is invertible if and only if $a(x) \neq 0$ for all $x \in \overline{\mathbb{R}}$. Thus $\mathcal{A}$ is an inverse closed subalgebra of $\mathcal{B}$.*

(c) *The operator $\Phi^{-1}\mathrm{Co}(a)\Phi$ with $a \in C_p(\overline{\mathbb{R}})$ belongs to $\mathrm{id}_{\mathcal{A}}\{R_0\}$ if and only if $a(-\infty) = a(+\infty) = 0$.*

(d) *If $\gamma \in \mathbb{C}$ satisfies (1.3), then the function s_γ given by (1.8) and the function r_γ defined by*

$$r_\gamma(x) := 1/\sinh[\pi(x + i/p + i\gamma)], \quad x \in \mathbb{R},$$

belong to $C_p(\overline{\mathbb{R}})$ and

$$S_\gamma = \Phi^{-1}\mathrm{Co}(s_\gamma)\Phi, \quad R_\gamma = \Phi^{-1}\mathrm{Co}(r_\gamma)\Phi.$$

Let us describe the quotient algebra

$$\mathcal{A}^\pi = (\mathcal{A} + \mathcal{K})/\mathcal{K}.$$

By [30, Proposition 4.2.14], a Mellin convolution operator is Fredholm on the space $L^p(\mathbb{R}_+, d\mu)$ if and only if it is invertible on this space. Hence, Theorem 4.2 implies the following.

Corollary 4.3. (a) *The algebra $\mathcal{A}^\pi$ is commutative and its maximal ideal space is homeomorphic to $\overline{\mathbb{R}}$.*

(b) *The Gelfand transform of the coset $(\Phi^{-1}\mathrm{Co}(a)\Phi)^\pi \in \mathcal{A}^\pi$ for $a \in C_p(\overline{\mathbb{R}})$ is given by*

$$\left[(\Phi^{-1}\mathrm{Co}(a)\Phi)^\pi\right]^\wedge(x) = a(x) \quad \textit{for} \quad x \in \overline{\mathbb{R}}.$$

4.4. Some operator relations

Lemma 4.4 ([14, Lemma 2.4], [20, Lemma 4.2]). *Let $1 < p < \infty$ and $\gamma, \delta \in \mathbb{C}$ be such that $0 < 1/p + \Re\gamma < 1$ and $0 < 1/p + \Re\delta < 1$. Then*

$$P_\delta^+ - P_\gamma^+ = P_\gamma^- - P_\delta^- = \frac{1}{2}\sinh[\pi i(\gamma - \delta)]R_\gamma R_\delta, \quad P_\gamma^- P_\delta^+ = -\frac{e^{i\pi(\delta-\gamma)}}{4}R_\gamma R_\delta.$$

4.5. Compactness of commutators of singular integral operators and functional operators

Fix $\alpha, \beta \in SOS(\mathbb{R}_+)$ and consider the Banach algebra of functional operators with shifts and slowly oscillating data defined by

$$\mathcal{FO}_{\alpha,\beta} := \mathrm{alg}_\mathcal{B}\{U_\alpha, U_\alpha^{-1}, U_\beta, U_\beta^{-1}, cI \, : \, c \in SO(\mathbb{R}_+)\}.$$

Lemma 4.5 ([17, Lemma 2.8]). *Let $\alpha, \beta \in SOS(\mathbb{R}_+)$. If $A \in \mathcal{FO}_{\alpha,\beta}$ and $B \in \mathcal{A}$, then $AB \simeq BA$.*

5. Allan–Douglas localization

5.1. The Allan–Douglas local principle

Let $\mathfrak{A}$ be a Banach algebra with identity. A subalgebra $\mathfrak{Z}$ of $\mathfrak{A}$ is said to be a central subalgebra of $\mathfrak{A}$ if $za = az$ for all $z \in \mathfrak{Z}$ and all $a \in \mathfrak{A}$.

The proof of the following result is contained, e.g., in [3, Theorem 1.35(a)].

Theorem 5.1 (Allan–Douglas). *Let $\mathfrak{A}$ be a Banach algebra with identity e and let $\mathfrak{Z}$ be a closed central subalgebra of $\mathfrak{A}$ containing e. Let $M(\mathfrak{Z})$ be the maximal ideal space of $\mathfrak{Z}$, and for $\omega \in M(\mathfrak{Z})$, let $\mathfrak{J}_\omega$ refer to the smallest closed two-sided ideal of $\mathfrak{A}$ containing the ideal ω. Then an element a is left (resp. right, two-sided) invertible in $\mathfrak{A}$ if and only if $a + \mathfrak{J}_\omega$ is left (resp. right, two-sided) invertible in the quotient algebra $\mathfrak{A}/\mathfrak{J}_\omega$ for all $\omega \in M(\mathfrak{Z})$.*

The algebra $\mathfrak{A}/\mathfrak{J}_\omega$ is referred to as the local algebra of $\mathfrak{A}$ at $\omega \in M(\mathfrak{Z})$.

5.2. Algebras of singular integral operators with shifts and algebras of operators of local type

Following [15, Section 6.3], we consider the following sets:

$$\mathcal{Z} := \text{alg}_{\mathcal{B}}\{I, S_0, cR_0, K : c \in SO(\mathbb{R}_+),\ K \in \mathcal{K}\},$$
$$\Lambda := \{A \in \mathcal{B} : AC - CA \in \mathcal{K} \text{ for all } C \in \mathcal{Z}\}.$$

By [15, Lemma 6.7(a)], the set Λ is a closed unital subalgebra of the algebra $\mathcal{B}$, which is usually called the algebra of operators of local type.

For $\alpha, \beta \in SOS(\mathbb{R}_+)$, put

$$\mathcal{F}_{\alpha,\beta} := \text{alg}_{\mathcal{B}}(\{S_0\} \cup \mathcal{FO}_{\alpha,\beta}).$$

By a minor modification of the proof of [15, Theorem 6.8] with the aid of Lemma 4.5, we get the following.

Theorem 5.2. *We have* $\mathcal{K} \subset \mathcal{Z} \subset \mathcal{F}_{\alpha,\beta} \subset \Lambda$.

5.3. Maximal ideal space of the algebra $\mathcal{Z}^\pi$

It follows from Theorem 5.2 that the quotient algebras $\mathcal{Z}^\pi := \mathcal{Z}/\mathcal{K}$ and $\Lambda^\pi := \Lambda/\mathcal{K}$ are well defined. Clearly, $\mathcal{Z}^\pi$ lies in the center of Λ^π.

Theorem 5.3 ([15, Theorem 6.11]). *For the commutative Banach algebra $\mathcal{Z}^\pi$ the following statements hold:*

(a) *the maximal ideal space $M(\mathcal{Z}^\pi)$ of $\mathcal{Z}^\pi$ is homeomorphic to the set*
$$\{-\infty, +\infty\} \cup (\Delta \times \mathbb{R});$$
(b) *any coset $Z^\pi \in \mathcal{Z}^\pi$ is of the form*
$$Z^\pi = (c_+P_0^+)^\pi + (c_-P_0^-)^\pi + \lim_{n\to\infty} \sum_{k=1}^{m_n} (c_{n,k}H_{n,k})^\pi \tag{5.1}$$
where $c_\pm \in \mathbb{C}$, $c_{n,k} \in SO(\mathbb{R}_+)$, $H_{n,k} \in \text{id}_{\mathcal{A}}\{R_0\}$, and $m_n \in \mathbb{N}$;

(c) *the Gelfand transform of the coset $Z^\pi \in \mathcal{Z}^\pi$ defined by* (5.1) *is given for a point $(\xi, x) \in \Delta \times \mathbb{R}$ by*
$$(Z^\pi)\hat{}(\xi, x) = c_+p_0^+(x) + c_-p_0^-(x) + \lim_{n\to\infty} \sum_{k=1}^{m_n} c_{n,k}(\xi)(H_{n,k}^\pi)\hat{}(x),$$
where $(H_{n,k}^\pi)\hat{}(x)$ is the Gelfand transform of a coset $H_{n,k}^\pi \in \mathcal{A}^\pi$, which is calculated in Corollary 4.3(b).

5.4. Semi-Fredholmness of operators of local type

Let $\mathcal{J}_{+\infty}^\pi$, $\mathcal{J}_{-\infty}^\pi$ and $\mathcal{J}_{\xi,x}^\pi$ for $(\xi, x) \in \Delta \times \mathbb{R}$ be the closed two-sided ideals of the Banach algebra Λ^π generated, respectively, by the maximal ideals

$$\mathcal{I}_{+\infty}^\pi := \text{id}_{\mathcal{Z}^\pi}\{(P_0^-)^\pi, (gR_0)^\pi : g \in SO(\mathbb{R}_+)\},$$
$$\mathcal{I}_{-\infty}^\pi := \text{id}_{\mathcal{Z}^\pi}\{(P_0^+)^\pi, (gR_0)^\pi : g \in SO(\mathbb{R}_+)\},$$
$$\mathcal{I}_{\xi,x}^\pi := \{Z^\pi \in \mathcal{Z}^\pi : (Z^\pi)\hat{}(\xi, x) = 0\}$$

of the algebra $\mathcal{Z}^\pi$, and let

$$\Lambda^\pi_{+\infty} := \Lambda^\pi/\mathcal{J}^\pi_{+\infty}, \quad \Lambda^\pi_{-\infty} := \Lambda^\pi/\mathcal{J}^\pi_{-\infty}, \quad \Lambda^\pi_{\xi,x} := \Lambda^\pi/\mathcal{J}^\pi_{\xi,x}$$

be the corresponding quotient algebras (see also [15, Section 6.6]).

Obviously, an operator $T \in \Lambda$ is left Fredholm (resp. right Fredholm) on the space $L^p(\mathbb{R}_+)$ if the coset $T^\pi := A + \mathcal{K}$ is left invertible (resp. right invertible) in the quotient Banach algebra $\Lambda^\pi \subset \mathcal{B}^\pi$. Applying now Theorem 5.1 with $\mathfrak{A} = \Lambda^\pi$ and $\mathfrak{Z} = \mathcal{Z}^\pi$, we immediately obtain the following.

Theorem 5.4. *An operator $T \in \Lambda$ is left (resp. right) Fredholm on the space $L^p(\mathbb{R}_+)$ if the following three conditions are fulfilled:*

(i) *the coset $T^\pi + \mathcal{J}^\pi_{+\infty}$ is left (resp. right) invertible in the quotient algebra $\Lambda^\pi_{+\infty}$;*
(ii) *the coset $T^\pi + \mathcal{J}^\pi_{-\infty}$ is left (resp. right) invertible in the quotient algebra $\Lambda^\pi_{-\infty}$;*
(iii) *for every $(\xi, x) \in \Delta \times \mathbb{R}$, the coset $T^\pi + \mathcal{J}^\pi_{\xi,x}$ is left (resp. right) invertible in the quotient algebra $\Lambda^\pi_{\xi,x}$.*

It follows from Theorems 4.2(d) and 5.2 that $N \in \mathcal{F}_{\alpha,\beta} \subset \Lambda$. Thus, Theorem 5.4 is applicable to N. Hence, our next aim is to study one-sided invertibility of the cosets $N^\pi + \mathcal{J}^\pi_{+\infty}$, $N^\pi + \mathcal{J}^\pi_{-\infty}$ and $N^\pi + \mathcal{J}^\pi_{\xi,x}$ in the corresponding local algebras $\Lambda^\pi_{+\infty}$, $\Lambda^\pi_{-\infty}$ and $\Lambda^\pi_{\xi,x}$ for all $(\xi, x) \in \Delta \times \mathbb{R}$.

6. Mellin pseudodifferential operators and their symbols

6.1. Mellin PDO's: overview

Mellin pseudodifferential operators are generalizations of Mellin convolution operators. Let $\mathfrak{a}$ be a sufficiently smooth function defined on $\mathbb{R}_+ \times \mathbb{R}$. The Mellin pseudodifferential operator (shortly, Mellin PDO) with symbol $\mathfrak{a}$ is initially defined for smooth functions f of compact support by the iterated integral

$$\begin{aligned}[\mathrm{Op}(\mathfrak{a})f](t) &= [\mathcal{M}^{-1}\mathfrak{a}(t,\cdot)\mathcal{M}f](t) \\ &= \frac{1}{2\pi}\int_{\mathbb{R}} dx \int_{\mathbb{R}_+} \mathfrak{a}(t,x)\left(\frac{t}{\tau}\right)^{ix} f(\tau)\frac{d\tau}{\tau} \quad \text{for} \quad t \in \mathbb{R}_+. \end{aligned} \tag{6.1}$$

Obviously, if $\mathfrak{a}(t,x) = a(x)$ for all $(t,x) \in \mathbb{R}_+ \times \mathbb{R}$, then the Mellin pseudodifferential operator $\mathrm{Op}(\mathfrak{a})$ becomes the Mellin convolution operator

$$\mathrm{Op}(\mathfrak{a}) = \mathrm{Co}(a).$$

In 1991 Rabinovich [27] (see also [28]) proposed to use Mellin pseudodifferential operators with C^∞ slowly oscillating symbols to study singular integral operators with slowly oscillating coefficients on L^p spaces. Namely, he considered symbols $\mathfrak{a} \in C^\infty(\mathbb{R}_+ \times \mathbb{R})$ such that

$$\sup_{(t,x)\in\mathbb{R}_+\times\mathbb{R}} \left|(t\partial_t)^j \partial_x^k \mathfrak{a}(t,x)\right|(1+x^2)^{k/2} < \infty \quad \text{for all} \quad j,k \in \mathbb{Z}_+ \tag{6.2}$$

and

$$\lim_{t\to s}\sup_{x\in\mathbb{R}}\big|(t\partial_t)^j\partial_x^k\mathfrak{a}(t,x)\big|(1+x^2)^{k/2}=0 \text{ for all } j\in\mathbb{N},\quad k\in\mathbb{Z}_+, \tag{6.3}$$

where $\mathbb{Z}_+=\mathbb{N}\cup\{0\}$ and $s\in\{0,\infty\}$. Here and in what follows ∂_t and ∂_x denote the operators of partial differentiation with respect to t and to x. Notice that (6.2) defines nothing but the Mellin version of the Hörmander class $S^0_{1,0}(\mathbb{R})$ (see, e.g., [24, Chap. 2, Section 1] for the definition of the Hörmander classes $S^m_{\varrho,\delta}(\mathbb{R}^n)$). If $\mathfrak{a}$ satisfies (6.2), then the Mellin PDO $\mathrm{Op}(\mathfrak{a})$ is bounded on the spaces $L^p(\mathbb{R}_+,d\mu)$ for $1<p<\infty$ (see, e.g., [33, Chap. VI, Proposition 4] for the corresponding Fourier PDO's). Condition (6.3) is the Mellin version of Grushin's definition of symbols slowly varying in the first variable (see, e.g., [11], [24, Chap. 3, Definition 5.11]).

The idea of application of Mellin PDO's with considered class of symbols was exploited in a series of papers by Rabinovich and coauthors (see [29, Sections 4.6–4.7] for the complete history up to 2004). On the other hand, the smoothness conditions imposed on slowly oscillating symbols are very strong. In particular, they are not applicable directly to the problem we are dealing with in the present paper.

In 2005 the second author [22] developed a Fredholm theory for the Fourier pseudodifferential operators with slowly oscillating symbols of limited smoothness in the spirit of Sarason's definition [31, p. 820] of slow oscillation adopted in the present paper (much less restrictive than in [28] and in the works mentioned in [29]). Necessary for our purposes results from [22] were translated to the Mellin setting, for instance, in [16] with the aid of the transformation defined by (4.1)–(4.2). For the convenience of readers, we reproduce the results required in what follows exactly in the same form as they were stated in [16], where more details on their proofs can be found.

6.2. Boundedness of Mellin PDO's

If a is an absolutely continuous function of finite total variation on $\mathbb{R}$, then its derivative belongs to $L^1(\mathbb{R})$ and

$$V(a)=\int_{\mathbb{R}}|a'(x)|\,dx$$

(see, e.g., [26, Chap. VIII, Sections 3 and 9; Chap. XI, Section 4]). The set $V(\mathbb{R})$ of all absolutely continuous functions of finite total variation on $\mathbb{R}$ forms a Banach algebra when equipped with the norm

$$\|a\|_V:=\|a\|_{L^\infty(\mathbb{R})}+V(a)=\|a\|_{L^\infty(\mathbb{R})}+\int_{\mathbb{R}}|a'(x)|\,dx. \tag{6.4}$$

Let $C_b(\mathbb{R}_+,V(\mathbb{R}))$ denote the Banach algebra of all bounded continuous $V(\mathbb{R})$-valued functions on $\mathbb{R}_+$ with the norm

$$\|\mathfrak{a}(\cdot,\cdot)\|_{C_b(\mathbb{R}_+,V(\mathbb{R}))}=\sup_{t\in\mathbb{R}_+}\|\mathfrak{a}(t,\cdot)\|_V.$$

As usual, let $C_0^\infty(\mathbb{R}_+)$ be the set of all infinitely differentiable functions of compact support on $\mathbb{R}_+$.

Theorem 6.1 ([16, Theorem 3.1]). *If $\mathfrak{a} \in C_b(\mathbb{R}_+, V(\mathbb{R}))$, then the Mellin pseudodifferential operator* $\mathrm{Op}(\mathfrak{a})$, *defined for functions $f \in C_0^\infty(\mathbb{R}_+)$ by the iterated integral* (6.1), *extends to a bounded linear operator on the space $L^p(\mathbb{R}_+, d\mu)$ and there is a number $C_p \in (0,\infty)$ depending only on p such that*

$$\|\mathrm{Op}(\mathfrak{a})\|_{\mathcal{B}(L^p(\mathbb{R}_+,d\mu))} \le C_p \|\mathfrak{a}\|_{C_b(\mathbb{R}_+,V(\mathbb{R}))}.$$

6.3. Products of Mellin PDO's

Consider the Banach subalgebra $SO(\mathbb{R}_+, V(\mathbb{R}))$ of the algebra $C_b(\mathbb{R}_+, V(\mathbb{R}))$ consisting of all $V(\mathbb{R})$-valued functions $\mathfrak{a}$ on $\mathbb{R}_+$ that slowly oscillate at 0 and ∞, that is,

$$\lim_{r\to s} \max_{t,\tau\in[r,2r]} \|\mathfrak{a}(t,\cdot) - \mathfrak{a}(\tau,\cdot)\|_{L^\infty(\mathbb{R})} = 0, \quad s \in \{0,\infty\}.$$

Let $\mathcal{E}(\mathbb{R}_+, V(\mathbb{R}))$ be the Banach algebra of all $V(\mathbb{R})$-valued functions $\mathfrak{a}$ in the algebra $SO(\mathbb{R}_+, V(\mathbb{R}))$ such that

$$\lim_{|h|\to 0} \sup_{t\in\mathbb{R}_+} \left\|\mathfrak{a}(t,\cdot) - \mathfrak{a}^h(t,\cdot)\right\|_V = 0$$

where $\mathfrak{a}^h(t,x) := \mathfrak{a}(t, x+h)$ for all $(t,x) \in \mathbb{R}_+ \times \mathbb{R}$.

Theorem 6.2 ([16, Theorem 3.3]). *If $\mathfrak{a}, \mathfrak{b} \in \mathcal{E}(\mathbb{R}_+, V(\mathbb{R}))$, then*

$$\mathrm{Op}(\mathfrak{a})\mathrm{Op}(\mathfrak{b}) \simeq \mathrm{Op}(\mathfrak{a}\mathfrak{b}).$$

Lemma 6.3 ([16, Lemma 3.4]). *If $\mathfrak{a}, \mathfrak{b}, \mathfrak{c} \in \mathcal{E}(\mathbb{R}_+, V(\mathbb{R}))$ are such that $\mathfrak{a}$ depends only on the first variable and $\mathfrak{c}$ depends only on the second variable, then*

$$\mathrm{Op}(\mathfrak{a})\mathrm{Op}(\mathfrak{b})\mathrm{Op}(\mathfrak{c}) = \mathrm{Op}(\mathfrak{a}\mathfrak{b}\mathfrak{c}).$$

6.4. Applications of Mellin pseudodifferential operators

We immediately deduce the following assertion from [14, Lemma 4.1].

Lemma 6.4. *Suppose $f \in SO(\mathbb{R}_+)$ and $\gamma \in \mathbb{C}$ satisfies* (1.3). *Then the functions*

$$\mathfrak{f}(t,x) := f(t), \quad \mathfrak{s}_\gamma(t,x) := s_\gamma(x), \quad \mathfrak{r}_\gamma(t,x) := r_\gamma(x), \quad (t,x) \in \mathbb{R}_+ \times \mathbb{R},$$

belong to the Banach algebra $\mathcal{E}(\mathbb{R}_+, V(\mathbb{R}))$.

Lemma 6.5. *If $\mathfrak{b} \in \mathcal{E}(\mathbb{R}_+, V(\mathbb{R}))$, then the operator $\Phi^{-1}\mathrm{Op}(\mathfrak{b})\Phi$ belongs to the algebra Λ.*

Proof. Let $c \in SO(\mathbb{R}_+)$. It follows from Lemma 6.4 that the functions

$$\mathfrak{c}_0(t,x) := c(t) r_0(x), \quad \mathfrak{s}_0(t,x) := s_0(x), \quad (t,x) \in \mathbb{R}_+ \times \mathbb{R},$$

belong to the algebra $\mathcal{E}(\mathbb{R}_+, V(\mathbb{R}))$. Since $\mathcal{E}(\mathbb{R}_+, V(\mathbb{R})) \subset C_b(\mathbb{R}_+, V(\mathbb{R}))$, Theorem 6.1 implies that $B := \Phi^{-1}\mathrm{Op}(\mathfrak{b})\Phi \in \mathcal{B}$. We infer from Theorem 6.2 that

$$\mathrm{Op}(\mathfrak{s}_0)\mathrm{Op}(\mathfrak{b}) - \mathrm{Op}(\mathfrak{b})\mathrm{Op}(\mathfrak{s}_0) \in \mathcal{K}(L^p(\mathbb{R}_+, d\mu)), \tag{6.5}$$

$$\mathrm{Op}(\mathfrak{c}_0)\mathrm{Op}(\mathfrak{b}) - \mathrm{Op}(\mathfrak{b})\mathrm{Op}(\mathfrak{c}_0) \in \mathcal{K}(L^p(\mathbb{R}_+, d\mu)). \tag{6.6}$$

On the other hand, by Theorem 4.2(d) and Lemma 6.3,

$$S_0 = \Phi^{-1}\mathrm{Co}(s_0)\Phi = \Phi^{-1}\mathrm{Op}(\mathfrak{s}_0)\Phi, \tag{6.7}$$

$$cR_0 = c\Phi^{-1}\mathrm{Co}(r_0)\Phi = \Phi^{-1}\mathrm{Op}(\mathfrak{c}_0)\Phi. \tag{6.8}$$

Combining (6.5)–(6.8), we conclude that $S_0B - BS_0, (cR_0)B - B(cR_0) \in \mathcal{K}$. Hence $B \in \Lambda$. □

Applying [14, Lemma 4.4] and making minor modifications in the proof of [16, Lemma 4.5], we get the following.

Lemma 6.6. *Let $\gamma \in \mathbb{C}$ satisfy (1.3). Suppose $\alpha \in SOS(\mathbb{R}_+)$, ω is its exponent function, and U_α is the associated isometric shift operator on $L^p(\mathbb{R}_+)$. Then the operator $U_\alpha R_\gamma$ can be realized as the Mellin pseudodifferential operator up to a compact operator:*

$$U_\alpha R_\gamma \simeq \Phi^{-1}\mathrm{Op}(\mathfrak{c})\Phi,$$

where the function $\mathfrak{c}$, given by

$$\mathfrak{c}(t,x) := e^{i\omega(t)x} r_\gamma(x) \quad \textit{for} \quad (t,x) \in \mathbb{R}_+ \times \mathbb{R},$$

belongs to the Banach algebra $\mathcal{E}(\mathbb{R}_+, V(\mathbb{R}))$.

7. Sufficient conditions for the semi-Fredholmness

7.1. One-sided invertibility in the quotient algebras $\Lambda^\pi_{+\infty}$ and $\Lambda^\pi_{-\infty}$

The next lemma shows that the operator N can be written as a paired operator with respect to the pair (P_0^+, P_0^-).

Lemma 7.1. *Let $1 < p < \infty$ and let $\gamma \in \mathbb{C}$ satisfy (1.3). Suppose a_k, b_k belong to $SO(\mathbb{R}_+)$ for all $k \in \mathbb{Z}$, α, β belong to $SOS(\mathbb{R}_+)$, the operators $A_+ \in W^{SO}_{\alpha,p}$ and $A_- \in W^{SO}_{\beta,p}$ are given by (1.5), and the operator N is given by (1.6). Then the operator N can be represented in each of the forms*

$$N = A_+P_0^+ + C_-P_0^- = C_+P_0^+ + A_-P_0^-, \tag{7.1}$$

where

$$C_+ := A_+ + 2\sinh(\pi i\gamma)e^{\pi i\gamma}(A_+ - A_-)P_\gamma^-,$$

$$C_- := A_- + 2\sinh(\pi i\gamma)e^{-\pi i\gamma}(A_+ - A_-)P_\gamma^+.$$

Moreover, all operators $DP_0^+ - P_0^+D$ and $DP_0^- - P_0^-D$, where D belongs to the set $\{A_+, A_-, C_+, C_-\}$, are compact on the space $L^p(\mathbb{R}_+)$.

Proof. Both representations follow from Lemma 4.4. The compactness of the commutators $DP_0^\pm - P_0^\pm D$ is a consequence of parts (a) and (d) of Theorem 4.2 and Lemma 4.5. □

The following statement generalizes [15, Theorem 8.1] from the case of two-sided invertible binomial functional operators A_+ and A_- to the case of one-sided invertible operators $A_+ \in W^{SO}_{\alpha,p}$ and $A_- \in W^{SO}_{\beta,p}$. This generalization is possible thanks to Theorem 3.3, although the proof follows the same lines as in [15].

Theorem 7.2. *Let $1 < p < \infty$ and let $\gamma \in \mathbb{C}$ satisfy (1.3). Suppose a_k, b_k belong to $SO(\mathbb{R}_+)$ for all $k \in \mathbb{Z}$, α, β belong to $SOS(\mathbb{R}_+)$, the operators $A_+ \in W^{SO}_{\alpha,p}$ and $A_- \in W^{SO}_{\beta,p}$ are given by (1.5), and the operator N is given by (1.6).*

(a) *If the operator A_+ is left (resp. right) invertible on the space $L^p(\mathbb{R}_+)$, then the coset $N^\pi + \mathcal{J}^\pi_{+\infty}$ is left (resp. right) invertible in the quotient algebra $\Lambda^\pi_{+\infty}$.*
(b) *If the operator A_- is left (resp. right) invertible on the space $L^p(\mathbb{R}_+)$, then the coset $N^\pi + \mathcal{J}^\pi_{-\infty}$ is left (resp. right) invertible in the quotient algebra $\Lambda^\pi_{-\infty}$.*

Proof. Recall that $\mathcal{FO}_{\alpha,\beta} \subset \mathcal{F}_{\alpha,\beta} \subset \Lambda$ in view of Theorem 5.2. By Lemma 7.1, the operator N is represented in each of the forms (7.1), where

$$A_+ \in W^{SO}_{\alpha,p} \subset \mathcal{FO}_{\alpha,\beta} \subset \mathcal{F}_{\alpha,\beta} \subset \Lambda, \quad A_- \in W^{SO}_{\beta,p} \subset \mathcal{FO}_{\alpha,\beta} \subset \mathcal{F}_{\alpha,\beta} \subset \Lambda,$$

and $C_+, C_- \in \mathcal{F}_{\alpha,\beta} \subset \Lambda$.

(a) Take $A_+ \in W^{SO}_{\alpha,p}$. If A_+ is left (resp. right) invertible in $\mathcal{B}$, then it follows from Theorem 3.3 that there exists a left (resp. right) inverse $A_+^{(-1)}$ of A_+ such that $A_+^{(-1)} \in W^{SO}_{\alpha,p} \subset \mathcal{F}_{\alpha,\beta} \subset \Lambda$. Hence the coset $A_+^\pi = A_+ + \mathcal{K}$ is left (resp. right) invertible in the quotient algebra Λ^π, which implies the left (resp. right) invertibility of the coset $A_+^\pi + \mathcal{J}^\pi_{+\infty}$ in the quotient algebra $\Lambda^\pi_{+\infty}$. Hence we infer from (7.1) that

$$\begin{aligned} N^\pi + \mathcal{J}^\pi_{+\infty} &= (A_+P_0^+ + C_-P_0^-)^\pi + \mathcal{J}^\pi_{+\infty} \\ &= A_+^\pi + [(C_- - A_+)P_0^-]^\pi + \mathcal{J}^\pi_{+\infty} = A_+^\pi + \mathcal{J}^\pi_{+\infty} \end{aligned}$$

because $(P_0^-)^\pi \in \mathcal{I}^\pi_{+\infty} \subset \mathcal{J}^\pi_{+\infty}$. Thus, the left (resp. right) invertibility of the operator A_+ in $\mathcal{B}$ implies the left (resp. right) invertibility of the coset $N^\pi + \mathcal{J}^\pi_{+\infty}$ in the quotient algebra $\Lambda^\pi_{+\infty}$. Part (a) is proved.

(b) The proof is analogous. □

7.2. Invertibility in the quotient algebras $\Lambda^\pi_{\xi,x}$ with $(\xi, x) \in \Delta \times \mathbb{R}$

By a literal repetition with minor modifications of the proof of [15, Lemma 7.4], we get the following.

Lemma 7.3. *Let $\gamma \in \mathbb{C}$ satisfy (1.3). Suppose $\alpha \in SOS(\mathbb{R}_+)$ and ω is its exponent function. If $\xi \in \Delta$ and*

$$\mathfrak{a}(t, x) := e^{i\omega(t)x}(r_\gamma(x))^2 \quad \text{for} \quad (t, x) \in (\Delta \cup \mathbb{R}_+) \times \mathbb{R},$$

then there exists a function $\mathfrak{b}_\xi \in \mathcal{E}(\mathbb{R}_+, V(\mathbb{R}))$ such that

$$\mathfrak{a}(t, x) - \mathfrak{a}(\xi, x) = (\omega(t) - \omega(\xi))\mathfrak{b}_\xi(t, x)r_\gamma(x) \quad \text{for} \quad (t, x) \in \mathbb{R}_+ \times \mathbb{R}.$$

Lemma 7.4. *Let $\gamma \in \mathbb{C}$ satisfy (1.3). Suppose α is a slowly oscillating shift, ω is its exponent function, and U_α is the associated isometric shift operator on $L^p(\mathbb{R}_+)$. If $(\xi, x) \in \Delta \times \mathbb{R}$, then*

$$(U_\alpha R_\gamma^2)^\pi - e^{i\omega(\xi)x}(r_\gamma(x))^2 I^\pi \in \mathcal{J}_{\xi,x}^\pi.$$

Proof. The proof is developed by analogy with [15, Lemma 8.3]. In view of Lemma 6.6,

$$U_\alpha R_\gamma \simeq \Phi^{-1}\mathrm{Op}(\mathfrak{c}_{\omega,\gamma})\Phi, \tag{7.2}$$

where $\mathfrak{c}_{\omega,\gamma} \in \mathcal{E}(\mathbb{R}_+, V(\mathbb{R}))$ is given by

$$\mathfrak{c}_{\omega,\gamma}(t,y) := e^{i\omega(t)y} r_\gamma(y) \quad \text{for} \quad (t,y) \in \mathbb{R}_+ \times \mathbb{R}.$$

On the other hand, in view of Theorem 4.2(d) and Lemma 6.4,

$$R_\gamma = \Phi^{-1}\mathrm{Co}(r_\gamma)\Phi = \Phi^{-1}\mathrm{Op}(\mathfrak{r}_\gamma)\Phi, \tag{7.3}$$

where $\mathfrak{r}_\gamma \in \mathcal{E}(\mathbb{R}_+, V(\mathbb{R}))$ is given by

$$\mathfrak{r}_\gamma(t,y) = r_\gamma(y) \quad \text{for} \quad (t,y) \in \mathbb{R}_+ \times \mathbb{R}.$$

It follows from (7.2)–(7.3) and Lemma 6.3 that

$$U_\alpha R_\gamma^2 \simeq \Phi^{-1}\mathrm{Op}(\mathfrak{a})\Phi, \tag{7.4}$$

where

$$\mathfrak{a}(t,y) = \mathfrak{c}_{\omega,\gamma}(t,y)\mathfrak{r}_\gamma(t,y) = e^{i\omega(t)y}(r_\gamma(y))^2 \quad \text{for} \quad (t,y) \in \mathbb{R}_+ \times \mathbb{R}.$$

Since $\mathfrak{a}(\cdot, y) \in SO(\mathbb{R}_+)$ for every $y \in \mathbb{R}$, taking the Gelfand transform of $\mathfrak{a}(\cdot, y)$, we obtain

$$\mathfrak{a}(t,y) = e^{i\omega(t)y}(r_\gamma(y))^2 \quad \text{for} \quad (t,y) \in (\Delta \cup \mathbb{R}_+) \times \mathbb{R}.$$

Fix $(\xi, x) \in \Delta \times \mathbb{R}$. Let us represent the function $\mathfrak{a}$ in the form

$$\begin{aligned} \mathfrak{a}(t,y) &= \mathfrak{a}(t,y) - \mathfrak{a}(\xi,y) + \mathfrak{c}_{\omega,\gamma}(\xi,y) r_\gamma(y) \\ &= [\mathfrak{a}(t,y) - \mathfrak{a}(\xi,y)] + [\mathfrak{c}_{\omega,\gamma}(\xi,y) - \mathfrak{c}_{\omega,\gamma}(\xi,x)]\mathfrak{r}_\gamma(t,y) \\ &\quad + \mathfrak{c}_{\omega,\gamma}(\xi,x)[\mathfrak{r}_\gamma(t,y) - \mathfrak{r}_\gamma(t,x)] + \mathfrak{a}(\xi,x), \end{aligned} \tag{7.5}$$

where $(t,y) \in \mathbb{R}_+ \times \mathbb{R}$. Further, we deduce from Lemma 7.3 that there exists a function $\mathfrak{b}_\xi \in \mathcal{E}(\mathbb{R}_+, V(\mathbb{R}))$ such that

$$\mathfrak{a} - \mathfrak{a}(\xi, \cdot) = (\omega - \omega(\xi))\mathfrak{b}_\xi \mathfrak{r}_\gamma.$$

Hence, we infer from the above equality and Lemmas 6.4 and 6.3 that

$$\begin{aligned} \Phi^{-1}\mathrm{Op}(\mathfrak{a} - \mathfrak{a}(\xi,\cdot))\Phi &= \Phi^{-1}\mathrm{Op}((\omega - \omega(\xi))\mathfrak{b}_\xi\mathfrak{r}_\gamma)\Phi \\ &= (\Phi^{-1}\mathrm{Op}(\omega - \omega(\xi))\Phi) \cdot (\Phi^{-1}\mathrm{Op}(\mathfrak{b}_\xi)\Phi) \cdot (\Phi^{-1}\mathrm{Op}(\mathfrak{r}_\gamma)\Phi). \end{aligned}$$

The latter equality, Theorem 6.2 and equality (7.3) imply that

$$\Phi^{-1}\mathrm{Op}(\mathfrak{a} - \mathfrak{a}(\xi,\cdot))\Phi \simeq (\omega - \omega(\xi))R_\gamma(\Phi^{-1}\mathrm{Op}(\mathfrak{b}_\xi)\Phi). \tag{7.6}$$

Applying Theorem 5.3(c) and Corollary 4.3(b), we obtain

$$([(\omega - \omega(\xi))R_\gamma]^\pi)\widehat{\ }(\xi,x) = (\omega(\xi) - \omega(\xi))r_\gamma(x) = 0.$$

Therefore $[(\omega-\omega(\xi))R_\gamma]^\pi \in \mathcal{I}^\pi_{\xi,x}$. On the other hand, since $\mathfrak{b}_\xi \in \mathcal{E}(\mathbb{R}_+, V(\mathbb{R}))$, we conclude from Lemma 6.5 that $\Phi^{-1}\mathrm{Op}(\mathfrak{b}_\xi)\Phi \in \Lambda$. Then, taking into account (7.6) and the definition of the ideal $\mathcal{J}^\pi_{\xi,x}$, we infer that

$$[\Phi^{-1}\mathrm{Op}(\mathfrak{a}-\mathfrak{a}(\xi,\cdot))\Phi]^\pi = [(\omega-\omega(\xi))R_\gamma(\Phi^{-1}\mathrm{Op}(\mathfrak{b}_\xi)\Phi)]^\pi \in \mathcal{J}^\pi_{\xi,x}. \tag{7.7}$$

Taking into account the definition of the norm (6.4) in the algebra $V(\mathbb{R})$, it is easy to see that the function $\mathfrak{c}_{\omega,\gamma}(\xi,\cdot)$ belongs to $V(\mathbb{R})$, where

$$\mathfrak{c}_{\omega,\gamma}(\xi,y) = e^{i\omega(\xi)y} r_\gamma(y) \quad \text{for} \quad y \in \mathbb{R}_+.$$

Then, by Stechkin's inequality (Theorem 4.1), $\mathfrak{c}_{\omega,\gamma}(\xi,\cdot) \in C_p(\overline{\mathbb{R}})$. Hence, it follows from Theorem 4.2(a) that $\Phi^{-1}\mathrm{Co}(\mathfrak{c}_{\omega,\gamma}(\xi,\cdot))\Phi \in \mathcal{A}$. By Theorem 5.3(c) and Corollary 4.3(b),

$$\begin{aligned}
&([\Phi^{-1}\mathrm{Op}(\mathfrak{c}_{\omega,\gamma}(\xi,\cdot)-\mathfrak{c}_{\omega,\gamma}(\xi,x))\Phi]^\pi)\widehat{\ }(\xi,x)\\
&\qquad = ([\Phi^{-1}\mathrm{Co}(\mathfrak{c}_{\omega,\gamma}(\xi,\cdot))\Phi]^\pi)\widehat{\ }(\xi,x) - ([\mathfrak{c}_{\omega,\gamma}(\xi,x)I]^\pi)\widehat{\ }(\xi,x)\\
&\qquad = e^{i\omega(\xi)x} r_\gamma(x) - e^{i\omega(\xi)x} r_\gamma(x) = 0.
\end{aligned}$$

Therefore

$$[\Phi^{-1}\mathrm{Op}(\mathfrak{c}_{\omega,\gamma}(\xi,\cdot)-\mathfrak{c}_{\omega,\gamma}(\xi,x))\Phi]^\pi \in \mathcal{I}^\pi_{\xi,x}.$$

By this observation, Lemma 6.3 and equality (7.3), we obtain

$$\begin{aligned}
&[\Phi^{-1}\mathrm{Op}((\mathfrak{c}_{\omega,\gamma}(\xi,\cdot)-\mathfrak{c}_{\omega,\gamma}(\xi,x))\mathfrak{r}_\gamma)\Phi]^\pi\\
&\qquad = [\Phi^{-1}\mathrm{Op}(\mathfrak{c}_{\omega,\gamma}(\xi,\cdot)-\mathfrak{c}_{\omega,\gamma}(\xi,x))\Phi]^\pi[\Phi^{-1}\mathrm{Op}(\mathfrak{r}_\gamma)\Phi]^\pi\\
&\qquad = [\Phi^{-1}\mathrm{Op}(\mathfrak{c}_{\omega,\gamma}(\xi,\cdot)-\mathfrak{c}_{\omega,\gamma}(\xi,x))\Phi]^\pi R^\pi_\gamma \in \mathcal{J}^\pi_{\xi,x}.
\end{aligned} \tag{7.8}$$

Finally, in view of (7.3), Theorem 5.3(c) and Corollary 4.3(b), we deduce that

$$([\Phi^{-1}\mathrm{Op}(\mathfrak{r}_\gamma - r_\gamma(x))\Phi]^\pi)\widehat{\ }(\xi,x) = ([R_\gamma - r_\gamma(x)I]^\pi)\widehat{\ }(\xi,x) = r_\gamma(x)-r_\gamma(x) = 0.$$

Hence

$$\mathfrak{c}_{\omega,\gamma}(\xi,x)[\Phi^{-1}\mathrm{Op}(\mathfrak{r}_\gamma - r_\gamma(x))\Phi]^\pi \in \mathcal{I}^\pi_{\xi,x} \subset \mathcal{J}^\pi_{\xi,x}. \tag{7.9}$$

Combining (7.4)–(7.5) with (7.7)–(7.9), we arrive at the relation

$$(U_\alpha R^2_\gamma)^\pi - e^{i\omega(\xi)x}(r_\gamma(x))^2 I^\pi = [\Phi^{-1}\mathrm{Op}(\mathfrak{a})\Phi - \mathfrak{a}(\xi,x)I]^\pi \in \mathcal{J}^\pi_{\xi,x},$$

which completes the proof. □

Now we are in a position to prove that condition (ii) of Theorem 1.1 is sufficient for the invertibility of the coset $N^\pi + \mathcal{J}^\pi_{\xi,x}$ in the quotient algebra $\Lambda^\pi_{\xi,x}$.

Theorem 7.5. *Let $1<p<\infty$ and let $\gamma \in \mathbb{C}$ satisfy* (1.3). *Suppose a_k, b_k belong to $SO(\mathbb{R}_+)$ for all $k \in \mathbb{Z}$, α, β belong to $SOS(\mathbb{R}_+)$, the operators $A_+ \in W^{SO}_{\alpha,p}$ and $A_- \in W^{SO}_{\beta,p}$ are given by* (1.5), *and the operator N is given by* (1.6). *If $n(\xi,x) \neq 0$ for some $(\xi,x) \in \Delta \times \mathbb{R}$, where the function n is defined by* (1.7)–(1.9), *then the coset $N^\pi + \mathcal{J}^\pi_{\xi,x}$ is two-sided invertible in the quotient algebra $\Lambda^\pi_{\xi,x}$.*

Proof. We follow the main lines of the proof of [15, Theorem 8.4].

Fix $(\xi, x) \in \Delta \times \mathbb{R}$ and consider the operators

$$H_\pm := \frac{p_\gamma^\pm(x)}{[r_\gamma(x)]^2} R_\gamma^2. \tag{7.10}$$

Then it follows from Theorem 5.3(c) and Corollary 4.3(b) that

$$(H_\pm^\pi)\widehat{\ }(\xi, x) = p_\gamma^\pm(x). \tag{7.11}$$

Therefore, taking into account Corollary 4.3(b) once again, we get

$$(P_\gamma^\pm - H_\pm)^\pi \in \mathcal{I}_{\xi,x}^\pi$$

and

$$\big[A_+(P_\gamma^+ - H_+) + A_-(P_\gamma^- - H_-)\big]^\pi \in \mathcal{J}_{\xi,x}^\pi,$$

whence

$$N^\pi + \mathcal{J}_{\xi,x}^\pi = (A_+H_+ + A_-H_-)^\pi + \mathcal{J}_{\xi,x}^\pi. \tag{7.12}$$

We know from Theorem 4.2(d),(a) that $H_\pm \in \mathcal{A}$. Hence, we infer from Lemma 4.5 that for all $k \in \mathbb{Z}$,

$$(U_\alpha^k H_+)^\pi = (H_+U_\alpha^k)^\pi, \quad (U_\beta^k H_-)^\pi = (H_-U_\beta^k)^\pi. \tag{7.13}$$

Taking into account (7.11), it is easy to see that for all $k \in \mathbb{Z}$,

$$\big[(a_kH_+)^\pi - (a_k(\xi)H_+)^\pi\big]\widehat{\ }(\xi, x) = 0, \quad \big[(b_kH_-)^\pi - (b_k(\xi)H_-)^\pi\big]\widehat{\ }(\xi, x) = 0.$$

Hence

$$(a_kH_+)^\pi - (a_k(\xi)H_+)^\pi,\ (b_kH_-)^\pi - (b_k(\xi)H_-)^\pi \in \mathcal{I}_{\xi,x}^\pi \subset \mathcal{J}_{\xi,x}^\pi. \tag{7.14}$$

Applying (7.13) and (7.14), we obtain for all $k \in \mathbb{Z}$,

$$(a_kU_\alpha^kH_+)^\pi - (a_k(\xi)U_\alpha^kH_+)^\pi = [(a_kH_+ - a_k(\xi)H_+)U_\alpha^k]^\pi \in \mathcal{J}_{\xi,x}^\pi, \tag{7.15}$$

$$(b_kU_\beta^kH_-)^\pi - (b_k(\xi)U_\beta^kH_-)^\pi = [(b_kH_- - b_k(\xi)H_-)U_\beta^k]^\pi \in \mathcal{J}_{\xi,x}^\pi. \tag{7.16}$$

Then it follows from (7.12) and (7.15)–(7.16) that

$$N^\pi + \mathcal{J}_{\xi,x}^\pi = \sum_{k\in\mathbb{Z}} \big(a_k(\xi)U_\alpha^kH_+ + b_k(\xi)U_\beta^kH_-\big)^\pi + \mathcal{J}_{\xi,x}^\pi. \tag{7.17}$$

In view of Theorem 5.3(c) and Corollary 4.3(b), it is easy to see that

$$\big([(r_\gamma(x))^{-2}R_\gamma^2 - I]^\pi\big)\widehat{\ }(\xi, x) = 0.$$

Hence

$$H_\pm^\pi - p_\gamma^\pm(x)I^\pi \in \mathcal{I}_{\xi,x}^\pi \subset \mathcal{J}_{\xi,x}^\pi. \tag{7.18}$$

By Lemmas 2.3, 2.6, and 7.4, we deduce for all $k \in \mathbb{Z}$ that

$$(U_\alpha^kR_\gamma^2)^\pi - e^{ik\omega(\xi)x}(r_\gamma(x))^2I^\pi,\ (U_\beta^kR_\gamma^2)^\pi - e^{ik\eta(\xi)x}(r_\gamma(x))^2I^\pi \in \mathcal{J}_{\xi,x}^\pi.$$

The above inclusions together with (7.10) imply for every $k \in \mathbb{Z}$ that

$$(U_\alpha^kH_+)^\pi - e^{ik\omega(\xi)x}p_\gamma^+(x)\,I^\pi \in \mathcal{J}_{\xi,x}^\pi, \tag{7.19}$$

$$(U_\beta^kH_-)^\pi - e^{ik\eta(\xi)x}p_\gamma^-(x)\,I^\pi \in \mathcal{J}_{\xi,x}^\pi. \tag{7.20}$$

Combining (7.18)–(7.20), we arrive at the equality

$$N^\pi + \mathcal{J}^\pi_{\xi,x} = n(\xi,x)I^\pi + \mathcal{J}^\pi_{\xi,x},$$

where $n(\xi,x)$ is given by (1.8)–(1.9). If $n(\xi,x) \neq 0$, then one can check straightforwardly that $(1/n(\xi,x))I^\pi + \mathcal{J}^\pi_{\xi,x}$ is the inverse of the coset $N^\pi + \mathcal{J}^\pi_{\xi,x}$ in the quotient algebra $\Lambda^\pi_{\xi,x}$. □

7.3. Proof of Theorem 1.1

The proof is analogous to that of [15, Theorem 1.2]. We know from Theorem 4.2(d) and Theorem 5.2 that $N \in \Lambda$. If condition (i) of Theorem 1.1 is fulfilled, that is, if the operators A_+ and A_- are left (resp. right) invertible, then by Theorem 7.2 the coset $N^\pi + \mathcal{J}^\pi_{+\infty}$ is left (resp. right) invertible in the quotient algebra $\Lambda^\pi_{+\infty}$ and the coset $N^\pi + \mathcal{J}^\pi_{-\infty}$ is left (resp. right) invertible in the quotient algebra $\Lambda^\pi_{-\infty}$. On the other hand, if condition (ii) of Theorem 1.1 holds, then in view of Theorem 7.5, the coset $N^\pi + \mathcal{J}^\pi_{\xi,x}$ is two-sided invertible in the quotient algebra $\Lambda^\pi_{\xi,x}$ for every pair $(\xi,x) \in \Delta \times \mathbb{R}$. Then, by Theorem 5.4, the operator $N \in \Lambda$ is left (resp. right) Fredholm. □

8. Semi-Fredholmness of weighted singular integral operators with coefficients being binomial functional operators

8.1. Criteria for the two-sided and strict one-sided invertibility of $aI - bU_\alpha$

Suppose $a, b \in SO(\mathbb{R}_+)$ and $\alpha \in SOS(\mathbb{R}_+)$. For $s \in \{0, \infty\}$, put

$$L_*(s; a, b) := \liminf_{t\to s}(|a(t)| - |b(t)|), \quad L^*(s; a, b) := \limsup_{t\to s}(|a(t)| - |b(t)|).$$

Fix a point $\tau \in \mathbb{R}_+$ and put

$$\tau_{-,\alpha} := \lim_{n\to-\infty} \alpha_n(\tau), \quad \tau_{+,\alpha} := \lim_{n\to+\infty} \alpha_n(\tau).$$

Then either $\tau_{-,\alpha} = 0$ and $\tau_{+,\alpha} = \infty$, or $\tau_{-,\alpha} = \infty$ and $\tau_{+,\alpha} = 0$. The points $\tau_{+,\alpha}$ and $\tau_{-,\alpha}$ are called attracting and repelling points of α, respectively.

We say that the triple $\{\alpha, a, b\}$ satisfies conditions (I1), (I2), (LI), (RI) if

(I1) $L_*(\tau_{-,\alpha}; a, b) > 0$ and $L_*(\tau_{+,\alpha}; a, b) > 0$ and $\inf_{t\in\mathbb{R}_+} |a(t)| > 0$;

(I2) $L^*(\tau_{-,\alpha}; a, b) < 0$ and $L^*(\tau_{+,\alpha}; a, b) < 0$ and $\inf_{t\in\mathbb{R}_+} |b(t)| > 0$;

(LI) $L^*(\tau_{-,\alpha}; a, b) < 0 < L_*(\tau_{+,\alpha}; a, b)$ and for every $t \in \mathbb{R}_+$ there is an integer k_t such that $b[\alpha_k(t)] \neq 0$ for $k < k_t$ and $a[\alpha_k(t)] \neq 0$ for $k > k_t$.

(RI) $L^*(\tau_{+,\alpha}; a, b) < 0 < L_*(\tau_{-,\alpha}; a, b)$ and for every $t \in \mathbb{R}_+$ there is an integer k_t such that $b[\alpha_k(t)] \neq 0$ for $k \geq k_t$ and $a[\alpha_k(t)] \neq 0$ for $k < k_t$.

Theorem 8.1 ([18, Theorems 1.1–1.2]). *Let $a, b \in SO(\mathbb{R}_+)$, $\alpha \in SOS(\mathbb{R}_+)$, and let the binomial functional operator A be given by*

$$A := aI - bU_\alpha.$$

(a) *The operator A is invertible on the Lebesgue space $L^p(\mathbb{R}_+)$ if and only if the triple $\{\alpha, a, b\}$ satisfies either condition* (I1), *or condition* (I2).
(b) *The operator A is strictly left invertible on the space $L^p(\mathbb{R}_+)$ if and only if the triple $\{\alpha, a, b\}$ satisfies condition* (LI).
(c) *The operator A is strictly right invertible on the space $L^p(\mathbb{R}_+)$ if and only if the triple $\{\alpha, a, b\}$ satisfies condition* (RI).

8.2. Sufficient conditions for the semi-Fredholmness

Combining Theorem 1.1 and Theorem 8.1, we arrive at the following.

Corollary 8.2. *Let $1 < p < \infty$ and let $\gamma \in \mathbb{C}$ satisfy* (1.3). *Suppose a, b, c, d belong to $SO(\mathbb{R}_+)$, α, β belong to $SOS(\mathbb{R}_+)$, and $\omega, \eta \in SO(\mathbb{R}_+)$ are the exponent functions of the shifts α, β, respectively. Consider the operator*

$$M := (aI - bU_\alpha)P_\gamma^+ + (cI - dU_\beta)P_\gamma^-$$

and the corresponding function m defined for $(\xi, x) \in (\mathbb{R}_+ \cup \Delta) \times \mathbb{R}$ by

$$m(\xi, x) := (a(\xi) - b(\xi)e^{i\omega(\xi)x})p_\gamma^+(x) + (c(\xi) - d(\xi)e^{i\eta(\xi)x})p_\gamma^-(x),$$

where the functions $p_\gamma^\pm$ are defined by (1.8).

(a) *If each of the triples $\{\alpha, a, b\}$ and $\{\beta, c, d\}$ satisfies either condition* (I1) *or condition* (I2) (*but not necessarily the same condition*), *and*
$$\inf_{x \in \mathbb{R}} |m(\xi, x)| > 0 \quad \text{for every} \quad \xi \in \Delta, \tag{8.1}$$
then the operator M is Fredholm on the space $L^p(\mathbb{R}_+)$.
(b) *If each of the triples $\{\alpha, a, b\}$ and $\{\beta, c, d\}$ satisfies only one of conditions* (I1), (I2) *and* (LI) (*but not necessarily the same condition*), *and condition* (8.1) *is fulfilled, then the operator M is left Fredholm on the space $L^p(\mathbb{R}_+)$.*
(c) *If each of the triples $\{\alpha, a, b\}$ and $\{\beta, c, d\}$ satisfies only one of conditions* (I1), (I2) *and* (RI) (*but not necessarily the same condition*), *and condition* (8.1) *is fulfilled, then the operator M is right Fredholm on the space $L^p(\mathbb{R}_+)$.*

Another (more involved) proof of Corollary 8.2(a), relying on criteria for the Fredholmness of Mellin pseudodifferential operators (see [22] and [16, Theorem 3.6]), is given in [19, Theorem 1.3]. The converse statement to Corollary 8.2(a) is proved in [20, Theorem 1.2]. The statements of parts (b) and (c) in Corollary 8.2 are new.

Acknowledgment

We would like to thank the anonymous referee for pointing out that Lemma 2.1 can be obtained from [13, Example 2.16].

Added in proof

We proved recently in [21, Theorem 1.1] that conditions (i) and (ii) of Theorem 1.1 are also necessary for the left (resp. right) Fredholmness of the operator N.

References

[1] A. Böttcher, Yu.I. Karlovich, and I.M. Spitkovsky, *Convolution Operators and Factorization of Almost Periodic Matrix Functions.* Operator Theory: Advances and Applications, vol. 131. Birkhäuser, Basel, 2002.

[2] A. Böttcher and B. Silbermann, *Introduction to Large Truncated Toeplitz Matrices.* Springer, New York, 1999.

[3] A. Böttcher and B. Silbermann, *Analysis of Toeplitz Operators.* Springer, Berlin, 2006.

[4] J.B. Conway, *A Course in Functional Analysis.* Springer, Berlin, 1990.

[5] V.D. Didenko and B. Silbermann, *Approximation of Additive Convolution-Like Operators. Real C*-Algebra Approach.* Birkhäuser, Basel, 2008.

[6] R. Duduchava, *Integral Equations with Fixed Singularities.* Teubner Verlagsgesellschaft, Leipzig, 1979.

[7] R. Duduchava, *On algebras generated by convolutions and discontinuous functions.* Integr. Equ. Oper. Theory **10** (1987) 505–530.

[8] G. Fernández-Torres and Yu.I. Karlovich, *Two-sided and one-sided invertibility of the Wiener type functional operators with a shift and slowly oscillating data.* Banach J. Math. Analysis **11** (2017), 554–590.

[9] G. Fernández-Torres and Yu.I. Karlovich, *Fredholmness of nonlocal singular integral operators with slowly oscillating data.* In: "Integral Methods in Science and Engineering, vol. 1", eds. C. Constanda et al. Birkhäuser, Cham, 2017, pp. 95–105.

[10] I. Gohberg and N. Krupnik, *One-Dimensional Linear Singular Integral Equations. I. Introduction.* Operator Theory: Advances and Applications, vol. 53. Birkhäuser, Basel, 1992.

[11] V.V. Grushin, *Pseudodifferential operators on* $\mathbb{R}^n$ *with bounded symbols.* Funct. Anal. Appl. **4** (1970) 202–212.

[12] R. Hagen, S. Roch, and B. Silbermann, *Spectral Theory of Approximation Methods for Convolution Equations.* Operator Theory: Advances and Applications, vol. 74. Birkhäuser, Basel, 1994.

[13] R. Hagen, S. Roch, and B. Silbermann, *C*-Algebras and Numerical Analysis.* Pure and Applied Mathematics, vol. 236. Marcel Dekker, New York, 2001.

[14] A.Yu. Karlovich, *Fredholmness and index of simplest weighted singular integral operators with two slowly oscillating shifts.* Banach J. Math. Anal. **9** (2015) 24–42.

[15] A.Yu. Karlovich, Yu.I. Karlovich, and A.B. Lebre, *Sufficient conditions for Fredholmness of singular integral operators with shifts and slowly oscillating data.* Integr. Equ. Oper. Theory **70** (2011) 451–483.

[16] A.Yu. Karlovich, Yu.I. Karlovich, and A.B. Lebre, *Fredholmness and index of simplest singular integral operators with two slowly oscillating shifts.* Operators and Matrices **8** (2014) 935–955.

[17] A.Yu. Karlovich, Yu.I. Karlovich, and A.B. Lebre, *On a weighted singular integral operator with shifts and slowly oscillating data.* Complex Anal. Oper. Theory **10** (2016) 1101–1131.

[18] A.Yu. Karlovich, Yu.I. Karlovich, and A.B. Lebre, *One-sided invertibility criteria for binomial functional operators with shift and slowly oscillating data.* Mediterr. J. Math. **13** (2016) 4413–4435.

[19] A.Yu. Karlovich, Yu.I. Karlovich, and A.B. Lebre, *The index of weighted singular integral operators with shifts and slowly oscillating data.* J. Math. Anal. Appl. **450** (2017), 606–630.

[20] A.Yu. Karlovich, Yu.I. Karlovich, and A.B. Lebre, *Necessary Fredholm conditions for weighted singular integral operators with shifts and slowly oscillating data.* J. Integr. Equ. Appl. **29** (2017), 365–399.

[21] A.Yu. Karlovich, Yu.I. Karlovich, and A.B. Lebre, *Criteria for $n(d)$-normality of weighted singular integral operators with shifts and slowly oscillating data.* Proc. London Math. Soc. **116** (2018) 997–1027.

[22] Yu.I. Karlovich, *An algebra of pseudodifferential operators with slowly oscillating symbols.* Proc. London Math. Soc. **92** (2006) 713–761.

[23] Yu.I. Karlovich, *Nonlocal singular integral operators with slowly oscillating data.* In "Operator Algebras, Operator Theory and Applications", eds. M.A. Bastos et al. Operator Theory: Advances and Applications, vol. 181, 2008, pp. 229–261.

[24] H. Kumano-go, *Pseudo-Differential Operators.* The MIT Press. Cambridge, MA, 1982.

[25] M.A. Naimark, *Normed Algebras.* Wolters-Noordhoff Publishing, Groningen, 1972.

[26] I.P. Natanson, *Theory of Functions of a Real Variable.* Frederick Ungar Publishing Co., New York, 1955.

[27] V.S. Rabinovich, *Singular integral operators on a composed contour with oscillating tangent and pseudodifferential Mellin operators.* Soviet Math. Dokl. **44** (1992) 791–796.

[28] V.S. Rabinovich, *Mellin pseudodifferential operators techniques in the theory of singular integral operators on some Carleson curves.* In "Differential and Integral Operators (Regensburg, 1995)", eds. I. Gohberg et al. Operator Theory: Advances and Applications, vol. 102, 1998, pp. 201–218.

[29] V.S. Rabinovich, S. Roch, and B. Silbermann, *Limit Operators and Their Applications in Operator Theory.* Operator Theory: Advances and Applications, vol. 150. Birkhäuser, Basel, 2004.

[30] S. Roch, P.A. Santos, and B. Silbermann, *Non-Commutative Gelfand Theories. A Tool-kit for Operator Theorists and Numerical Analysts.* Springer, Berlin, 2011.

[31] D. Sarason, *Toeplitz operators with piecewise quasicontinuous symbols.* Indiana Univ. Math. J. **26** (1977) 817–838.

[32] I.B. Simonenko and Chin Ngok Minh, *Local Approach to the Theory of One-Dimensional Singular Integral Equations with Piecewise Continuous Coefficients. Noethericity.* Rostov University Press, Rostov-on-Don, 1986 (in Russian).

[33] E.M. Stein, *Harmonic Analysis: Real-Variable Methods, Orthogonality, and Oscillatory Integrals.* Princeton Univ. Press, Princeton, NJ, 1993.

[34] K. Yosida, *Functional Analysis.* 6th ed. Springer, Berlin, 1980.

Alexei Yu. Karlovich
Centro de Matemática e Aplicações
Departamento de Matemática
Faculdade de Ciências e Tecnologia
Universidade Nova de Lisboa
Quinta da Torre
2829–516 Caparica, Portugal
e-mail: oyk@fct.unl.pt

Yuri I. Karlovich
Centro de Investigación en Ciencias
Instituto de Investigación en Ciencias Básicas y Aplicadas
Universidad Autónoma del Estado de Morelos
Av. Universidad 1001, Col. Chamilpa
C.P. 62209 Cuernavaca, Morelos, México
e-mail: karlovich@uaem.mx

Amarino B. Lebre
Centro de Análise Funcional, Estruturas Lineares e Aplicações
Departamento de Matemática
Instituto Superior Técnico
Universidade de Lisboa
Av. Rovisco Pais
1049–001 Lisboa, Portugal
e-mail: alebre@math.tecnico.ulisboa.pt

Operator Theory:
Advances and Applications, Vol. 267, 247–259

Factorization of Singular Integral Operators with a Carleman Backward Shift: The Vector Case

Amarino B. Lebre and Juan S. Rodríguez

Abstract. On the vector Lebesgue space on the unit circle L_p^n ($p \in (1,\infty), n \in \mathbb{N}$) we consider singular integral operators with a Carleman backward shift of linear fractional type, of the form

$$T_{A,B} = A\,P_+ + B\,P_-$$

with

$$A = aI + bU, \quad B = cI + dU,$$

where $a,b,c,d \in L_\infty^{n\times n}$, $P_\pm = \frac{1}{2}(I \pm S)$ are the Cauchy projectors in L_p^n defined componentwise, and U is an involutory shift operator associated with the given Carleman backward shift also defined componentwise. By generalization to the vector case ($n > 1$) of the previously obtained results for the scalar case ($n = 1$), it is shown that whenever a certain $2n \times 2n$ matrix function, associated with the original singular integral operator, admits a bounded factorization in L_p^{2n} the Fredholm characteristics of the paired operator $T_{A,B}$ can be obtained in terms of that factorization, in particular the dimensions of the kernel and of the cokernel.

Mathematics Subject Classification (2010). Primary 47G10; Secondary 47A68.

Keywords. Singular integral operators, singular integral operators with shifts, backward linear fractional shift, Carleman shift.

1. Introduction

Let $\mathbb{T}$ denote the unit circle in the complex plane, denote by $L_\infty := L_\infty(\mathbb{T})$ the algebra of essentially bounded functions on $\mathbb{T}$, and for $p \in [1,\infty)$ let $L_p := L_p(\mathbb{T})$ denote the Banach space of functions which are absolutely integrable to the pth power on $\mathbb{T}$.

This work was partially supported by the Fundação para a Ciência e a Tecnologia (Portuguese Foundation for Science and Technology) through the project UID/MAT/04721/2013 (Centro de Análise Funcional, Estruturas Lineares e Aplicações).

Two previous papers [4, 6] were devoted to the study of scalar singular integral operators with a linear fractional backward Carleman shift acting on the Lebesgue space $L_p := L_p(\mathbb{T})$, $p \in (1, \infty)$, of the general form

$$T_{A,B} = A\,P_+ + B\,P_- . \tag{1.1}$$

with

$$A = aI + bU, \quad B = cI + dU, \tag{1.2}$$

where $a, b, c, d \in L_\infty$, $P_\pm = \frac{1}{2}(I \pm S)$ are the Cauchy projectors in L_p associated with the singular integral operator with Cauchy kernel S, which is defined by

$$S\varphi(t) = \frac{1}{\pi i} P.V. \int_{\mathbb{T}} \frac{1}{\tau - t} \varphi(\tau)\, d\tau \,, \tag{1.3}$$

and U is an involutive operator on L_p ($U^2 = I$) which anticommutes with S,

$$US = -SU. \tag{1.4}$$

As a model of the operator U we considered a weighted shift operator associated with a linear fractional backward Carleman shift (see Section 2 for details).

The main results of [6] are the following: (1) under certain conditions a singular integral operator $T_{A,B}$ of the form (1.1) admits a factorization as a product of three operators with the outer factors in that representation being invertible; (2) such a representation can be obtained from a factorization of a matrix function $\mathcal{C} \in L_\infty^{2\times 2}$ associated with the original operator; (3) one can characterize both the kernel and the cokernel of the operator, which depend on the partial indices of the factorization of $\mathcal{C}$ as well as on two numbers in the set $\{-1, 1\}$, which can also be obtained from the factorization of $\mathcal{C}$.

From the results obtained in [6] it is clear that, even in the case where the coefficients are scalar functions, the operator $T_{A,B}$ has a vector behaviour, in the sense that this operator is equivalent, after extension, to a singular integral operator with (2×2) matrix coefficients (cf. relation (3.2)). Therefore, it is rather natural to ask what happens if we consider matrix coefficients from the beginning. Although the question is natural the answer to this question took some time and required the deduction of a special representation, presented in Proposition 3.2, which proved to be the missing piece of the puzzle.

In recent years a series of papers by V. Didenko and B. Silbermann were published devoted to the characterization of the kernel and cokernel of some Toeplitz plus Hankel operators (see [2] and the references therein). The classes considered are characterized by a special relation satisfied by the original coefficients and their actions under the shift. It is shown that, in the case of Fredholmness, the defect numbers can be obtained in terms of the generalized factorizations of two scalar functions in L_p. We do not pursue this line of research here, though in [6, Section 7] some Toeplitz plus Hankel operators were considered as an application of the results obtained.

Supposing that the coefficients $a, b, c, d \in L_\infty^{n\times n}$ and the operators $P_\pm$ and U are defined componentwise, we consider the operator $T_{A,B}$ given by (1.1) defined on L_p^n. We will show that the above-mentioned results from [6] are valid in the new context, with the matrix function $\mathcal{C}$ now belonging to $L_\infty^{2n\times 2n}$. Also, apart from to the partial indices of $\mathcal{C}$ we now need $2n$ numbers in the set $\{-1, 1\}$. This question is addressed in Section 3. In the next section we give some background to the subject, recalling in particular the main tool used throughout the paper, the notion of factorization of a matrix function.

2. Preliminaries

The study of singular integral operators and its connection with the factorization of matrix functions has a very long and rich history. Throughout that development and referring only to the case of operators without shift under consideration (i.e., $b = d = 0$ in (1.2)) acting on the Lebesgue space L_p^n, the concept of factorization of matrix functions evolved in a manner that tied it to the Fredholm property of the original operator. Let us recall the basic results in this context, which can be found in [1] or [9].

For $p \in (1, \infty)$ set $q = p/(p-1)$ and let $L_p^+ = \operatorname{im} P_+$, $L_p^- = \operatorname{im} P_- \oplus \mathbb{C}$. Let $m \in \mathbb{N}$. An invertible matrix function $\mathcal{C} \in L_\infty^{m\times m}$ is said to admit a generalized factorization in (or relative to) L_p^m if

(a) it can be represented in the form

$$\mathcal{C} = \mathcal{C}_+ \, \Lambda \, \mathcal{C}_-,$$

where $\mathcal{C}_+ \in (L_p^+)^{m\times m}, \mathcal{C}_+^{-1} \in (L_q^+)^{m\times m}, \mathcal{C}_- \in (L_q^-)^{m\times m}, \mathcal{C}_-^{-1} \in (L_p^-)^{m\times m}$, and $\Lambda = \operatorname{diag}\{t^{\kappa_1}, \dots, t^{\kappa_m}\}$, where $\kappa_1 \geq \cdots \geq \kappa_m$ are integers,

(b) The operator B acting according to the rule $B\varphi = \mathcal{C}_+ P_- \mathcal{C}_+^{-1}\varphi$ is either a bounded linear operator on L_p^m or a densely defined operator with a continuous extension to L_p^m.

The numbers κ_j, $j = 1, \dots, m$, which are uniquely determined by the matrix function $\mathcal{C}$ (when ordered as above), are called the partial indices of $\mathcal{C}$.

In what follows we shall restrict our analysis to the case where the generalized factorization of $\mathcal{C}$ is bounded, i.e., when the factors $\mathcal{C}_\pm$ belong to $L_\infty^{m\times m}$, in which case the operator B above is automatically bounded. This also allows us to give meaning to certain operators written in terms of the factors of the factorization of $\mathcal{C}$ (see below).

We have,

Theorem 2.1. *Let $\mathcal{A}, \mathcal{B} \in L_\infty^{m\times m}$ and consider in L_p^m the operator*

$$S_{\mathcal{A},\mathcal{B}} = \mathcal{A}\, P_+ + \mathcal{B}\, P_-.$$

Then $S_{\mathcal{A},\mathcal{B}}$ is a Fredholm operator on L_p^m if and only if the following conditions hold:

(i) *$\mathcal{A}, \mathcal{B}$ are invertible in $L_\infty^{m\times m}$,*
(ii) *$\mathcal{C} = \mathcal{A}^{-1}\mathcal{B}$ admits a generalized factorization in L_p^m.*

Moreover, when $S_{\mathcal{A},\mathcal{B}}$ is a Fredholm operator on L_p^m it can be factorized, by means of the factorization in L_p^m of the matrix function $\mathcal{C}$, in the form

$$S_{\mathcal{A},\mathcal{B}} = (\mathcal{A}\mathcal{C}_+)\, S_{I,\Lambda}\, S_{\mathcal{C}_+^{-1},\mathcal{C}_-}. \tag{2.1}$$

The operator on the left of the right-hand side of the above representation is just a multiplication operator, corresponding to a singular integral operator with equal coefficients $\mathcal{A}\mathcal{C}_+ = S_{\mathcal{A}\mathcal{C}_+,\mathcal{A}\mathcal{C}_+}$. Observe that this way of writing $S_{\mathcal{A},\mathcal{B}}$ is meaningless in the general case, but if we suppose that $\mathcal{C}$ has a bounded factorization, as we do, then both of the outer operators in the above representation, $\mathcal{A}\mathcal{C}_+$ and $S_{\mathcal{C}_+^{-1},\mathcal{C}_-}$, are bounded and invertible in L_p. Hence, the Fredholm characteristics of $S_{\mathcal{A},\mathcal{B}}$ coincide with those of the middle factor $S_{I,\Lambda}$. In fact, the following general result holds.

Theorem 2.2. *Let $\mathcal{A}$ and $\mathcal{B}$ be invertible in $L_\infty^{m\times m}$ and suppose that $\mathcal{C} = \mathcal{A}^{-1}\mathcal{B}$ admits a factorization in L_p^m, say $\mathcal{C} = \mathcal{C}_+\,\Lambda\,\mathcal{C}_-$. Then*

$$\dim\ker S_{\mathcal{A},\mathcal{B}} = \sum_{\kappa_j>0} \kappa_j, \qquad \operatorname{codim}\operatorname{im} S_{\mathcal{A},\mathcal{B}} = -\sum_{\kappa_j<0} \kappa_j.$$

A very special case occurs if all the partial indices are equal to zero (thus $\Lambda = I$), in which case the factorization $\mathcal{C}$ is said to be canonical, and corresponds to the invertibility of the singular integral operator $S_{\mathcal{A},\mathcal{B}}$.

On the unit circle we consider a Carleman backward shift α of linear fractional type, i.e., the function α is either of the form

$$\alpha(t) = \frac{t-\beta}{\overline{\beta}t-1}, \qquad |\beta|>1, t\in\mathbb{T}, \tag{2.2}$$

or

$$\alpha(t) = \frac{\tau}{t}, \qquad t,\tau\in\mathbb{T}. \tag{2.3}$$

The latter form with $\tau = 1$ can be seen as the limit of the first expression as $|\beta|$ tends to infinity. The shift function (2.2) can be factorized as follows:

$$\alpha = \alpha_+\, t^{-1}\, \alpha_- \tag{2.4}$$

with

$$\alpha_+(t) = \frac{t-\beta}{\lambda}, \qquad \alpha_-(t) = \frac{\lambda t}{\overline{\beta}t-1}, \tag{2.5}$$

where $\lambda = i\sqrt{|\beta|^2-1}$. The following property holds for arbitrary $t\in\mathbb{T}$:

$$\alpha_\pm(\alpha(t)) = \alpha_\mp(t).$$

The shift α has two fixed points $t_\pm$ on $\mathbb{T}$, given by $t_\pm = \frac{1\pm\lambda}{\overline{\beta}}$, and we have $\alpha_+(t_\pm) = \alpha_-(t_\pm) = \pm t_\pm$. When α is of the form (2.3) the factorization (2.4) holds with $\alpha_+ = \alpha_- \equiv \sqrt{\tau}$ and $t_\pm = \pm\sqrt{\tau}$, for any choice of the square root.

Associated with the shift function (2.2) or (2.3) we consider a weighted shift operator $U : L_p \to L_p$, defined by

$$(U\varphi)(t) = (\chi\,\varphi)(\alpha(t)) = \chi^{-1}(t)\,\varphi(\alpha(t)), \tag{2.6}$$

where

$$\chi(t) = t\alpha_-^{-1}(t), \quad t \in \mathbb{T}. \tag{2.7}$$

The following properties of this function will be used frequently below:

$$\text{(a) } \operatorname{ind}_{\mathbb{T}} \chi = 1, \quad \text{(b) } \chi\chi_\alpha \equiv 1, \quad \text{(c) } \chi(t_\pm) = \pm 1. \tag{2.8}$$

Here we have used the following notational convention:

$$x_\alpha = x \circ \alpha$$

for any function x defined on $\mathbb{T}$, a convention that we adopt throughout the paper. Taking into account the above-mentioned properties of the shift and related functions it is now straightforward to verify that the weighted shift operator U given by (2.6) is an involutive operator ($U^2 = I$) and that the anticommutation relation (1.4) holds.

For general information about singular integral operators with shift we refer the reader to [8] and [9].

3. Generalization to the vector case

In this section we generalize the results obtained in our previous work [4, 6] to the case of matrix-valued coefficients, that is we suppose that the coefficients a, b, c, d of the functional operators A and B in (1.2) belong to $L_\infty^{n\times n}$, $n > 1$. Supposing that the operators $P_\pm, U$ are defined componentwise in L_p^n all the notation used in those papers keeps its meaning with obvious changes. For instance, the constant matrix $\mathcal{E}$ there must be replaced by

$$\mathcal{E} = \begin{pmatrix} 0 & \mathcal{I}_n \\ \mathcal{I}_n & 0 \end{pmatrix}, \tag{3.1}$$

where $\mathcal{I}_n$ is the identity matrix of order n.

As in the scalar case the starting point is the following: based on the well-known operator relation

$$\frac{1}{2}\begin{pmatrix} I & I \\ U & -U \end{pmatrix}\begin{pmatrix} T_{A,B} & 0 \\ 0 & T_{\widetilde{A},\widetilde{B}} \end{pmatrix}\begin{pmatrix} I & U \\ I & -U \end{pmatrix} = S_{\mathcal{A},\mathcal{B}}, \tag{3.2}$$

where $\widetilde{A} = a\,I - b\,U$, $\widetilde{B} = c\,I - d\,U$, and

$$\mathcal{A} = \begin{pmatrix} a & d \\ b_\alpha & c_\alpha \end{pmatrix}, \quad \mathcal{B} = \mathcal{E}\mathcal{A}_\alpha\mathcal{E}, \tag{3.3}$$

to each singular integral operator with shift $T_{A,B}$ acting on L_p^n of the form (1.1) with coefficients in $L_\infty^{n\times n}$ we associate a singular integral operator without shift $S_{\mathcal{A},\mathcal{B}}$ acting on L_p^{2n} and we write

$$S_{\mathcal{A},\mathcal{B}} = \Phi(T_{A,B}) \tag{3.4}$$

to express that correspondence. Roughly speaking, doubling the dimension of the space is the price to pay for the elimination of the dependence on the shift. Notice that the map Φ is well defined in the class of singular integral operators $T_{A,B}$ with $L_\infty^{n\times n}$ coefficients, is injective, but is not a homomorphism. However, it enjoys a useful property: if $S_{\hat{\mathcal{A}},\hat{\mathcal{B}}} = \Phi(T_{\hat{A},\hat{B}})$, $S_{\check{\mathcal{A}},\check{\mathcal{B}}} = \Phi(T_{\check{A},\check{B}})$ and, moreover, $S_{\hat{\mathcal{A}},\hat{\mathcal{B}}}S_{\check{\mathcal{A}},\check{\mathcal{B}}} = S_{\mathcal{A},\mathcal{B}}$ with $S_{\mathcal{A},\mathcal{B}} = \Phi(T_{A,B})$, then $T_{\hat{A},\hat{B}}T_{\check{A},\check{B}} = T_{A,B}$.

Moreover, notice that the coefficients $\mathcal{A}$ and $\mathcal{B}$ of the singular integral operator (without shift) $S_{\mathcal{A},\mathcal{B}} = \Phi(T_{A,B})$ are connected in a special manner (cf. (3.3)), and this connection is responsible for all further results that we will derive, as we shall see.

Taking into account that the invertibility of $\mathcal{A}$ is necessary for the Fredholm property of $S_{\mathcal{A},\mathcal{B}}$, recalling Theorem 2.1, the analysis of such a singular integral operator $S_{\mathcal{A},\mathcal{B}}$ is closely related with the factorization properties of the matrix function $\mathcal{C} \in L_\infty^{2n\times 2n}$ defined by

$$\mathcal{C} = \mathcal{A}^{-1}\mathcal{B} = \mathcal{A}^{-1}\mathcal{E}\mathcal{A}_\alpha\mathcal{E}. \tag{3.5}$$

Observe that the matrix function $\mathcal{C}$ does not uniquely determine the matrix function $\mathcal{A}$, and that is the reason why we will refer to $\mathcal{A}$ as *a representative of* $\mathcal{C}$. The set of all $\mathcal{C} \in L_\infty^{2n\times 2n}$ for which there exists an invertible matrix function $\mathcal{A} \in L_\infty^{2n\times 2n}$ such that (3.5) holds will be denoted by $L_{\infty,\alpha}^{2n\times 2n}$.

A very important feature of matrix functions in $L_{\infty,\alpha}^{2n\times 2n}$ is that, whenever it is factorable in L_p^{2n}, there exists a factorization of it such that the outer factors are related in a particular way. This will be highlighted in Proposition 3.2 below, but first we introduce some notation and derive an auxiliary result.

When dealing with a factorization in L_p^{2n} of a matrix function $\mathcal{C} \in L_\infty^{2n\times 2n}$ it is sometimes convenient to consider the number $\ell \le 2n$ of partial indices which are pairwise distinct and we will write the central factor in the form

$$\Lambda = \operatorname{diag}\{t^{\kappa_1}\mathcal{I}_{d_1}, \dots, t^{\kappa_\ell}\mathcal{I}_{d_\ell}\}, \quad (\kappa_1 > \dots > \kappa_\ell) \tag{3.6}$$

where d_i is the multiplicity of the partial index κ_i, $i = 1, \dots, \ell$. The following result is the vector analogue of [6, Proposition 2.3] and its proof can be established similarly to the scalar case, bearing in mind well-known results about the connection between two different factorizations of the same matrix function (see [9]).

Proposition 3.1. *Let* $\mathcal{C} \in L_{\infty,\alpha}^{2n\times 2n}$, $n \in \mathbb{N}$, *and suppose that* $\mathcal{C} = \mathcal{C}_+\Lambda\mathcal{C}_-$ *is a generalized factorization of* $\mathcal{C}$ *in* L_p^{2n}. *Write* Λ *as in* (3.6) *and set*

$$\Lambda_\pm = \operatorname{diag}\{\alpha_\pm^{\kappa_1}\mathcal{I}_{d_1}, \dots, \alpha_\pm^{\kappa_\ell}\mathcal{I}_{d_\ell}\}, \quad \kappa_1 > \dots > \kappa_\ell, \ \sum_{j=1}^{\ell} d_j = 2n.$$

Then

$$\mathcal{H} = \Lambda_+(\mathcal{C}_-)_\alpha \mathcal{E}\mathcal{C}_+$$

is a polynomial matrix function satisfying

$$\mathcal{H}_\alpha = \Lambda_- \Lambda^{-1} \mathcal{H}^{-1} \Lambda \Lambda_-^{-1} \tag{3.7}$$

with the following block structure

$$\mathcal{H} = \begin{pmatrix} \mathcal{H}_1 & * & * \\ 0 & \ddots & * \\ 0 & 0 & \mathcal{H}_l \end{pmatrix}, \tag{3.8}$$

where each block $\mathcal{H}_i$, $i = 1, \ldots, \ell$, in the main diagonal is a non-singular constant matrix of order d_i, the multiplicity of the partial index κ_i, such that $\mathcal{H}_i^2 = \mathcal{I}_{d_i}$.

It should be noted that the blocks $\mathcal{H}_i$, $i = 1, \ldots, \ell$, are uniquely determined by the matrix function $\mathcal{C}$ up to similarity.

Now let us prepare the result that allows us to relate the outer factors of a factorization of $\mathcal{C}$ given by (3.5). This result is based on an idea of T. Ehrhardt presented in [3] and already used in [5]. Suppose that $\mathcal{C}$ satisfies the hypotheses of the above proposition, let $\mathcal{H}$ be the matrix function given there and let

$$\mathcal{L} = \operatorname{diag}\{\mathcal{H}_1, \ldots, \mathcal{H}_\ell\}\Lambda\Lambda_-^{-1} = \operatorname{diag}\{\chi^{\kappa_1}\mathcal{H}_1, \ldots, \chi^{\kappa_\ell}\mathcal{H}_\ell\}.$$

Taking into account that $\mathcal{H}_j = \mathcal{H}_j^{-1}$, $j = 1, \ldots, \ell$, due to (3.8), we may write

$$\mathcal{H}^{-1}\Lambda\Lambda_-^{-1} = (I + \mathcal{K})\mathcal{L}, \tag{3.9}$$

where $\mathcal{K}$ is a nilpotent polynomial matrix function. Further, each $\mathcal{H}_j$ is similar to a diagonal matrix with the numbers ± 1 in the diagonal, and therefore there exists a non-singular $2n \times 2n$ matrix $\mathcal{X}$ such that

$$\mathcal{D} = \mathcal{X}^{-1}\mathcal{L}\mathcal{X} = \operatorname{diag}\{\varepsilon_1\chi^{\kappa_1}, \ldots, \varepsilon_{2n}\chi^{\kappa_{2n}}\} \tag{3.10}$$

with $\varepsilon_j \in \{-1, 1\}$, $j = 1, \ldots, 2n$. As the blocks $\mathcal{H}_j$ are uniquely determined by the matrix function $\mathcal{C}$ up to similarity, the numbers ε_j, $j = 1, \ldots, 2n$, are uniquely determined by $\mathcal{C}$ up to reordering within the set of coefficients of the same power of χ.

Proposition 3.2. *Let $\mathcal{C}$ be given by (3.5), suppose that it satisfies the conditions of Proposition 3.1 and let $\mathcal{H}$ be the matrix function given there. Then $\mathcal{C}$ admits a representation of the form*

$$\mathcal{C} = \mathcal{X}_+\mathcal{D}_\mathcal{E}\mathcal{X}_-, \tag{3.11}$$

where $\mathcal{X}_+ = \mathcal{E}(\mathcal{X}_-^{-1})_\alpha\mathcal{E} \in (L_p^+)^{2n\times 2n}$ is given by $\mathcal{X}_+ = \mathcal{C}_+(I + \mathcal{K})^{\frac{1}{2}}\mathcal{X}$, for $\mathcal{X}$ and $\mathcal{K}$ as introduced above, and $\mathcal{D}_\mathcal{E} = \mathcal{D}\mathcal{E}$ with $\mathcal{D}$ given by (3.10).

Proof. We adapt the proof of [3, Theorem 3.2] to our case. Starting as in the scalar case, from the given factorization of $\mathcal{C}$ in L_p^{2n}, $\mathcal{C} = \mathcal{C}_+\Lambda\mathcal{C}_-$, we establish that

$$\mathcal{C} = \mathcal{C}_+\mathcal{H}^{-1}\Lambda\Lambda_-^{-1}\left(\mathcal{C}_+^{-1}\right)_\alpha\mathcal{E}, \tag{3.12}$$

where $\mathcal{H}$ is the polynomial matrix function referred to in Proposition 3.1. Note that, due to (3.12) and the properties of the factorization of the shift α, we have $\mathcal{H}^{-1}\Lambda\Lambda_-^{-1} = (\mathcal{H}^{-1}\Lambda\Lambda_-^{-1})_\alpha^{-1}$. Consequently, together with $\mathcal{H}^{-1}\Lambda\Lambda_-^{-1} = (I+\mathcal{K})\mathcal{L}$, we also have

$$\mathcal{H}^{-1}\Lambda\Lambda_-^{-1} = \mathcal{L}(I+\mathcal{K}_\alpha)^{-1}, \tag{3.13}$$

because $\mathcal{L}\mathcal{L}_\alpha = \mathcal{I}_{2n}$, due to $\chi\chi_\alpha \equiv 1$. Then $\mathcal{K}\mathcal{L} = \mathcal{L}\widetilde{\mathcal{K}}$ with $\widetilde{\mathcal{K}} = (I+\mathcal{K}_\alpha)^{-1} - I$. Using the method of induction one establishes that $\mathcal{K}^m\mathcal{L} = \mathcal{L}\widetilde{\mathcal{K}}^m$ for every $m \in \mathbb{N}$. The matrix functions $(I+\mathcal{K})^{1/2}$ and $(I+\widetilde{\mathcal{K}})^{1/2} = (I+\mathcal{K}_\alpha)^{-1/2}$ are well defined by series expansions, which are finite due to nilpotency of $\mathcal{K}$ and $\widetilde{\mathcal{K}}$. Using these expansions, it follows that

$$(I+\mathcal{K})^{1/2}\mathcal{L} = \mathcal{L}(I+\mathcal{K}_\alpha)^{-1/2}.$$

This, together with (3.9) and (3.13), implies that

$$\mathcal{H}^{-1}\Lambda\Lambda_-^{-1} = (I+\mathcal{K})^{1/2}\,\mathcal{L}\,(I+\mathcal{K}_\alpha)^{-1/2}$$

and, using the diagonalizing matrix $\mathcal{X}$ for $\mathcal{L}$, introduced before this proposition, we also get

$$\mathcal{H}^{-1}\Lambda\Lambda_-^{-1} = (I+\mathcal{K})^{1/2}\mathcal{X}\,\mathcal{D}\,\mathcal{X}^{-1}(I+\mathcal{K}_\alpha)^{-1/2},$$

where $\mathcal{D}$ is the diagonal matrix function (3.10). Thus, substituting this relation in (3.12), we obtain the representation of $\mathcal{C}$ as stated in the proposition. □

Proposition 3.2 constitutes the breakthrough needed to generalize our previous results. In fact, once this result is established, it is easy to see that the main theorem about the factorization of the operator $T_{A,B}$, [6, Theorem 4.2], maintains its validity in the new context with the same type of proof. This allows us to represent the given operator $T_{A,B}$ as a product of three operators in the same class where the outer operators in this representation are invertible.

Theorem 3.3. *Let α be the linear fractional backward shift (2.2), let U be the corresponding shift operator given by (2.6), and let $T_{A,B}$ be the singular integral operator with shift as defined in (1.1)–(1.2) with coefficients a, b, c, d belonging to the algebra $L_\infty^{n\times n}$. Suppose that the matrix functions $\mathcal{A}$ and $\mathcal{B}$, given by (3.3), are such that $\mathcal{A}$ is invertible in $L_\infty^{2n\times 2n}$ and that $\mathcal{C} = \mathcal{A}^{-1}\mathcal{B}$ admits a bounded factorization in L_p^{2n}. Then $T_{A,B}$ admits the factorization*

$$T_{A,B} = T_{A_1,B_1}\,T_{A_0,B_0}\,T_{A_2,B_2} \tag{3.14}$$

where each operator on the right-hand side is a singular integral operator with backward shift α of the form (1.1), the outer operators T_{A_j,B_j}, $j=1,2$, are invertible in L_p^n, and the central operator T_{A_0,B_0} has simple coefficients which can be obtained from a representative of the anti-block-diagonal matrix function $\mathcal{D}_\mathcal{E}$ for $\mathcal{D}$ given by (3.10).

It comes out from the proof of this theorem that the central operator T_{A_0,B_0}, which is therefore responsible for the Fredholm properties of the original operator $T_{A,B}$, is of a very special form. Indeed,

$$\Phi(T_{A_0,B_0}) = \mathcal{R}P_+ + \mathcal{E}\mathcal{R}_\alpha\mathcal{E}P_- = \mathcal{R}(P_+ + \mathcal{D}_\mathcal{E}P_-),$$

so that $\mathcal{R}$ is a representative of the anti-block diagonal matrix function $\mathcal{D}_\mathcal{E}$ with the diagonal matrix function $\mathcal{D}$ given by (3.10). Thus, as in the scalar case, we are left with the problem of finding a representative of the matrix function $\mathcal{D}_\mathcal{E}$ and to use it to characterize the central operator in the operator factorization (3.14). Let us now analyze these problems in the new context, but first we recall the result obtained in [6] for the scalar case.

Theorem 3.4 ([6, Theorem 6.2]). *Let $n = 1$. Under the conditions of Theorem* 3.3, *let $\mathcal{D}$ be the matrix function* (3.10), $\mathcal{D} = \operatorname{diag}\{\varepsilon_1\chi^{\kappa_1}, \varepsilon_2\chi^{\kappa_2}\}$. *Then*

$$\dim\ker T_{A,B} = \begin{cases} 0 & \text{if } \kappa_1 \le 0, \\ [\frac{\kappa_1}{2} + \frac{1-\varepsilon_1}{4}] & \text{if } \kappa_2 \le 0 \text{ and } \kappa_1 > 0, \\ [\frac{\kappa_1}{2} + \frac{1-\varepsilon_1}{4}] + [\frac{\kappa_2}{2} + \frac{1-\varepsilon_2}{4}] & \text{if } \kappa_2 > 0, \end{cases}$$

and

$$\operatorname{codim}\operatorname{im} T_{A,B} = \begin{cases} 0 & \text{if } \kappa_2 \ge 0, \\ [-\frac{\kappa_2}{2} + \frac{1+\varepsilon_2}{4}] & \text{if } \kappa_1 \ge 0 \text{ and } \kappa_2 < 0, \\ [-\frac{\kappa_2}{2} + \frac{1+\varepsilon_2}{4}] + [-\frac{\kappa_1}{2} + \frac{1+\varepsilon_1}{4}] & \text{if } \kappa_1 < 0. \end{cases}$$

We recall that the above numbers κ_1 and κ_2 ($\kappa_1 \ge \kappa_2$) are nothing but the partial indices of $\mathcal{C}$ in L^2_p and that ε_1 and ε_2 are the numbers on the main diagonal of $\mathcal{H}$ in (3.7) for the scalar case.

To proceed we will need to write some formulas where matrices of dimension $2n \times 2n$ are written in terms of blocks of matrices with dimension 2×2, which have a meaning in the context of the scalar case ($n = 1$). We then make the following notational convention: we put the subscript 0 in every term of dimension 2×2, which corresponds to the scalar case. A similar convention will be used for operators. According to this convention, for instance, we will denote by $\mathcal{E}_0$ the matrix corresponding to $\mathcal{E}$ in (3.1) for the scalar case.

Let us denote by $\widetilde{\mathcal{E}}$ the $2n \times 2n$ diagonal matrix with n 2×2 blocks all equal to $\mathcal{E}_0$. Clearly, the matrices $\mathcal{E}$ and $\widetilde{\mathcal{E}}$ are similar and so there exists a non-singular matrix $\mathcal{M}$ such that

$$\widetilde{\mathcal{E}} = \mathcal{M}^{-1}\mathcal{E}\mathcal{M}.$$

Now, suppose we are given a $2n \times 2n$ diagonal matrix function $\widetilde{\mathcal{D}}$ in the form of n blocks of 2×2 diagonal matrix functions,

$$\widetilde{\mathcal{D}} = \operatorname{diag}\{\mathcal{D}_0^{(1)}, \ldots, \mathcal{D}_0^{(n)}\},$$

and suppose in addition that each block $\mathcal{D}_0^{(j)}$ is of the form (3.10) for $n = 1$. Using the results of [6, Section 5], we know how to find a representative $\mathcal{R}_0^{(j)}$ of $(\mathcal{D}_0^{(j)})_{\mathcal{E}_0}$

with $(\mathcal{D}_0^{(j)})_{\mathcal{E}_0} = \mathcal{D}_0^{(j)}\mathcal{E}_0$, yielding

$$\mathcal{D}_0^{(j)} = (\mathcal{R}_0^{(j)})^{-1}\mathcal{E}_0(\mathcal{R}_0^{(j)})_\alpha, \quad j = 1, \dots, n.$$

Let $\widetilde{\mathcal{R}}$ be the $2n \times 2n$ matrix function with the n-block form

$$\widetilde{\mathcal{R}} = \operatorname{diag}\{\mathcal{R}_0^{(1)}, \dots, \mathcal{R}_0^{(n)}\}.$$

Then

$$\widetilde{\mathcal{D}} = \widetilde{\mathcal{R}}^{-1}\widetilde{\mathcal{E}}\widetilde{\mathcal{R}}_\alpha \tag{3.15}$$

and, consequently,

$$\widetilde{\mathcal{D}} = (\mathcal{M}\widetilde{\mathcal{R}})^{-1}\mathcal{E}(\mathcal{M}\widetilde{\mathcal{R}})_\alpha, \tag{3.16}$$

which means that $\mathcal{M}\widetilde{\mathcal{R}}$ is a representative of $\widetilde{\mathcal{D}}_{\mathcal{E}}$.

In general, the problem of finding a representative of the matrix function $\mathcal{D}_{\mathcal{E}}$ with $\mathcal{D}$ given by (3.10) can be solved by this procedure. It amounts to setting:

$$\widetilde{\mathcal{D}} = \mathcal{M}^{-1}\mathcal{D}\mathcal{M}.$$

In fact, from (3.16) it follows that

$$\mathcal{D} = (\mathcal{M}\widetilde{\mathcal{R}}\mathcal{M}^{-1})^{-1}\mathcal{E}(\mathcal{M}\widetilde{\mathcal{R}}\mathcal{M}^{-1})_\alpha = \mathcal{R}^{-1}\mathcal{E}\mathcal{R}_\alpha \tag{3.17}$$

and therefore $\mathcal{R} = \mathcal{M}\widetilde{\mathcal{R}}\mathcal{M}^{-1}$ is a representative of $\mathcal{D}_{\mathcal{E}}$.

In terms of operators we have

$$\begin{aligned} S_{\mathcal{R},\mathcal{E}\mathcal{R}_\alpha\mathcal{E}} &= \mathcal{R}(P_+ + \mathcal{D}_{\mathcal{E}}P_-) \\ &= \mathcal{R}(P_+ + \mathcal{M}\widetilde{\mathcal{D}}\widetilde{\mathcal{E}}\mathcal{M}^{-1}P_-) \\ &= \mathcal{R}\mathcal{M}(P_+ + \widetilde{\mathcal{D}}\widetilde{\mathcal{E}}P_-)\mathcal{M}^{-1} \\ &= \mathcal{M}\widetilde{\mathcal{R}}(P_+ + \widetilde{\mathcal{D}}\widetilde{\mathcal{E}}P_-)\mathcal{M}^{-1} = \mathcal{M}S_{\widetilde{\mathcal{R}},\widetilde{\mathcal{E}}\widetilde{\mathcal{R}}_\alpha\widetilde{\mathcal{E}}}\mathcal{M}^{-1}. \end{aligned}$$

Taking (3.15) into account, the operator $S_{\widetilde{\mathcal{R}},\widetilde{\mathcal{E}}\widetilde{\mathcal{R}}\widetilde{\mathcal{E}}} = \widetilde{\mathcal{R}}(P_+ + \widetilde{\mathcal{D}}_{\widetilde{\mathcal{E}}}P_-)$ is a direct sum of n scalar operators, namely

$$S_{\widetilde{\mathcal{R}},\widetilde{\mathcal{E}}\widetilde{\mathcal{R}}\widetilde{\mathcal{E}}} = \widetilde{\mathcal{R}}(P_+ + \widetilde{\mathcal{D}}_{\widetilde{\mathcal{E}}}P_-) = \bigoplus_{j=1}^{n} \mathcal{R}_0^{(j)}(P_+ + (\mathcal{D}_0^{(j)})_{\mathcal{E}_0}P_-) = \bigoplus_{j=1}^{n} S_{\mathcal{R}_0^{(j)},\mathcal{E}_0\mathcal{R}_0^{(j)}\mathcal{E}_0}.$$

Recalling that $S_{\mathcal{R},\mathcal{E}\mathcal{R}\mathcal{E}} = \Phi(T_{A_0,B_0})$, we conclude that T_{A_0,B_0} is equivalent to the direct sum of n scalar operators

$$T_{A_0,B_0} \sim \bigoplus_{j=1}^{n} T_{A_0^{(j)},B_0^{(j)}}$$

where $\Phi_0(T_{A_0^{(j)},B_0^{(j)}}) = S_{\mathcal{R}_0^{(j)},\mathcal{E}_0\mathcal{R}_0^{(j)}\mathcal{E}_0}$, $j = 1, \dots, n$, Φ_0 denoting the map (3.4) for the scalar case. Consequently, its Fredholm properties can be derived from those of these n scalar operators. Therefore we have proved the main result concerning the properties of the operator $T_{A,B}$ in the vector case.

Theorem 3.5. *Under the conditions of Theorem* 3.3, *we have*

$$\dim\ker T_{A,B} = \sum_{j=1}^{n} \dim\ker T_{A_0^{(j)},B_0^{(j)}}, \operatorname{codim}\operatorname{im} T_{A,B} = \sum_{j=1}^{n} \operatorname{codim}\operatorname{im} T_{A_0^{(j)},B_0^{(j)}},$$

where the dimension of the kernel and the codimension of the image of each $T_{A_0^{(j)},B_0^{(j)}}$, $j = 1,\ldots,n$, *are given by Theorem* 3.4.

Finally, whenever needed, the above results can also be used to completely characterize the kernel and the cokernel of the operator $T_{A,B}$, since that characterization for the scalar case is available in [6, Section 6].

In [6] we gave several examples of invertible scalar singular integral operators with shift, of the class under consideration, which differ in the regularity of the coefficients. These examples were constructed based on the knowledge of the representatives of anti-diagonal 2×2 matrix functions, which can have different nature depending on the parity of the partial indices of $\mathcal{C} \in L_\infty^{2\times 2}$ and on the sign of the determinant of the corresponding polynomial 2×2 matrix function $\det \mathcal{H} = \varepsilon_1\varepsilon_2$, as follows:

- a rational matrix function, if the partial indices of $\mathcal{C}$ are of the same parity and $\det \mathcal{H} = -1$;
- a continuous non-rational matrix function on $\mathbb{T}$ except at one point (one of the fixed points of the shift), if the partial indices of $\mathcal{C}$ are of distinct parity;
- a piecewise rational matrix function on $\mathbb{T}$ with two points of discontinuity, the fixed points of the shift α, if the partial indices of $\mathcal{C}$ are of the same parity and $\det \mathcal{H} = 1$.

We finishing by giving an example of an invertible singular integral operator with shift on L_p^2. According to the previous results, the defect numbers of such an operator are determined by the central factor $\mathcal{D}_\mathcal{E}$ of the factorization given in (3.11). So, we concentrate our attention on the case where the matrix function $\mathcal{C}$ reduces to the central factor $\mathcal{D}_\mathcal{E} = \mathcal{D}\mathcal{E}$. More precisely, suppose we want to identify the operator $T_{A,B}$ on L_p^2, $p \in (1,\infty)$, such that

$$\Phi(T_{A,B}) = \mathcal{R}(P_+ + \mathcal{D}_\mathcal{E} P_-),$$

where $\mathcal{R}$, is a representative of $\mathcal{D}_\mathcal{E}$ with

$$\mathcal{D} = \operatorname{diag}\{\chi,\chi,\chi,-\chi^{-1}\}, \quad \mathcal{E} = \begin{pmatrix} 0 & 0 & 1 & 0 \\ 0 & 0 & 0 & 1 \\ 1 & 0 & 0 & 0 \\ 0 & 1 & 0 & 0 \end{pmatrix}.$$

Note that the (decreasing) partial indices of $\mathcal{D}$ are the tuple $(1,1,1,-1)$. Using the method described above, we have

$$\widetilde{\mathcal{E}} = \mathcal{M}\mathcal{E}\mathcal{M} = \begin{pmatrix} 0 & 1 & 0 & 0 \\ 1 & 0 & 0 & 0 \\ 0 & 0 & 0 & 1 \\ 0 & 0 & 1 & 0 \end{pmatrix} = \begin{pmatrix} \mathcal{E}_0 & \vdots & 0 \\ \cdots & & \cdots \\ 0 & \vdots & \mathcal{E}_0 \end{pmatrix}$$

with

$$\mathcal{M} = \mathcal{M}^{-1} = \begin{pmatrix} 1 & 0 & 0 & 0 \\ 0 & 0 & 1 & 0 \\ 0 & 1 & 0 & 0 \\ 0 & 0 & 0 & 1 \end{pmatrix},$$

and

$$\widetilde{\mathcal{D}} = \mathcal{M}\mathcal{D}\mathcal{M} = \operatorname{diag}\{\widetilde{\mathcal{D}}_1, \widetilde{\mathcal{D}}_2\}$$

with

$$\widetilde{\mathcal{D}}_1 = \operatorname{diag}\{\mathcal{X}, \mathcal{X}\}, \quad \widetilde{\mathcal{D}}_2 = \operatorname{diag}\{\mathcal{X}, -\mathcal{X}^{-1}\}.$$

Using the results of [6, Section 5], we know representatives for the 2×2 anti-diagonal matrices $\widetilde{\mathcal{D}}_j\mathcal{E}_0$, $j = 1, 2$. In fact, setting

$$\widetilde{\mathcal{R}}_1 = \begin{pmatrix} \gamma & \gamma_\alpha \\ \mathcal{X}_\alpha\gamma_\alpha & \mathcal{X}_\alpha\gamma \end{pmatrix}, \quad \widetilde{\mathcal{R}}_2 = \begin{pmatrix} 1 & -\mathcal{X} \\ \mathcal{X}_\alpha & 1 \end{pmatrix},$$

where γ is the characteristic function of one of the two arcs of $\mathbb{T}$ connecting the two fixed points of the shift α, it is straightforward to verify that, for each $j = 1, 2$, the equation $(\widetilde{\mathcal{R}}_j)^{-1}\mathcal{E}_0(\widetilde{\mathcal{R}}_j)_\alpha = \widetilde{\mathcal{D}}_j$ holds, which means that $\widetilde{\mathcal{R}}_j$ is a representative of $\widetilde{\mathcal{D}}_j\mathcal{E}_0$. From (3.17), the representative $\mathcal{R}$ of $\mathcal{D}_\mathcal{E}$ is $\mathcal{R} = \mathcal{M}\operatorname{diag}\{\widetilde{\mathcal{R}}_1, \widetilde{\mathcal{R}}_2\}\mathcal{M}$.

In terms of operators, we have,

$$\begin{aligned}
\Phi(T_{A,B}) &= \mathcal{R}(P_+ + \mathcal{D}\mathcal{E}P_-) = \mathcal{R}P_+ + \mathcal{E}(\mathcal{R})_\alpha\mathcal{E}P_- \\
&= \mathcal{M}\begin{pmatrix} \widetilde{\mathcal{R}}_1(P_+ + \widetilde{\mathcal{D}}_1\mathcal{E}_0P_-) & 0 \\ 0 & \widetilde{\mathcal{R}}_2(P_+ + \widetilde{\mathcal{D}}_2\mathcal{E}_0P_-) \end{pmatrix}\mathcal{M} \\
&= \mathcal{M}\begin{pmatrix} \widetilde{\mathcal{R}}_1P_+ + \mathcal{E}_0(\widetilde{\mathcal{R}}_1)_\alpha\mathcal{E}_0P_-) & 0 \\ 0 & \widetilde{\mathcal{R}}_2P_+ + \mathcal{E}_0(\widetilde{\mathcal{R}}_2)_\alpha\mathcal{E}_0P_-) \end{pmatrix}\mathcal{M}
\end{aligned}$$

which means that T_{A,B_j} is (as expected) the direct sum of two scalar operators:

$$T_{A,B} = T_{A_1,B_1} \oplus T_{A_2,B_2}$$

with

$$T_{A_1,B_1} = \gamma(I + \mathcal{X}U)P_+ + \gamma_\alpha(\mathcal{X}I + U)P_-$$

and

$$T_{A_2,B_2} = (I + \mathcal{X}U)P_+ + (I - \mathcal{X}U)P_-.$$

Moreover, it follows from Theorem 3.4 that each of the scalar operators T_{A_1,B_1} and T_{A_2,B_2} is invertible in L_p, and thus $T_{A,B}$ is invertible in L_p^2. In fact, we have for the case of T_{A_1,B_1}: $\kappa_1 = \kappa_2 = \varepsilon_1 = \varepsilon_2 = 1$, and for the case of T_{A_2,B_2}: $\kappa_1 = \varepsilon_1 = 1$ and $\kappa_2 = \varepsilon_2 = -1$.

Observe then that in spite of the fact that the partial indices of $\mathcal{D}$ are all nonzero the resulting singular integral operator is invertible, being the direct sum of two scalar operators with different regularity of the coefficients: T_{A_1,B_1} with discontinuous coefficients (the points of discontinuity are the fixed points of the shift α) and T_{A_2,B_2} with polynomial and simple coefficients.

Acknowledgment

We would like to thank the anonymous referee for useful comments and remarks about some inaccuracies in the original version, which helped to improve the final version of this paper. We are also deeply grateful to our colleague Roger Picken, who took the unfortunate task of correcting our bad sentences in English, his mother tongue.

References

[1] K. Clancey and I. Gohberg, *Factorization of Matrix Functions and Singular Integral Operators.* Operator Theory: Advances and Applications, vol. 3. Birkhäuser, Basel, 1981.

[2] V.D. Didenko and B. Silbermann, *Generalized Toeplitz plus Hankel operators: kernel structure and defect numbers.* Complex Anal. Oper. Theory **10** 2016, 1351–1381.

[3] T. Ehrhardt, *Invertibility theory for Toeplitz plus Hankel operators and singular integral operators with flip.* J. Funct. Anal. **208** (2004), 64–106.

[4] V.G. Kravchenko, A.B. Lebre, and J.S. Rodríguez, *Factorization of singular integral operators with a Carleman shift via factorization of matrix functions: the anticommutative case.* Math. Nachr. **280** (2007), 1157–1175.

[5] V.G. Kravchenko, A.B. Lebre, and J.S. Rodríguez, *Matrix functions consimilar to the identity and singular integral operators.* Complex Anal. Oper. Theory **2** (2008), 593–615.

[6] V.G. Kravchenko, A.B. Lebre, and J.S. Rodríguez, *Factorization of singular integral operators with a Carleman backward shift: the case of bounded measurable coefficients.* J. Anal. Math. **107** (2009), 1–37.

[7] V.G. Kravchenko and G.S. Litvinchuk, *Introduction to the Theory of Singular Integral Operators with Shift.* Kluwer Academic Publishers, London, 1994.

[8] G.S. Litvinchuk, *Solvability Theory of Boundary Value Problems and Singular Integral Equations with Shift.* Kluwer Academic Publishers, London, 2000.

[9] G.S. Litvinchuk and I.M. Spitkovskii, *Factorization of Measurable Matrix Functions.* Operator Theory: Advances and Applications, vol. 25, Birkhäuser, Basel, 1987.

Amarino B. Lebre
Centro de Análise Funcional,
Estruturas Lineares e Aplicações
and
Departamento de Matemática
Instituto Superior Técnico
Universidade de Lisboa
Av. Rovisco Pais
1049–001 Lisboa, Portugal
e-mail: `alebre@math.tecnico.ulisboa.pt`

Juan S. Rodríguez
Centro de Análise Funcional,
Estruturas Lineares e Aplicações
and
Departamento de Matemática, F.C.T.
Universidade do Algarve
Campus de Gambelas
8005-139 Faro, Portugal
e-mail: `jsanchez@ualg.pt`

Operator Theory:
Advances and Applications, Vol. 267, 261–284

Extension-Restriction Theorems for Algebras of Approximation Sequences

Steffen Roch

Dedicated to Bernd Silbermann on the occassion of his 75th birthday.

Abstract. The C^*-algebra $\mathcal{S}(\mathsf{T}(C))$ of the finite sections discretization for Toeplitz operators with continuous generating functions is fairly well understood. Since its description in [3], this algebra serves both as a source of inspiration and as an archetypal example of an algebra generated by an discretization procedure. Moreover, it turns out that *every* separable C^*-algebra of approximation sequences has a common structure with $\mathcal{S}(\mathsf{T}(C))$ after an extension by compact sequences and a suitable fractal restriction. We explain what this statement means and give a proof.

Mathematics Subject Classification (2010). Primary 65J10; Secondary 46L99, 47N40.

Keywords. Finite sections discretization, Toeplitz operators, fractal restriction, compact extension, Silbermann algebras.

1. Introduction

Let H be a Hilbert space and $\mathcal{P} = (P_n)_{n\geq 1}$ a filtration on H, i.e., a sequence of orthogonal projections of finite rank that converges strongly to the identity operator on H. By $\mathcal{F}^{\mathcal{P}}$ we denote the set of all bounded sequences $(A_n)_{n\geq 1}$ of operators $A_n \in L(\operatorname{im} P_n)$, and by $\mathcal{G}^{\mathcal{P}}$ the set of all sequences $(A_n) \in \mathcal{F}^{\mathcal{P}}$ with $\|A_n\| \to 0$. Provided with the operations

$$(A_n) + (B_n) := (A_n + B_n), \quad (A_n)(B_n) := (A_n B_n), \quad (A_n)^* := (A_n^*)$$

and the norm $\|(A_n)\| := \sup \|A_n P_n\|$, $\mathcal{F}^{\mathcal{P}}$ becomes a unital C^*-algebra and $\mathcal{G}^{\mathcal{P}}$ is a closed ideal of $\mathcal{F}^{\mathcal{P}}$. The importance of the quotient algebra $\mathcal{F}^{\mathcal{P}}/\mathcal{G}^{\mathcal{P}}$ in numerical analysis stems from the fact that a coset $(A_n) + \mathcal{G}^{\mathcal{P}}$ is invertible in $\mathcal{F}^{\mathcal{P}}/\mathcal{G}^{\mathcal{P}}$ if and only if the operators A_n are invertible for all sufficiently large n and if the norms of the inverses are uniformly bounded. This is equivalent to saying that (A_n) is a stable sequence. Note in that connection that

$$\|(A_n) + \mathcal{G}^{\mathcal{P}}\|_{\mathcal{F}^{\mathcal{P}}/\mathcal{G}^{\mathcal{P}}} = \limsup \|A_n\|. \tag{1.1}$$

With every non-empty subset A of $L(H)$, we associate the smallest closed and symmetric subalgebra $\mathcal{S}^{\mathcal{P}}(\mathsf{A})$ of $\mathcal{F}^{\mathcal{P}}$ that contains all sequences $(P_nAP_n)_{n\geq 1}$ with $A \in \mathsf{A}$. The algebra $\mathcal{S}(\mathsf{T}(C))$ of the finite sections discretization for Toeplitz operators with continuous generating function arises exactly in this way. Here, H is the Hilbert space $l^2(\mathbb{Z}^+)$, P_n is the projection which sends the sequence $(x_0, x_1, \ldots)$ to $(x_0, \ldots, x_{n-1}, 0, 0, \ldots)$, and A is the C^*-algebra $\mathsf{T}(C)$ generated by all Toeplitz operators $T(a)$ with a being a continuous function on the complex unit circle $\mathbb{T}$. Recall that $T(a)$ is given by the matrix representation $(a_{i-j})_{i,j\geq 0}$ with respect to the standard basis of $l^2(\mathbb{Z}^+)$, where a_k denotes the kth Fourier coefficient of a. (We agree to omit the superscript $\mathcal{P}$ when the filtration is specified in this way.)

The sequences in $\mathcal{S}(\mathsf{T}(C))$ are completely described in the following theorem, discovered by A. Böttcher and B. Silbermann and first published in 1983 in their paper [3] on the convergence of the finite sections method for quarter plane Toeplitz operators (see also [4] and [6], Section 1.4.2). To state their results, let R_n stand for the operator on $l^2(\mathbb{Z}^+)$ which sends $(x_0, x_1, \ldots)$ to $(x_{n-1}, \ldots, x_0, 0, 0, \ldots)$. It is not hard to see that for each sequence $\mathbf{A} = (A_n) \in \mathcal{S}(\mathsf{T}(C))$, the strong limits $W(\mathbf{A}) :=$ s-lim A_nP_n and $\widetilde{W}(\mathbf{A}) :=$ s-lim $R_nA_nR_nP_n$ exist and that W and $\widetilde{W}$ are unital *-homomorphisms from $\mathcal{S}(\mathsf{T}(C))$ to $L(l^2(\mathbb{Z}^+))$ (actually, to $\mathsf{T}(C)$).

Theorem 1.1.

(a) *The algebra $\mathcal{S}(\mathsf{T}(C))$ consists of all sequences $(A_n)_{n\geq 1}$ of the form*

$$(A_n) = (P_nT(a)P_n + P_nKP_n + R_nLR_n + G_n) \tag{1.2}$$

where $a \in C(\mathbb{T})$, K and L are compact operators on $l^2(\mathbb{Z}^+)$, and $(G_n) \in \mathcal{G}$. The representation of a sequence $(A_n) \in \mathcal{S}(\mathsf{T}(C))$ in this form is unique.

(b) *For every sequence $\mathbf{A} \in \mathcal{S}(\mathsf{T}(C))$, the coset $\mathbf{A}+\mathcal{G}$ is invertible in the quotient algebra $\mathcal{S}(\mathsf{T}(C))/\mathcal{G}$ (equivalently, in $\mathcal{F}/\mathcal{G}$) if and only if the operators $W(\mathbf{A})$ and $\widetilde{W}(\mathbf{A})$ are invertible.*

Corollary 1.2. *The quotient algebra $\mathcal{S}(\mathsf{T}(C))/\mathcal{G}$ is *-isomorphic to the C^*-algebra of all pairs*

$$(T(a) + K, T(\tilde{a}) + L) \in L(l^2(\mathbb{Z}^+)) \times L(l^2(\mathbb{Z}^+)) \tag{1.3}$$

with $a \in C(\mathbb{T})$ and K, L compact. In particular, the mapping which sends the sequence (1.2) to the pair (1.3) is a *-homomorphism from $\mathcal{S}(\mathsf{T}(C))$ onto $\mathcal{S}(\mathsf{T}(C))/\mathcal{G}$ with kernel $\mathcal{G}$.

Now we can give a first idea of what the statement in the abstract that "an algebra has a common structure with $\mathcal{S}(\mathsf{T}(C))$" means. It is not hard to check that the set $\mathcal{J}$ of all sequences $(P_nKP_n + R_nLR_n + G_n)$ with K, L compact and $(G_n) \in \mathcal{G}$ forms a closed two-sided ideal of $\mathcal{S}(\mathsf{T}(C))$. By Corollary 1.2, the quotient $\mathcal{J}/\mathcal{G}$ is naturally isomorphic to the product $K(l^2(\mathbb{Z}^+)) \times K(l^2(\mathbb{Z}^+))$. Thus, this quotient has two natural irreducible representations $(K, L) \mapsto K$ and

$(K, L) \mapsto L$ which extend uniquely to irreducible representations of $\mathcal{S}(\mathsf{T}(C))/\mathcal{G}$. These extensions coincide (up to unitary equivalence) with the quotients $W^{\mathcal{G}}$ and $\widetilde{W}^{\mathcal{G}}$ of the homomorphisms W and $\widetilde{W}$ by $\mathcal{G}$ (note that $\mathcal{G}$ lies in the kernel of both W and $\widetilde{W}$), and these quotients have the property described in assertion (b) of Theorem 1.1.

To fix the latter property formally, we introduce the following notions. Given a unital C^*-algebra $\mathcal{A}$ and a family $\mathcal{W}$ of unital *-homomorphisms W from $\mathcal{A}$ to certain unital C^*-algebras, we say that $\mathcal{W}$ is *strictly spectral* for $\mathcal{A}$ if an element $a \in \mathcal{A}$ is invertible if and only if $W(a)$ is invertible for every $W \in \mathcal{W}$. We say that $\mathcal{W}$ is *spectral* for $\mathcal{A}$ if an element $a \in \mathcal{A}$ is invertible if and only if $W(a)$ is invertible for every $W \in \mathcal{W}$ and if $\sup_{W \in \mathcal{W}} \|(W(a))^{-1}\| < \infty$. Clearly, both notions coincide if $\mathcal{W}$ is a finite family.

We recall further that a C^*-algebra $\mathcal{A}$ is called *elementary* if it is isomorphic to the algebra of the compact operators on some Hilbert space, and that it is a *dual* algebra if it is isomorphic to a direct sum of elementary algebras. If $\mathcal{J}$ a closed ideal of $\mathcal{A}$ which is elementary (respective dual) when considered as a C^*-algebra, then we call $\mathcal{J}$ an *elementary* (respective a *dual*) ideal of $\mathcal{A}$. It is easy to see that every dual ideal $\mathcal{J}$ is generated (as a C^*-algebra) by its elementary ideals, $\mathcal{K}_t$ with $t \in T$, say. Since every closed ideal of $\mathcal{J}$ is also a closed ideal of $\mathcal{A}$, $\mathcal{J}$ can be identified with the smallest closed ideal of $\mathcal{A}$ which contains all elementary ideals $\mathcal{K}_t$. See [2] for an overview on dual algebras.

Now we can make the statement that a C^*-subalgebra $\mathcal{A}$ of $\mathcal{F}$ has a "common structure structure with $\mathcal{S}(\mathsf{T}(C))$" more precise: It means that $\mathcal{A}$ contains $\mathcal{G}$, that the quotient $\mathcal{A}/\mathcal{G}$ contains an ideal $\mathcal{J}$ which is dual, and that the extensions of the irreducible representations of $\mathcal{J}/\mathcal{G}$ to $\mathcal{A}/\mathcal{G}$ form a spectral family for $\mathcal{A}/\mathcal{G}$.

The motivation for this paper came from the observation made again and again over the years since Silbermann's paper [19] that many concrete algebras of approximation sequences have a "common structure with $\mathcal{S}(\mathsf{T}(C))$". This observation appeared (at least to the author) as a big miracle, since neither were the discretized operators very close to Toeplitz operators, nor had the used discretization procedures something in common with the fairly simple idea of taking finite sections. To get an impression, here is a very incomplete list of papers from different fields where this phenomenon occurs: [3, 5, 7–9, 11]. The explanation of that fact proposed in the present paper is that at least every *separable* subalgebra of $\mathcal{F}$ has a "common structure with $\mathcal{S}(\mathsf{T}(C))$", after *suitable extension by compact sequences and suitable fractal restriction.*

It should be mentioned that the proposed solution is by no means the only way to explain the "miracle": In [18], Santos and Silbermann singled out properties of the operator algebra to be discretized (e.g., generated by a shift) and of the discretization mapping (e.g., its positivity, a nice behavior of quasicommutators) which ensure that the resulting algebras have a common structure with $\mathcal{S}(\mathsf{T}(C))$.

The paper is organized as follows. The phrases "fractal restriction" and "extension by compact sequences" are explained in the following two sections, followed by a section which studies "fractal algebras of compact sequences". The goal of these sections is to introduce the language and to provide some facts for later reference without proofs. The heart of the paper is Section 5 where two versions of extension-restriction theorems are derived.

2. Fractal restriction

The idea behind the notion of a fractal algebra comes from a remarkable property of the algebra $\mathcal{S}(\mathsf{T}(C))/\mathcal{G}$: the structure of this algebra is determined by two representations W and $\widetilde{W}$, and these representations are defined by certain strong limits. A consequence of this "limit form" is that the operators $W(\mathbf{A})$ and $\widetilde{W}(\mathbf{A})$ can be determined from each subsequence of the sequence $\mathbf{A} \in \mathcal{S}(\mathsf{T}(C))$. This observation implies that whenever a *subsequence* of a sequence $\mathbf{A} \in \mathcal{S}(\mathsf{T}(C))$ is stable, then the operators $W(\mathbf{A})$ and $\widetilde{W}(\mathbf{A})$ are already invertible and, hence, the *full* sequence $\mathbf{A}$ is stable by Theorem 1.1.

One can state this observation in a slightly different way: every sequence in $\mathcal{S}(\mathsf{T}(C))$ can be recovered from each its subsequence up to a sequence tending to zero in the norm. In that sense, the essential information on a sequence in $\mathcal{S}(\mathsf{T}(C))$ is stored in each of its subsequences. Subalgebras of $\mathcal{F}$ with this property were called *fractal* in [17] in order to emphasize exactly this self-similarity aspect. We will recall some basic properties of fractal algebras that will be needed in what follows and start with the official definition of a fractal algebra. (Note that this definition also makes sense in a more general context when $\mathcal{F}$ is the direct product and $\mathcal{G}$ the direct sum of a sequence of C^*-algebras.)

Let $\mathfrak{S}_{\mathbb{N}}$ denote the set of all strictly increasing sequences $\eta : \mathbb{N} \to \mathbb{N}$. Given $\eta \in \mathfrak{S}_{\mathbb{N}}$, we let $\mathcal{F}_\eta$ denote the set of all subsequences $(A_{\eta(n)})$ of sequences (A_n) in $\mathcal{F}$. One can make $\mathcal{F}_\eta$ to a C^*-algebra in a natural way. The mapping $R_\eta : \mathcal{F} \to \mathcal{F}_\eta$, $(A_n) \mapsto (A_{\eta(n)})$ is called the *restriction* of $\mathcal{F}$ onto $\mathcal{F}_\eta$. For every subset $\mathcal{S}$ of $\mathcal{F}$, we abbreviate $R_\eta \mathcal{S}$ by $\mathcal{S}_\eta$. It is easy to see that $\mathcal{G}_\eta$ coincides with the ideal of the sequences in $\mathcal{F}_\eta$ which tend to zero in the norm.

Let $\mathcal{A}$ be a C^*-subalgebra of $\mathcal{F}$. A *-homomorphism W from $\mathcal{A}$ into a C^*-algebra $\mathcal{B}$ is called *fractal* if, for every $\eta \in \mathfrak{S}_{\mathbb{N}}$, there is a mapping $W_\eta : \mathcal{A}_\eta \to \mathcal{B}$ such that $W = W_\eta R_\eta|_{\mathcal{A}}$. Thus, the image of a sequence in $\mathcal{A}$ under a fractal homomorphism can be reconstructed from each of its (infinite) subsequences. It is not hard to check that W_η is a *-homomorphism again. The homomorphisms W and $\widetilde{W}$ defined in Section 1 are archetypal examples of a fractal homomorphism.

Definition 2.1.

(a) A C^*-subalgebra $\mathcal{A}$ of $\mathcal{F}$ is called *fractal* if the canonical homomorphism $\pi : \mathcal{A} \to \mathcal{A}/(\mathcal{A} \cap \mathcal{G})$, $\mathbf{A} \mapsto \mathbf{A} + (\mathcal{A} \cap \mathcal{G})$ is fractal.

(b) A sequence $\mathbf{A} \in \mathcal{F}$ is called *fractal* if the smallest C^*-subalgebra of $\mathcal{F}$ which contains the sequence $\mathbf{A}$ and the identity sequence is fractal.

Here are some equivalent characterizations of fractal algebras.

Theorem 2.2.

(a) *A C^*-subalgebra $\mathcal{A}$ of $\mathcal{F}$ is fractal if and only if the implication*

$$R_\eta(\mathbf{A}) \in \mathcal{G}_\eta \Rightarrow \mathbf{A} \in \mathcal{G} \tag{2.1}$$

holds for every sequence $\mathbf{A} \in \mathcal{A}$ and every $\eta \in \mathfrak{S}_\mathbb{N}$.

(b) *If $\mathcal{A}$ is a fractal C^*-subalgebra of $\mathcal{F}$, then $\mathcal{A}_\eta \cap \mathcal{G}_\eta = (\mathcal{A} \cap \mathcal{G})_\eta$ for every $\eta \in \mathfrak{S}_\mathbb{N}$.*

(c) *A C^*-subalgebra $\mathcal{A}$ of $\mathcal{F}$ is fractal if and only if the algebra $\mathcal{A} + \mathcal{G}$ is fractal.*

In many instances, the following theorem offers a comfortable way to check the fractality of a specific subalgebra of $\mathcal{F}$ (for example, the fractality of $\mathcal{S}(\mathsf{T}(C))$, where the corresponding homomorphism W is the product of fractal homomorphisms W and $\widetilde{W}$).

Theorem 2.3. *A unital C^*-subalgebra $\mathcal{A}$ of $\mathcal{F}$ is fractal if and only if there is a unital and fractal *-homomorphism W from $\mathcal{A}$ into a unital C^*-algebra $\mathcal{B}$ such that, for every sequence $\mathbf{A} \in \mathcal{A}$, the coset $\mathbf{A} + (\mathcal{A} \cap \mathcal{G})$ is invertible in $\mathcal{A}/(\mathcal{A} \cap \mathcal{G})$ if and only if $W(\mathbf{A})$ is invertible in $\mathcal{B}$.*

The following results from [17] give a first impression of the power of fractality.

Proposition 2.4. *Let $\mathcal{A}$ be a unital fractal C^*-subalgebra of $\mathcal{F}$ and $\mathbf{A} = (A_n) \in \mathcal{A}$. Then,*

(a) *the sequence $\mathbf{A}$ is stable if and only if it possesses a stable subsequence.*

(b) *the limit $\lim_{n\to\infty} \|A_n\|$ exists and is equal to $\|\mathbf{A} + \mathcal{G}\|$.*

Thus, the upper limit in (1.1) is in fact a limit if the sequence (A_n) belongs to a fractal algebra. A similar improvement can be observed for the convergence of certain spectral quantities. We will need the following notions.

Let $(M_n)_{n\in\mathbb{N}}$ be a sequence of non-empty compact subsets of the complex plane. The *upper limit* $\limsup M_n$ (also called the *partial limiting set*) resp. the *lower limit* $\liminf M_n$ (or the *uniform limiting set*) of the sequence (M_n) consists of all points $x \in \mathbb{C}$ which are partial limits resp. the limit of a sequence (m_n) of points $m_n \in M_n$. The upper and lower limits of a sequence (M_n) coincide if and only if this sequence converges with respect to the Hausdorff metric

$$h(L, M) := \max\{\max_{l\in L} \operatorname{dist}(l, M), \max_{m\in M} \operatorname{dist}(m, L)\}.$$

Recall in this connection that non-empty compact subsets of $\mathbb{C}$ form a complete metric space with respect to the Hausdorff distance and that the relatively compact subsets of this space are precisely its bounded subsets.

We denote the spectrum of an element a of a unital algebra $\mathcal{A}$ by $\sigma_\mathcal{A}(a)$ or $\sigma(a)$. Recall also that an element a of a C^*-algebra is called *normal* if $aa^* = a^*a$. Assertion (a) of the following result is the analog of the limsup-formula (1.1) for norms: assertion (b) its improvement in the presence of fractality.

Proposition 2.5.

(a) *If* $(A_n) \in \mathcal{F}$ *is a normal sequence, then*

$$\limsup \sigma(A_n) = \sigma_{\mathcal{F}/\mathcal{G}}((A_n) + \mathcal{G}).$$

(b) *If* (A_n) *is a normal sequence in a unital and fractal* C^**-subalgebra of* $\mathcal{F}$*, then*

$$\limsup \sigma(A_n) = \liminf \sigma(A_n). \tag{2.2}$$

(c) *A normal sequence* $(A_n) \in \mathcal{F}$ *is fractal if and only if* (2.2) *holds.*

Similar results hold for other spectral quantities, for example for the sequences of the condition numbers, the sets of the singular values, the ϵ-pseudospectra, and the numerical ranges of the A_n (see Chapter 3 in [6]). These results indicate that, given an (in general non-fractal) sequence in $\mathcal{F}$, it is of vital importance to single out (one of) its fractal subsequences. That this is indeed possible is a consequence of the following *fractal restriction theorem* first proved in [12]. The proof given there was based on Proposition 2.5 (c) and is rather involved. A much simpler proof employs the converse of Proposition 2.4 (b), which on its hand follows easily from Theorem 2.2 (a).

Theorem 2.6. *Let* $\mathcal{A}$ *be a separable* C^**-subalgebra of* $\mathcal{F}$*. Then there is an* $\eta \in \mathfrak{S}_{\mathbb{N}}$ *such that the restricted algebra* $\mathcal{A}_\eta$ *is fractal. In particular, every sequence in* $\mathcal{F}$ *possesses a fractal subsequence.*

One cannot expect that Theorem 2.6 holds for arbitrary C^*-subalgebras of $\mathcal{F}$; for example it is certainly not true for the algebra l^∞. On the other hand, non-separable fractal algebras exist: the finite sections algebra for Toeplitz operators with *piecewise continuous* generating function can serve as an example.

3. Compactness and essential fractality

The ideal $\mathcal{J}$ of the finite sections algebra $\mathcal{S}(\mathsf{T}(C))$ is evidently related with compact operators. We will make this relation precise by introducing an ideal $\mathcal{K}$ of $\mathcal{F}$ consisting of sequences of compact type. The role of this ideal in numerical analysis is comparable with the role of the ideal of the compact operators in operator theory. Throughout this section we assume that $\mathcal{F}$ is a direct product of matrix algebras $\mathbb{C}^{\delta(n)\times\delta(n)}$ with dimension function $\delta : \mathbb{N} \to \mathbb{N}$ tending to infinity.

Definition 3.1. Let $\mathcal{K}$ denote the smallest closed ideal of $\mathcal{F}$ which contains all sequences $(K_n) \in \mathcal{F}$ with $\operatorname{rank} K_n \le 1$ for every n. We refer to the elements of $\mathcal{K}$ as *compact sequences* and call a sequence in $\mathcal{F}$ a *Fredholm sequence* if it is invertible modulo the ideal $\mathcal{K}$.

Thus, a sequence $(A_n) \in \mathcal{F}$ belongs to $\mathcal{K}$ if and only if, for every $\varepsilon > 0$, there is a sequence $(K_n) \in \mathcal{F}$ such that

$$\sup_n \|A_n - K_n\| < \varepsilon \quad \text{and} \quad \sup_n \operatorname{rank} K_n < \infty. \tag{3.1}$$

It is a simple consequence of this fact that $\mathcal{G} \subseteq \mathcal{K}$.

We say that a sequence $\mathbf{A} \in \mathcal{F}$ has *finite essential rank* if it is the sum of a sequence in $\mathcal{G}$ and a sequence $(K_n) \in \mathcal{F}$ with $\sup_n \operatorname{rank} K_n < \infty$. If $\mathbf{A}$ is of finite essential rank, then there is a smallest integer $r \geq 0$ such that $\mathbf{A}$ can be written as $(G_n)+(K_n)$ with $(G_n) \in \mathcal{G}$ and $\sup_n \operatorname{rank} K_n = r$. We call this integer the *essential rank* of $\mathbf{A}$ and write $\operatorname{ess\,rank} \mathbf{A} = r$. If $\mathbf{A}$ is not of finite essential rank, then we put $\operatorname{ess\,rank} \mathbf{A} = \infty$. Thus, the sequences of essential rank 0 are just the sequences in $\mathcal{G}$.

In our running example, the intersection $\mathcal{S}(\mathsf{T}(C)) \cap \mathcal{K}$ is just the ideal $\mathcal{J}$, and the essential rank of the sequence $(P_n K P_n + R_n L R_n + G_n)$ equals $\operatorname{rank} K + \operatorname{rank} L$.

Both the compactness and the Fredholm property of a sequence $(A_n) \in \mathcal{F}$ can be characterized in terms of the asymptotic behavior of the singular values of the A_n; see Sections 4.2 and 5.1 in [14] for the following results. We denote the decreasingly ordered singular values of an $n \times n$ matrix A by

$$\|A\| = \Sigma_1(A) \geq \Sigma_2(A) \geq \cdots \geq \Sigma_n(A) \geq 0 \tag{3.2}$$

and set $\sigma_k(A) := \Sigma_{n-k+1}(A)$.

Theorem 3.2. *The following conditions are equivalent for a sequence $(K_n) \in \mathcal{F}$:*
(a) $\lim_{k\to\infty} \sup_{n\geq k} \Sigma_k(K_n) = 0$;
(b) $\lim_{k\to\infty} \limsup_{n\to\infty} \Sigma_k(K_n) = 0$;
(c) (K_n) *is compact.*

Corollary 3.3.
(a) *A sequence $(K_n) \in \mathcal{F}$ is of essential rank r if and only if*

$$\limsup_{n\to\infty} \Sigma_r(K_n) > 0 \quad \text{and} \quad \lim_{n\to\infty} \Sigma_{r+1}(K_n) = 0.$$

(b) *If $(K_n) \in \mathcal{K}$, then $\lim_{n\to\infty} \sigma_k(K_n) = 0$ for every k.*

Theorem 3.4. *The following conditions are equivalent for a sequence $(A_n) \in \mathcal{F}$:*
(a) *(A_n) is a Fredholm sequence.*
(b) *There are sequences $(B_n) \in \mathcal{F}$ and $(J_n) \in \mathcal{K}$ with $\sup_n \operatorname{rank} J_n < \infty$ such that $B_n A_n = I_n + J_n$ for all $n \in \mathbb{N}$.*
(c) *There is a $k \in \mathbb{N}$ such that $\liminf_{n\to\infty} \sigma_{k+1}(A_n) > 0$.*

The smallest non-negative integer k which satisfies condition (c) in the previous theorem is called the *α-number* $\alpha(\mathbf{A})$ of the Fredholm sequence $\mathbf{A} = (A_n)$. It corresponds to the kernel dimension of a Fredholm operator. Equivalently, $\alpha(\mathbf{A})$ is the smallest non-negative integer k for which there exist a sequence $(B_n) \in \mathcal{F}$ and a sequence $(J_n) \in \mathcal{K}$ of essential rank k such that $B_n A_n^* A_n = I_n + J_n$ for all $n \in \mathbb{N}$.

Next we extend the concept of fractality and define fractality with respect to the ideal of the compact sequences. For details see [16].

Definition 3.5. A C^*-subalgebra $\mathcal{A}$ of $\mathcal{F}$ is said to be *essentially fractal* (or *$\mathcal{K}$-fractal*) if the canonical homomorphism $\pi^{\mathcal{K}} : \mathcal{A} \to \mathcal{A}/(\mathcal{A} \cap \mathcal{K})$ is fractal and if $(\mathcal{A} \cap \mathcal{K})_\eta = \mathcal{A}_\eta \cap \mathcal{K}_\eta$ for each $\eta \in \mathfrak{S}_{\mathbb{N}}$.

In the same way one defines $\mathcal{J}$-fractality with respect to an arbitrary closed ideal $\mathcal{J}$ of $\mathcal{F}$. If $\mathcal{J} = \mathcal{G}$, then the second condition in Definition 3.5 is automatically satisfied by Theorem 2.2; thus, $\mathcal{G}$-fractality coincides with fractality in the sense of Definition 2.1.

The following result shows that essential fractality implies what one expects.

Theorem 3.6. *Let $\mathcal{A}$ be an essentially fractal and unital C^*-subalgebra of $\mathcal{F}$. A sequence in $\mathcal{A}$ is compact (Fredholm) if and only if one of its subsequences is compact (Fredholm).*

Corollary 3.7. *Let $\mathcal{A}$ be a fractal C^*-subalgebra of $\mathcal{F}$. If $(\mathcal{A} \cap \mathcal{K})_\eta = \mathcal{A}_\eta \cap \mathcal{K}_\eta$ for each $\eta \in \mathfrak{S}_{\mathbb{N}}$, then $\mathcal{A}$ is essentially fractal.*

Essential fractality has striking consequences for the behavior of the smallest singular values of a Fredholm sequence.

Theorem 3.8. *Let $\mathcal{A}$ be an essentially fractal and unital C^*-subalgebra of $\mathcal{F}$. A sequence $(A_n) \in \mathcal{A}$ is Fredholm if and only if there is a $k \in \mathbb{N}$ such that*

$$\limsup_{n\to\infty} \sigma_k(A_n) > 0. \tag{3.3}$$

Consequently, if a sequence (A_n) in an essentially fractal and unital C^*-subalgebra of $\mathcal{F}$ fails to be Fredholm, then

$$\lim_{n\to\infty} \sigma_k(A_n) = 0 \qquad \text{for each } k \in \mathbb{N}. \tag{3.4}$$

We call a sequence with that property *not normally solvable*, in analogy with the corresponding notion from operator theory.

Corollary 3.9. *Let $\mathcal{A}$ be an essentially fractal and unital C^*-subalgebra of $\mathcal{F}$. Then a sequence in $\mathcal{A}$ is either Fredholm or not normally solvable.*

Example. The finite sections algebra $\mathcal{S}(\mathsf{T}(C))$ for Toeplitz operators is essentially fractal, as follows easily from the description of $\mathcal{S}(\mathsf{T}(C))$ in Theorem 1.1 in combination with Corollary 3.7. The finite sections algebra $\mathcal{S}(\mathsf{BDO}(\mathbb{N}))$ for band-dominated operators examined in [14] is an example of an algebra which is essentially fractal but not fractal. Finally, the sequence (A_n) where

$$A_n := \begin{cases} \operatorname{diag}(0, 0, \dots, 0, 1) & \text{if } n \text{ is even} \\ \operatorname{diag}(0, 1, \dots, 1, 1) & \text{if } n \text{ is odd,} \end{cases}$$

is fractal, but not essentially fractal: its subsequence (A_{2n}) is compact, but (A_{2n+1}) is Fredholm.

As for fractality, there is an *essential fractal restriction theorem* (see [16]). Its proof is based on the fact that, for every separable C^*-subalgebra $\mathcal{A}$ of $\mathcal{F}$, there is a sequence η such that not only the sequence of the norms $(\|A_{\eta(n)}\|)_{n\geq 1} = (\Sigma_1(A_{\eta(n)}))_{n\geq 1}$ converges for every $(A_n) \in \mathcal{A}$ (which was basic in the proof of the fractal restriction theorem), but also *every* sequence $(\Sigma_k(A_{\eta(n)}))_{n\geq 1}$ of the singular values, for every $k \in \mathbb{N}$.

Theorem 3.10. *Let $\mathcal{A}$ be a separable C^*-subalgebra of $\mathcal{F}$. Then there is an $\eta \in \mathfrak{S}_{\mathbb{N}}$ such that the restricted algebra $\mathcal{A}_\eta$ is essentially fractal.*

4. Fractal algebras of compact sequences

Compact sequences in fractal algebras behave particularly well. To state these results, we need some more notions.

A non-zero element k of a C^*-algebra $\mathcal{A}$ is said to be of *algebraic rank one* if, for each $a \in \mathcal{A}$, there is a complex number μ such that $kak = \mu k$. We let $\mathcal{C}(\mathcal{A})$ stand for the smallest closed subalgebra of $\mathcal{A}$ which contains all elements of algebraic rank one. If such elements do not exist, we set $\mathcal{C}(\mathcal{A}) = \{0\}$. In any case, $\mathcal{C}(\mathcal{A})$ is a closed ideal of $\mathcal{A}$, the elements of which we call *compact*. The following theorem summarizes some well-known equivalent descriptions of $\mathcal{C}(\mathcal{A})$ (see Theorem 1.4.5 in [1], and recall the notion of a dual algebra from the introduction).

Theorem 4.1. *Let $\mathcal{A}$ be a unital C^*-algebra and $\mathcal{J}$ a closed ideal of $\mathcal{A}$. The following assertions are equivalent:*

(a) $\mathcal{J} = \mathcal{C}(\mathcal{J})$.
(b) *$\mathcal{J}$ is a dual algebra.*
(c) *The spectrum of every self-adjoint element of $\mathcal{J}$ is at most countable and has 0 as only possible accumulation point.*

Every dual ideal of a C^*-algebra comes with an associated lifting theorem (see [19] for a first version of that theorem and [6] for a proof).

Theorem 4.2 (Lifting theorem for dual ideals). *Let $\mathcal{A}$ be a unital C^*-algebra. For every element t of a set T, let $\mathcal{J}_t$ be an elementary ideal of $\mathcal{A}$ such that $\mathcal{J}_s\mathcal{J}_t = \{0\}$ whenever $s \neq t$, and let $W_t : \mathcal{A} \to L(H_t)$ denote the irreducible representation of $\mathcal{A}$ which extends the (unique up to unitary equivalence) irreducible representation of $\mathcal{J}_t$. Let further $\mathcal{J}$ stand for the smallest closed ideal of $\mathcal{A}$ which contains all ideals $\mathcal{J}_t$.*

(a) *An element $a \in \mathcal{A}$ is invertible if and only if the coset $a + \mathcal{J}$ is invertible in $\mathcal{A}/\mathcal{J}$ and if every operator $W_t(a)$ is invertible in $L(H_t)$.*
(b) *The separation property holds, i.e., $W_s(\mathcal{J}_t) = \{0\}$ whenever $s \neq t$.*
(c) *If $j \in \mathcal{J}$, then $W_t(j)$ is compact for every $t \in T$.*
(d) *If $a + \mathcal{J}$ is invertible in $\mathcal{A}/\mathcal{J}$, then all operators $W_t(a)$ are Fredholm and all but a finite number of them is invertible.*

A basic observation of [13] is that a compact sequence in a fractal algebra has the spectral property of Theorem 4.1 (c). The following result is thus an immediate consequence of that theorem. It implies in particular that every unital and fractal C^*-subalgebra of $\mathcal{F}$ which contains non-trivial compact sequences is subject to the lifting theorem.

Corollary 4.3. *Let $\mathcal{A}$ be a unital and fractal C^*-subalgebra of $\mathcal{F}$ which contains the ideal $\mathcal{G}$. Then the ideal $(\mathcal{A} \cap \mathcal{K})/\mathcal{G}$ of $\mathcal{A}/\mathcal{G}$ is a dual algebra.*

5. Restriction-extension theorems

5.1. Weights of elementary algebras of sequences

A projection in a C^*-algebra is a self-adjoint element p with $p^2 = p$. A closed ideal $\mathcal{J}$ of a C^*-algebra $\mathcal{A}$ *lifts projections*, if every projection in $\mathcal{A}/\mathcal{J}$ contains a representative which is a projection in $\mathcal{A}$. Closed ideals of C^*-algebras do not lift projections in general (take $\mathcal{A} = C([0, 1])$ and $\mathcal{J} = \{f \in \mathcal{A} : f(0) = f(1) = 0\}$). The following proposition states that elementary ideals of $\mathcal{F}/\mathcal{G}$ lift projections. More general, every dual ideal of a C^*-algebra owns the projection lifting property.

Proposition 5.1. *Let $\mathcal{B}$ be an elementary C^*-subalgebra of $\mathcal{F}/\mathcal{G}$.*

(a) *Every projection $p \in \mathcal{B}$ lifts to a sequence $(\Pi_n) \in \mathcal{F}$ of orthogonal projections, i.e., $(\Pi_n) + \mathcal{G} = p$.*

(b) *If p and q are rank one projections in $\mathcal{B}$ which lift to projections (Π_n^p) and (Π_n^q) in $\mathcal{F}$, respectively, then $\dim \operatorname{im} \Pi_n^p = \dim \operatorname{im} \Pi_n^q$ for all sufficiently large n.*

Thus, *for large n*, the numbers $\dim \operatorname{im} \Pi_n^p$ are uniquely determined by the algebra $\mathcal{B}$; they do neither depend on the rank one projection p nor on the choice of its lifting. For a precise formulation of that property, define an equivalence relation $\sim$ on the set of all sequences of non-negative integers by calling two sequences (α_n), (β_n) *equivalent* if $\alpha_n = \beta_n$ for all sufficiently large n. Then Proposition 5.1 states that the equivalence class which contains the sequence $(\dim \operatorname{im} \Pi_n^p)_{n \geq 1}$ is uniquely determined by the algebra $\mathcal{B}$. We denote this equivalence class by $\alpha^{\mathcal{B}}$ and call it the *weight* of the elementary algebra $\mathcal{B}$. We say that $\mathcal{B}$ is *of positive weight* if $\alpha^{\mathcal{B}}$ contains a sequence of positive numbers, and $\mathcal{B}$ is *of weight one* if $\alpha^{\mathcal{B}}$ contains the constant sequence of ones. Note that the weight is bounded if $\mathcal{B}$ is in $\mathcal{K}/\mathcal{G}$; in this case (Π_n^p) is a compact sequence of projections and therefore of finite essential rank.

5.2. Silbermann pairs and $\mathcal{J}$-Fredholm sequences

Next we are going to examine the consequences of the lifting theorem in the context of Silbermann pairs.

A *Silbermann pair* $(\mathcal{A}, \mathcal{J})$ consists of a unital C^*-subalgebra $\mathcal{A}$ of $\mathcal{F}$ and a closed ideal $\mathcal{J}$ of $\mathcal{A}$ such that $\mathcal{G} \subseteq \mathcal{J} \subseteq \mathcal{K}$, $\mathcal{G} \neq \mathcal{J}$ and $\mathcal{J}/\mathcal{G}$ is a dual algebra. Every sequence in $\mathcal{A}$ which is invertible modulo $\mathcal{J}$ is called a *$\mathcal{J}$-Fredholm* sequence. Note that every $\mathcal{J}$-Fredholm sequence is a Fredholm sequence in sense of Definition 3.1 (but a Fredholm sequence in $\mathcal{A}$ need not be $\mathcal{J}$-Fredholm because $\mathcal{J}$ may be properly contained in $\mathcal{A} \cap \mathcal{K}$).

Under the hypotheses of Corollary 4.3, $(\mathcal{A}, \mathcal{A} \cap \mathcal{K})$ is a Silbermann pair, and a sequence in $\mathcal{A}$ is $(\mathcal{A} \cap \mathcal{K})$-Fredholm if and only if it is Fredholm. The study of Silbermann pairs (in the special case when $\mathcal{J}/\mathcal{G}$ is an elementary subalgebra of $\mathcal{K}/\mathcal{G}$) was initiated by Silbermann in [20].

Let $(\mathcal{A}, \mathcal{J})$ be a Silbermann pair. Being dual by definition, the algebra $\mathcal{J}/\mathcal{G}$ is the direct sum of a family $(I_t)_{t \in T}$ of elementary algebras with associated bijective representations $W_t : I_t \to K(H_t)$. These representations extent to irreducible representations $\mathcal{A} \to L(H_t)$ which we denote by W_t again. In this context, the Lifting theorem 4.2 specifies as follows.

Theorem 5.2. *Let $(\mathcal{A}, \mathcal{J})$ be a Silbermann pair.*

(a) *A sequence $\mathbf{A} \in \mathcal{A}$ is stable if and only if it is $\mathcal{J}$-Fredholm and if the operators $W_t(\mathbf{A})$ are invertible for each $t \in T$.*
(b) *$W_s(I_t) = \{0\}$ whenever $s \neq t$.*
(c) *If $\mathbf{J} \in \mathcal{J}$, then $W_t(\mathbf{J})$ is a compact operator for every $t \in T$.*
(d) *If $\mathbf{A} \in \mathcal{A}$ is $\mathcal{J}$-Fredholm, then all operators $W_t(\mathbf{A})$ are Fredholm and all but a finite number of them is invertible.*

For every elementary ideal I_t, let (α_n^t) be a representative of the weight α^{I_t}. Assertion (d) of Theorem 5.2 implies that for every $\mathcal{J}$-Fredholm sequence $\mathbf{A} \in \mathcal{A}$, the sum

$$\alpha_n(\mathbf{A}) := \sum_{t \in T} \alpha_n^t \dim \ker W_t(\mathbf{A}) \tag{5.1}$$

is finite. This definition obviously depends on the choice of the representatives of the weights. But the equivalence class of the sequence $(\alpha_n(\mathbf{A}))$ modulo $\sim$ is uniquely determined, since only a finite number of items in the sum (5.1) is not zero.

A basic phenomenon of a $\mathcal{J}$-Fredholm sequence (A_n) is the following splitting property of the (increasingly ordered) singular values $\sigma_k(A_n)$ of A_n.

Theorem 5.3. *Let $(\mathcal{A}, \mathcal{J})$ be a Silbermann pair and $\mathbf{A} = (A_n) \in \mathcal{A}$ a $\mathcal{J}$-Fredholm sequence. Then $\mathbf{A}$ is a Fredholm sequence, and*

$$\lim_{n\to\infty} \sigma_{\alpha_n(\mathbf{A})}(A_n) = 0 \quad \textit{whereas} \quad \liminf_{n\to\infty} \sigma_{\alpha_n(\mathbf{A})+1}(A_n) > 0. \tag{5.2}$$

The proof makes use of results on lifting of families of mutually orthogonal projections and on generalized (or Moore–Penrose) invertibility. For details see [13].

Theorem 5.3 has some remarkable consequences. First note that the number

$$\alpha(\mathbf{A}) := \limsup_{n\to\infty} \alpha_n(\mathbf{A}) \tag{5.3}$$

is well defined and finite for every $\mathcal{J}$-Fredholm sequence $\mathbf{A} \in \mathcal{A}$. Since $(\alpha_n(\mathbf{A}))$ is a sequence of non-negative integers, it has a constant subsequence the entries of which are equal to $\alpha(\mathbf{A})$. Together with (5.2), this shows that

$$\liminf_{n\to\infty} \sigma_{\alpha(\mathbf{A})}(A_n) = 0 \quad \text{and} \quad \liminf_{n\to\infty} \sigma_{\alpha(\mathbf{A})+1}(A_n) > 0. \tag{5.4}$$

Corollary 5.4. *Let $(\mathcal{A}, \mathcal{J})$ be a Silbermann pair and $\mathbf{A} \in \mathcal{A}$ a $\mathcal{J}$-Fredholm sequence. Then $\mathbf{A}$ is a Fredholm sequence, and its α-number is given by* (5.3).

Many subalgebras of $\mathcal{F}$ which arise from concrete approximation methods have the property that every rank one projection in $\mathcal{J}/\mathcal{G}$ lifts to a sequence of projections of rank one (equivalently, that every elementary algebra I_t has weight one). In this case we call $(\mathcal{A}, \mathcal{J})$ a Silbermann pair of *local weight one*. For Silbermann pairs with this property, Theorem 5.3 and its Corollary 5.4 specify as follows.

Corollary 5.5. *Let $(\mathcal{A}, \mathcal{J})$ be a Silbermann pair of local weight one and $\mathbf{A} = (A_n) \in \mathcal{A}$ a $\mathcal{J}$-Fredholm sequence. Then*

$$\alpha(\mathbf{A}) = \sum_{t \in T} \dim \ker W_t(\mathbf{A}), \tag{5.5}$$

and the sequence $\mathbf{A}$ has the $\alpha(\mathbf{A})$-splitting property, i.e., the number of the singular values of A_n which tend to zero is $\alpha(\mathbf{A})$.

5.3. Spectral Silbermann pairs

A Silbermann pair $(\mathcal{A}, \mathcal{J})$ is called *spectral* or *strictly spectral* if the family $\{W_t\}_{t \in T}$ of the lifting homomorphisms of $(\mathcal{A}, \mathcal{J})$ is spectral or strictly spectral for the algebra $\mathcal{A}/\mathcal{G}$, respectively. For strictly spectral Silbermann pairs, the assertions of the lifting theorem can be completed as follows.

Theorem 5.6. *Let $(\mathcal{A}, \mathcal{J})$ be a strictly spectral Silbermann pair and $\mathbf{A} \in \mathcal{A}$. Then*

(a) *$\mathbf{A}$ is stable if and only if all operators $W_t(\mathbf{A})$ are invertible;*
(b) $\|\mathbf{A} + \mathcal{G}\|_{\mathcal{F}/\mathcal{G}} = \max_{t \in T} \|W_t(\mathbf{A})\|$.
(c) *$\mathbf{A}$ is $\mathcal{J}$-Fredholm if and only if all operators $W_t(\mathbf{A})$ are Fredholm and if there are only finitely many of them which are not invertible;*
(d) *$\mathbf{A} \in \mathcal{J}$ if and only if all operators $W_t(\mathbf{A})$ are compact and if, for each $\varepsilon > 0$, there are only finitely many of them with $\|W_t(\mathbf{A})\| > \varepsilon$.*

Proof. Assertion (a) is a re-formulation of the strict spectral condition. Assertion (b) is a consequence of (a) and of general properties of strictly spectral families of homomorphisms; see Theorem 5.39 in [6] (where strictly spectral families were called *sufficient*).

(c) The 'only if' part follows from the Lifting theorem 5.2 (d). Conversely, let $\mathbf{A} \in \mathcal{A}$ be a sequence for which all operators $W_t(\mathbf{A})$ are Fredholm and for which there is a finite subset T_0 of T which consists of all t such that $W_t(\mathbf{A})$ is not invertible. Then all operators $W_t(\mathbf{A}^*\mathbf{A})$ are Fredholm, and they are invertible if $t \notin T_0$. Let $t \in T_0$. Then $W_t(\mathbf{A}^*\mathbf{A})$ is a Fredholm operator of index 0. Hence, there is a compact operator K_t such that $W_t(\mathbf{A}^*\mathbf{A}) + K_t$ is invertible. Choose a sequence $\mathbf{K}_t \in \mathcal{J}$ with $W_t(\mathbf{K}_t) = K_t$ and $W_s(\mathbf{K}_t) = 0$ for $s \neq t$ (which is possible by the separation property in Theorem 5.2). Then the sequence $\mathbf{K} := \sum_{t \in T_0} \mathbf{K}_t$ belongs to the ideal $\mathcal{J}$, and all operators $W_t(\mathbf{A}^*\mathbf{A} + \mathbf{K})$ are invertible. By assertion (a), the sequence $\mathbf{A}^*\mathbf{A} + \mathbf{K}$ is stable. Similarly, one finds a sequence $\mathbf{L} \in \mathcal{J}$ such that $\mathbf{A}\mathbf{A}^* + \mathbf{L}$ is stable. Thus, the sequence $\mathbf{A}$ is invertible modulo $\mathcal{J}$, whence its $\mathcal{J}$-Fredholm property.

(d) The 'only if' part follows again from the Lifting theorem 5.2 (c). For the 'if' part, let $\mathbf{K} \in \mathcal{A}$ be a sequence such that, for every $\varepsilon > 0$, there are only finitely many $t \in T$ with $\|W_t(\mathbf{K})\| > \varepsilon$. For $n \in \mathbb{N}$, let T_n stand for the (finite) subset of

T which collects all t with $\|W_t(\mathbf{K})\| > 1/n$. For each $t \in T_n$, choose a sequence $\mathbf{K}^t \in \mathcal{J}$ with $W_t(\mathbf{K}^t) = W_t(\mathbf{K})$ and $W_s(\mathbf{K}^t) = 0$ for $s \neq t$ (employ the separation property in Theorem 5.2 again), and set $\mathbf{K}_n := \sum_{t \in T_n} \mathbf{K}^t$. Then $W_t(\mathbf{K} - \mathbf{K}_n) = 0$ for $t \in T_n$ and $W_t(\mathbf{K}_n) = 0$ for $t \notin T_n$. Hence, $\sup_{t \in T} \|W_t(\mathbf{K} - \mathbf{K}_n)\| \leq 1/n$ for every $n \in \mathbb{N}$. By Theorem 5.6 (b), the supremum in this estimate coincides with $\|\mathbf{K} - \mathbf{K}_n + \mathcal{G}\|_{\mathcal{F}/\mathcal{G}}$. Thus, $\mathbf{K}$ is the norm limit of a sequence in $\mathcal{J}$, whence $\mathbf{K} \in \mathcal{J}$. □

An example for a strictly spectral Silbermann pair is $(\mathcal{S}(\mathsf{T}(C)), \mathcal{J})$, with $\mathcal{J}$ specified as in the introduction. A sequence in $\mathcal{S}(\mathsf{T}(C))$ is Fredholm if and only if its strong limit is a Fredholm operator (note that $T(a)$ and $T(\tilde{a})$ are Fredholm operators only simultaneously). Equivalently, the sequence $\mathbf{A} := (P_nT(a)P_n + P_nKP_n + R_nLR_n + G_n)$ with $a \in C(\mathbb{T})$, K, L compact and $(G_n) \in \mathcal{G}$ is Fredholm if and only if $T(a)$ is a Fredholm operator. In this case,

$$\alpha(\mathbf{A}) = \dim \ker(T(a) + K) + \dim \ker(T(\tilde{a}) + L). \tag{5.6}$$

In particular, if $K = L = 0$ and if a is not the zero function, then

$$\begin{aligned}\alpha(\mathbf{A}) &= \dim \ker T(a) + \dim \ker T(\tilde{a}) \\ &= \max\{\dim \ker T(a), \dim \ker T(\tilde{a})\}.\end{aligned}$$

The second equality holds by a theorem of Simonenko and Coburn which states that one of the quantities $\dim \ker T(a)$ or $\dim \ker T(\tilde{a})$ is zero for each non-zero Toeplitz operator.

5.4. Silbermann algebras

We call a unital fractal C^*-subalgebra $\mathcal{A}$ of $\mathcal{F}$ a *Silbermann algebra* if $(\mathcal{A}, \mathcal{A} \cap \mathcal{K})$ is a spectral Silbermann pair; a Silbermann algebra $\mathcal{A}$ is called *strict* if the pair $(\mathcal{A}, \mathcal{A} \cap \mathcal{K})$ is strictly spectral. The following result is an immediate consequence of Theorem 5.6.

Theorem 5.7.

(a) *Let $\mathcal{A}$ be a Silbermann algebra. A sequence $\mathbf{A} \in \mathcal{A}$ is compact if and only if all operators $W_t(\mathbf{A})$ are compact and if, for each $\varepsilon > 0$, only a finite number of them has a norm greater than ε.*

Let $\mathcal{A}$ be a strict Silbermann algebra. A sequence $\mathbf{A} \in \mathcal{A}$ is

(b) *Fredholm if and only if all operators $W_t(\mathbf{A})$ are Fredholm and if only a finite number of them is not invertible;*

(c) *stable if and only if all operators $W_t(\mathbf{A})$ are invertible.*

The goal of this section is fractality and essential fractality properties of Silbermann algebras, which we prepare by two technical results. The first one describes the fractal algebras among the elementary subalgebras of $\mathcal{K}$.

Proposition 5.8. *Let $\mathcal{J}$ be a C^*-subalgebra of $\mathcal{K}$ which contains $\mathcal{G}$ properly and for which the quotient algebra $\mathcal{J}/\mathcal{G}$ is elementary. Then $\mathcal{J}$ is fractal if and only if $\mathcal{J}/\mathcal{G}$ is of positive weight.*

Proof. First let $\mathcal{J}/\mathcal{G}$ be of positive weight and let $\eta \in \mathfrak{S}_{\mathbb{N}}$. The mapping $W_\eta : \mathcal{J} \to \mathcal{F}_\eta/\mathcal{G}_\eta$, $\mathbf{J} \mapsto R_\eta \mathbf{J} + \mathcal{G}_\eta$ is a *-homomorphism which has the ideal $\mathcal{G}$ in its kernel. Since $\mathcal{J}/\mathcal{G}$ is elementary (thus, simple), either $\ker W_\eta = \mathcal{G}$ or $\ker W_\eta = \mathcal{J}$. The latter is impossible: the positivity of the weight implies that there is a sequence in $\mathcal{J}$ which consists of non-zero projections, and no restriction of that sequence can tend to zero in the norm. Thus, $\ker W_\eta = \mathcal{G}$, and the quotient homomorphism $W_\eta^\pi : \mathcal{J}/\mathcal{G} \to \mathcal{J}_\eta/\mathcal{G}_\eta$ which sends $\mathbf{J}+\mathcal{G}$ to $W_\eta(\mathbf{J})$ is a *-isomorphism between these algebras. Define

$$\pi_\eta : \mathcal{J}_\eta \to \mathcal{J}/\mathcal{G}, \quad R_\eta \mathbf{J} \mapsto (W_\eta^\pi)^{-1}(R_\eta \mathbf{J} + \mathcal{G}_\eta).$$

Then $\pi_\eta R_\eta$ is the canonical homomorphism from $\mathcal{J}$ onto $\mathcal{J}/\mathcal{G}$, whence the fractality of $\mathcal{J}$ follows.

Let now $\mathcal{J}$ be a fractal algebra and assume that $\mathcal{J}/\mathcal{G}$ is not of positive weight. Then the weight (α_n) of $\mathcal{J}/\mathcal{G}$ contains infinitely many zeros, say $\alpha_{\eta(n)} = 0$ for a certain $\eta \in \mathfrak{S}_{\mathbb{N}}$. Thus, every algebraic rank one projection p in $\mathcal{J}/\mathcal{G}$ lifts to a sequence (Π_n) of projections with $\Pi_{\eta(n)} = 0$ for all n. Then $(\Pi_n) \in \mathcal{G}$ by Theorem 2.2, i.e., $p = 0$. This is impossible, since $\mathcal{J}$ contains $\mathcal{G}$ properly. □

The following lemma is immediate from the definition of a fractal homomorphism.

Lemma 5.9. *Let $\mathcal{A}$ be a C^*-subalgebra of $\mathcal{F}$, $\mathcal{J}$ a closed ideal of $\mathcal{A}$ with $\mathcal{G} \subseteq \mathcal{J}$, $W : \mathcal{J} \to L(H)$ an irreducible representation of $\mathcal{J}$ with $\mathcal{G} \subseteq \ker W$, and $W' : \mathcal{A} \to L(H)$ the (unique) irreducible extension of W. If W is fractal, then W' is fractal.*

We say that a Silbermann pair $(\mathcal{A}, \mathcal{J})$ has *positive* or *constant local weight* if every elementary component of $\mathcal{J}/\mathcal{G}$ has a positive or constant weight, respectively. We call a Silbermann algebra $\mathcal{A}$ of *positive* or *constant local weight* if the Silbermann pair $(\mathcal{A}, \mathcal{A} \cap \mathcal{K})$ has the corresponding property. If $\mathcal{A}$ is a Silbermann algebra of constant local weight, then the numbers $\alpha_n(\mathbf{A})$ in (5.1) are independent of n.

Theorem 5.10. *Let $(\mathcal{A}, \mathcal{J})$ be a strictly spectral Silbermann pair of positive local weight. Then*

(a) *$\mathcal{A}$ is a fractal algebra,*
(b) *$\mathcal{J} = \mathcal{A} \cap \mathcal{K}$.*

Proof. (a) Let W_t be the lifting of an elementary component $\mathcal{J}_t$ of $\mathcal{J}$. We use the notation W_t also for the extensions of W_t to irreducible representations of $\mathcal{J}$ and $\mathcal{A}$. Since $(\mathcal{A}, \mathcal{J})$ is a strictly spectral Silbermann pair, the family $\{W_t\}_{t \in T}$ of the lifting homomorphisms of $\mathcal{J}/\mathcal{G}$ is strictly spectral for $\mathcal{A}/\mathcal{G}$. Moreover, all homomorphisms W_t are fractal. Indeed, since every elementary component of $\mathcal{J}$ is fractal by Proposition 5.8 and $\mathcal{G} \subseteq \ker W_t$, the representations W_t of $\mathcal{J}_t$ are fractal. Lemma 5.9 implies that the W_t are also fractal as representations of the algebra $\mathcal{A}$. Now the fractality of $\mathcal{A}$ comes from Theorem 2.3 (consider the homomorphism

$$W : \mathcal{A} \to \Pi_{t \in T} L(H_t), \quad a \mapsto (t \mapsto W_t(a))$$

which is unital and fractal and satisfies the invertibility condition in Theorem 2.3).

(b) $\mathcal{A}$ is fractal by assertion (a), so $(\mathcal{A} \cap \mathcal{K})/\mathcal{G}$ is a dual algebra by Corollary 4.3. Let $\{W_t\}_{t \in T}$ refer to the family of the lifting homomorphisms associated with that algebra. Since $\mathcal{J}/\mathcal{G}$ is a closed ideal of $(\mathcal{A} \cap \mathcal{K})/\mathcal{G}$, the lifting homomorphisms of $\mathcal{J}/\mathcal{G}$ form a subset $\{W_t\}_{t \in S}$ of $\{W_t\}_{t \in T}$, with S a non-empty subset of T.

Let $\mathbf{A} \in \mathcal{A}$ be invertible modulo $\mathcal{A} \cap \mathcal{K}$. By Theorem 5.6 (c), all operators $W_t(\mathbf{A})$ are Fredholm, and only a finite number of them is not invertible. Hence, all operators $W_t(\mathbf{A})$ with $t \in S$ are Fredholm, and all but finitely many of them are invertible. Again by By Theorem 5.6 (c), the sequence $\mathbf{A}$ is invertible modulo $\mathcal{J}$. Hence, invertibility modulo $\mathcal{A} \cap \mathcal{K}$ is equivalent to invertibility modulo $\mathcal{J}$ for sequences in $\mathcal{A}$, which implies $\mathcal{A} \cap \mathcal{K} = \mathcal{J}$. □

Corollary 5.11. *If $(\mathcal{A}, \mathcal{J})$ is a strictly spectral Silbermann pair of positive local weight, then $\mathcal{A}$ is a strict Silbermann algebra.*

Corollary 5.12. *Let $\mathcal{J}$ be a C^*-subalgebra of $\mathcal{F}$ with the following properties:*

(a) $\mathcal{G} \subset \mathcal{J} \subset \mathcal{K}$,
(b) *$\mathcal{J}/\mathcal{G}$ is a dual algebra,*
(c) *$\mathcal{J}$ is of positive local weight, and*
(d) *at least one elementary component of $\mathcal{J}/\mathcal{G}$ has infinite dimension.*

Then the minimal unitization $\mathbb{C}\mathbf{I} + \mathcal{J}$ of $\mathcal{J}$ is fractal.

Indeed, by Theorem 5.10, we have to show that $(\mathbb{C}\mathbf{I} + \mathcal{J}, \mathcal{J})$ is a strictly spectral Silbermann pair. But this follows easily from the lifting theorem. □

Next we turn our attention to relations between a strict Silbermann algebra $\mathcal{A}$ and its restriction $R_\eta \mathcal{A} = \mathcal{A}_\eta$.

Theorem 5.13. *Let $\mathcal{A}$ be a unital and fractal C^*-subalgebra of $\mathcal{F}$ which contains the ideal $\mathcal{G}$ and let $\eta \in \mathfrak{S}_{\mathbb{N}}$. Then the following assertions are equivalent:*

(a) *$\mathcal{A}$ is a strict Silbermann algebra.*
(b) *$(\mathcal{A}_\eta, (\mathcal{A} \cap \mathcal{K})_\eta)$ is a strictly spectral Silbermann pair of a positive local weight.*

Proof. First we show that if $\mathcal{A}$ is fractal, then $(\mathcal{A}_\eta, (\mathcal{A} \cap \mathcal{K})_\eta)$ is a Silbermann pair of a positive local weight. If $\mathcal{A}$ is fractal, its restriction $\mathcal{A}_\eta$ is fractal. Then, by Corollary 4.3, $(\mathcal{A}_\eta \cap \mathcal{K}_\eta)/\mathcal{G}_\eta$ is a dual algebra. Being a closed ideal of a dual algebra, the algebra $(\mathcal{A} \cap \mathcal{K})_\eta$ is dual itself, and this algebra consists of compact sequences in $\mathcal{F}_\eta$ only. Hence, $(\mathcal{A}_\eta, (\mathcal{A} \cap \mathcal{K})_\eta)$ is a Silbermann pair.

Let p be a minimal projection in $(\mathcal{A} \cap \mathcal{K})_\eta/\mathcal{G}_\eta$, and let $(P_{\eta(n)})$ be a lifting of p to a sequence of projections. Since $\mathcal{A}_\eta$ is fractal, this sequence is fractal. Theorem 2.2 (a) then implies that $P_{\eta(n)} \neq 0$ for all sufficiently large n. Hence, the weight of the elementary component containing p is positive.

Next we show that for every fractal algebra $\mathcal{A}$ and every $\eta \in \mathfrak{S}_{\mathbb{N}}$, there is a natural *-isomorphism

$$\xi_\eta : \mathcal{A}/\mathcal{G} \to \mathcal{A}_\eta/\mathcal{G}_\eta, \quad \mathbf{A} + \mathcal{G} \mapsto (R_\eta \mathbf{A}) + \mathcal{G}_\eta.$$

Indeed, ξ_η is well defined since $\mathbf{A} - \mathbf{B} \in \mathcal{G}$ implies that $(R_\eta \mathbf{A}) - (R_\eta \mathbf{B}) \in \mathcal{G}_\eta$, and ξ_η has a trivial kernel since $R_\eta \mathbf{A} \in \mathcal{G}_\eta$ implies $\mathbf{A} \in \mathcal{G}$ via Theorem 2.2. The isomorphism ξ_η maps $(\mathcal{A}\cap\mathcal{K})/\mathcal{G}$ onto $(\mathcal{A}\cap\mathcal{K})_\eta/\mathcal{G}_\eta$. Hence, both ideals are canonically isomorphic, which implies that if W is one of the lifting homomorphisms of the Silbermann pair $(\mathcal{A},\ \mathcal{A}\cap\mathcal{K})$, then $W\xi_\eta^{-1}$ is a lifting homomorphism of the pair $(\mathcal{A}_\eta, (\mathcal{A}\cap\mathcal{K})_\eta)$ and, conversely, if W_η is a lifting homomorphism of $(\mathcal{A}_\eta, (\mathcal{A}\cap\mathcal{K})_\eta)$, then $W_\eta\xi_\eta$ is a lifting homomorphism of $(\mathcal{A},\ \mathcal{A}\cap\mathcal{K})$.

Now it is easy to see that the strictly spectral hypotheses in (a) and (b) imply each other. For example, we verify that (b) $\Rightarrow$ (a); the reverse implication follows similarly. Let the Silbermann pair $(\mathcal{A}_\eta,\ (\mathcal{A}\cap\mathcal{K})_\eta)$ be strictly spectral, and let $\{W_t\}_{t\in T}$ refer to the family of the lifting homomorphisms of the Silbermann pair $(\mathcal{A},\ \mathcal{A}\cap\mathcal{K})$. Let $\mathbf{A}\in\mathcal{A}$ be a sequence for which all operators $W_t(\mathbf{A}+\mathcal{G})$ are invertible. Then all operators $W_t\xi_\eta^{-1}\xi_\eta(\mathbf{A}+\mathcal{G})$ with $t\in T$ are invertible. Since the $W_t\xi_\eta^{-1}$ run through the lifting homomorphisms of the pair $(\mathcal{A}_\eta,\ (\mathcal{A}\cap\mathcal{K})_\eta)$, and since this pair is strictly spectral, the restricted coset $\xi_\eta(\mathbf{A}+\mathcal{G})$ is invertible in $\mathcal{A}_\eta/\mathcal{G}_\eta$. Then $\mathbf{A}+\mathcal{G}$ is invertible in $\mathcal{A}/\mathcal{G}$, since ξ_η is an isomorphism. □

Corollary 5.14. *Any restriction of a strict Silbermann algebra is again a strict Silbermann algebra.*

Proof. Let $\mathcal{A}$ be a strict Silbermann algebra and $\eta\in\mathfrak{S}_\mathbb{N}$. Then $(\mathcal{A}_\eta,\ (\mathcal{A}\cap\mathcal{K})_\eta)$ is a strictly spectral Silbermann pair, and every elementary component of $(\mathcal{A}\cap\mathcal{K})_\eta)/\mathcal{G}_\eta$ has positive weight by Theorem 5.13. By Corollary 5.11, $\mathcal{A}_\eta$ is a strict Silbermann algebra. □

Corollary 5.15. *Any strict Silbermann algebra is essentially fractal.*

Indeed, the canonical homomorphism $\mathcal{A}\to\mathcal{A}/(\mathcal{A}\cap\mathcal{K})$ is fractal since $\mathcal{A}$ is fractal and $\mathcal{G}$ is in the kernel of that homomorphism. Further, $(\mathcal{A}\cap\mathcal{K})_\eta=\mathcal{A}_\eta\cap\mathcal{K}_\eta$ by Theorem 5.10 (b). □

Corollary 5.16. *Let* $\mathbf{A}$ *be a Fredholm sequence in a strict Silbermann algebra of constant local weight. Then every restriction* $\mathbf{A}_\eta$ *of* $\mathbf{A}$ *is Fredholm, and the sequences* $\mathbf{A}$ *and* $\mathbf{A}_\eta$ *have the same* α*-number.*

5.5. Forcing the spectral property

Again, let $\mathcal{F}:=\mathcal{F}^\delta$ be an algebra of matrix sequences with dimension function δ and $\mathcal{G}:=\mathcal{G}^\delta$ the associated ideal of zero sequences. We say that a C^*-subalgebra $\mathcal{A}^{\text{ext}}$ of $\mathcal{F}$ is an *extension* of a C^*-subalgebra $\mathcal{A}$ of $\mathcal{F}$ *by compact sequences* if there is a subset $\mathcal{K}'$ of the ideal $\mathcal{K}$ of the compact sequences in $\mathcal{F}$ such that $\mathcal{A}^{\text{ext}}$ is the smallest C^*-subalgebra of $\mathcal{F}$ which contains $\mathcal{A}$ and $\mathcal{K}'$. The goal of this section is to prove the following result.

Theorem 5.17. *Let* $\mathcal{A}$ *be a unital separable* C^**-subalgebra of* $\mathcal{F}$*. Then there is an extension* $\mathcal{A}^{\text{ext}}$ *of* $\mathcal{A}$ *by compact sequences and an* $\eta\in\mathfrak{S}_\mathbb{N}$ *such that the restriction* $\mathcal{A}^{\text{ext}}_\eta$ *is a Silbermann algebra.*

In other words, after extending $\mathcal{A}$ by a suitable set of compact sequences and then passing to a suitable restriction, we arrive at a spectral Silbermann pair $(\mathcal{A}^{\text{ext}}_\eta,\ \mathcal{A}^{\text{ext}}_\eta\cap\mathcal{K}_\eta)$.

Proof. Let $\mathcal{A}_0$ be a countable dense subset of $\mathcal{A}$ containing the identity sequence. Then the set $\mathcal{A}_0^* \mathcal{A}_0$ is also countable and dense in $\mathcal{A}$.

For each sequence $\mathbf{A} = (A_n)$ in $\mathcal{A}_0$, we write

$$A_n^* A_n = E_n^* \mathrm{diag}\,(\lambda_1(A_n^* A_n), \ldots, \lambda_{\delta(n)}(A_n^* A_n))E_n \tag{5.7}$$

with a unitary matrix E_n and increasingly ordered eigenvalues

$$0 \le \lambda_1(A_n^* A_n) \le \cdots \le \lambda_{\delta(n)}(A_n^* A_n)$$

of $A_n^* A_n$. For $l, r \in \mathbb{N}$, let $K_{l,r,n}$ be the $\delta(n) \times \delta(n)$-matrix which is zero if $\max\{l, r\} > \delta(n)$ and which has a 1 at the lrth entry and zeros at all other entries if $\max\{l, r\} \le \delta(n)$. The sequence $\mathbf{K}^{\mathbf{A},l,r}$ with entries $K_n^{\mathbf{A},l,r} := E_n^* K_{l,r,n} E_n$ (note that E_n depends on $\mathbf{A}$) is a sequence of matrices of rank at most 1, hence compact. Let $\mathcal{A}^{\text{ext}}$ stand for the smallest C^*-subalgebra of $\mathcal{F}$ which contains the algebra $\mathcal{A}$, the ideal $\mathcal{G}$, and all sequences $\mathbf{K}^{\mathbf{A},l,r}$ with $\mathbf{A} \in \mathcal{A}_0$ and $l, r \in \mathbb{N}$. This algebra is also separable. Hence, by Theorem 2.6 and Theorem 3.10, there is an $\eta \in \mathfrak{S}_{\mathbb{N}}$ such that the restriction $\mathcal{A}_\eta^{\text{ext}}$ is both fractal and essentially fractal. We claim that $\mathcal{A}_\eta^{\text{ext}}$ is a Silbermann algebra. Note that the sequences $\mathbf{K}^{\mathbf{A},r} := \sum_{l=1}^r \mathbf{K}^{\mathbf{A},l,l}$ with entries

$$K_n^{\mathbf{A},r} := E_n^* \mathrm{diag}\,(1, \ldots, 1, 0, \ldots, 0)E_n \tag{5.8}$$

where r ones followed by $\delta(n) - r$ zeros belong to $\mathcal{A}^{\text{ext}}$.

To simplify notation, we will assume that η is the identity mapping (otherwise replace δ by $\delta \circ \eta$ in what follows). Let $\mathcal{J} := \mathcal{A}^{\text{ext}} \cap \mathcal{K}$.

Since $\mathcal{A}$ is a C^*-subalgebra and $\mathcal{J}$ is a closed ideal of $\mathcal{A}^{\text{ext}}$, the algebraic sum $\mathcal{A} + \mathcal{J}$ is a C^*-subalgebra of $\mathcal{A}^{\text{ext}}$. This subalgebra contains $\mathcal{A}$, $\mathcal{G}$ and all sequences $\mathbf{K}^{\mathbf{A},l,r}$. Thus, $\mathcal{A}^{\text{ext}} = \mathcal{A} + \mathcal{J}$.

Since $\mathcal{A}^{\text{ext}}$ is fractal, the ideal $\mathcal{J}/\mathcal{G}$ is dual by Corollary 4.3. Let $(I_t)_{t \in T}$ denote the set of its elementary components and let $W_t : I_t \to L(H_t)$ stand for the irreducible representation associated with I_t. As before, we will denote an irreducible representation of I_t and its irreducible extensions to $\mathcal{A}^{\text{ext}}/\mathcal{G}$ and $\mathcal{A}^{\text{ext}}$ by the same symbol.

Let $\mathbf{A} \in \mathcal{A}_0$. We claim that the coset $\mathbf{K}^{\mathbf{A},1,1} + \mathcal{G} = \mathbf{K}^{\mathbf{A},1} + \mathcal{G}$ is an algebraic rank one projection in $(\mathcal{A}^{\text{ext}} \cap \mathcal{K})/\mathcal{G}$. Indeed, the entries $K_n^{\mathbf{A},1}$ are projection matrices of rank one. Hence, for every positive sequence $(B_n^* B_n) \in \mathcal{A}^{\text{ext}}$, there is a sequence (β_n) of complex numbers such that

$$K_n^{\mathbf{A},1} B_n^* B_n K_n^{\mathbf{A},1} = \beta_n K_n^{\mathbf{A},1} \quad \text{for every } n \in \mathbb{N}.$$

The sequence $(\beta_n K_n^{\mathbf{A},1})_{n \in \mathbb{N}}$ is fractal, and β_n is the largest singular value of $\beta_n K_n^{\mathbf{A},1}$. By Proposition 2.4 (b), the sequence (β_n) is convergent. Since every sequence in $\mathcal{A}^{\text{ext}}$ is a linear combination of four positive sequences, we conclude that, for every sequence $\mathbf{C} = (C_n) \in \mathcal{A}^{\text{ext}}$, there is a convergent sequence (γ_n) of complex numbers such that

$$K_n^{\mathbf{A},1} C_n K_n^{\mathbf{A},1} = \gamma_n K_n^{\mathbf{A},1} \quad \text{for every } n \in \mathbb{N}.$$

Put $\gamma := \lim_{n\to\infty}\gamma_n$. Then $\mathbf{K}^{\mathbf{A},1}\mathbf{C}\mathbf{K}^{\mathbf{A},1} - \gamma\mathbf{K}^{\mathbf{A},1} \in \mathcal{G}$, which proves the claim.

Since the elementary components of $\mathcal{J}/\mathcal{G}$ are generated by algebraic rank one projections, there is a $t(\mathbf{A}) \in T$ such that $\mathbf{K}^{\mathbf{A},1} + \mathcal{G} \in I_{t(\mathbf{A})}$. Since $I_{t(\mathbf{A})}$ is an ideal, the equality

$$\mathbf{K}^{\mathbf{A},l,r} = \mathbf{K}^{\mathbf{A},l,1}\mathbf{K}^{\mathbf{A},1,1}\mathbf{K}^{\mathbf{A},1,r}$$

implies that $\mathbf{K}^{\mathbf{A},l,r} + \mathcal{G} \in I_{t(\mathbf{A})}$ for every pair $l, r \in \mathbb{N}$. In particular, all cosets $\mathbf{K}^{\mathbf{A},r} + \mathcal{G}$ belong to $I_{t(\mathbf{A})}$. Since the cosets $\mathbf{K}^{\mathbf{A},l,l} + \mathcal{G}$ are linearly independent algebraic rank one projections and the representation $W_{t(\mathbf{A})}$ is irreducible, the operators $W_{t(\mathbf{A})}(\mathbf{K}^{\mathbf{A},l,l})$ form a linearly independent set of projection operators of rank one. Hence, the Hilbert space $H_{t(\mathbf{A})}$ has infinite dimension.

Since $\mathcal{A}_0$ is dense in $\mathcal{A}$, the sequences $\mathbf{A}+\mathbf{K}$ with $\mathbf{A} \in \mathcal{A}_0$ and $\mathbf{K} \in \mathcal{A}^{\text{ext}} \cap \mathcal{K}$ form a dense subset $\mathcal{A}_0^{\text{ext}}$ of $\mathcal{A}^{\text{ext}}$. Let $\mathbf{B} := \mathbf{A}+\mathbf{K}$ be a sequence of this form, for which $\mathbf{B}^*\mathbf{B}$ is not a Fredholm sequence (equivalently, $\mathbf{B}^*\mathbf{B}$ is not a $\mathcal{J}$-Fredholm sequence, since $\mathcal{J}$ contains all compact sequences in $\mathcal{A}^{\text{ext}}$). Then $\mathbf{A}^*\mathbf{A} = (A_n^*A_n)$ is not a Fredholm sequence, hence

$$\lim_{n\to\infty} \lambda_r(A_n^*A_n) = 0 \quad \text{for every } r \in \mathbb{N} \tag{5.9}$$

by (3.4) (recall (5.7) and remember that $\mathcal{A}^{\text{ext}}$ is essentially fractal after restriction). From (5.7)–(5.9) we conclude that $\mathbf{A}^*\mathbf{A}\mathbf{K}^{\mathbf{A},r} \in \mathcal{G}$ for every $r \in \mathbb{N}$, hence

$$W_{t(\mathbf{A})}(\mathbf{A}^*\mathbf{A})W_{t(\mathbf{A})}(\mathbf{K}^{\mathbf{A},r}) = 0 \quad \text{for every } r \in \mathbb{N}. \tag{5.10}$$

Since $(W_{t(\mathbf{A})}(\mathbf{K}^{\mathbf{A},r}))_{r\ge1}$ is an increasing sequence of orthogonal projections on $H_{t(\mathbf{A})}$, this sequence converges strongly, and its limit, P, is the orthogonal projection from $H_{t(\mathbf{A})}$ onto the closure of the linear span of the union of the ranges of the $W_{t(\mathbf{A})}(\mathbf{K}^{\mathbf{A},r})$ (see, for example, Theorem 4.1.2 in [10]). So we conclude from (5.10) that $W_{t(\mathbf{A})}(\mathbf{A}^*\mathbf{A})P = 0$. Thus, and by Theorem 5.2 (c),

$$W_{t(\mathbf{A})}(\mathbf{B}^*\mathbf{B})P = W_{t(\mathbf{A})}(\mathbf{A}^*\mathbf{A})P + W_{t(\mathbf{A})}(\mathbf{A}^*\mathbf{K} + \mathbf{K}^*\mathbf{A} + \mathbf{K}^*\mathbf{K})P$$

is a compact operator. Then $W_{t(\mathbf{A})}(\mathbf{B}^*\mathbf{B})$ cannot be invertible: otherwise, the projection P were compact, but the range of P has infinite dimension, which follows by the same arguments as the infinite dimensionality of $H_{t(\mathbf{A})}$.

Thus, whenever $\mathbf{B} \in \mathcal{A}_0^{\text{ext}}$ and $\mathbf{B}^*\mathbf{B}$ is not a Fredholm sequence, then one of the operators $W_t(\mathbf{B}^*\mathbf{B})$ is not invertible. Conversely, if all operators $W_t(\mathbf{B}^*\mathbf{B})$ with $t \in T$ are invertible, then $\mathbf{B}^*\mathbf{B}$ is a Fredholm sequence. By Theorem 5.2 (a) this implies that, whenever all operators $W_t(\mathbf{B}^*\mathbf{B})$ with $t \in T$ are invertible, then the sequence $\mathbf{B}^*\mathbf{B}$ is a stable. This fact holds for all sequences $\mathbf{B}$ in the dense subset $\mathcal{A}_0^{\text{ext}}$ of $\mathcal{A}^{\text{ext}}$, from which it is easy to conclude that the family $(W_t)_{t\in T}$ is spectral for $\mathcal{A}^{\text{ext}}$. □

It is not clear if one can make $\mathcal{A}^{\text{ext}}$ to a *strict* Silbermann algebra by a suitable restriction. The point is that the implication, obtained at the end of the previous proof, is verified only for sequences in a dense subset of $\mathcal{A}^{\text{ext}}$. Another open question is if there is a version of Theorem 5.17 which works without restriction if one already starts with a fractal and essentially fractal separable subalgebra $\mathcal{A}$ of $\mathcal{F}$.

5.6. Forcing local rank one

Our final goal is forcing the local rank one property by a suitable extension and restriction. The following result on restrictions of Silbermann pairs has been already shown as part of the proof of Theorem 5.13.

Proposition 5.18. *Let $(\mathcal{A}, \mathcal{J})$ be a Silbermann pair (in $\mathcal{F}$) and $\eta \in \mathfrak{S}_{\mathbb{N}}$. Then $(\mathcal{A}_\eta, \mathcal{J}_\eta)$ is a Silbermann pair (in $\mathcal{F}_\eta$).*

Let $\mathcal{J}/\mathcal{G} \cong \oplus_{t \in T} K(H_t)$ be the representation of the dual algebra $\mathcal{J}/\mathcal{G}$ as a direct sum of elementary components. Since these components are simple, every homomorphic image of $K(H_t)$ is either $\{0\}$ or isomorphic to $K(H_t)$. Given $\eta \in \mathfrak{S}_{\mathbb{N}}$, let T_η denote the set of all $t \in T$ which are *not* mapped to the zero ideal under the homomorphism

$$(A_n) + \mathcal{G} \mapsto (A_{\eta(n)}) + \mathcal{G}_\eta. \tag{5.11}$$

The following lemma says that the weights of $\mathcal{J}/\mathcal{G}$ behave naturally under the mapping (5.11), provided that $\mathcal{J}$ is fractal. Recall from Corollary 5.12 that the latter condition holds if all elementary components of $\mathcal{J}/\mathcal{G}$ have positive weight and if at least one of them has infinite dimension.

Lemma 5.19. *Let $\mathcal{J}$ be a fractal algebra and $t \in T_\eta$. Let $\mathcal{J}_{t,\eta}$ be the image of the elementary component $\mathcal{J}_t \cong K(H_t)$ of $\mathcal{J}/\mathcal{G}$ under the homomorphism* (5.11). *Then $\mathcal{J}_{t,\eta} \cong K(H_t)$ is elementary, and the weight of $\mathcal{J}_{t,\eta}$ is the restriction α_η of the weight α of $\mathcal{J}_t$.*

Proof. Let $(A_{\eta(n)})+\mathcal{G}_\eta$ be a projection of algebraic rank one in $\mathcal{J}_{t,\eta}$. The fractality of $\mathcal{J}$ ensures that the quotient homomorphism (5.11) is an isomorphism from $\mathcal{J}/\mathcal{G}$ to $\mathcal{J}_\eta/\mathcal{G}_\eta$ and, thus, from $\mathcal{J}_t$ to $\mathcal{J}_{t,\eta}$. Hence, the pre-image $(A_n) + \mathcal{G} \in \mathcal{J}_t$ of $(A_{\eta(n)})+\mathcal{G}_\eta$ is uniquely determined, and it is a projection of algebraic rank one. It is clear that if (Π_n) is a projection which lifts $(A_n)+\mathcal{G}$, then $(\Pi_{\eta(n)})$ is a projection which lifts $(A_{\eta(n)}) + \mathcal{G}_\eta$. This is the assertion. □

In the next result we will see how to force local constant weight (recall that the latter means that every elementary component of $\mathcal{J}/\mathcal{G}$ has constant weight).

Theorem 5.20. *Let $(\mathcal{A}, \mathcal{J})$ be a Silbermann pair and assume that $\mathcal{J}$ is separable and fractal. Then there is an $\eta \in \mathfrak{S}_{\mathbb{N}}$ such that the Silbermann pair $(\mathcal{A}_\eta, \mathcal{J}_\eta)$ has constant local weight.*

Proof. Since $\mathcal{J}$ is separable, the number of the elementary components of $\mathcal{J}/\mathcal{G}$ is at most countable. We enumerate these components by $\mathcal{J}^{(1)}$, $\mathcal{J}^{(2)}$, ... and denote the weight of $\mathcal{J}^{(i)}$ by $\alpha^{(i)}$. Every weight $\alpha^{(i)}$ is bounded and has, thus, a constant subsequence. This fact is used in the following construction.

Starting with the identity mapping η_0 on $\mathbb{N}$, there is a subsequence η_1 of η_0 such that the restriction $\alpha_{\eta_1}^{(1)}$ is a positive constant. We continue in this way and get, for every $k \geq 1$, a subsequence η_k of η_{k-1} such that the restriction $\alpha_{\eta_k}^{(k)}$ is a positive constant. If the number of elementary components of $\mathcal{J}/\mathcal{G}$ is finite, we let η be the last of the sequences η_k obtained in this way; otherwise we set $\eta(n) := \eta_n(n)$ for $n \in \mathbb{N}$. In each case, every restriction $\alpha_\eta^{(i)}$ is a positive constant. Then $(\mathcal{A}_\eta, \mathcal{J}_\eta)$ is a Silbermann pair by Proposition 5.18, and the local weights of that pair are just the η-restrictions of the weights of $(\mathcal{A}, \mathcal{J})$ by Lemma 5.19. Thus, $(\mathcal{A}_\eta, \mathcal{J}_\eta)$ is a Silbermann pair with constant local weight. □

Let $\mathcal{A}$ be a separable Silbermann algebra with a constant local weight. Again, we denote the elementary components of $(\mathcal{A} \cap \mathcal{K})/\mathcal{G}$ by $\mathcal{J}^{(1)}, \mathcal{J}^{(2)}, \ldots$ and write $\alpha^{(i)}$ for the weight of $\mathcal{J}^{(i)}$, which now can be identified with a non-negative integer. For every i, let $p_i \in \mathcal{J}^{(i)}$ be a minimal (hence, algebraic rank one) projection. Given p_i, choose a projection $(\Pi_n^{(i)})_{n\geq 1} \in \mathcal{A}$ which lifts p_i and which is specified such that $\dim \operatorname{im} \Pi_n^{(i)} = \alpha^{(i)}$ for all n, and write $\Pi_n^{(i)}$ as a sum $\sum_{k=1}^{\alpha^{(i)}} \Pi_n^{(i,k)}$ of projections with $\dim \operatorname{im} \Pi_n^{(i,k)} = 1$. For example, choose unitary matrices $E_n^{(i)}$ such that

$$\Pi_n^{(i)} = E_n^{(i)} \operatorname{diag}(1, \ldots, 1, 0, \ldots, 0)(E_n^{(i)})^*,$$

with $\alpha^{(i)}$ of ones followed by $\delta(n) - \alpha^{(i)}$ of zeros. Then set

$$\Pi_n^{(i,k)} := E_n^{(i)} \operatorname{diag}(0, \ldots, 0, 1, 0, \ldots, 0)(E_n^{(i)})^*,$$

with the 1 at the kth position. $(\Pi_n^{(i,k)})_{n\geq 1}$ is a sequence of rank one projections in $\mathcal{F}$. Let $\tilde{\mathcal{A}}$ be the smallest closed subalgebra of $\mathcal{F}$ which contains the algebra $\mathcal{A}$ and all sequences $(\Pi_n^{(i,k)})_{n\geq 1}$. Since $\tilde{\mathcal{A}}$ is again a separable C^*-algebra, the fractal restriction theorem applies.

Theorem 5.21. *Let $\mathcal{A}$ be a separable Silbermann algebra and let $\eta \in \mathfrak{S}_\mathbb{N}$ be such that $\tilde{\mathcal{A}}_\eta$ is fractal. Then the algebra $\tilde{\mathcal{A}}_\eta$ obtained in this way is a Silbermann algebra of local weight one.*

Proof. For brevity, we write $\mathcal{A}$ and $\tilde{\mathcal{A}}$ for $\mathcal{A}_\eta$ and $\tilde{\mathcal{A}}_\eta$, respectively, and set $\mathcal{J} := \mathcal{A} \cap \mathcal{K}$ and $\tilde{\mathcal{J}} := \tilde{\mathcal{A}} \cap \mathcal{K}$. Then $\tilde{\mathcal{A}}$ is both a fractal algebra and an extension of $\mathcal{A}$ by compact sequences. In particular,

$$\tilde{\mathcal{A}}/\tilde{\mathcal{J}} = (\mathcal{A} + \tilde{\mathcal{J}})/\tilde{\mathcal{J}} \cong \mathcal{A}/(\mathcal{A} \cap \tilde{\mathcal{J}}) = \mathcal{A}/(\mathcal{A} \cap \mathcal{K}) = \mathcal{A}/\mathcal{J}.$$

We are going to relate the irreducible representations of $\mathcal{A}$, associated with the elementary components of $\mathcal{J}/\mathcal{G}$, with the irreducible representations of $\tilde{\mathcal{A}}$, associated with the elementary components of $\tilde{\mathcal{J}}/\mathcal{G}$.

Let $W : \mathcal{J} \to L(K)$ be an irreducible representation of $\mathcal{J}$ associated with an elementary component of $\mathcal{J}/\mathcal{G}$. Let $x \in K$ be a cyclic vector. We extend W to an irreducible representation $\tilde{W}$ of $\mathcal{A}$ by setting $\tilde{W}(a)W(j)x := W(aj)x$ for $a \in \mathcal{A}$ and $j \in \mathcal{J}$.

Since $\mathcal{G}$ is in the kernel of W, the quotient of $\tilde{W}$ by $\mathcal{G}$ is well defined; we denote it by $\tilde{W}$ again. Then there is an irreducible representation $\pi : \tilde{\mathcal{J}} \to L(\tilde{H})$, a closed subspace H of $\tilde{H}$ which is invariant with respect to $\pi(\mathcal{J})$, and a unitary operator $U : H \to K$ such that

$$W(j) = U\pi(j)|_H U^* \quad \text{for } j \in \mathcal{J}. \tag{5.12}$$

Set $U^*x =: y \in H$. Then $y \neq 0$; hence y is cyclic for π, and we can extend π to an irreducible representation $\tilde{\pi} : \tilde{\mathcal{A}} \to L(\tilde{H})$ by

$$\tilde{\pi}(a)\pi(j)y := \pi(aj)y \quad \text{for } a \in \tilde{\mathcal{A}},\ j \in \tilde{\mathcal{J}}.$$

From (5.12) we conclude that $W(aj)x = U\pi(aj)|_H U^*x$ for $j \in \mathcal{J}$ and $a \in \mathcal{A}$; hence,

$$\begin{aligned}\tilde{W}(a)W(j)x &= W(aj)x \\ &= U\pi(aj)|_H U^*x \\ &= U\tilde{\pi}(a)\pi(j)|_H y \\ &= U\tilde{\pi}(a)|_H U^*U\pi(j)U^*x \\ &= U\tilde{\pi}(a)|_H U^*W(j)x.\end{aligned}$$

Since the vectors $W(j)x$ lie dense in K, we conclude that H is an invariant subspace for $\tilde{\pi}(\mathcal{A})$ and $\tilde{W}(a) = U\tilde{\pi}(a)|_H U^*$ for $a \in \mathcal{A}$. Note that every $\tilde{\pi}$ obtained in this way is an irreducible representation of $\tilde{\mathcal{J}}$ with $\mathcal{G}$ in its kernel; so it is associated with an elementary component of $\tilde{\mathcal{J}}/\mathcal{G}$.

Let $(I_t)_{t\in T}$ be the elementary components of $\mathcal{J}/\mathcal{G}$ and $W_t : \mathcal{A} \to L(K_t)$ be the lifting homomorphism associated with I_t. For every $t \in T$, we construct an extension of W_t to an irreducible representation of $\tilde{\mathcal{A}}$ as above. In general, this extension will not be unique. For a given irreducible representation $\tilde{\pi}$ of $\tilde{\mathcal{A}}$ obtained from an irreducible representation of $\tilde{\mathcal{J}}/\mathcal{G}$, we let $T_{\tilde{\pi}}$ denote the set of all $t \in T$ for which the above extension of W_t leads to $\tilde{\pi}$, i.e., for which there are a closed subspace H_t of $\tilde{H}$ invariant with respect to $\tilde{\pi}(\mathcal{A})$ and a unitary operator $U_t : H_t \to K_t$ such that

$$W_t(a) = U_t\tilde{\pi}(a)|_{H_t} U_t^* \quad \text{for } a \in \mathcal{A}. \tag{5.13}$$

We claim that the Hilbert spaces H_t with $t \in T_{\tilde{\pi}}$ are pairwise orthogonal. Let $s, t \in T_{\tilde{\pi}}$ and $s \neq t$. Inserting $a \in I_s$ into (5.13) we get $0 = W_t(a) = U_t\tilde{\pi}(a)|_{H_t} U_t^*$ whence $\tilde{\pi}(I_s)|_{H_t} = 0$. On the other hand, let $0 \neq x_s \in K_s$. Since $W_s : I_s \to L(K_s)$ is irreducible, x_s is algebraically cyclic, i.e., $W_s(I_s)x_s = K_s$. With $y_s := U_s x_s$ we obtain

$$H_s = U_s^*K_s = U_s^*W_s(I_s)x_s = \tilde{\pi}(I_s)U_s x_s = \tilde{\pi}(I_s)y_s.$$

Let now $z_s \in H_s$ and $z_t \in H_t$. By the previous line, there is a $j_s \in I_s$ such that $z_s = \tilde{\pi}(j_s)y_s$. Then

$$\langle z_s, z_t\rangle = \langle \tilde{\pi}(j_s)y_s, z_t\rangle = \langle y_s, \tilde{\pi}(j_s)^*z_t\rangle = \langle y_s, \tilde{\pi}(j_s^*)z_t\rangle = 0$$

because of $\tilde{\pi}(I_s)|_{H_t} = 0$ as observed above. This proves the claim.

Now we can finish the proof of the theorem. Let $\tilde{\mathbf{A}} \in \tilde{\mathcal{A}}$ be a sequence such that all operators $\tilde{\pi}(\tilde{\mathbf{A}})$ are invertible and the norms of their inverses are uniformly bounded by a constant C, where $\tilde{\pi}$ runs through the irreducible representations of $\tilde{\mathcal{A}}$ coming from the elementary components of $\tilde{\mathcal{J}}/\mathcal{G}$. Write $\tilde{\mathbf{A}}$ as $\mathbf{A} + \tilde{\mathbf{J}}$ where $\mathbf{A} \in \mathcal{A}$ and $\tilde{\mathbf{J}} \in \tilde{\mathcal{J}}$ (recall that $\tilde{\mathcal{A}}$ is an extension of $\mathcal{A}$ by compact sequences). Since $\tilde{\mathcal{J}}$ is dual, there are only finitely many π with $\|\pi(\tilde{\mathbf{J}})\| \geq 1/(2C)$. Thus, all operators $\tilde{\pi}(\mathbf{A})$ are Fredholm, only a finite number of them is not invertible, and the norms of their inverses are uniformly bounded.

From (5.13) we then conclude that all operators $W_t(\mathbf{A})$ are Fredholm, that only a finite number of them is not invertible, and that the norms of their inverses are uniformly bounded. Then all operators $W_t(\mathbf{A}^*\mathbf{A})$ are Fredholm with index 0, only a finite number of them is not invertible, and the norms of their inverses are uniformly bounded. Thus, there is a sequence $\mathbf{J} \in \mathcal{J}$ such that all operators $W_t(\mathbf{A}^*\mathbf{A}+\mathbf{J})$ are invertible and the norms of their inverses are uniformly bounded; the (finite) sum $\mathbf{J}$ of the (compact) orthogonal projections onto the kernels of $W_t(\mathbf{A}^*\mathbf{A})$ will do the job. Because $\mathcal{A}$ is a Silbermann algebra, the sequence $\mathbf{A}^*\mathbf{A}+\mathbf{J}$ is invertible. In particular, $\mathbf{A}^*\mathbf{A}$ is a Fredholm sequence. Analogously one shows that $\mathbf{A}\mathbf{A}^*$ is a Fredholm sequence. But then $\mathbf{A}$ itself is a Fredholm sequence in $\mathcal{A}$, which implies that $\mathbf{A} + \tilde{\mathbf{J}} = \tilde{\mathbf{A}}$ is a Fredholm sequence in $\tilde{\mathcal{A}}$, hence invertible modulo $\tilde{\mathcal{J}}$. By the lifting theorem, applied to the Silbermann pair $(\tilde{\mathcal{A}}, \tilde{\mathcal{J}})$, the sequence $\tilde{\mathbf{A}}$ is stable. Thus, $\tilde{\mathcal{A}}$ is a Silbermann algebra. □

If one starts with a strict Silbermann algebra $\mathcal{A}$, then this construction yields a strict Silbermann algebra $\tilde{\mathcal{A}}_\eta$ of local weight one.

5.7. A few examples

Block Toeplitz operators. As in the introduction, we consider Toeplitz operators $T(a)$ on $l^2(\mathbb{Z}^+)$ and their finite sections with respect to the filtration (P_n), but now the Toeplitz operators are generated by matrix-valued continuous functions $a : \mathbb{T} \to \mathbb{C}^{N\times N}$. We denote the related algebra of the (full) finite sections discretization for these operators by $\mathcal{S}(\mathsf{T}(C^{N\times N}))$. It is not hard to derive the analogues of Theorem 1.1 and Corollary 1.2 in this setting, with the sequences (R_nLR_n) replaced by the sequences $(R_nL_iR_n)$ when $n = kN+i$ with compact operators L_i, and with the homomorphism $\widetilde{W}$ replaced by the family of the N homomorphisms

$$\widetilde{W}_i(\mathbf{A}) := \text{s-lim}_{n\to\infty} R_{nN+i}A_{nN+i}R_{nN+i}P_{nN+i}$$

with $i \in \{0, 1, \ldots, N-1\}$. In particular, *every restriction* $\mathcal{S}_{\eta_i}(\mathsf{T}(C^{N\times N}))$ *with* $\eta_i(n) := nN + i$ *is a strict Silbermann algebra* (and has, in that sense, the same structure as $\mathcal{S}(\mathsf{T}(C)))$.

One-sided almost Mathieu operators. These are the operators

$$H_{\alpha,\lambda,\theta} := V_{-1} + V_1 + aI : l^2(\mathbb{Z}^+) \to l^2(\mathbb{Z}^+)$$

where V_1 and V_{-1} stand for the forward and backward shift operator, respectively, and where $a(n) := \lambda \cos 2\pi(n\alpha + \theta)$ with real parameters α, λ and θ. Further we write AP_α for the smallest closed subalgebra of $L(l^2(\mathbb{Z}^+))$ which contains all operators $H_{\alpha,\lambda,\theta}$ with arbitrary λ and θ (but with fixed α) and all compact operators. The algebra of the finite sections discretization of AP_α with respect to the same filtration (P_n) as before is denoted by $\mathcal{S}(\mathsf{AP}_\alpha)$.

We only deal with the non-periodic case when $\alpha \in (0, 1)$ is irrational. Then we write α as a continued fraction with nth approximant p_n/q_n such that

$$|\alpha - p_n/q_n| < q_n^{-2}.$$

It has been shown in [11, Theorem 5.1] that the restriction $\mathcal{S}_\eta(\mathsf{AP}_\alpha)$ with $\eta(n) := q_n$ is a strict Silbermann algebra (with exactly the same structure as $\mathcal{S}(\mathsf{T}(C))$).

Operators in Cuntz algebras. For $N \geq 2$, let O_N denote the smallest C^*-subalgebra of $L(l^2(\mathbb{Z}^+))$ which contains the operators

$$S_i : (x_k)_{k\geq 0} \mapsto (y_k)_{k\geq 0} \quad \text{with} \quad y_k := \begin{cases} x_r & \text{if } k = rN + i \\ 0 & \text{else} \end{cases} \tag{5.14}$$

with $i = 0, \ldots, N-1$. These operators are isometries, and they satisfy the Cuntz axiom

$$S_0S_0^* + \cdots + S_{N-1}S_{N-1}^* = I. \tag{5.15}$$

Since the (abstract) Cuntz algebra $\mathcal{O}_N$ is simple, $\mathcal{O}_N$ is *-isomorphic to the (concrete = represented) algebra O_N. We use the same filtration (P_n) as before and consider the smallest closed subalgebra $\mathcal{S}(\mathsf{O}_N)$ of $\mathcal{F} = \mathcal{F}^\mathcal{P}$ which contains all finite sections sequences (P_nAP_n) with $A \in \mathsf{O}_N$. It turns out that the algebra $\mathcal{S}(\mathsf{O}_N)$ fails to be fractal, and it is a main result of [15] that $\eta(n) := N^n$ is a right choice to make the restricted algebra $\mathcal{S}_\eta(\mathsf{O}_N)$ fractal.

References

[1] W. Arveson, *An Invitation to C^*-Algebras.* Springer, 1976.

[2] M.C.F. Berglund, *Ideal C^*-algebras.* Duke Math. J. **40** (1973), 241–257.

[3] A. Böttcher, B. Silbermann, *The finite section method for Toeplitz operators on the quarter-plane with piecewise continuous symbols.* Math. Nachr. **110** (1983), 279–291.

[4] A. Böttcher, B. Silbermann, *Introduction to Large Truncated Toeplitz Matrices.* Springer, 1999.

[5] T. Ehrhardt, S. Roch, B. Silbermann, *Finite section method for Toeplitz and singular integral operators of operator valued piecewise quasicontinuous functions.* Singular Integral Operators and Related Topics (eds. A. Böttcher, I. Gohberg), Operator Theory: Adv. Appl. **90**, Birkhäuser, 1996, 204–243.

[6] R. Hagen, S. Roch, B. Silbermann, *C^*-Algebras and Numerical Analysis.* Marcel Dekker, Inc., 2001.

[7] P. Junghanns, R. Kaiser, D. Potts, *Collocation quadrature methods and fast summation for Cauchy singular equations with fixed singularities.* Lin. Alg. Appl. **491** (2016), 187–238.

[8] P. Junghanns, B. Silbermann, *Local theory of the collocation method for the approximate solution of singular integral equations, I.* Integral Equations Oper. Theory **7** (1984), 791–807.

[9] H. Mascarenhas, B. Silbermann, *Sequences of variable-coefficient Toeplitz matrices and their singular values.* J. Functional Anal. **270** (2016), 4, 1479–1500.

[10] G.J. Murphy, *C^*-Algebras and Operator Theory.* Academic Press, Inc., 1990.

[11] V.S. Rabinovich, S. Roch, B. Silbermann, *Finite sections of band-dominated operators with almost periodic coefficients.* Modern Operator Theory and Applications (eds. Y.M. Erusalimsky et al.), Operator Theory: Adv. Appl. **170**, Birkhäuser, 2006, 205–228.

[12] S. Roch, *Algebras of approximation sequences: Fractality.* Problems and Methods in Mathematical Physics, Oper. Theory: Adv. Appl. **121**, Birkhäuser, 2001, 471–497.

[13] S. Roch, *Algebras of approximation sequences: Fredholm theory in fractal algebras.* Studia Math. **150** (2002), 1, 53–77.

[14] S. Roch, *Finite sections of band-dominated operators.* Memoirs AMS **191**, 895, Providence, R.I., 2008.

[15] S. Roch, *Spatial discretization of Cuntz algebras.* Houston J. Math. **36** (2010), 4, 1097–1132.

[16] S. Roch, *Arveson dichotomy and essential fractality.* Operator Theory, Pseudo-Differential Equations, and Mathematical Physics (eds. Y.I. Karlovich, L. Rodino, B. Silbermann, I.M. Spitkovski), Oper. Theory: Adv. Appl. **228**, Birkhäuser, 2013, 325–342.

[17] S. Roch, B. Silbermann, *C^*-algebra techniques in numerical analysis.* J. Oper. Theory **35** (1996), 2, 241–280.

[18] P.A. Santos, B. Silbermann, *An approximation theory for operators generated by shifts.* Numerical Functional Anal. Optimization **27** (2006), 451–484.

[19] B. Silbermann, *Lokale Theorie des Reduktionsverfahrens für Toeplitzoperatoren.* Math. Nachr. **104** (1981), 137–146.

[20] B. Silbermann, *Fredholm theory and numerical linear algebra.* Recent Advances in Operator Algebra and Its Applications, Oper. Theory: Adv. Appl. **160**, Birkhäuser, 2005, 403–411.

Steffen Roch
Technische Universität Darmstadt
Fachbereich Mathematik
Schlossgartenstrasse 7
D-64289 Darmstadt, Germany
e-mail: `roch@mathematik.tu-darmstadt.de`

Operator Theory:
Advances and Applications, Vol. 267, 285–315

Toeplitz and Hankel Algebras – Axiomatic and Asymptotic Aspects

Steffen Roch and Bernd Silbermann

Abstract. In 1983, the authors introduced a Banach algebra of – as they called them – Toeplitz-like operators. This algebra is defined in an axiomatic way; its elements are distinguished by the existence of four related strong limits. The algebra is in the intersection of Barria and Halmos' asymptotic Toeplitz operators and of Feintuch's asymptotic Hankel operators.

In the present paper, we start with repeating and extending this approach and introduce Toeplitz and Hankel operators in an abstract and axiomatic manner. In particular, we will see that our abstract Toeplitz operators can be characterized both as shift invariant operators and as compressions. Then we show that the classical Toeplitz and Hankel operators on the spaces $H^p(\mathbb{T})$, $l^p(\mathbb{Z}_+)$, and $L^p(\mathbb{R}_+)$ are concrete realizations of our abstract Toeplitz operators. Finally we generalize some results by Didas on derivations on Toeplitz and Hankel algebras to the axiomatic context.

Mathematics Subject Classification (2010). Primary 47L80; Secondary 47B35.
Keywords. Toeplitz-like operators, abstract Toeplitz and Hankel operators, Toeplitz algebras, quasicommutator ideals.

1. Introduction

When an operator A on a Banach space X bears the name *Toeplitz operator* then it is usually distinguished by one of the following properties:

- A is the compression of a "nice" (e.g., a normal) operator onto a non-trivial complementable closed subspace of X, or
- A owns a kind of shift-invariance.

For a concrete example, we recall the definition of the classical Toeplitz operators (as well as of their close relatives, the Hankel operators) on the Hardy space. Let $1 < p < \infty$. For a function $f \in L^p(\mathbb{T})$, the Lebesgue space over the unit circle

$\mathbb{T} := \{z \in \mathbb{C} : |z| = 1\}$ in the complex plane, we let

$$\hat{f}_n = \frac{1}{2\pi} \int_0^{2\pi} f(e^{i\varphi}) e^{-in\varphi} \, d\varphi$$

denote its nth Fourier coefficient. It is well known that the Riesz projection

$$P : \sum_{n \in \mathbb{Z}} \hat{f}_n e^{in\varphi} \to \sum_{n \in \mathbb{Z}_+} \hat{f}_n e^{in\varphi},$$

is bounded on $L^p(\mathbb{T})$; its range is called the *Hardy space* $H^p = H^p(\mathbb{T})$. We still set $Q := I - P$ and introduce the reflection operator

$$J : L^p(\mathbb{T}) \to L^p(\mathbb{T}), \; (Jf)(t) := t^{-1} f(t^{-1}).$$

Then $J^2 = I$ and $JPJ = Q$.

Let $a \in L^\infty(\mathbb{T})$. We write aI for the operator $f \mapsto af$ of multiplication by a on $L^p(\mathbb{T})$. Then the *Toeplitz operator* $T(a)$ and the *Hankel operator* $H(a)$ on H^p are defined as the compressions

$$T(a) := PaI|_{\mathrm{im}\,\mathrm{P}} = PaI|_{H^p} \quad \text{and} \quad H(a) := PaJ|_{\mathrm{imP}} = PaJ|_{H^p},$$

respectively. In particular, $T(a)$ is the compression of the "nice" multiplication operator aI onto the Hardy space. The function a is called the *generating function* of both $T(a)$ and $H(a)$. The assignment $a \mapsto T(a)$ is one-to-one, whereas $a \mapsto H(a)$ is not. Toeplitz and Hankel operators possess completely different properties, but are nevertheless closely related by

$$T(ab) = T(a)T(b) + H(a)H(\tilde{b}), \quad H(ab) = T(a)H(b) + H(a)T(\tilde{b}) \tag{1.1}$$

where $\tilde{c}(t) := c(1/t)$ for $c \in L^\infty(\mathbb{T})$. Note that $JcJ = \tilde{c}I$.

The function system $\{\chi_n\}_{n \in \mathbb{Z}_+}$, $\chi_n(t) := t^n$ $(t \in \mathbb{T})$, forms a Schauder basis of H^p. The matrix representations of $T(a)$ and $H(a)$, $a \in L^\infty(\mathbb{T})$, with respect to this basis are given by

$$(\hat{a}_{j-k})_{j,k=0}^\infty, \quad (\hat{a}_{j+k+1})_{j,k=1}^\infty,$$

respectively, from which we conclude the other of the distinguishing properties of Toeplitz operators, as follows. For every positive integer n, define

$$V_n : H^p \to H^p, \; f \mapsto \chi_n f \quad \text{and} \quad V_{-n} : H^p \to H^p, \; f \mapsto P(\chi_{-n} f).$$

Then A is a Toeplitz operator on H^p if and only if it is shift-invariant in the sense that

$$A = V_{-n} A V_n \quad \text{for all positive } n \in \mathbb{Z}. \tag{1.2}$$

This is the starting point for several lines of research. Many work has been done to understand algebras generated by Toeplitz (and Hankel) operators. To mention at least one result in that direction, let $\mathcal{A}$ be a closed subalgebra of $L^\infty(\mathbb{T})$. We denote by $\mathrm{alg}\, T(\mathcal{A})$ the smallest closed subalgebra of $L(H^p)$, the Banach algebra of the bounded linear operators on H^p, which contains all Toeplitz operators with generating function in $\mathcal{A}$. Further we write $\mathrm{qc}\,(T)$ for the *quasicommutator ideal* of $\mathrm{alg}\, T(\mathcal{A})$, i.e., for the smallest closed two-sided ideal of $\mathrm{alg}\, T(\mathcal{A})$ which contains all operators of the form $T(ab) - T(a)T(b)$ with $a, b \in \mathcal{A}$.

Theorem 1.1.

(a) *The algebra* $\mathrm{alg}\, T(\mathcal{A})$ *decomposes into the direct sum*

$$\mathrm{alg}\, T(\mathcal{A}) = T(\mathcal{A}) \oplus \mathrm{qc}\,(T).$$

(b) *The sequence*

$$\{0\} \to \mathrm{qc}\,(T) \to \mathrm{alg}\, T(\mathcal{A}) \to \mathcal{A} \to \{0\}$$

is short exact. Here, the second arrow stands for the embedding and the third one for the canonical mapping from $\mathrm{alg}\, T(\mathcal{A})$ *onto the quotient algebra* $\mathrm{alg}\, T(\mathcal{A})/\mathrm{qc}\,(T)$*, which is isomorphic to* $\mathcal{A}$.

Further results along these lines, also for other kinds of Toeplitz operators, can be found in [4, Section 2.40].

Barria and Halmos [2] took (1.2) as a starting point to initiate the study of asymptotic Toeplitz operators. They call an operator $A \in L(H^2)$ an *asymptotic Toeplitz operator* if the sequence $(V_{-n}AV_n)_{n\geq 0}$ converges strongly[1]. If the strong limit of that sequence exists, then it is necessarily a Toeplitz operator, whence the notation. They also proved that every operator A in $\mathrm{alg}\, TH(L^\infty)$, the smallest closed subalgebra of $L(H^2)$ which contains all Toeplitz and Hankel operators with generating functions in L^∞, is asymptotically Toeplitz. Since the strong limit of the operators $V_{-n}AV_n$ is a Toeplitz operator $T(\varphi)$ with generating function $\varphi \in L^\infty$, they arrived at a symbol map $\mathrm{smb} : A \mapsto \varphi$. It is proved in [2] that the restriction of smb to $\mathrm{alg}\, TH(L^\infty)$ is a contractive *-homomorphism from $\mathrm{alg}\, TH(L^\infty)$ onto L^∞. This symbol map obviously fulfills

$$\mathrm{smb}\, T(a) = a \quad \text{for every } a \in L^\infty.$$

For the Toeplitz algebra $\mathrm{alg}\, T(L^\infty) \subset L(H^2)$, this result was already discovered by R.G. Douglas [10] using different methods.

Using a similar strong limit, Feintuch [11] introduced *asymptotic Hankel operators.* Both "asymptotic approaches" have the disadvantage that the sets of all asymptotic Toeplitz (resp. Hankel) operators do not form algebras (see [2], Example 13).

In 1983, we published a preprint [19] where we independently of [2] introduced a class of asymptotic Toeplitz operators (called Toeplitz-like operators in what follows). Besides the strong limit by Barria and Halmos, we required the existence of three other strong limits (one of which is Feintuch's, but note that [11] appeared only in 1989) so that the set of these operators actually forms a Banach algebra. The motivation for introducing that algebra came from the study of the finite section method for operators belonging to the algebra generated by

[1]Other kinds of convergence, e.g., norm, weak and Cesaro convergence, are considered as well. Our focus is on strong convergence.

Toeplitz and Hankel operators with piecewise continuous generating functions. In 1985 we published our papers [20], [21] where we presented the results and related applications.

The approach of [19] offers moreover the possibility to introduce Toeplitz and Hankel operators in an abstract and axiomatic manner. Especially, the analogs of (1.1) and of Theorem 1.1 can be proved for this class of abstract Toeplitz and Hankel operators. In the present paper, we start with repeating and extending this approach. In particular, we will see that our abstract Toeplitz operators can be characterized both as shift invariant operators and as compressions. Then we show that the classical Toeplitz and Hankel operators on the spaces $H^p(\mathbb{T})$, $l^p(\mathbb{Z}_+)$, and $L^p(\mathbb{R}_+)$ are concrete realizations of our abstract Toeplitz operators. Finally we generalize some results by Didas [8] on derivations on Toeplitz and Hankel algebras to the axiomatic context.

2. Toeplitz-like operators

Here we introduce, in an axiomatic way, the algebra $\mathsf{TL}(X)$ of the Toeplitz-like operators on a Banach space X. For full proofs and some more details we refer to [19, 20] where this algebra was introduced and studied for the first time. A few proofs are sketched here to make the presentation more accessible.

2.1. The axioms

Let X be a Banach space and write $L(X)$ for the Banach algebra of the bounded linear operators on X. The identity operator on X is denoted by I. Further let $G \neq \{0\}$ be a subgroup of the additive group of the real numbers and G_+ the semigroup of its non-negative elements. Our construction of the algebra of the Toeplitz-like operators is based on two families, $\mathcal{V} = (V_t)_{t\in G}$ and $\mathcal{R} = (R_t)_{t\in G}$, of bounded linear operators on X which are specified and related by the following axioms:

(T_1) The mappings $t \mapsto V_t$ and $t \mapsto V_{-t}$ are semigroup homomorphisms on G_+. In particular, $V_s V_t = V_{s+t}$ and $V_{-s}V_{-t} = V_{-(s+t)}$ for all $s, t \in G_+$.

(T_2) $V_{-t}V_t = I$ but $V_t V_{-t} \neq I$ for all $t \in G_+ \setminus \{0\}$. Thus, for positive t, the operators V_t are invertible only from the left-hand side.

By (T_1) and (T_2), $V_0 = I$. Set $Q_t := V_t V_{-t}$ and $P_t := I - Q_t$ for $t \in G$.

(T_3) $R_t^2 = P_t$ and $V_s R_t = R_{s+t} P_t$ for $s, t \in G$.

(T_4) $V_{-t} \to 0$ strongly as $t \to \infty$ and $\sup_{t\in G_+}\{\|V_t\|, \|R_t\|\} < \infty$.

The following properties are easily verified.

Lemma 2.1.

(a) *$P_t^2 = P_t$ and $Q_t^2 = Q_t$ for $t \in G$.*
(b) *$P_{-t} = 0$ and $Q_{-t} = I$ for $t \in G_+$.*
(c) *$P_t \to I$ and $Q_t \to 0$ strongly as $t \to \infty$.*

Thus, P_t and Q_t are complementary projections.

Lemma 2.2. *For all* $s, t \in G$,

(a) $V_s V_t = V_{s+t}$ *whenever* $s \leq 0$ *or* $t \geq 0$.
(b) $V_s V_t = Q_s V_{s+t} = V_{s+t} Q_{-t}$.
(c) $P_s V_{s+t} = V_{s+t} P_{-t}$.
(d) $P_s P_t = P_{\min\{s,t\}}$ *and* $Q_s Q_t = Q_{\max\{s,t\}}$.

Lemma 2.3. *For all* $s, t \in G$,

(a) $R_t P_t = P_t R_t = R_t$.
(b) $R_t = 0$ *for* $t \leq 0$.
(c) $R_t V_{-s} = P_t R_{s+t}$.
(d) $R_s R_t = V_{s-t} P_t = P_s V_{s-t}$.

Notice also that

$$\sup_{t \in G}\{\|V_t\|, \|P_t\|, \|Q_t\|, \|R_t\|\} =: M < \infty. \tag{2.1}$$

We will see several concrete examples of operators V_t and R_t satisfying axioms (T_1)–(T_4) in Section 5. Note also that if the operators V_t and R_t satisfy these axioms on a Banach space X, then the $n \times n$ diagonal matrices

$$\operatorname{diag}(V_t, V_t, \ldots, V_t) \quad \text{and} \quad \operatorname{diag}(R_t, R_t, \ldots, R_t)$$

satisfy these axioms in place of V_t and R_t on the direct sum X^n of n copies of X.

2.2. The algebra $\mathsf{TL}(X)$ of the Toeplitz-like operators

Given families $(V_t)_{t \in G}$ and $(R_t)_{t \in G}$ of operators subject to axioms (T_1)–(T_4), let $\mathsf{TL}(X)$ (with TL for *Toeplitz-like*) stand for the set of all operators $A \in L(X)$ for which the sequences $(V_{-t} A V_t)$, $(R_t A R_t)$, $(V_{-t} A R_t)$ and $(R_t A V_t)$ converge strongly as $t \to \infty$. Their strong limits are denoted by $\mathcal{T}(A)$, $\widetilde{\mathcal{T}}(A)$, $\mathcal{H}(A)$ and $\widetilde{\mathcal{H}}(A)$, respectively.

Example.

(a) The operators V_t and R_t belong to $\mathsf{TL}(X)$, and

$$\mathcal{T}(V_t) = V_t, \quad \widetilde{\mathcal{T}}(V_t) = V_{-t}, \quad \mathcal{H}(V_t) = R_t, \quad \widetilde{\mathcal{H}}(V_t) = 0.$$

The operators R_t lie in the kernel of each of the four strong limits.

(b) If $V_t \to 0$ and $R_t \to 0$ weakly as $t \to \infty$, then every compact operator K on X belongs to $\mathsf{TL}(X)$, and

$$\mathcal{T}(K) = \widetilde{\mathcal{T}}(K) = \mathcal{H}(K) = \widetilde{\mathcal{H}}(K) = 0.$$

The following is Theorem 1.1 in [20].

Theorem 2.4.

(a) $\mathsf{TL}(X)$ *is a norm-closed subalgebra of* $L(X)$ *which contains the identity operator.*

(b) *The following identities hold for all* A, $B \in \mathsf{TL}(X)$

$$\mathcal{T}(AB) = \mathcal{T}(A)\mathcal{T}(B) + \mathcal{H}(A)\widetilde{\mathcal{H}}(B),$$
$$\widetilde{\mathcal{T}}(AB) = \widetilde{\mathcal{T}}(A)\widetilde{\mathcal{T}}(B) + \widetilde{\mathcal{H}}(A)\mathcal{H}(B),$$
$$\mathcal{H}(AB) = \mathcal{H}(A)\widetilde{\mathcal{T}}(B) + \mathcal{T}(A)\mathcal{H}(B),$$
$$\widetilde{\mathcal{H}}(AB) = \widetilde{\mathcal{T}}(A)\widetilde{\mathcal{H}}(B) + \widetilde{\mathcal{H}}(A)\mathcal{T}(B).$$

Theorem 2.5.

(a) $\mathcal{T}$, $\widetilde{\mathcal{T}}$, $\mathcal{H}$ *and* $\widetilde{\mathcal{H}}$ *are bounded linear operators on* $\mathsf{TL}(X)$ *which map this algebra into itself.*

(b) *The composition of any two of the operators* $\mathcal{T}$, $\widetilde{\mathcal{T}}$, $\mathcal{H}$ *and* $\widetilde{\mathcal{H}}$ *belongs to* $\{\mathcal{T}, \widetilde{\mathcal{T}}, \mathcal{H}, \widetilde{\mathcal{H}}, 0\}$. *In particular,*

$\circ$	$\mathcal{T}$	$\widetilde{\mathcal{T}}$	$\mathcal{H}$	$\widetilde{\mathcal{H}}$
$\mathcal{T}$	$\mathcal{T}$	$\widetilde{\mathcal{T}}$	0	0
$\widetilde{\mathcal{T}}$	$\widetilde{\mathcal{T}}$	$\mathcal{T}$	0	0
$\mathcal{H}$	$\mathcal{H}$	$\widetilde{\mathcal{H}}$	0	0
$\widetilde{\mathcal{H}}$	$\widetilde{\mathcal{H}}$	$\mathcal{H}$	0	0.

Proof. The linearity of the operators in (a) is evident; their boundedness is a consequence of (2.1). So let us verify the first column of the table for example. A basic ingredient are the identities collected in the following lemma.

Lemma 2.6. *Let* $A \in \mathsf{TL}(X)$ *and* $s \in G_+$. *Then*

(a) $V_{-s}\mathcal{T}(A)V_s = \mathcal{T}(A)$,
(b) $R_s\mathcal{T}(A)R_s = P_s\widetilde{\mathcal{T}}(A)P_s$,
(c) $V_{-s}\mathcal{T}(A)R_s = \mathcal{H}(A)P_s$,
(d) $R_s\mathcal{T}(A)V_s = P_s\widetilde{\mathcal{H}}(A)$.

Now let $A \in \mathsf{TL}(X)$ and write

$$\mathcal{T}(A) = V_{-t}AV_t + C_t \quad \text{with } C_t \to 0 \text{ strongly as } t \to \infty.$$

Then, for every fixed $s \in G_+$,

$$V_{-s}\mathcal{T}(A)V_s = V_{-s}V_{-t}AV_tV_s + V_{-s}C_tV_s = V_{-s-t}AV_{s+t} + V_{-s}C_tV_s.$$

Letting t go to ∞ we arrive at identity (a) in Lemma 2.6, which on its hand implies that $\mathcal{T}(\mathcal{T}(A)) = \mathcal{T}(A)$. In other words, $\mathcal{T} \circ \mathcal{T} = \mathcal{T}$.

Analogously, using (T_3) and Lemma 2.3 (c) we write

$$R_s\mathcal{T}(A)R_s = R_sV_{-t}AV_tR_s = P_sR_{s+t}AR_{s+t}P_s + R_sC_tR_s,$$

and passage to the strong limit as $t \to \infty$ yields

$$R_s\mathcal{T}(A)R_s = P_s\widetilde{\mathcal{T}}(A)P_s.$$

Letting s go to ∞ we obtain $\widetilde{\mathcal{T}}(\mathcal{T}(A)) = \widetilde{\mathcal{T}}(A)$ or $\widetilde{\mathcal{T}} \circ \mathcal{T} = \widetilde{\mathcal{T}}$. Similarly,

$$V_{-s}\mathcal{T}(A)R_s = V_{-s-t}AR_{s+t}P_s + V_{-s}C_tR_s$$

gives $V_{-s}\mathcal{T}(A)R_s = \mathcal{H}(A)P_s$, whence $\mathcal{H}(\mathcal{T}(A)) = \mathcal{H}(A)$ and, finally,

$$R_sT(A)V_s = P_sR_{s+t}AV_{s+t} + R_sC_tV_s$$

implies $R_s\mathcal{T}(A)V_s = P_s\widetilde{\mathcal{H}}(A)$ and $\widetilde{\mathcal{H}}(\mathcal{T}(A)) = \widetilde{\mathcal{H}}(A)$. The other entries of the table can be checked analogously. □

Theorem 2.7.

(a) *$\mathcal{T}$ is a bounded projection on* $\mathsf{TL}(X)$, *and* $\operatorname{im}\mathcal{T}$ *is a closed subspace of* $L(X)$.
(b) $\mathsf{TL}(X) = \operatorname{im}\mathcal{T} \oplus \ker\mathcal{T}$.
(c) $\ker\mathcal{T}$ *is a closed two-sided ideal of* $\mathsf{TL}(X)$ *and a left-sided ideal of* $L(X)$.

It is simple consequence of the last two identities in Theorem 2.4 that $\ker\mathcal{H}$ and $\ker\widetilde{\mathcal{H}}$ are closed subalgebras of $\mathsf{TL}(X)$.

2.3. Compact operators in $\mathsf{TL}(X)$

As mentioned in Example (b) in Section 2.2, the compact operators belong to $\mathsf{TL}(X)$ if $V_s \to 0$ and $R_s \to 0$ weakly. We will see now that the converse is also true.

Theorem 2.8. *The following assertions are equivalent:*

(a) *all operators of rank one are in* $\mathsf{TL}(X)$;
(b) $V_s \to 0$ *and* $R_s \to 0$ *weakly;*
(c) *all compact operators are in* $\mathsf{TL}(X)$.

Proof. The only implication that needs a proof is (a) ⇒ (b). We prepare this proof by a few facts.

Fact 1. *The R_s do not converge strongly.* Indeed, suppose the R_s converge strongly to some operator $R \in L(X)$. Then

$$V_{-s}AR_s \to 0AR = 0 \quad \text{strongly for every } A \in L(X).$$

In particular, $\mathcal{H}(A) = 0$ for every $A \in \mathsf{TL}(X)$. But then $R_s = \mathcal{H}(V_s) = 0$, hence $P_s = R_s^2 = 0$ for every $s \in G_+$, which contradicts the strong convergence of P_s to the identity.

Fact 2. *Every rank one operator in* $\mathsf{TL}(X)$ *lies in* $\ker\mathcal{T}$. Indeed, let $Kx := \langle x, y\rangle z$ with $y \in X^*$ and $z \in X$. Then

$$V_{-s}KV_sx = \langle V_sx, y\rangle V_{-s}z \to 0$$

since $V_{-s} \to 0$ strongly and $\sup\|V_s\| < \infty$ by axiom (T_4).

Fact 3. *There is a constant $c > 0$ such that*

$$c^{-1}\|P_sx\| \le \|R_sx\| \le c\|P_sx\| \quad \textit{for all } x \in X.$$

Indeed, with $c := \sup\|R_s\|$ we obtain

$$\|R_sx\| = \|R_sP_sx\| \le c\|P_sx\| = c\|R_s^2x\| \le c^2\|R_sx\|.$$

Now to the proof of the implication. By Fact 1, there is a $z \in X$ such that the $R_s z$ do not converge. Clearly, $z \neq 0$. Consider the rank one operators

$$K_y x := \langle x,\, y\rangle z \quad \text{with } y \in X^*.$$

They belong to $\mathsf{TL}(X)$ by assumption and are, hence, in $\ker \mathcal{T}$ by Fact 2. Since $\ker \mathcal{T} = \ker \widetilde{\mathcal{T}}$ by Theorem 2.5, we conclude that

$$R_s K_y R_s x = \langle R_s x,\, y\rangle R_s z \to 0 \quad \text{for all } x \in X,\, y \in X^*. \tag{2.2}$$

Since $\|P_s z\| \to \|z\|$ and by Fact 3, we get

$$\|R_s z\| \geq c^{-1}\|P_s z\| \geq (2c)^{-1}\|z\|$$

for s sufficiently large. Hence, (2.2) implies that $\langle R_s x,\, y\rangle \to 0$ for all $x \in X$ and $y \in X^*$; in other words: $R_s \to 0$ weakly.

The proof for V_s runs similarly. Now we use the inclusion $\ker \mathcal{T} \subseteq \ker \widetilde{\mathcal{H}}$ by Theorem 2.5 to conclude that

$$R_s K_y V_s x = \langle V_s x,\, y\rangle R_s z \to 0 \quad \text{for all } x \in X,\, y \in X^*$$

from which we obtain the weak convergence $V_s \to 0$ as before. □

It is not by accident that both sequences in assertion (b) of the previous theorem tend weakly to zero.

Proposition 2.9. *Let (T_1)–(T_4) be satisfied. Then*

$$V_s \to 0 \textit{ weakly} \quad \Leftrightarrow \quad R_s \to 0 \textit{ weakly}.$$

Proof. We need the following fact, which is easy to check: If the sequence (A_t) is uniformly bounded and if $\langle A_t x,\, y\rangle \to \langle Ax,\, y\rangle$ for all x in a dense subset of X and for all $y \in X^*$, then $A_t \to A$ weakly. Using this fact, we can prove the assertion as follows.

First assume that $V_s \to 0$ weakly. Then, for every $r \in G_+$ and $x \in X$,

$$\langle R_t P_r x,\, y\rangle = \langle V_{t-r} R_r x, y\rangle \to 0 \quad \text{as } t \to \infty$$

by (T_3). Since $P_t \to I$ strongly, the elements of the form $P_r x$ with $r \in G_+$ and $x \in X$ lie dense in X. Thus, we conclude from the fact that $R_t \to 0$ weakly.

Conversely, assume that $R_s \to 0$ weakly. Then, for every $r \in G_+$ and $x \in X$,

$$\langle V_t P_r x,\, y\rangle = \langle R_{t-r} R_r x, y\rangle \to 0 \quad \text{as } t \to \infty$$

by Lemma 2.3 (d). Now the fact implies that $V_t \to 0$ weakly. □

3. Abstract Toeplitz and Hankel operators

3.1. Symbols

Let $\mathcal{S}mb$ denote the quotient algebra $\mathsf{TL}(X)/\ker \mathcal{T}$, which is correctly defined by Theorem 2.7, and let smb refer to the canonical homomorphism from $\mathsf{TL}(X)$ onto $\mathcal{S}mb$. We call smb A the *symbol* of A. The symbol of the identity operator is denoted by e. The symbol of an operator carries important information. For example, we mention the following inverse closedness result from [20], the proof of which makes use of the fact that $\ker \mathcal{T}$ is a left-sided ideal of $L(X)$.

Theorem 3.1. *Let $A \in \mathsf{TL}(X)$ be invertible in $L(X)$. Then A is invertible in $\mathsf{TL}(X)$ if and only if $\operatorname{smb} A$ is invertible in $\mathcal{S}mb$.*

By Theorem 2.7 (a), every coset $q = \operatorname{smb} A \in \mathcal{S}mb$ contains exactly one operator from $\operatorname{im} \mathcal{T}$, namely $\mathcal{T}(A)$. We call this operator the *abstract Toeplitz operator* with symbol q and denote it by $\mathsf{T}(q)$. Similarly, we call $\mathsf{H}(q) := \mathcal{H}(\mathsf{T}(q))$ the *abstract Hankel operator* with symbol q. The mappings

$$\mathsf{T} : \mathcal{S}mb \to \operatorname{im} \mathcal{T},\ q \mapsto \mathsf{T}(q) \quad \text{and} \quad \mathsf{H} : \mathcal{S}mb \to \operatorname{im} \mathcal{H},\ q \mapsto \mathsf{H}(q) \tag{3.1}$$

are linear by construction. Whereas T is also a bijection, H fails to be injective in general. We will see in Corollary 3.20 that these mappings are bounded.

There is a natural involution $q \mapsto \tilde{q}$ on the symbol algebra $\mathcal{S}mb$ defined by $\tilde{q} := \operatorname{smb} \widetilde{\mathcal{T}}(\mathsf{T}(q))$.

Lemma 3.2.

(a) *The mapping $q \mapsto \tilde{q}$ is an automorphism of $\mathcal{S}mb$ with $\tilde{\tilde{q}} = q$.*

(b) *If $M = 1$ in (2.1), then $\sim$ is an isometry and $\|q\|_{\mathcal{S}mb} = \|\mathsf{T}(q)\|_{L(X)}$ for all $q \in \mathcal{S}mb$.*

It is not hard to see that $\widetilde{\mathcal{T}}(\mathsf{T}(q)) = \mathsf{T}(\tilde{q})$ and $\widetilde{\mathcal{H}}(\mathsf{T}(q)) = \mathsf{H}(\tilde{q})$. The following lemmas show that abstract Toeplitz and Hankel operators behave as the concrete Toeplitz and Hankel operators introduced in the introduction and deserve, hence, their name.

Lemma 3.3. *Let $p, q \in \mathcal{S}mb$. Then*

$$\mathsf{T}(pq) = \mathsf{T}(p)\mathsf{T}(q) + \mathsf{H}(p)\mathsf{H}(\tilde{q}), \quad \mathsf{H}(pq) = \mathsf{H}(p)\mathsf{T}(\tilde{q}) + \mathsf{T}(p)\mathsf{H}(q).$$

Proof. Choose A and B in $\mathsf{TL}(X)$ such that $\operatorname{smb} A = p$ and $\operatorname{smb} B = q$. Then $pq = \operatorname{smb}(AB)$, and the operators $\mathsf{T}(p)$, $\mathsf{T}(q)$ and $\mathsf{T}(pq)$ coincide with $\mathcal{T}(A)$, $\mathcal{T}(B)$ and $\mathcal{T}(AB)$, respectively, by definition. Since $A - \mathsf{T}(p)$ and $B - \mathsf{T}(q)$ lie in $\ker \mathcal{T}$, we conclude from Theorem 2.5 (b) that $\mathcal{H}(A) = \mathcal{H}(\mathsf{T}(p)) = \mathsf{H}(p)$ and $\widetilde{\mathcal{H}}(B) = \widetilde{\mathcal{H}}(\mathsf{T}(q)) = \mathsf{H}(\tilde{q})$. Hence, the assertion of the lemma is a consequence of the identities in Theorem 2.4 (b). □

Lemma 3.4. *Let $q_1, \ldots, q_n \in \mathcal{S}mb$ and set $A := \prod_{i=1}^n \mathsf{T}(q_i)$ and $q := \prod_{i=1}^n q_i$. Then*

$$\mathcal{T}(A) = \mathsf{T}(q), \quad \widetilde{\mathcal{T}}(A) = \mathsf{T}(\tilde{q}), \quad \mathcal{H}(A) = \mathsf{H}(q), \quad \widetilde{\mathcal{H}}(A) = \mathsf{H}(\tilde{q}).$$

There are several algebras that can be associated with abstract Toeplitz and Hankel operators. The simplest ones are the *Toeplitz algebras* $\operatorname{alg} \mathsf{T}(\mathcal{S})$ where $\mathcal{S}$ is a closed subalgebra of the symbol algebra $\mathcal{S}mb$. By definition, $\operatorname{alg} \mathsf{T}(\mathcal{S})$ is the smallest closed subalgebra of $L(X)$ which contains all abstract Toeplitz operators $\mathsf{T}(c)$ with $c \in \mathcal{S}$. Clearly, $\operatorname{alg} \mathsf{T}(\mathcal{S})$ is a closed subalgebra of $\mathsf{TL}(X)$.

Similarly, given closed subalgebras $\mathcal{S}_1$ and $\mathcal{S}_2$ of the symbol algebra $\mathcal{S}mb$, let $\operatorname{alg}\mathsf{TH}(\mathcal{S}_1, \mathcal{S}_2)$ stand for the smallest closed subalgebra of $L(X)$ which contains all abstract Toeplitz operators $\mathsf{T}(c_1)$ with $c_1 \in \mathcal{S}_1$ and all abstract Hankel operators $\mathsf{H}(c_2)$ with $c_2 \in \mathcal{S}_2$. If $\mathcal{S}_1 = \mathcal{S}_2 =: \mathcal{S}$, then we simply write $\operatorname{alg}\mathsf{TH}(\mathcal{S})$ in place of $\operatorname{alg}\mathsf{TH}(\mathcal{S}, \mathcal{S})$. Algebras of this form are often referred to as *Hankel algebras.* By Theorem 2.5, $\operatorname{alg}\mathsf{TH}(\mathcal{S}_1, \mathcal{S}_2) \subseteq \mathsf{TL}(X)$. We will see in Section 5.1 that, in general, $\operatorname{alg}\mathsf{TH}(\mathcal{S}mb)$ is properly contained in $\mathsf{TL}(X)$.

A third natural candidate, with a strong coupling between Toeplitz and Hankel operators, is the *Toeplitz-plus-Hankel algebra* $\operatorname{alg}\mathsf{TH}^+(\mathcal{S})$ which is the smallest closed subalgebra of $L(X)$ which contains all Toeplitz-plus-Hankel operators $T(c) + H(c)$ with $c \in \mathcal{S}$. Clearly, $\operatorname{alg}\mathsf{TH}^+(\mathcal{S}) \subsetneq \operatorname{alg}\mathsf{TH}(\mathcal{S})$.

3.2. Quasicommutators and decompositions

The goal of this section is to understand the ideal $\ker \mathcal{T}$ as a quasicommutator ideal. A general approach to this circle of ideas is as follows. Let $\mathcal{A}$ and $\mathcal{B}$ be Banach algebras and $D : \mathcal{A} \to \mathcal{B}$ a linear and bounded mapping. We write $\operatorname{alg} D(\mathcal{A})$ for the smallest closed subalgebra of $\mathcal{B}$ which contains all elements $D(a)$ with $a \in \mathcal{A}$. We say that $D(\mathcal{A})$ generates $\mathcal{B}$ (as a Banach algebra) if $\operatorname{alg} D(\mathcal{A}) = \mathcal{B}$.

The *quasicommutator ideal* generated by D is a measure for the deviation of D from being multiplicative (i.e., an algebra homomorphism). This quasicommutator ideal, denoted by $\operatorname{qc}(D)$, is the smallest closed ideal of $\operatorname{alg} D(\mathcal{A})$ which contains all quasi-commutators $D(ab) - D(a)D(b)$ with $a, b \in \mathcal{A}$. Clearly, if D is multiplicative then $\operatorname{qc}(D) = \{0\}$, whereas $\operatorname{qc}(D) = \operatorname{alg} D(\mathcal{A})$ if this algebra is simple and D is not multiplicative. The following lemma provides an equivalent description of $\operatorname{qc}(D)$ is terms of higher quasi-commutators. The proof is an easy exercise.

Lemma 3.5. *Let $\mathcal{A}$, $\mathcal{B}$ be Banach algebras and $D : \mathcal{A} \to \mathcal{B}$ a bounded linear mapping such that $D(\mathcal{A})$ generates $\mathcal{B}$. Then $\operatorname{qc}(D)$ is equal to*

$$\operatorname{clos\,span}_{\mathcal{B}} \{D(a_1 \cdots a_n) - D(a_1) \cdots D(a_n) : n \in \mathbb{N},\ a_1, \ldots, a_n \in \mathcal{A}\}.$$

In what follows we consider bounded linear mappings $D : \mathcal{A} \to \mathcal{B}$ which own a homomorphic left inverse, i.e., we suppose there is a bounded homomorphism $W : \mathcal{B} \to \mathcal{A}$ such that $W(D(a)) = a$ for all $a \in A$. Mappings D with that property are often called *discretizations* or *quantizations*, in order to emphasize their role in numerical analysis or operator theory, respectively (see, e.g., Section 6.5.3 in [14]).

Theorem 3.6. *Let $\mathcal{A}$ and $\mathcal{B}$ be Banach algebras, $D : \mathcal{A} \to \mathcal{B}$ a bounded linear mapping, and $W : \mathcal{B} \to \mathcal{A}$ a bounded homomorphism such that $W(D(a)) = a$ for every $a \in \mathcal{A}$. Then*

(a) *D is injective, $D(\mathcal{A})$ is a closed subspace of $\mathcal{B}$, and W is surjective,*
(b) *the algebra $\mathcal{B}$ decomposes into the direct sum*

$$\mathcal{B} = D(\mathcal{A}) \oplus \ker W, \tag{3.2}$$

(c) *if $D(\mathcal{A})$ generates $\mathcal{B}$, then $\ker W = \operatorname{qc}(D)$.*

(d) *if D and W are contractions, then D is an isometry, the canonical projection $P : \mathcal{B} \to D(\mathcal{A})$ associated with the decomposition $\mathcal{B} = D(\mathcal{A}) \oplus \ker W$ has norm 1, and*

$$\|D(a)\| = \min_{k \in \ker W} \|D(a) + k\| \quad \text{for every } a \in \mathcal{A}. \tag{3.3}$$

Proof. (a) First we show that $D(\mathcal{A})$ is a closed subspace of $\mathcal{B}$. Let $D(a_n)$ be a sequence in $D(\mathcal{A})$ which converges to $b \in \mathcal{B}$. Then the $W(D(a_n)) = a_n$ form a Cauchy sequence in $\mathcal{A}$ which converges to an element $a \in \mathcal{A}$. Since D is bounded, the $D(a_n)$ converge to $D(a)$; hence $b = D(a) \in D(\mathcal{A})$, and $D(\mathcal{A})$ is closed. The other assertions follow immediately from $W \circ D = \mathrm{id}_{\mathcal{A}}$.

(b) Writing $b \in \mathcal{B}$ as $b = D(W(b)) + (b - D(W(b))$ and noting that

$$W(b - D(W(b)) = W(b) - W(D(W(b))) = 0$$

we obtain $\mathcal{B} = D(\mathcal{A}) + \ker W$. To see that this sum is direct, assume that $D(a) \in \ker W$. Then $a = W(D(a)) = 0$, hence $D(a) = 0$.

(c) Since $W(D(ab) - D(a)D(b)) = W(D(ab)) - W(D(a))W(D(b)) = 0$ and $\mathrm{qc}\,(D)$ is a closed ideal of $\mathcal{B}$, the inclusion $\mathrm{qc}\,(D) \subseteq \ker W$ is clear (and it holds without the additional assumption in (c)).

For the reverse inclusion, let $k \in \ker W$. Since $D(\mathcal{A})$ generates $\mathcal{B}$, there is a sequence of elements $k_n = \sum_i \prod_j D(a_{ijn})$ with $a_{ijn} \in \mathcal{A}$ such that $\|k - k_n\| \to 0$. Using Lemma 3.5, we can write k_n as

$$k_n = D(\sum_i \prod_j a_{ijn}) + q_n = D(W(k_n)) + q_n = D(W(k_n - k)) + q_n$$

with $q_n \in \mathrm{qc}\,(D)$ (note that $W(k) = 0$). Then

$$\begin{aligned} \|k - q_n\| &\le \|k - k_n\| + \|k_n - q_n\| \le \|k - k_n\| + \|D(W(k_n - k))\| \\ &\le (1 + \|D\|\,\|W\|)\,\|k - k_n\| \ \to\ 0. \end{aligned}$$

Thus, k can be approximated as closely as desired by elements in $\mathrm{qc}\,(D)$. Since the quasi-commutator ideal is closed, the assertion follows.

(d) Let $a \in \mathcal{A}$ and $k \in \ker W$. Then $\|a\| = \|W(D(a))\| \le \|D(a)\| \le \|a\|$ and

$$\|D(a)\| = \|D(W(D(a)))\| = \|D(W(D(a) + k))\| \le \|D(a) + k\|,$$

which implies assertion (d). □

The conditions in Theorem 3.6 are also necessary, as the following proposition shows.

Proposition 3.7. *Let the Banach algebra $\mathcal{B} = \mathcal{M} \oplus \mathcal{J}$ be the direct sum of a closed subspace $\mathcal{M}$ and a closed ideal $\mathcal{J}$. Then there are a Banach algebra $\mathcal{A}$, a bounded linear mapping $D : \mathcal{A} \to \mathcal{M}$ such that $D(\mathcal{A}) = \mathcal{M}$, and a bounded homomorphism $W : \mathcal{B} \to \mathcal{A}$ which is a left inverse for D. Moreover, $\mathcal{J} = \ker W$, and $\mathcal{J} = \ker W = \mathrm{qc}\,(D)$ if $\mathcal{M}$ generates $\mathcal{B}$.*

Proof. Let $P : \mathcal{B} \to \mathcal{M}$ be the canonical projection associated with the decomposition $\mathcal{B} = \mathcal{M} \oplus \mathcal{J}$, set $\mathcal{A} := \mathcal{B}/\mathcal{J}$, and write W for the canonical homomorphism from $\mathcal{B}$ onto $\mathcal{A}$. Then the mapping $D : \mathcal{A} \to \mathcal{M}$, $a + \mathcal{J} \mapsto P(a)$, is correctly defined, linear and surjective, and

$$W \circ D : a + \mathcal{J} \mapsto P(a) \mapsto P(a) + \mathcal{J} = a + \mathcal{J}$$

is the identity mapping on $\mathcal{A}$. Since $\mathcal{M}$ is closed and D is inverse to the (evidently bounded and surjective) mapping $W|_{\mathcal{M}} : \mathcal{M} \to \mathcal{A}$, D is bounded by the bounded inverse mapping theorem. The equality $\mathcal{J} = \ker W = \mathrm{qc}\,(D)$ is a consequence of Theorem 3.6. □

3.3. Algebras of abstract Toeplitz and Hankel operators

Next we are going to apply these general results in the context of (abstract) Toeplitz and Hankel algebras. We start with the Toeplitz algebra $\mathrm{alg}\,\mathsf{T}(\mathcal{S})$, with $\mathcal{S}$ a closed subalgebra of $\mathcal{S}mb$. Slightly modifying the above notation, we write $\mathrm{qc}_{\mathsf{T}}\,(\mathcal{S})$ (instead of $\mathrm{qc}\,(\mathsf{T}|_{\mathcal{S}})$) for the quasicommutator ideal of the mapping $\mathsf{T} : \mathcal{S}mb \to L(X)$ restricted to $\mathcal{S}$. Thus, $\mathrm{qc}_{\mathsf{T}}\,(\mathcal{S})$ is the smallest closed ideal of $\mathrm{alg}\,\mathsf{T}(\mathcal{S})$ which contains all operators $\mathsf{T}(cd) - \mathsf{T}(c)\mathsf{T}(d)$ with $c, d \in \mathcal{S}$. Then the restriction of the symbol homomorphism smb to $\mathrm{alg}\,\mathsf{T}(\mathcal{S})$ is a left inverse to the linear mapping T, restricted to $\mathcal{S}$. In this context, Theorem 3.6 specifies as follows.

Corollary 3.8.

(a) $\mathrm{qc}_{\mathsf{T}}\,(\mathcal{S}) = \ker(\mathrm{smb}\,|_{\mathrm{alg}\,\mathsf{T}(\mathcal{S})}) = \ker(\mathcal{T}|_{\mathrm{alg}\,\mathsf{T}(\mathcal{S})})$.

(b) *The algebra* $\mathrm{alg}\,\mathsf{T}(\mathcal{S})$ *decomposes into the direct sum*

$$\mathrm{alg}\,\mathsf{T}(\mathcal{S}) = \mathsf{T}(\mathcal{S}) \oplus \mathrm{qc}_{\mathsf{T}}\,(\mathcal{S}).$$

(c) *The sequence*

$$\{0\} \to \mathrm{qc}_{\mathsf{T}}\,(\mathcal{S}) \to \mathrm{alg}\,\mathsf{T}(\mathcal{S}) \to \mathcal{S} \to \{0\},$$

with the second arrow standing for the natural embedding and the third one for the restriction of the symbol mapping, is exact.

In general, the Hankel algebras $\mathrm{alg}\,\mathsf{TH}(\mathcal{S}_1, \mathcal{S}_2)$ do not arise from a discretization mapping in the above sense; in particular there is no natural quasicommutator ideal. So we deal with the kernel of $\mathcal{T}$ instead of the quasicommutator ideal. Theorem 2.7 specifies as follows to this context.

Corollary 3.9.

(a) $\mathrm{alg}\,\mathsf{TH}(\mathcal{S}_1, \mathcal{S}_2) = \mathsf{T}(\mathcal{S}_1) \oplus \ker(\mathcal{T}|_{\mathrm{alg}\,\mathsf{TH}(\mathcal{S}_1,\mathcal{S}_2)})$.

(b) *The sequence*

$$\{0\} \to \ker(\mathcal{T}|_{\mathrm{alg}\,\mathsf{TH}(\mathcal{S}_1,\mathcal{S}_2)}) \to \mathrm{alg}\,\mathsf{TH}(\mathcal{S}_1, \mathcal{S}_2) \to \mathcal{S}_1 \to \{0\},$$

where the second arrow stands for the embedding and the third one for the restriction of the symbol mapping, is exact.

In this context, there is an equivalent description of the kernel of $\mathcal{T}$ (see [8] for the H^2-setting).

Lemma 3.10. *Let* $\mathcal{S}_1 \subseteq \mathcal{S}_2$. *Then* $\ker(\mathcal{T}|_{\operatorname{alg}\mathsf{TH}(\mathcal{S}_1,\mathcal{S}_2)})$ *is the smallest closed ideal of* $\operatorname{alg}\mathsf{TH}(\mathcal{S}_1, \mathcal{S}_2)$ *which contains all Hankel operators* $\mathsf{H}(q)$ *with* $q \in \mathcal{S}_2$.

Proof. Abbreviate $\ker(\mathcal{T}|_{\operatorname{alg}\mathsf{TH}(\mathcal{S}_1,\mathcal{S}_2)})$ by $\mathcal{J}_1$ and the closed ideal generated by $\mathsf{H}(\mathcal{S}_2)$ by $\mathcal{J}_2$. Since every Hankel operator is in $\ker\mathcal{T}$, it is immediate that $\mathcal{J}_2 \subseteq \mathcal{J}_1$. For the reverse inclusion, let $A \in \mathcal{J}_1$. Then A can be represented as a norm limit

$$A = \lim_{n\to\infty}\left(\sum_{i=1}^{\alpha_n} A_i^{(n)}\mathsf{H}(b_i^{(n)})C_i^{(n)} + \sum_{i=1}^{\gamma_n}\prod_{j=1}^{\delta_n}\mathsf{T}(d_{ij}^{(n)})\right)$$

with symbols $d_{ij}^{(n)} \in \mathcal{S}_1$ and $b_i^{(n)} \in \mathcal{S}_2$ and with operators $A_i^{(n)}, C_i^{(n)} \in \mathsf{TH}(\mathcal{S}_1, \mathcal{S}_2)$. Since both A and all Hankel operators are in $\ker\mathcal{T}$ and $\mathcal{T}$ is continuous, we conclude from Lemma 3.4 that

$$0 = \mathcal{T}(A) = \lim_{n\to\infty}\mathsf{T}\left(\sum_{i=1}^{\gamma_n}\prod_{j=1}^{\delta_n} d_{ij}^{(n)}\right).$$

Hence, $A = A - \mathcal{T}(A)$ is equal to

$$\lim_{n\to\infty}\left(\sum_{i=1}^{\alpha_n} A_i^{(n)}\mathsf{H}(b_i^{(n)})C_i^{(n)} + \sum_{i=1}^{\gamma_n}\prod_{j=1}^{\delta_n}\mathsf{T}(d_{ij}^{(n)}) - \mathsf{T}\left(\sum_{i=1}^{\gamma_n}\prod_{j=1}^{\delta_n} d_{ij}^{(n)}\right)\right).$$

The first item is evidently in $\mathcal{J}_2$; the second one is in the quasicommutator ideal of $\mathsf{T}(\mathcal{S}_1)$ by Lemma 3.5. This ideal is generated by the quasicommutators

$$\mathsf{T}(a)\mathsf{T}(b) - \mathsf{T}(ab) = -\mathsf{H}(a)\mathsf{H}(\tilde{b})$$

which are in $\mathcal{J}_2$ because of $\mathcal{S}_1 \subseteq \mathcal{S}_2$. □

In contrast to the algebras $\operatorname{alg}\mathsf{TH}(\mathcal{S}_1, \mathcal{S}_2)$, the Toeplitz-plus-Hankel algebras $\operatorname{alg}\mathsf{TH}^+(\mathcal{S})$ again arise from a discretization mapping, namely from the restriction of the mapping

$$\mathsf{T} + \mathsf{H} : \mathcal{S}mb \to \mathsf{LT}(X), \quad s \mapsto \mathsf{T}(s) + \mathsf{H}(s)$$

to $\mathcal{S}$, again with smb as a homomorphic left inverse. We denote the related quasicommutator ideal by $\mathrm{qc}_{T+H}(\mathcal{S})$, i.e., $\mathrm{qc}_{T+H}(\mathcal{S})$ is the smallest closed ideal of $\operatorname{alg}\mathsf{TH}^+(\mathcal{S})$ which contains all operators

$$\mathsf{T}(cd) + \mathsf{H}(cd) - (\mathsf{T}(c) + \mathsf{H}(c))(\mathsf{T}(d) + \mathsf{H}(d)) \quad \text{with } c, d \in \mathcal{S}.$$

Corollary 3.11.

(a) $\mathrm{qc}_{T+H}(\mathcal{S}) = \ker(\mathrm{smb}|_{\operatorname{alg}\mathsf{TH}^+(\mathcal{S})}) = \ker(\mathcal{T}|_{\operatorname{alg}\mathsf{TH}^+(\mathcal{S})})$.

(b) *The algebra* $\operatorname{alg}\mathsf{TH}^+(\mathcal{S})$ *decomposes into the direct sum*

$$\operatorname{alg}\mathsf{TH}^+(\mathcal{S}) = \{\mathsf{T}(c) + \mathsf{H}(c) : c \in \mathcal{S}\} \oplus \mathrm{qc}_{T+H}(\mathcal{S}).$$

(c) *The sequence*

$$\{0\} \to \mathrm{qc}_{T+H}(\mathcal{S}) \to \operatorname{alg}\mathsf{TH}^+(\mathcal{S}) \to \mathcal{S} \to \{0\},$$

with the second arrow standing for the natural embedding and the third one for the restriction of the symbol mapping, is exact.

We conclude by a few remarks on commutators. The *commutator ideal* com $\mathcal{B}$ of a Banach algebra $\mathcal{B}$ is the smallest closed ideal of $\mathcal{B}$ which contains all commutators $ab - ba$ with $a, b \in \mathcal{B}$. It is easy to see that in the context of Theorem 3.6 and if $D(\mathcal{A})$ generates $\mathcal{B}$, the commutator ideal of $\mathcal{B}$ is already generated by commutators $D(c)D(d) - D(d)D(c)$ with $c, d \in \mathcal{A}$. Thus, if $\mathcal{A}$ is commutative, then com $\mathcal{B}$ is contained in qc $(\mathcal{A})$, which follows from

$$D(c)D(d) - D(d)D(c) = (D(c)D(d) - D(cd)) - (D(d)D(c) - D(dc)).$$

Note also that the commutator ideal of a closed subalgebra $\mathcal{A}$ of $\mathsf{LT}(X)$ with commutative symbol algebra smb $\mathcal{A}$ is contained in ker $\mathcal{T}$.

The above remarks apply in particular to Toeplitz and Toeplitz-plus-Hankel algebras. For the commutator ideal of Hankel algebras, we have the following.

Lemma 3.12. *If* $\mathcal{S}_1$ *is commutative, then* com alg $\mathsf{TH}(\mathcal{S}_1, \mathcal{S}_2)$ *is contained in* ker $\mathcal{T}$.

Proof. Let $A, B \in \operatorname{alg} TH(\mathcal{S}_1, \mathcal{S}_2)$. By Corollary 3.9, we write $A = \mathsf{T}(a) + K$ and $B = \mathsf{T}(b) + L$ with $a, b \in \mathcal{S}_1$ and $K; L \in \ker \mathcal{T}$. Then

$$AB - BA = \mathsf{T}(a)\mathsf{T}(b) - \mathsf{T}(b)\mathsf{T}(a) + M$$

with $M \in \ker \mathcal{T}$. The assertion follows since

$$\mathcal{T}(\mathsf{T}(a)\mathsf{T}(b) - \mathsf{T}(b)\mathsf{T}(a)) = \mathsf{T}(ab - ba) = 0$$

by Lemma 3.4 and since $\mathcal{S}_1$ is commutative. □

3.4. Continuous symbols

The operators V_t belong to the algebra $\mathsf{TL}(X)$ as we observed in Example 2.2. Let alg $(\mathcal{V})$ stand for the smallest closed subalgebra of $\mathsf{TL}(X)$ which contains all of these operators.

Lemma 3.13. *Let* $\mathcal{A}$ *be a closed subalgebra of* $\mathsf{TL}(X)$ *which contains* $\mathcal{V}$. *Then the following sets are equal for every* $t_0 \in G_+ \setminus \{0\}$:

(a) *the smallest closed ideal of* $\mathcal{A}$ *which contains all operators* P_t, $t \in G$;
(b) *the smallest closed ideal of* $\mathcal{A}$ *which contains* P_{t_0}.

Proof. Let $\mathcal{I}$ stand for the smallest closed ideal of $\mathcal{A}$ which contains P_{t_0}. Clearly, $\mathcal{I}$ is contained in the ideal described in (a). For the reverse inclusion, we have to show that $P_t \in \mathcal{I}$ for every $t \in G_+$. Let $k \in \mathbb{Z}_+$ and $k \geq 2$. Then

$$\begin{aligned} P_{kt_0} - P_{(k-1)t_0} &= V_{(k-1)t_0}V_{-(k-1)t_0} - V_{kt_0}V_{-kt_0} \\ &= V_{(k-1)t_0}(I - V_{t_0}V_{-t_0})V_{-(k-1)t_0} \\ &= V_{(k-1)t_0}P_{t_0}V_{-(k-1)t_0} \in \mathcal{I}. \end{aligned}$$

Summing up we conclude that $P_{kt_0} \in \mathcal{I}$ for every $k \in \mathbb{Z}_+$. Now, given $t \in G_+$, choose $k \in \mathbb{Z}_+$ such that $t < kt_0$. Then, by Lemma 2.2 (a),

$$\begin{aligned} V_{-(kt_0-t)} P_{kt_0} V_{kt_0-t} &= I - V_{-(kt_0-t)} V_{kt_0} V_{-kt_0} V_{kt_0-t} \\ &= I - V_t V_{-t} = P_t, \end{aligned}$$

whence $P_t \in \mathcal{I}$. □

In particular we see that if one of the projections P_t with $t \in G_+ \setminus \{0\}$ is compact, then each of these projections is compact.

Let $\mathcal{I}(\mathcal{P})$ denote the smallest closed ideal of $\operatorname{alg}(\mathcal{V})$ which contains (one or all of) the projections P_t. Further write φ_t for the symbol of V_t and $\mathcal{C}$ for the smallest closed subalgebra of the symbol algebra $\mathcal{S}mb$ which contains all symbols φ_t. We call $\mathcal{C}$ the *algebra of the continuous symbols.* This notion is inspired by Gohberg and Feldman's text [13]. Note that $\varphi_s \varphi_t = \varphi_{s+t}$ for all $s, t \in G$ and that φ_0 is the identity element of $\mathcal{S}mb$.

Lemma 3.14.

(a) $\operatorname{alg}(\mathcal{V}) = \operatorname{alg}(\mathsf{T}(\mathcal{C}))$.

(b) $\mathcal{I}(\mathcal{P}) = \ker(\mathcal{T}|_{\operatorname{alg}(\mathcal{V})}) = \operatorname{qc}_{\mathsf{T}}(\mathcal{C})$.

Proof. Assertion (a) is a consequence of $\mathsf{T}(\varphi_t) = V_t$. The second equality in (b) comes from Corollary 3.8. The first equality is then evident since the quasicommutators $P_t = I - V_t V_{-t}$ belong to the kernel of $\mathcal{T}$, and they generate the quasicommutator ideal $\operatorname{qc}_{\mathsf{T}}(\mathcal{C})$. □

Whereas the symbol algebra $\mathcal{S}mb$ need not to be commutative, its subalgebra of the continuous symbols is commutative. Moreover, the following holds.

Lemma 3.15. *$\mathcal{C}$ is a closed subalgebra in the center of $\mathcal{S}mb$.*

Proof. Let $q \in \mathcal{S}mb$ and $t \in G_+$. By Lemma 2.6 (a), $V_{-t}\mathsf{T}(q)V_t = \mathsf{T}(q)$. Passing to symbols we obtain $\varphi_{-t} q \varphi_t = q$ or, equivalently, $q\varphi_t = \varphi_t q$ for $t \in G_+$. Since $\varphi_{-t} = (\varphi_t)^{-1}$, this equality holds for negative t as well. □

3.5. Abstract Toeplitz operators as compressions

We infer from Lemma 2.6 (a) that the elements of $\operatorname{im}\mathcal{T}$, hence all abstract Toeplitz operators, are shift invariant. We will see now that abstract Toeplitz operators also arise as compressions of abstract Laurent operators, for which we have to embed X into a larger space. This construction will also shed new light onto the algebra $\mathsf{TL}(X)$ since it identifies the four strong limits $\mathcal{T}$, $\widetilde{\mathcal{T}}$, $\mathcal{H}$ and $\widetilde{\mathcal{H}}$ as components of a single strong limit.

Let the Banach space X, the group G and the operators V_t, R_t, P_t and Q_t with $t \in G$ be as in Section 2.1. Let X^2 stand for the direct sum $X \oplus X$, provided with the norm $\|(x_1, x_2)\| = \|x_1\|_X + \|x_2\|_X$ (or with another norm making X^2 to a Banach space into which X is isometrically embedded). Thinking of the elements of X^2 as column vectors, we identify operators in $L(X^2)$ with 2×2 matrices with entries in $L(X)$.

For $t \in G_+$, define operators $U_{\pm t}$ on X^2 by

$$U_t := \begin{pmatrix} V_t & R_t \\ 0 & V_{-t} \end{pmatrix} \quad \text{and} \quad U_{-t} := \begin{pmatrix} V_{-t} & 0 \\ R_t & V_t \end{pmatrix}. \tag{3.4}$$

Proposition 3.16. *The operators U_t form a commutative group, i.e., $U_s U_t = U_t U_s = U_{s+t}$ for all $s,\, t \in G$.*

Proof. We prepare the proof by establishing the identities

$$V_s V_t + R_s R_{-t} = V_{s+t} \quad \text{for all } s,\, t \in G \tag{3.5}$$

$$V_s R_t + R_s V_{-t} = R_{s+t} \quad \text{for all } s,\, t \in G. \tag{3.6}$$

From Lemmas 2.2 (b) and 2.3 (d) we know that $R_s R_{-t} = P_s V_{s+t}$ and $V_s V_t = Q_s V_{s+t}$. Summing up gives (3.5). For (3.6), recall from Axiom (T_3) and Lemma 2.3 (c) that $V_s R_t = R_{s+t} P_t$ and $R_s V_{-t} = P_s R_{s+t}$. Since $V_{-t} = V_{-t} Q_t$, the last identity implies that $R_s V_{-t} = P_s R_{s+t} Q_t$. It thus remains to show that

$$P_s R_{s+t} Q_t = R_{s+t} Q_t \quad \text{or, equivalently,} \quad Q_s R_{s+t} Q_t = 0,$$

which is further equivalent to $V_{-s} R_{s+t} V_t = 0$. Now we have

$$\begin{aligned} V_{-s} R_{s+t} V_t &= R_t P_{s+t} V_t && \text{by Axiom } (T_3) \\ &= R_t P_t P_{s+t} V_t && \text{by Lemma 2.3 (a)} \\ &= R_t P_{s+t} P_t V_t && \text{by Lemma 2.2 (d)} \\ &= 0 && \text{since } P_t V_t = 0. \end{aligned}$$

The identities (3.5) and (3.6) together imply that

$$\begin{pmatrix} V_s & R_s \\ R_{-s} & V_{-s} \end{pmatrix} \begin{pmatrix} V_t & R_t \\ R_{-t} & V_{-t} \end{pmatrix} = \begin{pmatrix} V_{s+t} & R_{s+t} \\ R_{-s-t} & V_{-s-t} \end{pmatrix}$$

for all $s,\, t \in G$, from which the assertion follows (note that $R_t = 0$ for $t \le 0$). □

The simple identity

$$U_{-t} \begin{pmatrix} A & 0 \\ 0 & 0 \end{pmatrix} U_t = \begin{pmatrix} V_{-t} A V_t & V_{-t} A R_t \\ R_t A V_t & R_t A R_t \end{pmatrix},$$

holding for every $A \in L(X)$ and $t \in G_+$, implies the relation between the two approaches to Toeplitz-like operators.

Theorem 3.17. *Let $A \in L(X)$. Then the strong limit*

$$\mathcal{L}(A) := \text{s-lim}_{t \to \infty} U_{-t} \begin{pmatrix} A & 0 \\ 0 & 0 \end{pmatrix} U_t$$

exists if and only if $A \in \mathsf{TL}(X)$. In this case,

$$\mathcal{L}(A) = \begin{pmatrix} \mathcal{T}(A) & \mathcal{H}(A) \\ \widetilde{\mathcal{H}}(A) & \widetilde{\mathcal{T}}(A) \end{pmatrix}. \tag{3.7}$$

Corollary 3.18.

(a) *The mapping* $\mathcal{L} : \mathsf{TL}(X) \to L(X^2)$ *is a bounded homomorphism.*
(b) $U_{-t}\mathcal{L}(A)U_t = \mathcal{L}(A)$ *for every* $A \in \mathsf{TL}(X)$ *and* $t \in G$.

Proof. Assertion (a) comes from the definition of $\mathcal{L}$ and Theorem 2.4 (b); assertion (b) can be proved in the same as Lemma 2.6 (likewise: it follows from that lemma). □

Define mappings

$$E : X \to X^2,\ x \mapsto (x,\ 0)^T, \qquad E^{-1} : X^2 \to X,\ (x,\ y)^T \mapsto x.$$

Then, by (3.7),

$$E^{-1}\mathcal{L}(A)E = \mathcal{T}(A) \qquad \text{for every } A \in \mathsf{LT}(X). \tag{3.8}$$

In other words, if $P \in L(X^2)$ stands for the projection $(x,\ y)^T \mapsto (x,\ 0)^T$, then every abstract Toeplitz operator can be viewed as the compression $PA|_{\mathrm{imP}}$ of an *abstract Laurent operator* $A \in \operatorname{im}\mathcal{L}$.

Theorem 3.19.

(a) $\operatorname{im}\mathcal{L}$ *is a closed subalgebra of* $L(X^2)$.
(b) *The mapping* $A \mapsto \operatorname{smb}(E^{-1}AE)$ *is a bounded isomorphism from* $\operatorname{im}\mathcal{L}$ *onto* $\mathcal{S}mb = \operatorname{smb}\mathsf{TL}(X)$.

Proof. (a) Let $\mathcal{L}(A_n)$ be a sequence in $\operatorname{im}\mathcal{L}$ which converges in $L(X^2)$. Then, by (3.8), $(\mathcal{T}(A_n))$ is a Cauchy sequence in $\operatorname{im}\mathcal{T}$. Since $\operatorname{im}\mathcal{T}$ is closed by Theorem 2.7 (a), there is a operator $A \in \mathsf{LT}(X)$ such that $\mathcal{T}(A_n) \to \mathcal{T}(A)$ in the operator norm. The first column in the table in Theorem 2.5 shows that $\mathcal{L}(\mathcal{T}(B)) = \mathcal{L}(B)$ for every $B \in \mathsf{TL}(X)$. Hence, $\mathcal{L}(A_n) = \mathcal{L}(\mathcal{T}(A_n))$ converges in the norm to $\mathcal{L}(A) = \mathcal{L}(\mathcal{T}(A))$. Thus, $\operatorname{im}\mathcal{L}$ is a closed subspace of $L(X^2)$. It is also an algebra by Corollary 3.18.
(b) The mapping $A \mapsto \operatorname{smb}(E^{-1}AE)$ sends $\mathcal{L}(A)$ to $\operatorname{smb}\mathcal{T}(A)$ by (3.8). This mapping is a homomorphism because $\mathcal{T}(A) - \mathcal{T}(A)\mathcal{T}(B) \in \ker\mathcal{T}$ and smb is a homomorphism. We are going to show that this mapping is injective. Let $\operatorname{smb}(E^{-1}\mathcal{L}(A)E) = \operatorname{smb}\mathcal{T}(A) = 0$ for an operator $A \in \mathsf{TL}(X)$. Then $\mathcal{T}(A) \in \ker\mathcal{T}$; hence, $\mathcal{T}(A) = 0$. The first column in the table in Theorem 2.5 then shows that $\mathcal{L}(A) = \mathcal{L}(\mathcal{T}(A)) = 0$. □

It is not hard to see that the inverse of the mapping $A \mapsto \operatorname{smb}(E^{-1}AE)$ in Theorem 3.19 is explicitly given by

$$(\operatorname{smb}\mathsf{TL}(X) =)\quad \mathcal{S}mb \to \operatorname{im}\mathcal{L}, \quad q \mapsto \mathcal{L}(\mathsf{T}(q)).$$

So we have two ways to think of the symbol of a Toeplitz-like operator: first, as an element of the quotient algebra $\mathsf{TL}(X)/\ker\mathcal{T}$; second as an operator in $\mathcal{L}(\mathsf{TL}(X))$ acting on X^2.

Corollary 3.20. *The mappings* (3.1) *are bounded.*

Proof. We shall demonstrate this for the mapping $\mathsf{T} : \mathcal{S}mb \to \operatorname{im} \mathcal{T}$, $q \mapsto \mathsf{T}(q)$. Let $\mu : \mathcal{S}mb \to \operatorname{im} \mathcal{L}$ denote the inverse of the bounded isomorphism in Theorem 3.19 (b). Then μ is bounded by Banach's bounded inverse theorem; hence, the mapping

$$\mathcal{S}mb \to \operatorname{im} \mathcal{T}, \quad q \mapsto \mathsf{T}(q) = P\mu(q)|_X$$

is bounded. □

Remark 3.21. With the above notation and results, it becomes evident that Theorem 2.7 (b) can be viewed as a special case of Theorem 3.6, with $D : \operatorname{im} \mathcal{L} \to \mathsf{TL}(X)$, $A \mapsto E^{-1}AE$ as the discretization mapping and $\mathcal{L}$ as its homomorphic left inverse. □

4. Laurent-like operators

4.1. The axioms

In the previous sections, we started with two families $\mathcal{V}$ and $\mathcal{R}$ of operators to define first an algebra of Toeplitz-like operators and related abstract Toeplitz and Hankel operators, and then a group of shifts on a larger space, together with the related abstract Laurent operators. We will see now that one can also go the way around.

This approach works for Banach spaces X' which are a direct sum of two closed subspaces *of equal size.* Formally this means that there are operators P, Q, $J \in L(X')$ such that

$$P^2 = P,\ Q^2 = Q,\ P + Q = I \quad \text{and} \quad J^2 = I,\ JPJ = Q.$$

Then X' is equal to the direct sum $\operatorname{im} P \oplus \operatorname{im} Q$, the mapping $J|_{\operatorname{imP}} : \operatorname{im} P \to \operatorname{im} Q$ is a linear isomorphism and, with $X := \operatorname{im} P$, there is a linear isomorphism

$$X' \to X^2, \quad x \mapsto (Px,\ JQx) = (Px,\ PJx).$$

In what follows we simply assume that X' is already of the form X^2 for a certain Banach space X and that P, Q and J are given by

$$P : (x,\ y) \mapsto (x,\ 0), \quad Q : (x,\ y) \mapsto (0,\ y) \quad \text{and} \quad J : (x,\ y) \mapsto (y,\ x).$$

Again we write the elements of X^2 as column vectors and identify operators in $L(X^2)$ with 2×2 matrices with entries in $L(X)$.

Let $G \neq \{0\}$ and G_+ be as before, and let $(U_t)_{t \in G}$ be a bounded family of operators $U_t \in L(X^2)$ subject to the following axioms:

(L_1) The mapping $t \mapsto U_t$ is a group isomorphism on G. In particular, $U_s U_t = U_{s+t}$ for all s, $t \in G$.
(L_2) $U_0 = I$ and $JU_tJ = U_{-t}$ for every $t \in G$.
(L_3) $PU_sPU_tP = PU_{s+t}P$ for all s, $t \in G_+$.
(L_4) $U_{-t}PU_t \to I$ strongly as $t \to \infty$.

The first condition in axiom (L_2) ensures that all operators U_t are invertible in $L(X^2)$ and that $U_t^{-1} = U_{-t}$. By calling an invertible operator $U \in L(X^2)$ *J-unitary*, we can rephrase the second condition in axiom (L_2) as the J-unitarity of the U_t. It is easy to see that if U is J-unitary, then there are operators $V_\pm$, $R_\pm \in L(X)$ such that

$$U = \begin{pmatrix} V_+ & R_+ \\ R_- & V_- \end{pmatrix} \quad \text{and} \quad U^{-1} = \begin{pmatrix} V_- & R_- \\ R_+ & V_+ \end{pmatrix}.$$

In particular, there are operators V_t, $R_t \in L(X)$ such that

$$U_t = \begin{pmatrix} V_t & R_t \\ R_{-t} & V_{-t} \end{pmatrix} \quad \text{and} \quad U_t^{-1} = U_{-t} = \begin{pmatrix} V_{-t} & R_{-t} \\ R_t & V_t \end{pmatrix}$$

for all $t \in G_+$.

Lemma 4.1. *The so-defined operators $V_{\pm t}$ and $R_{\pm t}$ are subject to axioms (T_1)–(T_4).*

Proof. The group property of the U_t implies that

$$V_s V_t + R_s R_{-t} = V_{s+t}, \quad V_s R_t + R_s V_{-t} = R_{s+t} \quad \text{for } s,\, t \in G. \tag{4.1}$$

Since $V_s V_t = V_{s+t}$ by (L_3), the first identity in (4.1) implies that

$$R_s R_{-t} = 0 \quad \text{for } s,\, t \in G_+. \tag{4.2}$$

Further we conclude from (L_4) that

$$\begin{pmatrix} V_{-t} & R_{-t} \\ R_t & V_t \end{pmatrix} \begin{pmatrix} I & 0 \\ 0 & 0 \end{pmatrix} \begin{pmatrix} V_t & R_t \\ R_{-t} & V_{-t} \end{pmatrix} = \begin{pmatrix} V_{-t}V_t & V_{-t}R_t \\ R_{-t}V_t & R_t^2 \end{pmatrix} \to \begin{pmatrix} I & 0 \\ 0 & I \end{pmatrix}$$

as $t \to \infty$, whence $R_t^2 \to I$ as $t \to \infty$. Together with (4.1) this implies that

$$0 = R_s^2 R_{-t} \to R_{-t} \quad \text{as } s \to \infty.$$

Thus, $R_{-t} = 0$ for $t \in G_+$, and the matrix representations of U_t and U_{-t} are upper and lower triangular, respectively. With this information it is easy to check that the $(V_{-t})_{t\in G_+}$ own the semigroup property (which gives (T_1) and that $V_{-t}V_t = I$ for $t \in G_+$ and $V_t V_{-t} \to 0$ as $t \to \infty$. Then $V_{-t}V_tV_{-t} = V_{-t} \to 0$ as $t \to \infty$, i.e., (T_4) holds. Another consequence of $V_tV_{-t} \to 0$ is that $V_tV_{-t} \neq I$ respective $P_t := I - V_tV_{-t} \neq 0$ for large positive t. It follows as in the proof of Lemma 3.13 that then $P_t \neq 0$ for all positive t, i.e., $V_tV_{-t} \neq I$ for all positive t, which finishes the proof of (T_2).

It remains to check (T_3). Let $t \in G_+$. The identity $V_tV_{-t} + R_t^2 = I$ is nothing but the south-east corner of 2×2-matrix identity $U_{-t}U_t = I$. Further, $R_tV_t = 0$ by the south-west corner of that identity. Hence, $R_tQ_t = 0$ and $R_tP_t = P_tR_t = R_t$. Now the second identity in (T_3) follows easily by multiplying the second identity in (4.1) from the right-hand side by P_t. □

Note that, conversely, the operators defined by (3.4) satisfy the axioms (L_1)–(L_4).

4.2. Laurent-like operators

Let X^2 and the operators U_t be as in Section 4.1. We let $\mathsf{LL}(X^2)$ stand for the set of all operators $A \in L(X^2)$ for which the two strong limits

$$\mathcal{L}_\pm(A) := \text{s-lim}_{t\to\pm\infty} U_{-t}AU_t$$

exist. We call the elements of $\mathsf{LL}(X^2)$ *Laurent-like operators.*

Theorem 4.2.

(a) $\mathsf{LL}(X^2)$ *is a closed subalgebra of* $L(X^2)$.

(b) *An operator* $A \in L(X^2)$ *is in* $\mathsf{LL}(X^2)$ *if and only if*

$$A = \begin{pmatrix} B & C \\ D & E \end{pmatrix} \quad \text{with} \quad B, C, D, E \in \mathsf{TL}(X) \quad \text{and} \quad C, D \in \ker \mathcal{T}. \tag{4.3}$$

(c) *If* A *is of the form* (4.3), *then*

$$\mathcal{L}_+(A) = \begin{pmatrix} \mathcal{T}(B) & \mathcal{H}(B) \\ \widetilde{\mathcal{H}}(B) & \widetilde{\mathcal{T}}(B) \end{pmatrix} \quad \text{and} \quad \mathcal{L}_-(A) = \begin{pmatrix} \widetilde{\mathcal{T}}(E) & \widetilde{\mathcal{H}}(E) \\ \mathcal{H}(E) & \mathcal{T}(E) \end{pmatrix}. \tag{4.4}$$

Proof. Assertion (a) is standard. Let A be of the form (4.3). Then, for $t \in G_+$,

$$U_{-t}AU_t = \begin{pmatrix} V_{-t}BV_t & V_{-t}BR_t + V_{-t}CV_{-t} \\ R_tBV_t + V_tDV_t & R_tBR_t + V_tDR_t + R_tCV_{-t} + V_tEV_{-t} \end{pmatrix}.$$

The operators $V_{-t}BV_t$, $V_{-t}BR_t$, R_tBV_t and R_tBR_t converge strongly to $\mathcal{T}(B)$, $\mathcal{H}(B)$, $\widetilde{\mathcal{H}}(B)$ and $\widetilde{\mathcal{T}}(B)$ as $t \to \infty$, respectively, because $B \in \mathsf{TL}(X)$. Since $V_{-t} \to 0$ strongly, the operators $V_{-t}CV_{-t}$ and V_tEV_{-t} converge strongly to 0. Finally, the operators V_tDV_t and V_tDR_t converge strongly to 0 because $D \in \ker \mathcal{T}$ and by Lemma 1.11 in [20]. This proves the first assertion in (c); the second one follows analogously.

It remains to verify the "only if" implication in (b). From (L_4) and the J-unitarity of the U_t we conclude that P and Q belong to $\mathsf{LL}(X^2)$. Thus, if

$$A = \begin{pmatrix} B & C \\ D & E \end{pmatrix} \in \mathsf{LL}(X^2),$$

then

$$\begin{pmatrix} B & 0 \\ 0 & 0 \end{pmatrix}, \begin{pmatrix} 0 & C \\ 0 & 0 \end{pmatrix}, \begin{pmatrix} 0 & 0 \\ D & 0 \end{pmatrix}, \begin{pmatrix} 0 & 0 \\ 0 & E \end{pmatrix} \in \mathsf{LL}(X^2).$$

From

$$U_{-t} \begin{pmatrix} B & 0 \\ 0 & 0 \end{pmatrix} U_t = \begin{pmatrix} V_{-t}BV_t & V_{-t}BR_t \\ R_tBV_t & R_tBR_t \end{pmatrix}$$

we conclude that $B \in \mathsf{TL}(X)$. Further, the equality

$$U_{-t} \begin{pmatrix} 0 & 0 \\ D & 0 \end{pmatrix} U_t = \begin{pmatrix} 0 & 0 \\ V_tDV_t & V_tDR_t \end{pmatrix}$$

implies that the operators V_tDV_t and V_tDR_t converge strongly. Since $V_{-t} \to 0$ strongly, we conclude that

$$DV_t = V_{-t}V_tDV_t \to 0 \quad \text{and} \quad DR_t = V_{-t}V_tDR_t \to 0$$

strongly as $t \to \infty$. But then

$$V_{-t}DV_t \to 0, \quad R_tDV_t \to 0, \quad V_{-t}DR_t \to 0, \quad R_tDR_t \to 0$$

strongly as $t \to \infty$. Hence, $D \in \mathsf{TL}(X)$, and each of the four strong limits is zero. In particular, $D \in \ker \mathcal{T}$. Working with $\mathcal{L}_-$ instead of $\mathcal{L}_+$, we obtain that C, $E \in \mathsf{TL}(X)$ and $C \in \ker \mathcal{T}$. □

5. Examples of Toeplitz-like operators

Here we present a few concrete Banach spaces and examine the related algebras of Toeplitz-like operators. A main objective is to identify the abstract Toeplitz and Hankel operators $\mathsf{T}(a)$ and $\mathsf{H}(a)$ with classical (concrete) Toeplitz and Hankel operators acting on these spaces. For a general acquaintance with (concrete) Toeplitz and Hankel operators see, e.g., [4, 9, 16, 18].

5.1. Toeplitz-like operators on H^p

Let H^p, $1 < p < \infty$, be the Hardy space introduced in Section 1, and let $G = \mathbb{Z}$. Let $V_0 := I$ and, for $n \in \mathbb{Z}_+ \setminus \{0\}$, define operators $V_{\pm n}$ on H^p by

$$V_n f := \chi_n f \quad \text{and} \quad V_{-n} f := \chi_{-n} \left(f - \sum_{i=0}^{n-1} \hat{f}_i \chi_i \right),$$

respectively. Clearly, $V_{-n}V_n = I$ and $V_{-n} \to 0$ strongly as $n \to \infty$. Note also that

$$\ker V_{-n} = \left\{ f \in H^p : f = \sum_{i=0}^{n-1} \hat{f}_i \chi_i \right\}.$$

Finally, we define operators R_n on H^p by

$$R_n f := \sum_{i=0}^{n-1} \hat{f}_{n-1-i} \chi_i = JQ\chi_{-n}P$$

if $n \in \mathbb{Z}_+ \setminus \{0\}$ and $R_n = 0$ else. Then all axioms of Section 2.1 are satisfied, and the corresponding algebra $\mathsf{TL}(H^p)$ is well defined. Moreover, $V_n \to 0$ and $R_n \to 0$ weakly. We let U_n stand for the related operators on $(H_p)^2$, defined as in Section 3.5.

As already mentioned, the space $H^p \subset L^p(\mathbb{T})$ is equal to $\operatorname{im} P$, with P the classical Riesz projection. Set $Q := I - P$. Thus, $L^p(\mathbb{T}) = \operatorname{im} P \oplus \operatorname{im} Q$, whereas $(H^p)^2 = \operatorname{im} P \oplus \operatorname{im} P$, which we may write as

$$L^p(\mathbb{T}) = \begin{pmatrix} \operatorname{im} P \\ \operatorname{im} Q \end{pmatrix} \quad \text{and} \quad (H^p)^2 = \begin{pmatrix} \operatorname{im} P \\ \operatorname{im} P \end{pmatrix}.$$

With respect to this identification, we may further identify operators from $L^p(\mathbb{T})$ to $(H^p)^2$ with 2×2-matrices. It is easy to see that the operators

$$\eta := \begin{pmatrix} P & 0 \\ 0 & JQ \end{pmatrix} : \begin{pmatrix} \operatorname{im} P \\ \operatorname{im} Q \end{pmatrix} \to \begin{pmatrix} \operatorname{im} P \\ \operatorname{im} P \end{pmatrix}$$

and

$$\eta^{-1} := \begin{pmatrix} P & 0 \\ 0 & JP \end{pmatrix} : \begin{pmatrix} \operatorname{im} P \\ \operatorname{im} P \end{pmatrix} \to \begin{pmatrix} \operatorname{im} P \\ \operatorname{im} Q \end{pmatrix}$$

are inverse to each other and that $\eta^{-1} U_n \eta = \chi_n I$ for all $n \in \mathbb{Z}$. More precisely, consider the mapping

$$\rho : L^p(\mathbb{T}) \to L^p_2(\mathbb{T}) := \begin{pmatrix} \operatorname{im} P \\ \operatorname{im} Q \end{pmatrix}, \quad h \mapsto \begin{pmatrix} Ph \\ Qh \end{pmatrix}$$

which is an isomorphism with inverse

$$\rho^{-1} : L^p_2(\mathbb{T}) \to L^p(\mathbb{T}), \quad \begin{pmatrix} Ph \\ Qh \end{pmatrix} \mapsto Ph + Qh = h.$$

The norm on $L^p_2(\mathbb{T})$ is chosen in such a way that ρ becomes an isometry. A simple computation gives for $h \in L^p(\mathbb{T})$ and $n \in \mathbb{Z}_+$ that

$$\chi_n h = (P+Q)\chi_n (P+Q)h$$

equals

$$\chi_n h = \rho^{-1} \begin{pmatrix} P\chi_n P & P\chi_n Q \\ Q\chi_n P & Q\chi_n Q \end{pmatrix} \rho h.$$

Using the relations $P\chi_n P = V_n$, $Q\chi_n P = 0$,

$$P\chi_n Q = PJJ\chi_n JPJ = PJ\chi_{-n} PJ = JQ\chi_{-n} PJ = R_n J$$

(holding by the definition of R_n) and

$$Q\chi_n Q = JPJ\chi_n JPJ = JP\chi_{-n} PJ = JV_{-n} J,$$

one gets

$$\chi_n h = \rho^{-1} \begin{pmatrix} V_n & R_n J \\ 0 & JV_{-n} J \end{pmatrix} \rho h.$$

On the other hand,

$$\rho^{-1} \eta^{-1} U_n \eta \rho = \rho^{-1} \begin{pmatrix} V_n & R_n J \\ 0 & JV_{-n} J \end{pmatrix} \rho$$

which, together with the previous identity, implies $\chi_n h = \rho^{-1}\eta^{-1} U_n \eta \rho h$. In the same manner one can check that, for $h \in L^p(\mathbb{T})$ and $n \in \mathbb{Z} \setminus \mathbb{Z}_+$,

$$\chi_{-n} h = \rho^{-1} \eta^{-1} U_{-n} \eta \rho h.$$

Now let $A \in \mathsf{TL}(H^p)$ and $n \in \mathbb{Z}$. From $U_{-n} \mathcal{L}(A) U_n = \mathcal{L}(A)$ we conclude that

$$\chi_{-n} (\rho^{-1} \eta^{-1} \mathcal{L}(A) \eta \rho) \chi_n I = \rho^{-1} \eta^{-1} \mathcal{L}(A) \eta \rho.$$

Hence, $\rho^{-1}\eta^{-1}\mathcal{L}(A)\eta\rho$ is the operator of multiplication by some function $q \in L^\infty(\mathbb{T})$ (see [5]). This gives

$$\eta^{-1} \mathcal{L}(A) \eta = \begin{pmatrix} PqP & PqQ \\ QqP & QqQ \end{pmatrix}$$

which finally implies

$$\mathcal{L}(A) = \begin{pmatrix} T(q) & H(q) \\ H(\tilde{q}) & T(\tilde{q}) \end{pmatrix}.$$

In particular, $\mathcal{T}(A)$, $\widetilde{\mathcal{T}}(A)$, $\mathcal{H}(A)$ and $\widetilde{\mathcal{H}}(A)$ coincide with the familiar (concrete) Toeplitz and Hankel operators $T(q)$, $T(\tilde{q})$, $H(q)$ and $H(\tilde{q})$ on H^p as defined in Section 1.

As a consequence, we observe that the corresponding symbol algebra $\mathcal{S}mb$ is a subalgebra of $L^\infty(\mathbb{T})$. In fact, $\mathcal{S}mb = L^\infty(\mathbb{T})$, which is an immediate consequence of the identities

$$V_{-n}T(a)V_n = T(a), \qquad R_nT(a)R_n = P_nT(\tilde{a}), \tag{5.1}$$

$$V_{-n}T(a)R_n = H(a)P_n, \quad R_nT(a)V_n = P_nH(\tilde{a}) \tag{5.2}$$

for all $a \in L^\infty(\mathbb{T})$ and $n \in \mathbb{Z}_+$. Let us check the last identity of (5.2), for example. By the first relation in (1.1) we have

$$I = T(\chi_n\chi_{-n}) = T(\chi_n)T(\chi_{-n}) + H(\chi_n)H(\chi_n).$$

Hence, $H(\chi_n)^2 = I - Q_n = P_n$. Further,

$$R_nT(a)V_n = R_nT(a\chi_n) = R_n(T(\chi_n)T(a) - H(\chi_n)H(\tilde{a})).$$

Since $R_n = H(\chi_n)$ and $R_nT(\chi_n) = 0$, we conclude that $R_nT(a)V_n = P_nH(\tilde{a})$, as desired. Note that one could also start with (5.1), (5.2) in order to derive that $\eta^{-1}\mathcal{L}(T(a))\eta = aI$.

Let $\mathcal{C}$ be a closed subalgebra of $L^\infty(\mathbb{T})$. Then, in the present context, Corollary 3.9 specifies to

$$\operatorname{alg} TH(\mathcal{C}) = T(\mathcal{C}) \oplus \ker(\mathcal{T}|_{\operatorname{alg} TH(\mathcal{C})}).$$

This result was recently proved by Didas [8, Theorem 2.4] for $p = 2$ and under the assumption that $\mathcal{C}$ is an inner subalgebra of $L^\infty(\mathbb{T})$ in the sense of [17]. For example, $C(\mathbb{T})$ is inner, and so is every closed subalgebra of $L^\infty(\mathbb{T})$ which strictly contains H^∞. Moreover, Didas showed that $\ker(\mathcal{T}|_{\operatorname{alg} TH(\mathcal{C})})$ equals the commutator ideal of $\operatorname{alg} TH(\mathcal{C})$ under this additional assumption.

At the end of Section 2.2 we remarked that $\ker \mathcal{H}$ and $\ker \widetilde{\mathcal{H}}$ are subalgebras of $\mathsf{TL}(X)$. In the setting of this section, we can describe these algebras more precisely as

$$\ker \mathcal{H} = \{T(a) : a \in H^\infty\} \oplus \ker \mathcal{T}, \quad \ker \widetilde{\mathcal{H}} = \{T(a) : a \in \overline{H^\infty}\} \oplus \ker \mathcal{T}.$$

Remark 5.1. The construction of the algebra $\mathsf{TL}(X)$ works as well on weighted Hardy spaces $H^p(w)$ and the above results remain valid in this setting provided that the weight w is symmetric (i.e., $w(t) = w(1/t)$) and that the Riesz projection P is bounded on $L^p(w)$ (which holds if w is a Muckenhoupt weight, for example if w is a power weight; see [3], Chapter 2). In this setting, the operators V_n and R_n can be defined as above; they are uniformly bounded and all requirements made in Section 2.1 are satisfied.

5.2. Toeplitz-like operators on l^p

Here we are going to employ the alternative approach of Section 3.5 to define Toeplitz-like operators on the classical sequence spaces $l^p(\mathbb{Z}_+)$, $1 \le p < \infty$. Consider the reflection operator $J : (x_n)_{n\in\mathbb{Z}} \mapsto (y_n)_{n\in\mathbb{Z}}$ where $y_n = x_{-n-1}$ and the discrete Riesz projection $P : (x_n)_{n\in\mathbb{Z}} \mapsto (y_n)_{n\in\mathbb{Z}}$ where $y_n = x_n$ if $n \ge 0$ and $y_n = 0$ else. Clearly, $J^2 = I$ and $JPJ = I - P =: Q$.

Let $a \in L^\infty(\mathbb{T})$. On $l_0(\mathbb{Z})$, the linear space of the finitely supported sequences, we consider the Laurent operator $L(a)$ defined by

$$(L(a)x)_k := \sum_{m\in\mathbb{Z}} \hat{a}_{k-m} x_m.$$

Note that only finitely many items in this sum do not vanish. The function a is called a *multiplier* on $l^p(\mathbb{Z})$ if

$$\|L(a)\| := \sup\{\|L(a)x\|_p : x \in l_0(\mathbb{Z}),\ \|x\|_p\} < \infty.$$

If a is a multiplier, then $L(a)$ extends to a bounded linear operator on $l^p(\mathbb{Z})$ which we denote by $L(a)$ again. The set M^p of all multipliers on $l^p(\mathbb{Z})$ forms a Banach algebra under the norm $\|a\|_{M^p} := \|L(a)\|_{L(l^p(\mathbb{Z}))}$ (see [4, Chapter 2]). In particular, every function $a \in L^\infty(\mathbb{T})$ with bounded total variation belongs to M^p, and there is a constant c_p independent of a such that Stechkin's inequality

$$\|a\|_{M^p} \le c_p(\|a\|_\infty + \operatorname{Var}(a))$$

holds. Note also that $JL(a)J = L(\tilde{a})$ for every $a \in M^p$.

Write l^p for $l^p(\mathbb{Z}_+) = \operatorname{im} P$ and let $a \in M^p$. The operators

$$T(a) : l^p \to l^p,\ f \mapsto PL(a)f, \quad H(a) : l^p \to l^p,\ f \mapsto PL(a)QJf$$

are called the (concrete) *Toeplitz* and *Hankel operator* with generating function a, respectively. These operators are bounded if $a \in M^p$. It is also easy to see that if $T(a)f = g$ and $H(a)f = h$ then

$$g_j = \sum_{k=0}^{\infty} \hat{a}_{j-k} f_k, \quad h_j = \sum_{k=0}^{\infty} \hat{a}_{j+k+1} f_k.$$

Let $\{e_j\}_{j\in\mathbb{Z}_+}$ stand for the standard basis on l^p, i.e., the jth entry of e_j is 1 whereas all other entries are 0. The l^p-version of a Theorem by Brown and Halmos (Theorem 2.7 (b) in [4]) tells us that if $A \in L(l^p)$ and if there is a sequence $(a_n)_{n\in\mathbb{Z}}$ in $\mathbb{C}$ such that

$$\langle Ae_j, e_k\rangle = a_{j-k} \quad \text{for all } j,\, k \in \mathbb{Z}_+,$$

then there exists a multiplier $a \in M^p$ such that $A = T(a)$ and a_n is the nth Fourier coefficient of a. Here, as usual, $\langle x, y\rangle := \sum x_n \overline{y_n}$ for $x \in l^p$ and $y \in l^q$ with $p^{-1} + q^{-1} = 1$.

Next we are going to define the related algebra $\mathsf{TL}(l^p)$ of Toeplitz-like operators. Set $G = \mathbb{Z}$ and define

$$V_n := T(\chi_n) \quad \text{and} \quad R_n := H(\chi_n) \quad \text{for } n \in \mathbb{Z} \tag{5.3}$$

with $\chi_n(t) = t^n$ as before. One easily checks that these operators satisfy axioms (T_1)–(T_4); hence, the algebra $\mathsf{TL}(l^p)$ with respect to these families is well defined. Moreover, if $1 < p < \infty$, then $V_n \to 0$ and $R_n \to 0$ weakly as $n \to \infty$. Note that this claim fails for $p = 1$. Indeed, consider the bounded linear functional

$$g : l^1(\mathbb{Z}_+) \to \mathbb{C}, \quad x = (x_i)_{i \in \mathbb{Z}_+} \mapsto \sum_{i \in \mathbb{Z}_+} x_i.$$

Then $g(V_n x) = g(x)$ for all $x \in l^1(\mathbb{Z}_+)$ and $n \in \mathbb{Z}_+$; hence, the V_n cannot converge weakly to zero. Now the assertion for the R_s comes from Proposition 2.9.

If $A \in \mathsf{TL}(l^p)$ and $\mathcal{T}(A) = A$, then $V_{-1} A V_1 = A$. Thus, A satisfies the hypotheses of the Brown/Halmos theorem, and there is a multiplier $a \in M^p$ such that $A = T(a)$. Thus, every abstract Toeplitz (or Hankel) operator on l^p is a concrete Toeplitz (or Hankel) operator as defined above. The l^p-version of Theorem 2.7 implies

$$\mathsf{TL}(l^p) / \ker \mathcal{T} \cong M^p.$$

Remark 5.2. An equivalent treatment of Toeplitz-like operators on *weighted* l^p-spaces fails: the operators (5.3) are not longer uniformly bounded then. This is in strong contrast to the H^p-setting in the previous section. □

Our final goal is to provide some peculiar properties of the algebra $\mathsf{TL}(X)$, its ideal $\ker \mathcal{T}$ and its subalgebras $\operatorname{alg} T(L^\infty)$ and $\operatorname{alg} TH(L^\infty)$. In particular, we are going to show that

(a) $\mathsf{TL}(l^2)$ is not a C^*-algebra.
(b) $\ker \mathcal{T} \subset \mathsf{TL}(l^2)$ is not a two-sided ideal of $L(l^2)$.
(c) $\operatorname{alg} TH(L^\infty)$ is a proper subalgebra of $\mathsf{TL}(l^2)$.
(d) $\operatorname{alg} T(L^\infty)$ is a proper subalgebra of $\operatorname{alg} TH(L^\infty)$.

For (a), assume that $\mathsf{TL}(l^2)$ is a C^*-algebra. Then $\ker \mathcal{T}$ is a closed *symmetric* ideal of $\mathsf{TL}(l^2)$ and, hence, a closed *symmetric* left-sided ideal of $L(l^2)$. So we have BC, $B^*C^* \in \ker \mathcal{T}$ for every $B \in L(l^2)$ and $C \in \ker \mathcal{T}$ by the left ideal property, whence $(B^*C^*)^* = CB \in \ker \mathcal{T}$ by the symmetry. Thus, $\ker \mathcal{T}$ turns out to be a closed *two-sided* ideal of $L(l^2)$. Moreover, $\ker \mathcal{T}$ contains non-compact operators (for example, the Hankel operator which generates the Hilbert matrix), but not every bounded linear operator on l^2 is in $\ker \mathcal{T}$ (for example, $I \notin \ker \mathcal{T}$). This contradicts the well-known fact (first observed by Calkin [6]; see also [12]) that the only non-trivial closed two-sided ideal of $L(l^2)$ is the ideal of the compact operators.

The same arguments show (b), and (c) follows because $\operatorname{alg} TH(L^\infty)$ is a symmetric closed subalgebra of $\mathsf{TL}(l^2)$, whereas $\mathsf{TL}(l^2)$ fails to be symmetric by (a). Finally, (d) comes from the existence of Hankel operators which do not belong to $\operatorname{alg} T(L^\infty)$ as first observed by S. Axler and proved in Power [18].

The following explicit example of an operator $A \in \mathsf{TL}(l^2)$ for which $A^* \notin \mathsf{TL}(l^2)$ is due to our former colleague Hans-Jürgen Fischer.

Example. Let the operator A be given by the matrix representation $A = (a_{ij})_{i,j=0}^{\infty}$ where

$$a_{ij} = \begin{cases} \frac{1}{\sqrt{2^i}} & \text{if } 0 \le j \le 2^i - 1, \\ 0 & \text{if } j \ge 2^i. \end{cases}$$

We show that A is a bounded operator on l^2. Let $x = (x_i)_{i \ge 0} \in l^2$ and $\|x\| = 1$. Then

$$\|Ax\|^2 = \sum_{n=0}^{\infty} \frac{1}{2^n} \left| \sum_{k=0}^{2^n-1} x_k \right|^2 .$$

For $k \in \mathbb{Z}^+$, set $\lambda_k := (k+1)^{-\frac{1}{4}}$. Then, by the Cauchy–Schwarz inequality,

$$\left| \sum_{k=0}^{2^n-1} x_k \right|^2 = \left| \sum_{k=0}^{2^n-1} \lambda_k \lambda_k^{-1} x_k \right|^2 \le \left(\sum_{k=0}^{2^n-1} \lambda_k^2 \right) \left(\sum_{k=0}^{2^n-1} \left| \frac{x_k}{\lambda_k} \right|^2 \right).$$

The estimate

$$\sum_{k=0}^{2^n-1} \lambda_k^2 = \sum_{k=0}^{2^n-1} (k+1)^{-\frac{1}{2}} \le 2\sqrt{2^n} - 1$$

then gives

$$\begin{aligned} \|Ax\|^2 &\le \sum_{n=0}^{\infty} \frac{2\sqrt{2^n} - 1}{2^n} \left(\sum_{k=0}^{2^n-1} \left| \frac{x_k}{\lambda_k} \right|^2 \right) \\ &\le 2 \sum_{n=0}^{\infty} \sum_{k=0}^{2^n-1} \frac{1}{\sqrt{2^n}} \left| \frac{x_k}{\lambda_k} \right|^2 \\ &= 2 \sum_{k=0}^{\infty} \sum_{n \ge \log_2(k+1)} \frac{1}{\sqrt{2^n}} \left| \frac{x_k}{\lambda_k} \right|^2 . \end{aligned}$$

Using

$$\sum_{n \ge \log_2(k+1)} \frac{1}{\sqrt{2^n}} = \frac{1}{\sqrt{k+1}} \frac{1}{1 - 1/\sqrt{2}} = \frac{1}{\sqrt{k+1}} \sqrt{2}(\sqrt{2}+1),$$

we arrive at

$$\|Ax\|^2 \le 2\sqrt{2}(\sqrt{2}+1) \sum_{k=0}^{\infty} \frac{1}{\sqrt{k+1}} \left| \frac{x_k}{\lambda_k} \right|^2 = 2\sqrt{2}(\sqrt{2}+1) \sum_{k=0}^{\infty} |x_k|^2,$$

whence the boundedness of A and the estimate $\|A\|^2 \le 2\sqrt{2}(\sqrt{2}+1)$. From

$$\|AV_n e_k\|^2 = \|Ae_{n+k}\|^2 = 2^{-\log_2(n+k)} \quad \text{for } k,\, n \ge 0$$

and

$$\|AR_n e_k\|^2 = \|Ae_{n-k}\|^2 = 2^{-\log_2(n-k)} \quad \text{for } n > k \ge 0$$

we further conclude that $AV_n \to 0$ and $AR_n \to 0$ strongly as $n \to \infty$. Hence, $A \in \mathsf{TL}(l^2)$ (and $A \in \ker \mathcal{T}$). Suppose that $A^* \in \mathsf{TL}(l^2)$. Then $(V_{-n}A^*V_n)$ converges strongly to some operator $\mathcal{T}(A^*)$ by definition. On the other hand, $(V_{-n}A^*V_n) =$

$(V_{-n}AV_n)^* \to 0$ weakly because $V_{-n}AV_n \to 0$ strongly. Hence, $\mathcal{T}(A^*) = 0$. Then, by Lemma 1.11 in [20], $A^*V_n \to 0$ strongly, which is impossible since $\|A^*V_n e_0\| = 1$ for all $n \in \mathbb{Z}^+$. $\square$

Remark 5.3. It is an open question if $\operatorname{alg} TH(L^\infty)$ is a proper subalgebra of

$$\mathsf{TL}^*(l^2) := \{A \in \mathsf{TL}(l^2) : A^* \in \mathsf{TL}(l^2)\}.$$

5.3. Toeplitz-like operators on $L^p(\mathbb{R}_+)$

Throughout this section, we let L^p and $L^p_\pm$ stand for the classical Lebesgue spaces on $\mathbb{R}$, $\mathbb{R}_+ = [0, \infty)$ and $\mathbb{R}_- = \mathbb{R} \setminus \mathbb{R}_+$, respectively. Every function $a \in L^\infty$ induces a bounded operator $f \mapsto af$ on L^p which we denote by $m(a)$. In particular, if χ_+ and χ_- stand for the characteristic functions of $\mathbb{R}_+$ and $\mathbb{R}_-$, then we let P and Q denote the operators $m(\chi_+)$ and $m(\chi_-)$, respectively. The ranges of P and Q can be identified with L^p_+ and L^p_-. We will also need the flip operator J on L^p which is defined by $(Jf)(t) = f(-t)$. Clearly, $J^2 = I$ and $Jm(a)J = m(\tilde{a})$ where $\tilde{a} = a(-t)$ for $a \in L^\infty$.

We write the Fourier transform on L^1 in the form

$$(Ff)(x) = \frac{1}{\sqrt{2\pi}} \int_{\mathbb{R}} f(t)e^{ixt}\, dt, \quad x \in \mathbb{R},$$

and use the same letter F to denote the continuous extension of F to a unitary operator on L^2. Note that $F^{-1} = JF$.

Let $a \in L^\infty$. Then the operator $M(a) := F^{-1}m(a)F : L^2 \to L^2$ is bounded and has norm $\|a\|_\infty$. For $1 \le p < \infty$, we let $M^p(\mathbb{R})$ stand for the collection of all functions $a \in L^\infty$ owning the following property: Whenever $f \in L^2 \cap L^p$, then $M(a) \in L^p$, and there is a constant c_p independent of f such that $\|M(a)f\|_p \le c_p\|f\|_p$ for all $f \in L^2 \cap L^p$. If $a \in M^p(\mathbb{R})$, then $M(a) : L^2 \cap L^p \to L^p$ extends to a bounded operator on L^p, called the operator of convolution by a. We denote this operator by $M(a)$ again. The elements of $M^p(\mathbb{R})$ are also called L^p-multipliers.

An operator A on L^p is called *translation invariant* if $U_{-s}AU_s = A$ for all $s \in \mathbb{R}$ where $(U_s f)(t) := f(t-s)$. If $a \in M^p(\mathbb{R})$, then the convolution operator $M(a)$ is translation invariant. A theorem of Hörmander [15] (see also 9.2 in [4]) states that the converse is also true: every translation invariant operator on L^p is an operator of convolution by a certain L^p-multiplier.

If $a \in L^\infty$ is of finite total variation $V(a)$, then $a \in M^p(\mathbb{R})$, and the Stechkin inequality

$$\|M(a)\|_{L(L^p)} \le C_p(\|a\|_\infty + V(a))$$

holds. Here, C_p is a constant independent of a which can be chosen concretely as $C_p = \|M(\chi_+ - \chi_-)\|_{L(L^p)}$. Note also that $JM(a)J = M(\tilde{a})$. Further basic properties of $M^p(\mathbb{R})$ can be found in [4, Chapter 9], for example.

Let $1 \le p < \infty$. With every multiplier $a \in M^p(\mathbb{R})$, we associate the *Wiener–Hopf integral operator* $W(a)$ and the *Hankel integral operator* $H_{\mathbb{R}}(a)$ by

$$W(a) := PM(a)|_{L^p_+} : L^p_+ \to L^p_+ \quad \text{and} \quad H_{\mathbb{R}}(a) := PM(a)QJ|_{L^p_+} : L^p_+ \to L^p_+,$$

respectively. These operators are bounded on L^p_+, and the following analogues of (1.1) hold for all L^p-multipliers a, b:

$$W(ab) = W(a)W(b) + H_{\mathbb{R}}(a)H_{\mathbb{R}}(\tilde{b}), \quad H_{\mathbb{R}}(ab) = W(a)H_{\mathbb{R}}(b) + H_{\mathbb{R}}(a)W(\tilde{b}).$$

For $s \in \mathbb{R}$, define $\omega_s(x) := e^{isx}$. Then $\omega_s \in M^p(\mathbb{R})$ for every $p \in [1, \infty)$ and $M(\omega_s)$ is nothing but the translation operator U_s. Now set $V_s := W(\omega_s)$ for $s \in G$, $R_s := H_{\mathbb{R}}(\omega_s)$ for $s \in \mathbb{R}_+$ and $R_s = 0$ for $s \in \mathbb{R}_-$. Then the axioms in Section 2.1 are satisfied, thus, the algebra $\mathsf{TL}(L^p_+)$ is well defined. Moreover, the operators V_s and R_s tend weakly to 0 as $s \to \infty$ if $1 < p < \infty$. For $p = 1$, neither the V_s nor the R_s tend weakly to zero which follows by arguments similar to those for $X = l^1$.

As before, one can easily prove that for $s \geq 0$ and $a \in M^p(\mathbb{R})$,

$$V_{-s}W(a)V_s = W(a), \quad R_sW(a)R_s = P_sW(\tilde{a})P_s,$$

$$V_{-s}W(a)R_s = H_{\mathbb{R}}(a)P_s, \quad R_sW(a)V_s = P_sH_{\mathbb{R}}(\tilde{a}).$$

What results is that every Wiener–Hopf operator $W(a)$ belongs to $\mathsf{TL}(L^p_+)$ and that

$$\mathcal{T}(W(a)) = W(a), \quad \widetilde{\mathcal{T}}(W(a)) = W(\tilde{a}),$$

$$\mathcal{H}(W(a)) = H_{\mathbb{R}}(a), \quad \widetilde{\mathcal{H}}(W(a)) = H_{\mathbb{R}}(\tilde{a}).$$

Conversely, we are going to show that if $A \in \mathsf{TL}(L^p_+)$ and $\mathcal{T}(A) = A$, then there exists a multiplier $a \in M^p(\mathbb{R})$ such that $A = W(a)$. We proceed as in Section 5.1. Slightly abusing the notation, we let U_s stand for the related operators on $(L^p_+)^2$, defined as in Section 3.5. As already mentioned, the space $L^p_+ \subset L^p$ is equal to $\operatorname{im} P$. Thus, $L^p = \operatorname{im} P \oplus \operatorname{im} Q$, whereas $(L^p_+)^2 = \operatorname{im} P \oplus \operatorname{im} P$, which we may write as

$$L^p = \begin{pmatrix} \operatorname{im} P \\ \operatorname{im} Q \end{pmatrix} \quad \text{and} \quad (L^p_+)^2 = \begin{pmatrix} \operatorname{im} P \\ \operatorname{im} P \end{pmatrix}.$$

With respect to this identification, we may further identify operators from L^p to $(L^p_+)^2$ with 2×2-matrices. It is easy to see that the operators

$$\eta := \begin{pmatrix} P & 0 \\ 0 & JQ \end{pmatrix} : \begin{pmatrix} \operatorname{im} P \\ \operatorname{im} Q \end{pmatrix} \to \begin{pmatrix} \operatorname{im} P \\ \operatorname{im} P \end{pmatrix}$$

and

$$\eta^{-1} := \begin{pmatrix} P & 0 \\ 0 & JP \end{pmatrix} : \begin{pmatrix} \operatorname{im} P \\ \operatorname{im} P \end{pmatrix} \to \begin{pmatrix} \operatorname{im} P \\ \operatorname{im} Q \end{pmatrix}$$

are inverse to each other and that $\eta^{-1}U_s\eta = M(\omega_s)$ for all $s \in \mathbb{R}$.

Now let $A \in \mathsf{TL}(L^p_+)$. From $U_{-s}\mathcal{L}(A)U_s = \mathcal{L}(A)$ for $s \in \mathbb{R}_+$ we conclude that

$$M(\omega_{-s})(\eta^{-1}\mathcal{L}(A)\eta)M(\omega_s) = \eta^{-1}\mathcal{L}(A)\eta.$$

This equality implies that $\eta^{-1}\mathcal{L}(A)\eta$ is translation invariant. By Hörmander's theorem, there is a multiplier $a \in M^p(\mathbb{R})$ such that

$$\eta^{-1}\mathcal{L}(A)\eta = M(a) = \begin{pmatrix} PM(a)P & PM(a)Q \\ QM(a)P & QM(a)Q \end{pmatrix},$$

which implies

$$\mathcal{L}(A) = \begin{pmatrix} W(a) & H_{\mathbb{R}}(a) \\ H_{\mathbb{R}}(\tilde{a}) & W(\tilde{a}) \end{pmatrix}.$$

In particular, $\mathcal{T}(A)$, $\widetilde{\mathcal{T}}(A)$, $\mathcal{H}(A)$ and $\widetilde{\mathcal{H}}(A)$ coincide with the familiar (concrete) Wiener–Hopf and Hankel operators $W(a)$, $W(\tilde{q})$, $H_{\mathbb{R}}(a)$ and $H_{\mathbb{R}}(\tilde{a})$ as defined in the beginning of this section.

We thus arrived at the following theorem which can be viewed as the analogue of the Brown/Halmos theorem in the L^p_+-setting:

Theorem 5.4.
(a) *If* $a \in M^p(\mathbb{R})$, *then* $W(a)$ *belongs to* $\mathsf{TL}(L^p_+)$.
(b) *If* $A \in \mathsf{TL}(L^p_+)$ *and* $A = \mathcal{T}(A)$, *then there exists a multiplier* $a \in M^p(\mathbb{R})$ *such that* $A = W(a)$.

Corollary 5.5. $\mathsf{TL}(L^p_+)/\ker \mathcal{T} \cong M^p(\mathbb{R})$.

In particular we conclude that $M^p(\mathbb{R})$ actually forms a Banach algebra.

6. Derivations

Let $\mathcal{A}$ and $\mathcal{B}$ be Banach algebras. A *derivation* on $\mathcal{A}$ is a (not necessarily bounded) linear operator $\mathsf{D} : \mathcal{A} \to \mathcal{B}$ such that $\mathsf{D}(ab) = \mathsf{D}(a)b + a\mathsf{D}(b)$ for all a, $b \in \mathcal{A}$. Every element $c \in \mathcal{A}$ induces a derivation $a \mapsto ca - ac$ on $\mathcal{A}$.

Chernoff [7] showed that if X is an infinite-dimensional Banach space and $\mathcal{A}$ a closed subalgebra of $L(X)$ which contains the compact operators, then every derivation $\mathsf{D} : \mathcal{A} \to L(X)$ is bounded, and there is an operator $C \in L(X)$ such that $\mathsf{D}(A) = CA - AC$ for all $A \in \mathcal{A}$.

Theorem 6.1. *Let* $\mathcal{A}$ *be a closed subalgebra of* $\mathsf{LT}(X)$ *and suppose that the symbol algebra* $\operatorname{smb}\mathcal{A}$ *is commutative and semisimple. Then every bounded derivation* $\mathsf{D} : \mathcal{A} \to \mathcal{A}$ *maps* $\mathcal{A}$ *into* $\ker \mathcal{T}$.

Proof. Let π denote the canonical homomorphism from $\mathcal{A}$ onto $\mathcal{A}/\operatorname{com}\mathcal{A} =: \mathcal{A}^\pi$, with $\operatorname{com}\mathcal{A}$ again referring to the commutator ideal of $\mathcal{A}$. From

$$\begin{aligned} \mathsf{D}(AB - BA) &= (\mathsf{D}(A)B + A\mathsf{D}(B)) - (\mathsf{D}(B)A + B\mathsf{D}(A)) \\ &= (\mathsf{D}(A)B - B\mathsf{D}(A)) + (A\mathsf{D}(B) - \mathsf{D}(B)A) \end{aligned}$$

we conclude that D maps $\operatorname{com}\mathcal{A}$ into $\operatorname{com}\mathcal{A}$. Hence, the quotient mapping

$$\mathsf{D}^\pi : \mathcal{A}^\pi \to \mathcal{A}^\pi, \ \pi(A) \mapsto \pi(\mathsf{D}(A))$$

is a well-defined bounded derivation on $\mathcal{A}^\pi$. Since $\mathcal{A}^\pi = \mathcal{A}/\operatorname{com}\mathcal{A}$ is commutative, D^π maps the algebra $\mathcal{A}^\pi$ into its Jacobson radical $\operatorname{rad}\mathcal{A}^\pi$ by the Singer–Wermer theorem (see [22]).

Since $\operatorname{smb}\mathcal{A}$ is commutative, the commutator ideal $\operatorname{com}\mathcal{A}$ is contained in $\mathcal{A} \cap \ker \mathcal{T}$, as we observed before Lemma 3.12. Hence, π maps $\mathcal{A} \cap \ker \mathcal{T}$ onto a closed ideal $(\mathcal{A} \cap \ker \mathcal{T})^\pi$ of $\mathcal{A}^\pi$. Let ω stand for the canonical homomorphism from $\mathcal{A}^\pi$ onto $\mathcal{A}^\pi/(\mathcal{A}\cap\ker\mathcal{T})^\pi$. Then $\omega\circ\pi$ is a homomorphism on $\mathcal{A}$ with kernel $\mathcal{A}\cap\ker\mathcal{T}$. Hence, the quotient algebra $\mathcal{A}^\pi/(\mathcal{A}\cap\ker\mathcal{T})^\pi$ is isomorphic to the algebra $\operatorname{smb}\mathcal{A}$. Since $\operatorname{smb}\mathcal{A}$ is semisimple by assumption, we conclude that $\operatorname{rad}\mathcal{A}^\pi \subseteq (\mathcal{A}\cap\ker\mathcal{T})^\pi$. Thus, D^π maps $\mathcal{A}^\pi$ into $(\mathcal{A}\cap\ker\mathcal{T})^\pi$, which implies that D maps $\mathcal{A}$ into $\mathcal{A}\cap\ker\mathcal{T}$. □

Combining this result with Chernoff's theorem, we arrive at the following.

Corollary 6.2. *Let $\mathcal{A}$ be as in Theorem* 6.1 *and suppose that $\mathcal{A}$ contains the compact operators. Then every derivation $\mathsf{D}:\mathcal{A}\to\mathcal{A}$ is bounded, there is an operator $T\in L(X)$ such that $\mathsf{D}(A)=TA-AT$, and D maps into* $\ker(\mathcal{T}|_\mathcal{A})$.

Note that if $\mathcal{A}$ contains the compact operators then $\mathsf{TL}(X)$ contains the compact operators, which happens if and only if $V_s\to 0$ and $R_s\to 0$ weakly as $s\to\infty$ by Theorem 2.8.

Didas [8] derived Corollary 6.2 in the context of (concrete) Toeplitz and Hankel operators acting on the Hardy space H^2 on the unit disk (as defined in the introduction). Moreover, assuming that $\mathcal{C}$ is an inner subalgebra of $L^\infty(\mathbb{T})$ which strictly contains H^∞, he showed that every derivation on $\operatorname{alg} TH(\mathcal{C})$ maps into the *commutator* ideal of that algebra. The strict containment of H^∞ in $\mathcal{C}$ implies that $\operatorname{alg} TH(\mathcal{C})$ contains the compact operators and that the commutator ideal equals $\operatorname{alg} TH(\mathcal{C})\cap\ker\mathcal{T}$ (see [9]).

If $1<p<\infty$ and $\mathcal{S}$ is a closed subalgebra of the symbol algebra $L^\infty(\mathbb{T})$ which contains $C(\mathbb{T})$, then the Toeplitz algebra $\operatorname{alg}\mathsf{T}(\mathcal{S})\subset L(H^p)$ contains all compact operators. An analogous result holds for concrete Toeplitz algebras on l^p and $L^p(\mathbb{R})$, $1<p<\infty$, if $\mathcal{S}$ is a closed subalgebra of corresponding multiplier algebras which contains all functions in the Wiener algebra over $\mathbb{T}$ and $\mathbb{R}$, respectively.

References

[1] J. Barria, *On Hankel operators not in the Toeplitz algebra.* Proc. Amer. Math. Soc. **124** (1996), 1507–1511.

[2] J. Barria and P.R. Halmos, *Asymptotic Toeplitz operators.* Trans. Amer. Math. Soc. **273** (1982), 621–630.

[3] A. Böttcher and Yu.I. Karlovich, *Carleson Curves, Muckenhoupt Weights, and Toeplitz Operators.* Birkhäuser, Basel, 1997.

[4] A. Böttcher and B. Silbermann, *Analysis of Toeplitz Operators.* 2nd ed., Springer, Berlin, 2006).

[5] A. Brown and P.R. Halmos, *Algebraic properties of Toeplitz operators.* J. Reine Angew. Math. **213** (1964), 89–102.

[6] J.W. Calkin, *Two-sided ideals and congruences in the ring of bounded operators in Hilbert space.* Ann. Math. **42** (1941), 839–873.

[7] P.R. Chernoff, *Representations, automorphisms, and derivations of some operator algebras.* J. Funct. Anal. **12** (1973), 275–289.

[8] M. Didas, *On the structure of Hankel algebras.* Integr. Equ. Oper. Theory **80** (2014), 511–525.

[9] R.G. Douglas, *Banach Algebra Techniques in Operator theory.* Academic Press, New York 1972.

[10] R.G. Douglas, *Banach algebra techniques in the theory of Toeplitz operators.* CBMS Regional Conf. Ser. Math. **15**, Amer. Math. Soc., Providence, R. I., 1973.

[11] A. Feintuch, *On asymptotic Toeplitz and Hankel operators.* Oper. Theory: Adv. Appl. **41**, Birkhäuser 1989, 241–254.

[12] I.C. Gohberg, A.S. Markus, and I.A. Feldman, *Normally solvable operators and ideals associated with them.* Amer. Math. Soc. Transl. **61** (1967), 63–84.

[13] I. Gohberg and I. Feldman, *Convolution Equations and Projection Methods for Their Solution.* Nauka, Moskva 1971 (Russian; English transl.: Am. Math. Soc. Transl. of Math. Monographs **41**, Providence, R. I., 1974).

[14] R. Hagen, S. Roch, and B. Silbermann, *C^*-Algebras and Numerical Analysis.* Marcel Dekker, Inc., 2001.

[15] L. Hörmander, *Estimates for translation invariant operators in L^p spaces.* Acta Math. **104** (1960), 93–140.

[16] V. Peller, *Hankel Operators and Their Applications.* Springer, Berlin, 2003.

[17] S.C. Power, *C^*-algebras generated by Hankel operators and Toeplitz operators.* J. Funct. Anal. **31** (1979), 52–68.

[18] S.C. Power, *Hankel operators on Hilbert space.* Bull. London Math. Soc. **12** (1980), 422–442.

[19] S. Roch and B. Silbermann, *Toeplitz-like operators, quasicommutator ideals, numerical analysis I.* Preprint P-MATH-22/83, Institut für Mathematik der AdW, Berlin 1983.

[20] S. Roch and B. Silbermann, *Toeplitz-like operators, quasicommutator ideals, numerical analysis I.* Math. Nachr. **120** (1985), 141–173.

[21] S. Roch and B. Silbermann, *Toeplitz-like operators, quasicommutator ideals, numerical analysis II.* Math. Nachr. **134** (1987), 245–255.

[22] M. Singer and J. Wermer, *Derivations on commutative normed algebras.* Math. Ann. **129** (1955), 260–264.

Steffen Roch
Technische Universität Darmstadt
Fachbereich Mathematik
Schlossgartenstrasse 7
D-64289 Darmstadt, Germany
e-mail: `roch@mathematik.tu-darmstadt.de`

Bernd Silbermann
Technische Universität Chemnitz
Fakultät für Mathematik
D-09107 Chemnitz, Germany
e-mail: `silbermn@mathematik.tu-chemnitz.de`

Operator Theory:
Advances and Applications, Vol. 267, 317–358

More Than 40 Years of Algebraic Techniques in Numerical Analysis

Pedro A. Santos

Abstract. Algebraic techniques in Numerical Analysis represents the use of Banach and C*-algebra theory to help answer questions and solve problems in Numerical Analysis. On the forty-fourth anniversary of the seminal work by Kozak relating stability of an approximation sequence with invertibility in a suitably chosen Banach algebra, the author presents a review of the main issues, history, results and a few of the remaining challenges facing this interdisciplinary area of research, on the frontier between Operator Theory, Banach and C*-algebras, and Numerical Analysis.

Mathematics Subject Classification (2010). Primary 65-02, 65J10; Secondary 46H99, 47B35, 47L99.

Keywords. Algebraic methods, numerical analysis, history.

1. Introduction

Let X, Y be Banach spaces and denote by $\mathcal{B}(X,Y)$ the Banach algebra of all bounded linear operators on X with values in Y, with the usual norm

$$\|A\| := \sup_{x\neq 0} \|Ax\|/\|x\|.$$

We write $\mathcal{B}(X)$ for $\mathcal{B}(X,X)$.

Given $A \in \mathcal{B}(X,Y)$, consider the equation

$$Ax = y \tag{1}$$

with $y \in Y$ given, and $x \in X$ unknown. If the dimension of X is very large or infinite, it is usually not possible to solve such an equation directly. To solve the equation (1) numerically by a direct method, formally one can specify a sequence of simpler (*e.g.*, acting on a finite-dimensional space, or being an integral operators over a compact interval) operators A_τ, with τ in some (eventually partially[1]) ordered unbounded set $\mathbb{I}$, which "approximate" A, in the following way. Let $P_\tau \in \mathcal{B}(X)$ and $R_\tau \in \mathcal{B}(Y)$ be projections such that $P_\tau, R_\tau \to I$ strongly and let A_τ be a sequence of bounded operators $A_\tau : \mathrm{Im}\ P_\tau \to \mathrm{Im}\ R_\tau$. Put $y_\tau = R_\tau y$ and replace the equation $Ax = y$ by the sequence of the (simpler) equations $A_\tau x_\tau = y_\tau$. The crucial question is if this method *applies*, *i.e.*, if there is a τ_0 such that the equations $A_\tau x_\tau = y_\tau$ possess unique solutions for every right-hand side y for $\tau \geq \tau_0$, and if the sequence $(x_\tau)_{\tau\geq\tau_0}$ converges to a solution x of the original equation $Ax = y$. Note that the notion of applicability implies that A is surjective.

One could think that the invertibility of the operator A and strong convergence of A_τ to A would be necessary and sufficient for the approximation method to apply. But, unless the operators A_τ approximate A in the operator norm, that is not always the case, as the following simple example shows. We will consider X and Y as the space $l^2_\mathbb{Z}$ of the double infinite sequences $(u_j)_{j=-\infty}^{+\infty}$ which are square summable:

$$\sum_{j=-\infty}^{+\infty} |u_j|^2 < \infty, \quad \|u\| := \left(\sum_{j=-\infty}^{+\infty} |u_j|^2 \right)^{1/2}$$

and the respective canonical basis. Two simple operators on $l^2_\mathbb{Z}$ are the forward shift defined as $(Su)_j := u_{j-1}$ and its inverse. They are unitary operators. Another operator acting on $l^2_\mathbb{Z}$ is the finite section projection P_n, $n \geq 0$, defined by

$$(P_n u_j) := \begin{cases} u_j & \text{if } -n \leq j \leq n \\ 0 & \text{if } j < -n \text{ or } j > n. \end{cases}$$

Given any operator A on $l^2_\mathbb{Z}$, the sequence $(A_n)_{n>0}$ defined by $A_n := P_n A P_n$ approximates A in the strong sense, that is, $\|(A - A_n)u\| \to 0$ for any fixed $u \in l^2_\mathbb{Z}$. Approximating the operator A by the operator A_n defined above is applying the *Finite Section Method* to A.

Example 1.1. Let $A = I - 2S$. This is an invertible operator, as can easily be seen by writing

$$A = -2S(I - \frac{1}{2}S^{-1}).$$

The inverse is given by the Neumann series

$$A^{-1} = \left(I - \frac{1}{2}S^{-1}\right)^{-1} (-2S)^{-1} = \left(\sum_{k=0}^{+\infty} \left(\frac{1}{2}S^{-1}\right)^k\right) \left(-\frac{1}{2}S^{-1}\right)$$

$$= -\sum_{k=0}^{+\infty} 2^{-k-1} S^{-k-1}.$$

[1] In order not to overburden the notation and proofs with technical details, we will consider in general the index set to be $\mathbb{Z}^+$ or $\mathbb{R}^+$. But most results carry on to more general index sets, as can be seen for instance in [55, Chapter 4].

Representing A and A^{-1} as infinite matrices using the canonical basis, one obtains the Laurent matrices

$$A = \begin{bmatrix} \ddots & \ddots & \ddots & \ddots & \ddots & \ddots & \\ \ddots & 1 & 0 & 0 & 0 & 0 & \ddots \\ \ddots & -2 & 1 & 0 & 0 & 0 & \ddots \\ \ddots & 0 & -2 & \underline{1} & 0 & 0 & \ddots \\ \ddots & 0 & 0 & -2 & 1 & 0 & \ddots \\ \ddots & 0 & 0 & 0 & -2 & 1 & \ddots \\ & \ddots & \ddots & \ddots & \ddots & \ddots & \end{bmatrix}$$

and

$$A^{-1} = -\begin{bmatrix} \ddots & \ddots & \ddots & \ddots & \ddots & \ddots & \\ \ddots & 0 & 2^{-1} & 2^{-2} & 2^{-3} & 2^{-4} & \ddots \\ \ddots & 0 & 0 & 2^{-1} & 2^{-2} & 2^{-3} & \ddots \\ \ddots & 0 & 0 & \underline{0} & 2^{-1} & 2^{-2} & \ddots \\ \ddots & 0 & 0 & 0 & 0 & 2^{-1} & \ddots \\ \ddots & 0 & 0 & 0 & 0 & 0 & \ddots \\ & \ddots & \ddots & \ddots & \ddots & \ddots & \end{bmatrix},$$

where the 00-entry is underlined. Now the operator $A_n = P_n A P_n$ is represented as the $(2n+1) \times (2n+1)$ finite matrix

$$A_n = \begin{bmatrix} 1 & 0 & 0 & \dots & 0 \\ -2 & 1 & 0 & \ddots & 0 \\ 0 & -2 & 1 & \ddots & \vdots \\ \vdots & \ddots & \ddots & \ddots & 0 \\ 0 & 0 & \dots & -2 & 1 \end{bmatrix}$$

with inverse

$$(A_n)^{-1} = \begin{bmatrix} 1 & 0 & 0 & \dots & 0 \\ 2 & 1 & 0 & \ddots & 0 \\ 4 & 2 & 1 & \ddots & \vdots \\ \vdots & \ddots & \ddots & \ddots & 0 \\ 2^{2n} & \dots & 4 & 2 & 1 \end{bmatrix}.$$

So in this case A and its finite section are both invertible (for any $n > 0$) but the inverses $(A_n)^{-1}$ do not approximate A^{-1} in any way.

Other types of mismatch between an infinite-dimensional operator and its finite-dimensional approximations also occur when one considers the operator spectrum. In the above example, for instance, $\sigma(I - 2S)$ is a circle of radius 2 centered on the point 1, while $\sigma(A_n) = \{1\}$.

The applicability of an approximation method $A_\tau x = y_\tau$ to an operator A is closely related to the stability of the operator sequence (A_τ), *i.e.*, to the existence of an index $\tau_0 \in \mathbb{I}$, the invertibility of the operators A_τ for $\tau > \tau_0$ and to the uniform boundedness of the norms of their inverses. The result is due to Polski:

Proposition 1.2 (Polski, 1963). *If $A_\tau \to A$ strongly, then the approximation method applies to A if and only if the sequence (A_τ) is stable and the operator A is invertible.*

Proof. ($\Rightarrow$) If the approximation method applies to A then $A_\tau^{-1} R_\tau y$ converges for all $y \in Y$. Then, by the Uniform Boundedness Principle we obtain that $\sup_{\tau > \tau_0} \|(A_\tau^{-1}) R_\tau\| < \infty$. Regarding the invertibility write

$$\|P_\tau x - A_\tau^{-1} R_\tau A x\| \le \|A_\tau^{-1} R_\tau\| \, \|A_\tau P_\tau x - Ax\|.$$

Because of the stability and the strong convergence of $A_\tau P_\tau$ to A, the left-hand side goes to zero as $\tau \to \infty$. If x is in the kernel of A, this implies in particular that $P_\tau x \to 0$, that is, $x = 0$. So the kernel of A is trivial. Because the range of A is Y by definition of applicability of an approximation method, we conclude that A is invertible.

($\Leftarrow$) Let x_τ denote the solution of the approximate equation $A_\tau P_\tau x_\tau = R_\tau y$ and x that of the original equation. If the sequence (A_τ) is stable, then

$$\begin{aligned}\|x - x_\tau\|_X &= \|(P_\tau + I - P_\tau)x - (A_\tau)^{-1} R_\tau y\|_X \\ &= \|(A_\tau)^{-1} A_\tau P_\tau x + (I - P_\tau)x - (A_\tau)^{-1} R_\tau A x\|_X \\ &\le \|(A_\tau)^{-1} R_\tau\| \|A_\tau P_\tau x - Ax\|_X + \|(I - P_\tau)x\|_X \to 0. \qquad \square\end{aligned}$$

We have thus a necessary and sufficient condition for an approximation method to apply, the stability of its approximation sequence. One can say that the above statement can be resumed in the expression *applicability = approximation + stability.*

There are some instances where the stability problem is easy to solve. For example, let H be a Hilbert space, A a positive definite operator, and $(P_n)_{n \ge 1}$ be a sequence of self-adjoint projections which converges strongly to the identity operator. By the positive definiteness, there is a positive constant C such that

$$\langle Ax, x\rangle \ge C \|x\|^2 \quad \text{for all } x \in H.$$

Then it is immediate that

$$\langle P_n A P_n x_n, x_n\rangle = \langle A P_n x_n, P_n x_n\rangle \ge C \|x_n\|^2 \quad \text{for all } x_n \in \operatorname{Im} P_n,$$

and the operators $A_n : \text{Im } P_n \to \text{Im } P_n$ are invertible for each $n \geq 1$ with the norms of their inverses not greater that $1/C$. So, the sequence $(P_n A P_n)$ is stable and one even has $\tau_0 = 1$. Outside of these particular cases, the stability problem for more general types of operator proves to be a very delicate one. The insight by Polsky does not yet allow us to understand what characterizes operators and approximating methods for which there is applicability. The road to obtain those results was a long one, and started in the second part of the 20th century.

This paper presents an historical view of the major steps of this long path with some emphasis on the author's almost 25 years contact and contribution to this vast area of mathematical research. The objective is to present a readable and interesting account for the non-specialist, which allow him or her to understand the main issues and have pointers to a more in-depth look at some of the topics. We start in the next section by describing the grounding ideas, and then proceed dedicating some space to each generalization and use of algebraic techniques to tackle new problems. Finally we present a couple of possible future research directions on this already mature subject.

2. The pioneering works

During the sixties and seventies of last century, there was a lot of interest in the finite section method (FSM) for Toeplitz operator matrices. We introduce here the context and definition of such operators.

2.1. The context and the problem

Let $\mathbb{T}$ represent the unit circle on the complex plane and let $L^p(\mathbb{T})$, $1 \leq p < \infty$ denote the Banach space of (the classes of) complex-valued measurable functions a on $\mathbb{T}$ where the Lebesgue integral

$$\int_0^{2\pi} |a(e^{\mathrm{i}\theta})|^p \, d\theta \quad =: \|a\|_p^p$$

is finite, endowed with the norm $\|\cdot\|_p$. Moreover, let $L^\infty(\mathbb{T})$ denote the space of all essentially bounded measurable functions, with norm $\|a\|_\infty := \text{ess sup}_{t\in\mathbb{T}}|a(t)|$. For $1 < p < q < \infty$ one has

$$L^\infty(\mathbb{T}) \subset L^q(\mathbb{T}) \subset L^p(\mathbb{T}) \subset L^1(\mathbb{T}).$$

Given $a \in L^1(\mathbb{T})$, it is possible to define its Fourier coefficients a_n $(n \in \mathbb{Z})$ by

$$a_n := \frac{1}{2\pi} \int_0^{2\pi} a(e^{\mathrm{i}\theta}) e^{-\mathrm{i}n\theta} \, d\theta.$$

There is an isometry between the Hilbert space $L^2(\mathbb{T})$ and the double infinite sequence space $l^2_{\mathbb{Z}}$. A function a belongs to $L^2(\mathbb{T})$ if and only if $\sum_{n\in\mathbb{Z}} |a_n|^2 < \infty$ and in that case $\|a\|_2^2 = 2\pi \sum_{n\in\mathbb{Z}} |a_n|^2$. Moreover, the set of the trigonometric polynomials $\chi_n(t) := t^n$, $\{\chi^n\}_{n\in\mathbb{Z}}$ for $t \in \mathbb{T}$, forms an orthogonal basis in $L^2(\mathbb{T})$.

The sequence spaces $l^p_{\mathbb{Z}}$ (resp l^p), $1 \leq p < \infty$, are defined as the Banach space of all sequences $\varphi = (\varphi_n)_{n\in\mathbb{Z}}$ (resp $\varphi = (\varphi_n)_{n\in\mathbb{Z}^+}$) of complex numbers for which

$$\sum_{n\in\mathbb{Z}} |\varphi_n|^p < \infty \qquad \left(\text{resp. } \sum_{n\in\mathbb{Z}^+} |\varphi_n|^p < \infty \right)$$

with $\mathbb{Z}^+ := \{n \in \mathbb{Z} : n \geq 0\}$. Given a function $a \in L^\infty(\mathbb{T})$, the multiplication operator $M(a) : L^2(\mathbb{T}) \to L^2(\mathbb{T})$, $u \mapsto au$ can be written in the $l^2_{\mathbb{Z}}$ space as a convolution $(u_n) \mapsto ((a * u)_n)$ with $(a * u)_n := \sum_{k\in\mathbb{Z}} a_{n-k}u_k$. That operator can thus be represented as a double infinite matrix with constant entries along the diagonals and is known as the Laurent operator with symbol a. Not all $a \in L^\infty(\mathbb{T})$ generate bounded operators in l^p for $p \neq 2$. The set of all $a \in L^1(\mathbb{T})$ that do generate bounded Laurent operators is denoted by M^p and its elements are called multipliers. For a detailed construction of the corresponding operators for $p \neq 2$ see [26, Chapters 1 and 2]. The Wiener algebra W is constituted by all all $a \in L^1(\mathbb{T})$ whose Fourier coefficients belong to $l^1_{\mathbb{Z}}$. We have that $W \subset M^p$ for all p.

The discrete Riesz projection $P : l^p_{\mathbb{Z}} \to l^p$ is defined as $(u_n)_{n\in\mathbb{Z}} \mapsto (u_n)_{n\in\mathbb{Z}^+}$. One can also think of the Riesz projection as acting on $l^p_{\mathbb{Z}}$ and putting to zero all elements of the sequence with negative indices. Given $a \in M^p$, then the Toeplitz operator on l^p generated by a is defined by

$$T(a) : l^p \to l^p, \qquad u \mapsto P(a * u) \quad (= PM(a)Pu\ )$$

and can be represented as a one-sided infinite matrix

$$[a_{ij}]_{i,j\geq 0}, \qquad a_{ij} := a_{i-j}.$$

In fact, infinite Toeplitz and Laurent matrices form one of the simplest class of infinite-dimensional operators one can think of that can be non-compact, and in general, non-self-adjoint or positive definite. And the FSM for matrices is one of the simplest finite-dimensional strongly convergent approximation methods.

Besides applicability of the FSM, related subjects were also studied by mathematicians in the mid 20th century, mainly in the USA and USSR. So, for instance Baxter [5] obtained in 1963 results on the applicability of the FSM for Toeplitz operators with generating function in the Wiener algebra, while also working on determinant asymptotics. Determinant asymptotics of Toeplitz matrices are an older subject, with several seminal results obtained by Szegő [150, 151] since 1915, and then generalized by several authors until now [2, 26, 39] (see also Section 3.4).

From 1965 to 1971 Gohberg and Feldman developed a theory on the applicability of the FSM for convolution type operators (including both the discrete and continuous forms, paired operators, multi-dimensional case, etc) with continuous generating function, which was published as a book [46]. Widom [155, 156] studied the asymptotic inversion of convolution operators, to obtain estimates for determinants, and behavior of the spectrum.

A result by Kozak from 1973, would completely change the perspective, after being combined with an unrelated observation from Widom. Kozak, following a suggestion from his thesis adviser Dybin [26, Section 7.11] observed that the stability problem for a sequence was equivalent to an invertibility problem in a suitably constructed Banach algebra [81].

Let $\mathcal{E}$ be the set of all bounded sequences $\mathbb{A} := (A_\tau)_{\tau\in\mathbb{I}}$ of operators $A_\tau \in \mathcal{B}(X)$. Provided with the operations

$$(A_\tau) + (B_\tau) := (A_\tau + B_\tau), \quad \alpha(A_\tau) := (\alpha A_\tau), \quad (A_\tau)(B_\tau) := (A_\tau B_\tau)$$

and with the norm $\|(A_\tau)\| := \sup_\tau \|A_\tau\|_{\mathcal{B}(X)}$, the set $\mathcal{E}$ becomes a unital Banach algebra. Note that the constant sequences (A) are included in $\mathcal{E}$ for every operator $A \in \mathcal{B}(X)$. If X is a Hilbert space, then we can equip $\mathcal{E}$ with the involution $(A_\tau)^* := (A_\tau^*)$ which makes $\mathcal{E}$ a C^*-algebra.

Let $\mathcal{G}$ be the set of all sequences $(A_\tau) \in \mathcal{E}$ with $\lim_{\tau\to\infty} \|A_\tau\| = 0$. This set forms a closed two-ideal in $\mathcal{E}$.

Theorem 2.1 (Kozak). *A sequence $\mathbb{A} \in \mathcal{E}$ is stable if and only if $\mathbb{A} + \mathcal{G}$ is invertible in $\mathcal{E}/\mathcal{G}$.*

Proof. ($\Rightarrow$) If $\mathbb{A} = (A_\tau)_{\tau\in\mathbb{I}}$ and there exists a τ_0 such that A_τ is invertible for $\tau > \tau_0$ and the norms of the inverses are uniformly bounded, then one can define a sequence of operators B_τ by

$$B_\tau := \begin{cases} A_\tau^{-1} & \text{if } \tau > \tau_0 \\ 0 & \text{if } \tau \le \tau_0 \end{cases}.$$

Defining $\mathbb{B} := (B_\tau)_{\tau\in\mathbb{I}} \in \mathcal{E}$, we have that $\mathbb{B} + \mathcal{G}$ is the inverse of $\mathbb{A} + \mathcal{G}$ in $\mathcal{E}/\mathcal{G}$.
($\Leftarrow$) There exists $\mathbb{B} := (B_\tau)_{\tau\in\mathbb{I}} \in \mathcal{E}$ such that

$$B_\tau A_\tau = I + G_\tau \qquad \text{and} \qquad A_\tau B_\tau = I + G'_\tau$$

with $\|G_\tau\|$, $\|G'_\tau\| \to 0$. There is a τ_0 such that for $\tau > \tau_0$ both $\|G_\tau\|$ and $\|G'_\tau\|$ are less than $1/2$. Then $I + G_\tau$ and $I + G'_\tau$ are invertible and A_τ has both a right and left inverse. Moreover,

$$\|A_\tau^{-1}\| \le \|(I + G_\tau)^{-1}\| \|B_\tau\| \le 2\|B_\tau\|. \qquad \square$$

With this algebraization of the stability problem, Kozak arrived at studying invertibility in a non-commutative Banach algebra. The other main achievement of [81] was that Kozak generalized Simonenko's local principle, used for the investigation of the Fredholm property of operators, to a local principle for exploring invertibility in general Banach algebras. This generalization of Simonenko's local principle was powerful enough to treat the finite section method for multidimensional convolution operators with continuous symbols. Several attempts to extend Kozak's method to non-continuous symbols failed because the sequences that resulted were not of "local type", a key feature needed to use Simonenko's local principle (see, e.g., [120, Section 2.5]).

Subsequently other general local principles were introduced by several authors. They are all based on the happy circumstance that if a Banach algebras has a non-trivial center, then the invertibility problem in this algebra can be reduced to a collection of invertibility problems in simpler related algebras. A concise survey of this topic is in [120, Chapter 2], with proofs. Below we present the version which is known as local principle by Allan/Douglas. A different but equivalent (in the complex Banach algebra setting) formulation exists in the form of the Gohberg–Krupnik local principle introduced in [49].

Theorem 2.2 (Allan/Douglas). *Let $\mathcal{A}$ be a Banach algebra with identity e and let $\mathcal{C}$ be closed central subalgebra of A containing e. Let $M(\mathcal{C})$ be the maximal ideal space of $\mathcal{C}$, and for $\omega \in M(\mathcal{C})$, let $\mathcal{I}_\omega$ refer to the smallest closed two-sided ideal of $\mathcal{A}$ containing the ideal ω. Then an element $a \in \mathcal{A}$ is invertible in $\mathcal{A}$ if and only if $a + \mathcal{I}_\omega$ is invertible in the quotient algebra $\mathcal{A}/\mathcal{I}_\omega$ for all $\omega \in M(\mathcal{C})$.*

In the context of the above theorem, if $\mathcal{I}_\omega = \mathcal{A}$ one considers $a+\mathcal{I}_\omega$ invertible for all $a \in \mathcal{A}$.

To apply a local principle like the one described above, a specific structure in the algebra is needed, namely it must possess a central subalgebra. In order to describe the search to find such a subset in the sequence algebra $\mathcal{E}$ when considering finite sections of Toeplitz operators, one needs to introduce additional concepts like that of an Hankel operator. Let $Q := I - P$ and define the flip operator $J : l^p_{\mathbb{Z}} \to l^p_{\mathbb{Z}}$ by the rule $(Ju)_n = u_{-n-1}$. Note that $J^2 = I$ and $PJ = JQ$. The Hankel operator $H(a)$ on l^p is defined as the operator

$$H(a) := PM(a)QJ$$

and can be represented as a one-sided infinite matrix

$$[a_{ij}]_{i,j>0}, \qquad a_{ij} := a_{i+j-1}.$$

This Hankel operator "uses" only the positive-indexed Fourier coefficients of a. The negative-indexed coefficients of a are used in $H(\tilde{a})$ where $\tilde{a}(t) = a(1/t)$, that is, $M(\tilde{a}) = JM(a)J$, $H(\tilde{a}) = JQM(a)P$. The following proposition relates the Toeplitz and Hankel operators with the products of their symbols [26, Chapter 2]. In this form, (3) and (4) appeared for the first time in Widom's paper [157].

Proposition 2.3. *Let $a, b \in M^p$. Then $ab \in M^p$ and*

$$M(ab) = M(a)M(b), \tag{2}$$

$$T(ab) = T(a)T(b) + H(a)H(\tilde{b}), \tag{3}$$

$$H(ab) = T(a)H(b) + H(a)T(\tilde{b}). \tag{4}$$

Proof. We just write the proof of (3):

$$\begin{aligned} T(ab) &= PM(ab)P = PM(a)M(b)P = PM(a)PM(b)P + PM(a)QM(b)P \\ &= T(a)T(b) + PM(a)QJJQM(b)P = T(a)T(b) + H(a)H(\tilde{b}). \end{aligned}$$ □

A result by Hartman [62] from 1958 implies that if a belongs to C_p, the closure of W in M^p, then the operators $H(a)$ and $H(\tilde{a})$ are compact. When a and b are piecewise continuous functions[2], the respective Hankel operators are not compact in general, but the product $H(a)H(b)$ is compact if a and b have no common points of discontinuity [26, Proposition 6.29].

To return now to the main narrative, beginning with Baxter's paper referred above, a lot of further proofs of the applicability of the finite section method to invertible Toeplitz operators $T(a)$, $a \in C_p$, or a belonging to some special classes in the l^2 setting were presented (see, for instance, [31, 46, 104]). One proof is based on Widom's formula [157]:

$$T_n(ab) = T_n(a)T_n(b) + P_nH(a)H(\tilde{b})P_n + W_nH(\tilde{a})H(b)W_n, \tag{5}$$

where $T_n(a) = P_nT(a)P_n$, $P_n(u_0, u_1, \ldots) := (u_0, u_1, \ldots, u_{n-1}, 0, 0, \ldots)$ and $W_n(u_0, u_1, \ldots) := (u_{n-1}, \ldots, u_0, 0, 0, \ldots)$. Clearly, (5) is the counterpart of formula (3) for finite sections of Toeplitz operators $T(a)$. Note that (5) can be used to give asymptotic inverses B_n for a stable sequence $(T_n(a))$, $a \in C_p$, that is

$$\|B_nT_n(a) - P_n\| \to 0, \qquad \|T_n(a)B_n - P_n\| \to 0 \quad \text{as } n \to \infty.$$

Indeed, in this case one can check that a is invertible, set $B_n = T_n(a^{-1}) + P_nK(a)P_n + W_nK(\tilde{a})W_n$, and notice that the invertibility of $T(a)$ implies the invertibility of $T^{-1}(\tilde{a})$ and $\tilde{a}^{-1}$, where $K(a) = T(a)^{-1} - T(a^{-1})$ and $K(\tilde{a}) = T^{-1}(\tilde{a}) - T(\tilde{a}^{-1})$ are compact operators (see for instance [25]).

In 1977, Verbickiĭ and Krupnik published the paper [153] where they proved that if a is a piecewise continuous function with only one discontinuity, having a mild additional multiplier property, then the finite section method is applicable to $T(a)$ in l^p if and only if $T(a)$ and $T(\tilde{a})$ are invertible.

Using the above two results, around 1980, Böttcher and Silbermann developed a technique of separation of discontinuities. This technique allowed them to reduce problems for Toeplitz operators with finitely many discontinuities in the symbols to problems for compactly perturbed Toeplitz operators whose symbols have only one discontinuity. In this connection the following result from [46] is of importance.

Proposition 2.4 (Gohberg, Feldman). *Let A be a bounded operator and K a compact operator. If the finite section method applies to A and $A + K$ is invertible, then the finite section method applies to $A + K$.*

Combining their separation technique with (3), the Verbickiĭ–Krupnik result for a single jump, and Proposition 2.4, Böttcher and Silbermann [22] were able to extend the Verbickiĭ–Krupnik result to symbols with finitely many jumps.

2.2. Silbermann's insight

The new result using the separation technique would only be published in 1982 but, already in 1980, Silbermann had started to look for a way to generalize it

[2]A function $a \in L^\infty$ is said to be piecewise continuous, in short $a \in PC(\mathbb{T})$, if a has well-defined one-sided limits at all points $t \in \mathbb{T}$.

for a countable number of discontinuities and also for systems of Toeplitz operators, that is, for generators in the set $PC_p(\mathbb{T}) := PC \cap M_{<p>}$, with PC the set of piecewise continuous functions and $M_{<p>}$ the set of all functions from $L^\infty(\mathbb{T})$ that belong to all $M_{\hat{p}}$ for $\hat{p}$ in a neighbourhood of p [3] [12]. He thought to use the algebraization technique from Kozak's thesis with the local principle of Gohberg and Krupnik. But there was a problem. The algebra $\mathcal{E}/\mathcal{G}$ resulting from the application of Kozak's method, with $\mathcal{E}$ defined by

$$\mathcal{E} = \{(A_n)_{n\in\mathbb{Z}^+} : A_n : \text{Im } P_n \to \text{Im } P_n\},$$

seemed to have non-useful center for applying the local principle (it is connected with each index of the sequence and isomorphic to l^∞ – the local algebras are either trivial or isomophic to $\mathcal{E}/\mathcal{G}$ again). *Something must commute* were his thoughts at the time, as he recalled many years later [14]. In fact, for operators in $\mathcal{B}(l^p)$, if $f \in C(\mathbb{T})$ and $a \in PC_p(\mathbb{T})$, then $T(a)T(f)-T(f)T(a)$ is compact and invertibility in the Calkin algebra generated by the cosets of the form $T(a)+\mathcal{K}$, $a \in PC_p(\mathbb{T})$ can be studied using local principles because there exists a central subalgebra $\mathcal{C} = \{T(f)+\mathcal{K} : f \in C(\mathbb{T})\}$. Applying the idea to finite sections of Toeplitz operators one obtains from Widom's formula

$$T_n(a)T_n(f) - T_n(f)T_n(a) = P_nK_1P_n + W_nK_2W_n$$

with K_1, K_2 compact operators. Silbermann's incredible fruitful idea was then to check if the set

$$\mathcal{J} := \big\{(P_nK_1P_n + W_nK_2W_n + G_n),\ K_{1,2} \text{ compact}, (G_n) \in \mathcal{G}\big\}$$

would form a closed ideal of $\mathcal{E}$ or some related subalgebra, and if it was possible to relate invertibility in $\mathcal{E}/\mathcal{G}$ with invertibility in the new algebra. It turned out that in both cases his ingenious intuition was right. The set $\mathcal{J}$ is an ideal if there are operators A and $\tilde{A} \in \mathcal{B}(l^p)$ such that the sequences $(A_n) \in \mathcal{E}$ have strong limits

$$\begin{aligned} &\operatorname*{s-lim}_{n\to\infty} A_n = A, && \operatorname*{s-lim}_{n\to\infty} W_nA_nW_n = \tilde{A}, \\ &\operatorname*{s-lim}_{n\to\infty} A_n^* = A^*, && \operatorname*{s-lim}_{n\to\infty} (W_nA_nW_n)^* = \tilde{A}^*, \end{aligned} \tag{6}$$

where A^* and $\tilde{A}^*$ are bounded operators on l^q, with $\frac{1}{p}+\frac{1}{q}=1$. We will denote the set of the sequences that possess such strong limits as $\mathcal{F}$. The limit operators depend on the sequence (A_n) under consideration. These sequences form a closed Banach subalgebra $\mathcal{F}$ of $\mathcal{E}$, and one just needs now to relate invertibility in $\mathcal{E}/\mathcal{G}$ with invertibility in $\mathcal{F}/\mathcal{J}$. The proposition below, which appears in [142], is the first of what in this context were latter called *lifting* theorems. Its proof makes use of the interplay between multiplication by compact operators and weak/strong/norm convergence. For a modern and general perspective on lifting theorems, see [120].

Proposition 2.5 (Silbermann). *A sequence $\mathbb{A} \in \mathcal{F}$ is stable if and only if the corresponding operators A and $\tilde{A}$ are invertible, and the coset $\mathbb{A}+\mathcal{J}$ is invertible in $\mathcal{F}/\mathcal{J}$.*

---

[3]The use of $M_{<p>}$ over M_p facilitates some technical details in the proofs.

Invertibility in $\mathcal{F}/\mathcal{J}$ (in fact, in specific subalgebras like alg $\{(T_n(a))+\mathcal{J},\ a \in PC_p\}$) could then be studied using the local principle of Gohberg and Krupnik, in analogy with studying invertibility in the Calkin algebra for regular operators. In fact, in the case under study, invertibility of the coset $\mathbb{A}+\mathcal{J}$ in $\mathcal{F}/\mathcal{J}$ was already implied by the invertibility of the two operators A and $\tilde{A}$. So, Silbermann obtained the elegant result:

Theorem 2.6 (Silbermann). *If $T(a)$ is a bounded Toeplitz operator on $[l^p]^n$ with symbol $a \in [PC_p(\mathbb{T})]^{n\times n}$, then the FSM applies to $T(a)$ if and only if $T(a)$ and $T(\tilde{a})$ are invertible.*

Silbermann was not only able to prove the desired result regarding the applicability of the FSM to Toeplitz operators (or even systems of Toeplitz operators) with a countable number of discontinuities in their symbol, but the fundamental idea, finding a compact-like sequence ideal, would form the basis of a whole new branch in theoretic numerical analysis. The approach taken by Silbermann to study stability was later systematized by Böttcher, Silbermann and Roch into the following steps (see Figure 1).

Let $\mathcal{A}$ be an algebra containing sequences of operators for which one wants to obtain stability criteria.

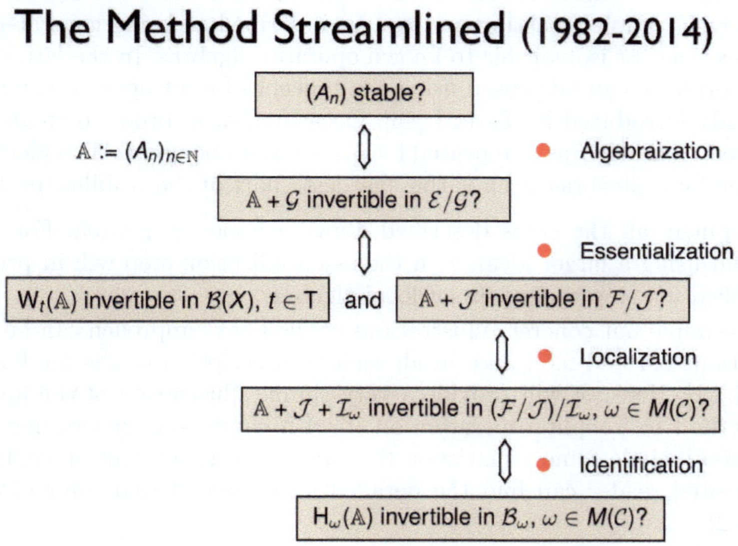

FIGURE 1. The algebraic approach to stability problems

1. **Algebraization:** Find a unital Banach algebra $\mathcal{E}$ which contains $\mathcal{A}$ and a closed ideal $\mathcal{G}\subset\mathcal{E}$ such that the stability problem becomes equivalent to an invertibility problem in the quotient algebra $\mathcal{E}/\mathcal{G}$.

2. **Essentialization:** Find a unital subalgebra $\mathcal{F}$ of $\mathcal{E}$ which contains $\mathcal{A}$ and $\mathcal{G}$, and a closed ideal $\mathcal{J}$ of $\mathcal{F}$ such that the algebra $\mathcal{A}^{\mathcal{J}} := (\mathcal{A} + \mathcal{J})/\mathcal{J}$ has a large center.

   To be able to use the latter property, one has to guarantee that no essential information is lost regarding the invertibility of a coset of a sequence $\mathbb{A} \in \mathcal{F}$ in the algebra $\mathcal{E}/\mathcal{G}$ and the invertibility of the same sequence modulo the ideal $\mathcal{J}$. This is done employing inverse-closedness results and a lifting theorem, which relates invertibility in those algebras with the invertibility of a collection of limit operators, indexed by a set T.

3. **Localization:** Use a local principle to translate the invertibility problem in the algebra $\mathcal{A}^{\mathcal{J}}$ to a family of simpler invertibility problems in local algebras. In this step one must, if using Allan/Douglas' local principle, to find a suitable central subalgebra and its maximal ideal space M. The central subalgebra is connected with the ideal $\mathcal{J}$ of the previous step.

   Then one uses the local principle to translate the invertibility problem from $\mathcal{F}/\mathcal{J}$ to the family of local algebras $\mathcal{A}_w^{\mathcal{J}}$, with $w \in M$.

4. **Identification:** Find necessary and sufficient conditions for the invertibility of elements in the local algebras.

   In some cases, we are in the situation that the local algebras have a relatively simple structure and can be described as being generated by two projections, or isomorphic to known operator algebras. In the latter case the isomorphism can be proved using the concept of limit operators, which were already introduced by Favard [43]. Occasionally, in order to treat more involved local algebras, a repeated localization can be useful. The identification of the local algebras often is the "hardest" part of the stability problem.

Of course, all the steps described above are closely related. For instance, finding and using a larger ideal $\mathcal{J}$ in the essentialization step will in principle in the identification step wield simpler local algebras.

Two simple but concrete illustrations of the above approach can be found in [120, Sections 6.4 and 6.5], and an alternative description of the method in [56, Section 6.1.1]. Here we will provide a very simple illustration of the application of the method to Toeplitz operators on l^2 with continuous generating function. We will also include some remarks on the case of piecewise continuous functions. The interested reader can find the detailed treatment of that later case in [56, Section 4.2].

2.3. Finite Section Method for Toeplitz operators with continuous generating function

Let $a \in C(\mathbb{T})$ or $PC(\mathbb{T})$, and consider the FSM applied to $T(a) : l^2 \to l^2$. That is, to solve the infinite system of equations

$$T(a)x = y$$

we consider systems of n equations in n unknowns, given by

$$T_n(a)x_n = P_n y$$

by selecting the first n lines and n columns of the infinite matrix $T(a)$. The projection P_n is the one defined after equation (5), selecting the first n entries of a vector. We would like then to study the stability of the sequences $(T_n(a))_{n\in\mathbb{Z}^+}$. The algebras we will consider are then exactly the algebras generated by those sequences, $\mathcal{C} := \operatorname{alg}\{(T_n(a))_{n\in\mathbb{Z}^+}, a \in C(\mathbb{T})\}$, or the larger $\mathcal{A} := \operatorname{alg}\{(T_n(a))_{n\in\mathbb{Z}^+}, a \in PC(\mathbb{T})\}$. In both cases, the first two steps of the method have no differences.

2.3.1. Algebraization. Let $\mathcal{E}$ be the algebra defined by

$$\mathcal{E} = \{(A_n)_{n\in\mathbb{Z}^+} : A_n : \text{Im } P_n \to \text{Im } P_n\},$$

with pointwise defined operations and involution, and the supremum norm. The algebra $\mathcal{E}$ is a Banach (in fact C*) algebra containing $\mathcal{A}$ and its subset $\mathcal{G}$ of all sequences of operators tending in the norm to zero forms a closed two-sided ideal. Kozak's result then implies that a sequence in $\mathcal{E}$ is stable if and only if its coset is invertible in $\mathcal{E}/\mathcal{G}$. But it is not easy in general to check invertibility in $\mathcal{E}/\mathcal{G}$, even for the sequences belonging to $\mathcal{C}$ or $\mathcal{A}$.

2.3.2. Essentialization. Consider now the subset $\mathcal{F}$ of $\mathcal{E}$ formed by all sequences $\mathbb{A} = (A_n)$ such that the following four strong limits exist:

$$\begin{aligned} \operatorname*{s-lim}_{n\to\infty} A_n &= \mathsf{W}_0(\mathbb{A}), & \operatorname*{s-lim}_{n\to\infty} W_n A_n W_n &= \mathsf{W}_1(\mathbb{A}), \\ \operatorname*{s-lim}_{n\to\infty} A_n^* &= \mathsf{W}_0(\mathbb{A})^*, & \operatorname*{s-lim}_{n\to\infty} (W_n A_n W_n)^* &= \mathsf{W}_1(\mathbb{A})^*. \end{aligned} \tag{7}$$

The set $\mathcal{F}$ forms a closed subalgebra of $\mathcal{E}$ containing $\mathcal{A}$. Note that the above strong limits are in reality unital algebra homomorphisms from $\mathcal{F}$ into $\mathcal{B}(l^2)$ and that $\mathsf{W}_{0,1}(\mathcal{G}) = 0$, which means they induce well-defined unital homomorphisms acting from $\mathcal{F}/\mathcal{G}$ into $\mathcal{B}(l^2)$. Define the subset

$$\mathcal{J} := \{(P_n K_0 P_n + W_n K_1 W_n + G_n),\ K_{0,1} \text{ compact}, (G_n) \in \mathcal{G}\}$$

of $\mathcal{F}$. One can prove that in fact $\mathcal{J}$ is a closed two-sided ideal of $\mathcal{F}$. To illustrate the interplay between the several concepts (strong and norm convergence, compact operators, etc) we will show here that it is a left ideal. For that, given sequences $(A_n) \in \mathcal{F}$ and $(J_n) \in \mathcal{J}$ one wants to check that $(A_n)(J_n)$ is in $\mathcal{J}$.

$$A_n(P_n K_0 P_n + W_n K_1 W_n + G_n) = A_n P_n K_0 P_n + A_n W_n K_1 W_n + A_n G_n.$$

The sequence $(A_n G_n)$ is clearly in $\mathcal{G}$ and thus in $\mathcal{J}$. We just have to concern ourselves with the other parts. Let us start with the sequence $(A_n P_n K_0 P_n)$. One can write

$$(A_n P_n K_1 P_n) = (P_n \mathsf{W}_0(\mathbb{A}) K_0 P_n) + (P_n(A_n - \mathsf{W}_0(\mathbb{A}))K_0 P_n)$$

and the first part of the left-hand side is clearly in $\mathcal{J}$. The second part is in $\mathcal{G}$ because

$$\|P_n(A_n - \mathsf{W}_0(\mathbb{A}))K_0 P_n\| \le \|(A_n - \mathsf{W}_0(\mathbb{A}))K_0\| \to 0$$

as n goes to infinity due to the strong convergence of $A_n - \mathsf{W}_0(\mathbb{A})$ to zero and the compactness of K_0. Likewise, for the sequence $(A_n W_n K_1 W_n)$,

$$(A_n W_n K_1 W_n) = (W_n \mathsf{W}_1(\mathbb{A}) K_1 W_n) + (W_n(W_n A_n W_n - \mathsf{W}_1(\mathbb{A}))K_1 P_n)$$

because $W_n^2 = P_n$ and so the same above reasoning can be made. The result is proved.

Now it is necessary to relate invertibility in $\mathcal{F}/\mathcal{G}$ with invertibility in $\mathcal{F}/\mathcal{J}$. This is done using the special properties of the homomorphisms. Usually one uses a general lifting theorem. For this example, and also to give some insight on the lifting theorem itself, I will give a sketch of a proof.

Theorem 2.7 (Lifting). *An element $\mathbb{A} + \mathcal{G}$ is invertible in $\mathcal{F}/\mathcal{G}$ if and only if the operators $\mathsf{W}_0(\mathbb{A})$ and $\mathsf{W}_1(\mathbb{A})$ are invertible and the coset $\mathbb{A} + \mathcal{J}$ is invertible in $\mathcal{F}/\mathcal{J}$.*

Proof. ($\Rightarrow$) It is clear that if $\mathbb{A} + \mathcal{G}$ is invertible then $\mathbb{A} + \mathcal{J}$ is invertible in $\mathcal{F}/\mathcal{J}$ and, being $\mathsf{W}_{0,1}$ unital homomorphisms, $\mathsf{W}_0(\mathbb{A})$ and $\mathsf{W}_1(\mathbb{A})$ are also invertible. The difficulty of the proof is in the other direction.
($\Leftarrow$) Consider the kernels of the homomorphisms $\mathsf{W}_{0,1}$. They are ideals of $\mathcal{F}$ and it is not difficult to see that (note that the sequence of operators (W_n) tends weakly to zero)

$$\mathcal{J} \cap \mathrm{Ker}\ \mathsf{W}_0 \cap \mathrm{Ker}\ \mathsf{W}_1 = \mathcal{G}. \tag{8}$$

Because $\mathcal{F}/\mathrm{Ker}\ \mathsf{W}_{0,1}$ is isomorphic to $\mathsf{W}_{0,1}(\mathcal{F})$, and the $\mathsf{W}_{0,1}(\mathbb{A})$ are invertible, there exist sequences $\mathbb{C}_0$ and $\mathbb{C}_1$ in $\mathcal{F}$ such that

$$\mathbb{A}\mathbb{C}_0 + \mathbb{K}_0 = \mathbb{I} \quad \text{and} \quad \mathbb{A}\mathbb{C}_1 + \mathbb{K}_1 = \mathbb{I}$$

where $\mathbb{I} := (P_n)$ is the identity sequence in $\mathcal{F}$ and $\mathbb{K}_{0,1} \in \mathrm{Ker}\ \mathsf{W}_{0,1}$. And because $\mathbb{A} + \mathcal{J}$ is invertible in $\mathcal{F}/\mathcal{J}$ we also have that there exists a sequence $\mathbb{B} \in \mathcal{F}$ such that $\mathbb{A}\mathbb{B} = \mathbb{I} + \mathbb{J}$, with $\mathbb{J} \in \mathcal{J}$. Then the product

$$(\mathbb{A}\mathbb{C}_0 - \mathbb{I})(\mathbb{A}\mathbb{C}_1 - \mathbb{I})(\mathbb{A}\mathbb{B} - \mathbb{I}) = \mathbb{A}\mathbb{B}' - \mathbb{I}$$

for some $\mathbb{B}' \in \mathcal{F}$ and it is in $\mathcal{G}$ by (8). Thus $\mathbb{A} + \mathcal{G}$ is right-invertible. The proof for left-invertibility is the same. □

2.3.3. Localization. In this step, one considers the smallest closed algebra formed by the sequences in $\mathcal{C}$ and in $\mathcal{J}$. To apply Allan/Douglas' local principle we must find a useful central subalgebra of

$$\mathcal{C}^J := (\mathcal{C} + \mathcal{J})/\mathcal{J} \subset \mathcal{F}/\mathcal{J}.$$

In fact, in the simple case of continuous functions only, the algebra $\mathcal{C}^J$ is commutative, as can easily be seen using Widom's formula (5) and Hartman's result on compactness of Toeplitz operators with continuous generator functions:

$$\begin{aligned} &T_n(a)T_n(b) - T_n(b)T_n(a) \\ &\quad = P_n\big(H(b)H(\tilde{a}) - H(a)H(\tilde{b})\big)P_n + W_n\big(H(\tilde{b})H(a) - H(\tilde{a})H(b)\big)W_n. \end{aligned}$$

For commutative algebras the local principle is equivalent to the usual Gelfand theory. The local algebras are all isomorphic to the complex field $\mathbb{C}$. The maximal ideal space of $\mathcal{C}^{\mathcal{J}}$ coincides with the maximal ideal space of $C(\mathbb{T})$ that is, with $\mathbb{T}$.

Things get more interesting if one considers the case of piecewise continuous functions. In that case the formula above gives that $\mathcal{C}^{\mathcal{J}}$ is a central subalgebra of

$$\mathcal{A}^J := (\mathcal{A} + \mathcal{J})/\mathcal{J} \subset \mathcal{F}/\mathcal{J}.$$

So, each point $w \in \mathbb{T}$ corresponds to a maximal ideal of $\mathcal{C}^{\mathcal{J}}$ which we denote by the same symbol, and let $\mathcal{I}_w$ be the smallest closed two-sided ideal of $\mathcal{A}^J$ containing w. Applying Allan/Douglas's local principle, invertibility of an element $(T_n(a)) + \mathcal{J}$ in $\mathcal{A}^J$ is equivalent to the invertibility of all the local cosets $(T_n(a)) + \mathcal{J} + \mathcal{I}_w$, $w \in \mathbb{T}$.

2.3.4. Identification. In the case of continuous functions only, for each $w \in \mathbb{T}$, the coset $(T_n(a)) + \mathcal{J} + w \in \mathcal{A}^{\mathcal{J}}/w$ coincides with

$$(T_n(a(w))) + \mathcal{J} + w = a(w)\mathbb{I} + \mathcal{J} + w,$$

which means in some sense that, locally, the continuous generating function behaves as a constant function. The cosets of $\mathbb{A} = (T_n(a))$ are thus invertible in each local algebra $\mathcal{A}^{\mathcal{J}}/w$ if and only if $a(w) \neq 0$. So, the coset $(T_n(a)) + \mathcal{J}$ is invertible in $\mathcal{F}/\mathcal{J}$ if and only if a is invertible in $C(\mathbb{T})$. The invertibility of the generator a is a consequence of the Fredholm property of the Toeplitz operator $T(a)$, which in our case already appears as $\mathsf{W}_0(\mathbb{A})$. So, in the case under consideration, the condition of invertibility in $\mathcal{F}/\mathcal{J}$ gives no extra conditions for the applicability of the FSM, besides the two strong limits. On the other hand, it is known that for (scalar) Toeplitz operators, $T(a)(= \mathsf{W}_0(\mathbb{A}))$ is invertible if and only if $T(\tilde{a})(= \mathsf{W}_1(\mathbb{A}))$ is invertible. We can conclude that the FSM is applicable for Toeplitz operators $T(a) : l^2 \to l^2$ with a continuous if and only if $T(a)$ is invertible.

Consider now a piecewise continuous function a. At the points w where a is continuous, the picture is as above: we obtain a scalar local algebra, with only multiples of the identity. This happens because the coset $(T_n(b_w)) + \mathcal{J}$ with the function b_w defined by $b_w(t) := a(w) - a(t)$ is in $\mathcal{I}_w$ ($b_w(w) = 0$) and so belongs to the 0 coset in the local algebra. If a has a discontinuity at the point w we can use a similar reasoning. But now, the local behaviour will be piecewise constant. Let χ_w be the piecewise constant function on $\mathbb{T}$ with the value 1 on the arc from w to $-w$ (with respect to the common orientation of $\mathbb{T}$) and 0 otherwise. Now the coset $(T_n(a_w)) + \mathcal{J} + \mathcal{I}_w$ with a_w defined by

$$a_w := a(w^+)\chi_w + a(w^-)(1 - \chi_w)$$

coincides with the original $(T_n(a)) + \mathcal{J} + \mathcal{I}_w$ because $a_w - a$ is a function continuous at w and taking the value 0 there, and so $(T_n(a_w - a)) + \mathcal{J} \in \mathcal{I}_w$. The above reasoning signifies that the local algebra at w is generated by the identity and the coset $(T_n(\chi_w)) + \mathcal{J} + \mathcal{I}_w$. Obtaining invertibility conditions for the local algebras, as refered in the previous section, is usually the hardest part, and we will not enter into details here. The idea in this case is that, being a singly generated algebra, it is commutative and it can be described via Gelfand theory, and the maximal ideal space of a singly generated algebra is homeomorphic to the spectrum of its generator. The spectrum in this case can be shown to be the interval $[0, 1]$ (see [56, Proposition 4.26] for details). Thus, the coset $(T_n(a_w)) + \mathcal{J} + \mathcal{I}_w$ is invertible in $\mathcal{A}^{\mathcal{J}}/\mathcal{I}_w$ if and only if the line segment in the complex plane joining $a(w^+)$ to $a(w^-)$ does not contain the point 0. In fact, in this case also, that condition is a necessary condition for the invertibility of the operator $T(a) = \mathsf{W}_0(\mathbb{A})$, and so the end result is as above.

3. Extension and application

Direct applications of the new method to powers of Toeplitz operators, sums and products and elements of the algebra generated by those types of operators appeared almost immediately in [107, 108, 122]. As if to show the hard road ahead, there was an overlook when applying the complex machinery of Gohberg–Krupnik's local principle in [107, 108] that rendered the general result obtained in those papers false. The issue was detected and solved by Andreas Rathsfeld (see [55, Section 4.7] for more details). A first abstract and general view of these results was published in 1985 [123].

Incidentally, a concrete invertible Toeplitz operator for which the Finite Section Method is not applicable was found by Treil in 1987 [152]. That example shows the applicability problem has indeed no simple solution to be found.

The new approach to stability problems pioneered by Kozak and given wide application possibilities by Silbermann would over the next 35 years be extended in several dimensions and applied to a great variety of problems. It would also reflect back into Banach and C*-algebra theory, by supplying that field with many new examples and new problems.

3.1. Convolution operator algebras

The new approach's first application, from the original 1981 paper [142], was heavily inspired by the study of the Calkin algebra connected to Toeplitz operators and their commutation properties. To apply the method to other types of operators, the structure of the algebras they generated and their commutation properties had to be known. Continuous convolution operators were an obvious early candidate for application, and thus a great deal of study of the algebras generated by operators of that type (singular integral and Wiener–Hopf operators) followed. Certain operators on a contour $\Gamma \subset \mathbb{C}$ can be formally defined as

$$(Au)(t) := \lambda u(t) + \int_\Gamma k(t-s)u(s)\,ds$$

where $\lambda \in \mathbb{C}$ and k can be understood to be either a function or a distribution, depending on Γ. In particular, when $k(x) = -\frac{1}{\pi \mathrm{i}}\frac{1}{x}$ we obtain the Singular Integral Operator (SIO) on Γ. With $\Gamma = \mathbb{R}$ we have a convolution, acting on $L^p(\mathbb{R})$ for instance, and its restriction to $L^p(\mathbb{R}^+)$ is called a Wiener–Hopf operator. If Γ is the unit circle $\mathbb{T}$ in the complex plane, one defines the Hardy spaces H^p of the functions with only positive indexed Fourier coefficients and the continuous version of the Riesz projection is

$$P : L^p \to H^p, \qquad P\left(\sum_{i=-\infty}^{+\infty} a_i t^i\right) \mapsto \sum_{i=0}^{+\infty} a_i t^i$$

with the Toeplitz operator defined as

$$T_a : H^p \to H^p, \qquad u \mapsto P(au).$$

For the theory of these types of operators see, e.g., [26] and the references cited therein.

Silbermann applied his newly discovered method to the FSM for singular integral equations in [143]. Böttcher [6, 7] was the first to prove criteria for the applicability of the finite section method to Wiener–Hopf integral operators on $L^p(\mathbb{R}^+)$. He used Silbermann's localization technique to reduce the problem to case of a symbol with a single jump. While in the discrete case, that is, for Toeplitz operators on ℓ^p whose symbol has a single jump, the analogous problem had been solved earlier by Verbickiĭ and Krupnik [153], Böttcher had to find a new approach to settle the L^p case. He observed in particular for the first time that it is not only two operators but rather a whole family of operators whose invertibility is necessary and sufficient for the applicability of the finite section method. This was the first appearance of the lentiform massive local spectra in the theory of singular integral operators [13].

The corresponding result for Toeplitz operators on the unit circle followed in 1986 [20], where, for a positive integer n, the finite section of a Toeplitz operator T_a is defined as the compression of T_a to the $(n+1)$-dimensional space of trigonometric polynomials of degree at most n.

Many other results followed (e.g., [21, 125, 127], see also the following sections) using, systematizing and extending results from authors like Gohberg and Krupnik [47, 48, 82], Simonenko and Min′ [149], Sarason [138], Duduchava [37], etc, sometimes leading to fruitful collaboration with those authors.

There was also an effort to understand the relation and differences between the three local principles (Simonenko's [147, 148], Gohberg–Krupnik's [49] and Allan/Douglas' [1, 35]) which culminated in [19]. The developments and collaborations described above fed into applications of the algebraic techniques in many cases.

3.2. Different operators, function and symbol spaces

Some operators related to Toeplitz operators like band-dominated operators were a natural venue for the new technique [101, 102]. Banded matrices are matrices where the non-zero entries are situated on a band around the main diagonal. The concept can be extended to other type of operators, that are then named *banded operators*. Band-dominated operators are norm limits of sequences of band operators and a generalization of Toeplitz operators with symbols in C_p (see [113] for an overview of the theory for operators in l^2). Other suitable candidates with some relation to Toeplitz or Wiener–Hopf operators were Hankel operators and other operators derived from sums and products of simpler multiplication and convolution operators [116, 118]. In fact, algebraic techniques are ideal to study properties of operators that result from sums and products of simpler building blocks. For operators acting in Hilbert spaces, the corresponding algebras are C*-algebras which makes the problems somewhat easier to tackle, due to the nice norm and inverse-closedness properties of such algebras (see, e.g., [120, Section 1.2] for a short discussion on this subject or [11] which condenses very well many of the issues in the C*-algebra setting). So some results were first obtained for L^2 spaces for instance, and only years later for L^p with $p \neq 2$.

Besides the above-mentioned pioneer works of Böttcher [6, 7] on $L^p(\mathbb{R})$, Roch and Silbermann [125] developed a symbol calculus for finite sections of singular integral operators with flip and piecewise continuous coefficients.

Starting in 1995, as part of his PhD thesis, the author tried to find conditions for the applicability of the FSM for algebras that included Wiener–Hopf plus Hankel operators acting on $L^2(\mathbb{R}^+)$. The main difficulties lay in the essentialization and the identification steps. Regarding the essentialization step, simple adaptations of the before used homomorphisms did not work because of the flip J, defined by $(Ju)(t) = u(-t)$, which appears in the Hankel operators, and it was not clear what these new homomorphisms would look like. Finally it was possible to find a new family of matrix homomorphisms that could be used. After several months of fruitless search the inspiration came when looking at the relations expressed on Section 7.13 of the first edition of [26], which is treating the discrete case without explicit Hankel operators in the algebra. It crossed the author's mind that similar strong limits in the continuous case might be compatible with the flip, and thus be well defined on the whole algebra generated by Wiener–Hopf plus Hankel operators, and that those strong limits could be arranged as a single matrix-operator-valued homomorphism.

When including the flip, the central subalgebra must be constituted by operators defined by even functions. This implies a "double point" localization at non-fixed points of the flip (which are 0 and ∞). A standard technique for dealing with this problem is doubling the dimension (see, e.g., [120, Section 1.1.5]), ending in operator matrices.

In the identification step, a complicated local algebra at infinity appeared, generated by four non-commuting idempotents, which had to be analysed with novel techniques [118].

Also, the approximation algebra building blocks were made of the constant sequences of multiplication operators (aI) and convolution operators ($W^0(b)$) for suitable generators a and b, the flip (J), together with the FSM projection sequence $(P_\tau)_{\tau\in\mathbb{I}}$, acting on the whole space $L^2(\mathbb{R})$. Before, a more restricted definition was used considering as generators for instance the sequences $(P_\tau W(a) P_\tau)$ acting in Im (P_τ). This different approach has the advantage of treating the sequence algebra as a generalization of the usual operator algebra for those operators, and the stability analysis as an extension of the operator algebra's Fredholm theory. The disadvantage is the appearance of complicated local algebras at infinity. In fact, Roch had use a similar idea a few years earlier [109] in an unpublished work, but that preprint was not known to the author at the time.

Other examples of Toeplitz-related operator sequences for which the algebraic methods apply and were more recently focus of attention are the variable coefficient Toeplitz matrices [92, 146] and random Jacobi operators [87]. Variable coefficient Toeplitz matrices are of the form

$$[a_{ij}]_{i,j=0}^{n}, \qquad a_{ij} := a_{i-j}(i/n, j/n),$$

where the $a_k(x, y)$ are the Fourier coefficients of a function $t \mapsto a(x, y, t)$ with $a : [0, 1] \times [0, 1] \times \mathbb{T} \to \mathbb{C}$. They appear in discretizations of ordinary differential equations with variable coefficients, and many other situations.

3.2.1. Beyond the C*-case. The generalization of the above results to Wiener–Hopf and Hankel operators in $L^p(\mathbb{R})$ for $1 < p < \infty$ turned out to be much harder than expected. As a first step in 2010 the author, together with Steffen Roch and Bernd Silbermann, managed to study the applicability of the FSM to operators belonging to the closed subalgebra of $\mathcal{B}(L^p(\mathbb{R}))$ generated by operators of multiplication by piecewise continuous functions on $\dot{\mathbb{R}}$ and operators of convolution by piecewise continuous Fourier multipliers [119]. This involved the use of certain intermediate algebras to guarantee inverse-closedness properties. When trying to extend the results of [119] to algebras with flip and to combine the usual homomorphisms with the doubling of the dimension, the authors were not able to overcome the occurring inverse-closedness problem.

In 2012, the author together with Steffen Roch [115] introduced two new families of homomorphisms, which were suitable to describe Banach operator algebras that include the flip *and* to tackle the inverse-closedness problem. These homomorphisms possess, in some sense, a built-in doubling of dimension. Besides studying their properties, these homomorphisms were used to give a description of the Calkin algebra generated by convolution, multiplication and the flip operators which is alternative to the description in [120, Section 5.7] and is interesting in its own right. The application of these homomorphisms to sequence algebras and, thus, to study the stability was the subject the next paper [116], published in 2014.

The type of generating function of the convolution operators can also be extended. The original result involved Toeplitz operators with piecewise continuous symbols and systems of equations with such operators. Alexei Karlovich, Helena Mascarenhas and the author [75] considered the applicability of the FSM to operators belonging to the algebra generated by operators of multiplication by piecewise continuous functions and convolution operators with symbols in the algebra generated by piecewise continuous and slowly oscillating functions. The space the operators act on was $L^p(\mathbb{R})$, which raised some non-trivial issues regarding inverse-closedness of the subalgebras that the usual methods to solve those type of problems use. This latter work built on the previous work of Bastos, Bravo, and

Yuri Karlovich on the compactness of commutators involving convolutions with slowly oscillating and piecewise continuous matrix functions [4]. In both of the above works there was a gap in the proof of the full result on the analysis of the local algebra related to the point ∞ [76]. That gap was resolved with the help of Markus Seidel [121, 140].

Lindner [86] considered approximations in the limit cases l^1, L^1, l^∞ and L^∞ for band-dominated operators. One of the difficulties is that for $p = \infty$ the FSM is no longer strongly convergent. Some problems and consequences of this setting are described in the next section.

We must mention also the book by Didenko and Silbermann [32] where they use a real (not complex) C^*-algebra approach to obtain conditions for the convergence of approximation methods for additive convolution-like operators.

3.2.2. Changing the domain. The multi-dimensional case (which originated the original paper from Kozak [80]) was further tackled by Böttcher and Silbermann in [23], extending the previous results to the FSM for Toeplitz operators on $l^2 \times l^2$ with piecewise continuous symbols defined on the torus. For this they needed to use a variant of the local principle called bilocalization. Böttcher [6] considered two-dimensional Wiener–Hopf integral operators in $L^p(\mathbb{R}^2)$, Hagen, Roch and Silbermann [54] explored spline approximation methods for multi-dimensional pseudodifferential operators on $L^p(\mathbb{R}^n)$ (see also next section) and Mascarenhas and Silbermann [90, 91] studied finite sections of convolution type operators on cones on $L^p(\mathbb{R}^2)$, with symbols in the Wiener algebra.

In their work on singular integral operators, Gohberg and Krupnik had considered contours in $\mathbb{C}$ besides the real line [47, 48, 51, 52]. The study of the corresponding numerical approximations followed for instance in [33, 55] where special care had to be taken on corner points, namely in some cases introducing using multiindiced sequences. The effect of the extra index is to cut-off part of the contour nearby the corner, simplifying the stability conditions.

3.3. Other approximation methods

While being very important in theoretical terms, the FSM for operators acting in spaces of non-periodic functions defined on sets like $\mathbb{R}^n$ (e.g., $L^p(\mathbb{R})$) is only used in practice as a first step in the numerical solution, in order to replace unbounded domains by compact ones, because the resulting approximations are still infinite-dimensional operators. Thus immediate and natural applications of the algebraic methods would be to polynomial and spline approximation. It is a strength of the new method that it could be applied easily to these different settings. This happens because they all share an underlying structure (see Section 3.5).

3.3.1. Polynomial approximation. Consider the unit circle $\mathbb{T}$ and the set of all trigonometric polynomials of degree less or equal to n:

$$\Pi_n := \left\{ u_n(t) = \sum_{k=-n}^{n} \beta_k t^k, \quad t = e^{\mathrm{i}\theta}, \beta_k \in \mathbb{C} \right\}. \tag{9}$$

Given a bounded function f defined on the unit circle $\mathbb{T}$, the Lagrange interpolation operator L_n onto Π_n is uniquely defined by the conditions

$$L_n f(t_j) = f(t_j), \quad t_j := e^{2\pi i j}/(2n+1), \quad j = -n, \ldots, n,$$

and is a projection. For $u \in L^p(\mathbb{T})$, the Galerkin projection of u onto Π_n is defined as

$$P_n u(t) := \sum_{k=-n}^{n} \beta_k t^k, \qquad \beta_k = \frac{1}{2\pi}\int_0^{2\pi} u(e^{\mathbf{i}s})e^{-\mathbf{i}ks}\, ds.$$

If S is the Cauchy singular integral operator defined on $\mathbb{T}$,

$$(Su)(t) := \frac{1}{\pi \mathbf{i}}\int_{\mathbb{T}} \frac{u(s)}{s-t} ds, \qquad t \in \mathbb{T},$$

which is a bounded operator on $L^p(\mathbb{T})$ for $1 < p < \infty$, one can define the projections $P = (I+S)/2$ and $Q = (I-S)/2$ and consider the equation

$$(aP + bQ)u = f, \qquad f \in L^p(\mathbb{T}) \tag{10}$$

where a and b are known bounded functions on $\mathbb{T}$. The solution u of Equation (10) can be approximated by $u_n := \sum_{k=-n}^{n} \beta_k t^k$ where the β_k are the solutions of the finite system

$$a(t_j)\sum_{k=0}^{n} t_j^k \beta_k + b(t_j)\sum_{k=-n}^{-1} t_j^k \beta_k = f_j, \quad j = -n, \ldots, -1, 0, 1, \ldots, n$$

where f_j can be either $f(t_j)$ if f is bounded, or $(P_n f)(t_j)$. The above is equivalent to the operator equation

$$L_n(aP + bQ)P_n u_n = L_n f \quad (\text{or } = P_n f).$$

Already in 1982 Junghanns and Silbermann [71] applied the new algebraic techniques to study the convergence of the above-described collocation method for one-dimensional singular integral equations on the unit circle. The results were published in English [72] in 1984 and included the treatment of systems of such equations with piecewise continuous coefficients which may have a countably infinite number of discontinuities. Incidentally, this is one of the first appearances of the new technique in English language. It is interesting to note that English started becoming the lingua franca of mathematics only in the last 40 years. Before, it was common to publish research papers written in the national languages and, in particular, it was common in the former German Democratic Republic to publish in either German or Russian.

For singular integral equations on a interval it is possible to use polynomial approximation based on orthogonal (Chebyshev or other) polynomials. Again the structure of the underlying algebras is similar to the previous cases. The applicability of the finite section (Galerkin) method for the Cauchy SIO with piecewise continuous coefficients was obtained in [70] and that of the collocation method was obtained in [68, 69, 73, 74]. Operators involving Mellin components were treated in [67]. More recent work on this subject appeared in [64, 65]. In all cases there were considered Hilbert function spaces.

The non-Hilbert case appears with delay over the Hilbert case. For instance, in [66] Junghanns et al presented necessary and sufficient conditions for the stability of collocation methods applied to Cauchy singular integral equations on an interval for weighted L^p spaces. They also studied the behavior of the approximation numbers and the k-splitting property of the respective sequence of discretized operators (see Section 3.4.2). The main impact of the change to the non C*-case, besides the added technical difficulty, was the appearance of additional conditions in terms of a family of limit operators indexed on the points of the interval, in a similar way to what had already be observed by Böttcher years earlier [7].

3.3.2. Spline approximation algebraic methods. We will illustrate the matter with a simple example of splines for functions defined on the real line, but the principle is the same for other domains. Consider functions on $L^p(\mathbb{R}^+)$. Let $(\Delta^n)_{n\in\mathbb{N}}$ denote a sequence of meshes (or partitions) in $\mathbb{R}^+$, $\Delta^n = \{x_k^{(n)} : 0 \le k \le n\}$ such that $0 = x_0^{(n)} \le x_1^{(n)} \le \cdots \le x_n^{(n)} = \infty$. To the sequence of meshes $(\Delta^n)_{n\in\mathbb{N}}$ we associate a sequence of piecewise polynomial spline spaces

$$\mathfrak{S}^n = \{u \in L^2(\mathbb{R}^+) : u|_{]x_{k-1}^{(n)}, x_k^{(n)}[} \in \mathbb{P}_d, u|_{]x_{n-1}^{(n)}, \infty[} = 0\},$$

where k runs from 1 to $n-1$ and $\mathbb{P}_d$ represents the set of polynomials of degree less than or equal to d (d fixed). The above definition is for the so-called maximum defect splines, that is, there are no restrictions on the smoothness of the spline functions on the mesh points. It is of course possible to define splines with continuity or other restrictions. On the other end of the scale are the smoothest splines, where the function u must be continuously differentiable a number of times, depending on d ($u \in C^{d-1}$).

Let P^n be a projection from $L_2(\mathbb{R}^+)$ onto S^n, which is a closed subspace of $L_2(\mathbb{R}^+)$.

It is now possible to consider the approximation method

$$P^n A P^n u_n = P^n v, \quad u_n \in \mathrm{Im}\,(P^n) \tag{11}$$

for approximating the equation

$$Au = v, \quad u, v \in L^2(\mathbb{R}^+).$$

In general, one can choose the projection P^n to be the orthogonal projection, which in this case corresponds to the Galerkin projection, a collocation projection (in a subspace of continuous functions), or any other projections associated with various numerical approximation methods. As long as the projection satisfies certain conditions on the convergence, it is possible to build the algebras of approximation sequences and apply algebraic techniques to obtain conditions for the stability of those sequences. It is also possible to use (infinite) uniform meshes, or substitute the maximum defect spline spaces for smoother splines.

Algebraic techniques in the case of spline approximation appeared first without algebras. Inspired by the results obtained by Silbermann in the algebraic setting described above and its own earlier work [98] that involved implicitly a discrete commutator property, Prössdorf [93, 94] in 1984 defined a localization principle applicable to spline approximation sequences directly, by proving the commutator properties, without needing an algebraic structure. But indeed, the algebraic structure is implicit. This variant of the algebraic method, including the introduction of associated operators, corresponding to the identification step of the algebraic method, continued to be used by Prössdorf and some co-workers (e.g., [30, 95–97, 99]).

Silbermann and his student Roland Hagen first managed to obtain results using algebraic techniques for the finite element collocation method applied to singular integral operators on the unit circle [57] and on the torus [58]. In [59] the analysis was extended to the qualocation method, which is similar to the Galerkin method, but with the exact inner products in the Galerkin method replaced by certain quadrature rules. This method is a compromise between the Galerkin and collocation methods, aiming to achieve the benefits of the first at a cost comparable to the second method. In [60], Hagen and Silbermann make a bridge with the work of Prössdorf and Rathsfeld [97] by reframing the problem using a C*-algebra and showing that it falls on the template for the algebraic techniques.

Overcoming the singular behavior of spline approximation methods for one-dimensional singular integral operators caused by discontinuities in the coefficients or by non-smooth geometries of the underlying curves was the object of [110] where Steffen Roch showed the singularities can be cut off by a modification of the approximation method which is closely related to the finite or infinite section method for discrete Toeplitz operators. Also in that year Hagen, Roch and Silbermann [54] tried to obtain a general formulation of stability problems for spline approximation methods in $L^p(\mathbb{R}^n)$. Those results were generalized and extendend to a wide set of approximation methods with regard to singular integral operators and some types of Mellin operators in the book "Spectral theory of approximation methods for convolution equations" [55] by the same authors published in 1995.

Regarding Wiener–Hopf operators, Roch [111] treated the infinite Galerkin method (*i.e.*, the Galerkin method in which the spline space has infinite dimension) with smoothest splines for Wiener–Hopf operators with piecewise continuous symbols. In the case of a uniform mesh and maximum defect finite-dimension spline approximations for Wiener–Hopf plus Hankel operators on the real line, the stability problem was solved in [133] by the author. Several new ideas were introduced: an infinite spline projection and a finite section projection were added as independent members of an algebra that also included the operators being approximated as constant elements. The combination of the two projections generated the finite-dimensional spline projection. This allowed the use of both the techniques for infinite Galerkin method from [55] and the new essentialization homomorphisms developed ealier [118]. A main problem that was solved was related to some of the local algebras, which were generated by four elements instead of the usual two or three. The solution was to carefully analyse the relations between the generators and extend the knowledge about these types of algebras. Finally in [135], Silbermann and the author tackled non-uniform meshes in $L^2(\mathbb{R}^+)$.

On another line of research, Didenko and Silbermann studied approximation methods for the Muskhelishvili equation on a simple closed curve Γ with a finite number of corner points. Approximate solutions of the equation were constructed as piecewise constant functions by the Galerkin and quadrature methods [34]. Since the curve is not smooth, the integral operators appearing in this equation are non-compact, which generates additional difficulties with the Fredholm property of the corresponding operator and the stability of the approximation methods.

In [134], the author studied the applicability of maximum defect polynomial (Galerkin) spline approximation methods with graded meshes to Wiener–Hopf operators with matrix-valued piecewise continuous generating function defined on $\dot{\mathbb{R}}$. The continuous symbol case for L^p was already studied in [42] and [100, Chapter 5], using factorization techniques. This work extended the L^2 piecewise continuous case in [135]. While the Banach algebra techniques were in some sense more classical, this work involved calculating norm estimates related to the approximations used, in order to be able to assert that the commutators indeed belong to the correct ideals.

3.3.3. Non-strongly convergent approximation sequences. When the sequence of projections P_τ is strongly convergent, most proofs involving the characterization of sequences as belonging to the ideal $\mathcal{J}$ make use of the close relationship between strong convergence and the ideal of compact operators. In this context, the observation that the multiplication of a strongly convergent sequence by a compact operator gives uniform convergence plays a fundamental role. For non-strongly convergent projections, as it is the case for the finite section projections in $l^\infty_\mathbb{Z}$ or $L^\infty(\mathbb{R})$, S. Roch and B. Silbermann [126] (see also [100, 4.36 et seq.]) initiated the idea of changing the above-mentioned connection between compactness and strong convergence to a definition. Starting with the projection sequence $\mathcal{P} := (P_n)_{n\in\mathbb{Z}^+}$, they substituted the usual compact operators by compact-like operators related to that sequence. The $\mathcal{P}$-compact operators will be those operators $K \in \mathcal{B}(X)$ such that

$$\|KP_\tau - K\| \text{ and } \|P_\tau K - K\| \text{ tend to zero as } \tau \to \infty.$$

This new approach with the adapted ideal of $\mathcal{P}$-compact operators happens to be useful also in the cases $1 < p < \infty$, and is at the heart of recent results [88, 89, 141].

In [88] the author, together with Helena Mascarenhas and Markus Seidel managed to use the new approach to study, together with the stability of operator sequences, a more general Fredholm theory in sequence algebras, including the splitting property (see Section 3.4 below) and an index formula in the case of $\mathcal{P}$-strongly converging finite section methods. This setting covered, in particular, convolution type operators in $L^p(\mathbb{R})$, with $p \in [1,\infty]$. A new class $\mathcal{Q}$ of quasi-banded operators was introduced, extending the previously known notion of band-dominated operators. It includes all convolutions with generating function continuous at all finite points of $\mathbb{R}$, but does not include the general piecewise continuous case. Results on its inverse and Fredholm closedness, and results on the stability of the FSM for such operators were obtained. This new context allowed for the treatment of the FSM for convolution type operators with symbols in a wide variety of function spaces. It was also possible to combine the convolution operators with multiplication operators by functions of again a variety of spaces and to obtain relatively simple conditions for stability. The results cover the cases $p \in \{1, \infty\}$, in which all Fourier convolution operators are contained in $\mathcal{Q}$. By using a special transformation, the flip J could also be included. An interesting observation is that the general results obtained by this method do not need the use of local principles. Without local principles, it was possible to cover convolution operators whose symbol has a jump at infinity.

The techniques pioneered in [88] were added to and reworked into a new approach in [89], which the authors named "Rich Approach". It completed the picture by describing also non-stable sequences in terms of the asymptotic behavior of their approximation numbers and Fredholm indices. It also covered much larger classes of data, including almost periodic, even bounded uniformly continuous multipliers b, piecewise continuous as well as operators of multiplication with almost arbitrary functions a having restrictions only on the behavior at infinity.

The new approach involved the introduction of the concept of rich sequences, where the existence of strong (or other) limits for the whole sequence is not necessary, as long as there are subsequences with certain properties. Using rich sequences with the concept of $\mathcal{P}$-convergence made constructions and proofs much simpler. In the previous works, when considering certain function spaces as generators for the convolution and multiplication operators, the localization had to be done over involved commutative subalgebras of sequences arising from functions whose definition was technical and the characterization of their maximal ideal spaces required hard proofs. This was now replaced by the much simpler algebra of the continuous functions. The approach used allowed to build upon much simpler (and probably even more general) definitions of the multiplier algebras.

An interesting point is that the role of the homomorphisms changed. While in the classic method the W homomorphisms served for the construction a sufficiently large ideal $\mathcal{J}$ such that the respective quotient has a large center, and other family H of homomorphisms provided the desired representatives for the identification of the local cosets, in this work only the W homomorphisms were used in the identification of the local cosets, whereas the H homomorphisms were used to build the ideal corresponding to $\mathcal{J}$.

3.4. Asymptotics, behavior of eigenvalues, approximation numbers and numerical ranges

Asymptotics of Toeplitz matrices were one of the main drivers for the development of the algebraic techniques described in this work. For instance, already in 1958, Widom was interested in the distribution of the eigenvalues for Hermitian Toeplitz matrices [154] and in 1974, his aim to calculate asymptotic determinants was the approximation of the inverses of (large) finite matrices by the known inverse of their infinite counterparts [156] following on the pioneer works of Szegő, Hartwig and Fisher, and others [63, 150, 151]. The above-mentioned paper [157] that presented for the first time formula (5) had the aim of extending some of the earlier results to Block Toeplitz matrices. The topic of Toeplitz determinants, which inspired the discovery of the algebraic techniques in numerical analysis, was touched also by those techniques. Already in 1985, Böttcher and Silbermann [24] proved the Fisher–Hartwig conjecture on Toeplitz determinants for a large class of symbols and after the fall of the Berlin wall, normal academic contacts between Widom in the USA and Böttcher and Silbermann in (now) Germany developed [27] further. See also Böttcher's nice expository paper [9].

Torsten Ehrhardt, a student from Silbermann, continued the work on the subject of Toeplitz Determinants with Fisher–Hartwig Singularities, publishing a series of works starting in 1997 ([3, 41]). In 2003 he gave a proof of the Szegő–Widom limit theorem using an algebraic approach [38].

The author also made a modest contribution to this topic by proving, together with A. Karlovich, a generalization of the Szegő strong limit theorem for Toeplitz determinants with the symbol having a nonstandard smoothness [78]. There it was assumed that the symbol belonged to the Wiener algebra and, moreover, the sequences of Fourier coefficients of the symbol with negative and nonnegative indices belonged to certain weighted Orlicz classes.

A book that contains a good overview of the issues and research done on Asymptotics and spectral properties of banded Toeplitz matrices up to the early 2000s is [17]. As was mentioned earlier, banded matrices are matrices where the non-zero entries are situated on a band around the main diagonal.

Work has continued on the subject, as is shown by [130], where the authors present a unified C*-algebra approach to classical Szegő limit theorems that applies to general (including non-selfadjoint) Toeplitz operators and in particular obtain results for operators with almost periodic diagonals. Other example is a recent generalization of the Szegő's strong limit theorem for certain classes of band-dominated operators [39] or perturbed operators [15].

3.4.1. Spectral approximation. The basic question in Spectral Approximation is whether the spectra of the operators in a convergent approximating sequence $(A_\tau)_{\tau\in\mathbb{I}}$ do approximate the spectrum of the limit operator A, $\sigma(A)$, in some sense. For that, one can define the partial limiting set of a sequence of subsets $(U_\tau)\subset\mathbb{C}$ as the set of all partial limits of all sequences (λ_τ) of $\lambda_\tau\in U_\tau$, $\tau\in\mathbb{I}$ and denote it by $p\text{-}\lim_{\tau\to\infty} U_\tau$. Then the question above can be more precisely stated as whether

$$p\text{-}\lim_{\tau\to\infty}\sigma(A_\tau)=\sigma(A).$$

In the case of Hermitian Toeplitz operators acting in l^2 or L^2 the answer is affirmative. In general it is not. A simple counter-example is given for instance by the unilateral forward shift V on l^2, defined by $(Vu)_0 = 0, (Vu)_j = u_{j-1}, j \in \mathbb{N}$. The spectrum of V is the closed unit disk, but approximations $A_n = P_n A P_n$ are represented as the $n+1 \times n+1$ finite matrices

$$A_n = \begin{bmatrix} 0 & 0 & 0 & \dots & 0 \\ 1 & 0 & 0 & \ddots & 0 \\ 0 & 1 & 0 & \ddots & \vdots \\ \vdots & \ddots & \ddots & \ddots & 0 \\ 0 & 0 & \dots & 1 & 0 \end{bmatrix}$$

with spectrum $\{0\}$. In fact, since the 1960 work of Schmidt and Spitzer [139] it is known that in the case of a trigonometric polynomial a, while the spectrum of the associated Laurent operators corresponds with the range of a, the limiting set of the spectra of the finite sections of that operator is a union of analytic arcs that are "inside" the closed curve $a(\mathbb{T}) \subset \mathbb{C}$. But while we have a negative answer for non-Hermitian operators to the spectral approximation question when considering the spectrum, the picture changes when one considers the ϵ-pseudospectrum of the operators. This notion was first introduced by Landau [85] in 1975. The ϵ-pseudospectrum is defined as the set

$$\sigma_\epsilon(A) := \{\lambda \in \mathbb{C} : \|(A - \lambda I)^{-1}\| \geq 1/\epsilon\}.$$

In 1992 Reichel and Trefethen [105] showed that for some classes of non-real symbols, the pseudospectra of the corresponding discrete convolution operators would mimic exactly the pseudospectra of the limit operator. Inspired by those results Böttcher in 1994 used algebraic techniques to extend this result to convolution operators on $L^2(\mathbb{R})$ for a much wider class of symbols [8]. His results also carried over to the discrete case. Later, Böttcher, Grudsky and Silbermann considered the case $1 < p < \infty$ for continuous (including matrix) symbols [18]. Roch [112] generalised the results for almost periodic symbols and included Numerical Ranges. In [56, Chapter 3] Hagen, Roch and Silbermann went on to prove that for the C*-case, fractality of the approximating algebra (see Section 3.5) guarantees convergence of the pseudospectra and also of the numerical ranges.

3.4.2. Approximation numbers. Given an operator A acting on the Banach space X, the lower approximation numbers from the right and from the left,

$$s_k^r(A) := \inf\{\|A - T\| : T \in \mathcal{B}(X), \dim \text{Ker } T \geq k\},$$
$$s_k^l(A) := \inf\{\|A - T\| : T \in \mathcal{B}(X), \dim \text{Coker } T \geq k\}.$$

measure the distance to (or degree of) injectivity and surjectivity, respectively. It is clear that

$$0 = s_0^{r/l}(A) \leq s_1^{r/l}(A) \leq s_2^{r/l}(A) \leq \cdots$$

for both right and left approximation numbers and if dim $X = n$ then $s_n^{r/l}(A) = \|A\|$. It is also not difficult to show (see, for instance, [10]), that

$$s_1^l(A) = s_1^r(A) = \begin{cases} 1/\|A^{-1}\| & \text{if } A \text{ is invertible} \\ 0 & \text{if } A \text{ is not invertible.} \end{cases} \tag{12}$$

Also notice that in case of X being a Hilbert space, these approximation numbers coincide with the lower singular values of A or A^*, respectively, that is, the square roots of the eigenvalues of A^*A and AA^*. In the early 1990s, Silbermann and Roch started [128, 144] to use algebraic techniques to extend Widom's earlier results [158] on the limiting set of singular values. A little later, by using the C*-algebra concept of Moore–Penrose invertibility in the approximation sequence algebra, they were able to derive a certain behavior of the singular values of the approximating matrices. They defined (A_n) to be a Moore–Penrose sequence if the coset $(A_n)+\mathcal{G}$ would be Moore–Penrose invertible as an element of the algebra $\mathcal{E}/\mathcal{G}$ and showed that in this case the set of the singular values of A_n would split into two parts. One of these parts remains bounded away from zero by a positive constant for all n, and the other part tends to zero as $n \to \infty$. Otherwise, $s_k^{l/r}(T_n(a)) \to 0$ as $n \to 0$ for each fixed k. The results were quite general, showing the power of C*-algebra techniques. When applied to the finite sections of a Toeplitz operator, the number k of singular values going to zero was given by

$$k = \dim \operatorname{Ker} T(a) + \dim \operatorname{Ker} T(\tilde{a}). \tag{13}$$

The results were written in preprint form in 1996, but were only published in 1998 [129].

Böttcher tried to extend these results to the Banach algebra case and in 1997 managed to prove the scalar version of the splitting property for some Toeplitz matrices with piecewise continuous symbol in l^p [10]. He settled the cases where either the limit operator $T(a)$ would be not normally solvable in l^p or l^q, with $1/p + 1/q = 1$, or the operator would be Fredholm in the whole scale of spaces l^r for $r \in [p, q]$. In those cases, the result is the same as for $p = 2$. The third case, when $T(a)$ is normally solvable in l^p and l^q but not on some l^r with $r \in [p, q]$, was only solved in 2006 by Rogozhin and Silbermann [131, 132]. They were able to consider the splitting property for all elements in the algebra of approximating sequences. In the third case above, the k-splitting property still occurs, but the number k must consider also the kernels of the non-invertible limit operators associated the local algebras at each point $t \in \mathbb{T}$.

Already in [129] the concept of a Fredholm sequence was introduced as a analogy to the notion of Fredholm operator. Broadly, Fredholm sequences are those sequences in $\mathcal{F}$ that are invertible modulo the ideal $\mathcal{J}$ (see Section 2.2). For Fredholm sequences the third case above cannot occur as all local representatives are invertible. When studying concrete approximation methods, it is possible to rely on this framework to extract the splitting property for Fredholm sequences.

Silbermann and Seidel [141] revisited this whole framework in the context of non-strongly convergent approximating sequences. In [88] the author, together with H. Mascarenhas and M. Seidel, building on those results, formulated and proved results on the splitting property for the large class of rich sequences. There, to a sequence $\mathbb{A}$ we associate numbers $\alpha(\mathbb{A})$ and $\beta(\mathbb{A})$ as the splitting number k corresponding to the right (resp. left) approximation numbers of the sequence $\mathbb{A}$ and were able to relate those numbers to the kernels and cokernels of the limit operators of the sequence.

3.5. Structure of the approximation algebras

Algebraic methods being applicable in such a wide range of spaces, operators and numerical methods, as we have seen in the sections above, hints at the existence of an underlying common structure in the occurring algebras. It is clear that the approximating algebras are related to the original Toeplitz operator algebra, which can be thought of as the algebra generated by the unilateral shift operator V.

In the case of C*-algebras, Coburn showed in the sixties [29] that the C*-algebra generated by any non-unitary isometry is *-isomorphic to the C*-algebra $\operatorname{alg}\{V\}$, where $V : l^2 \to l^2$ is the unilateral shift of multiplicity one. That algebra is exactly the algebra $\mathcal{T}(C) := \operatorname{alg}\{T(c) : c \in C(\mathbb{T})\}$ with $T(c) : l^2 \to l^2$. This algebra includes the ideal $\mathcal{K}$ of all compact operators and it was also known that the Calkin algebra $\mathcal{T}(C)/\mathcal{K}$ is isometrically isomorphic to the C*-algebra of the continuous functions $C(\mathbb{T})$.

With the above framework in mind, Böttcher and Silbermann in [23], when considering the convergence of the finite sections method for quarter plane Toeplitz operators, described for the first time the structure of the algebras of approximating sequences. They considered the algebra

$$\mathcal{A}(\mathcal{T}(C)) := \operatorname{alg}\{(P_nT(c)P_n)_{n\in\mathbb{Z}^+},\ c \in C(\mathbb{T})\}$$

containing all discretizations of Toeplitz operators with continuous generating function and obtained the following result:

Theorem 3.1. *The algebra $\mathcal{A}(\mathcal{T}(C))$ consists of all sequences of the form*

$$(A_n) = (P_nT(c)P_n + P_nK_0P_n + W_nK_1W_n + G_n) \tag{14}$$

with $K_{0,1}$ compact operators and $(G_n) \in \mathcal{G}$ and this representation is unique for each element. Moreover, the strong limits (6) exist for the elements of $\mathcal{A}(\mathcal{T}(C))$ and for each sequence $\mathbb{A} \in \mathcal{A}(\mathcal{T}(C))$ the coset $\mathbb{A} + \mathcal{G}$ is invertible in $\mathcal{A}/\mathcal{G}$ if and only if the strong limit operators $A = T(c) + K_0$ and $\tilde{A} = T(\tilde{c}) + K_1$ are invertible.

The above result implies that the quotient algebra $\mathcal{A}(\mathcal{T}(C))$ is $*$-isomorphic to the C*-algebra of all pairs

$$(T(c) + K_0, T(\tilde{c}) + K_1) \in \mathcal{B}(l^2) \times \mathcal{B}(l^2).$$

with c continuous, and $K_{0,1}$ compact operators.

As in the case of Silbermann's original paper [142], the set

$$\mathcal{J} = \{P_nK_0P_n + W_nK_1W_n + G_n,\ K_{0,1} \text{ compact}, G_n \in \mathcal{G}\}$$

is a closed two-sided ideal of $\mathcal{A}(\mathcal{T}(C))$. Write W_0 and W_1 for the $*$-homomorphisms corresponding to the strong limits in (6), such that

$$\mathsf{W}_0(A_n) := \operatorname*{s-lim}_{n\to\infty} A_n \quad \text{and} \quad \mathsf{W}_1(A_n) := \operatorname*{s-lim}_{n\to\infty} W_n A_n W_n.$$

It can be easily checked that they are in fact $*$-homomorphisms. Then $\mathcal{F} \subset \mathcal{E}$ is the sequence algebra where those homomorphisms are well defined. Introduce then the sets in $\mathcal{F}$

$$\mathcal{K}_0 = \{(P_n K P_n) + \mathcal{G},\ K \in \mathcal{K}\}$$
$$\mathcal{K}_1 = \{(W_n K W_n) + \mathcal{G},\ K \in \mathcal{K}\}.$$

They can be checked to be ideals of $\mathcal{F}/\mathcal{G}$, and the following structure emerges:

1. $\mathcal{G} \in \operatorname{Ker} \mathsf{W}_{0,1}$;
2. the quotient homomorphisms $\mathcal{F}/\mathcal{G} \to \mathcal{B}(l^2)$, $\mathbb{A} + \mathcal{G} \mapsto \mathsf{W}_{0,1}(\mathbb{A})$ are $*$-isomorphisms between $\mathcal{K}_{0,1}$ and $\mathcal{K}$;
3. $\mathcal{K}_0 \in \operatorname{Ker} \mathsf{W}_1$ and $\mathcal{K}_1 \in \operatorname{Ker} \mathsf{W}_0$;
4. an element $\mathbb{A} \in \mathcal{A}$ is invertible in $\mathcal{F}$ if and only if the operators $\mathsf{W}_{0,1}(\mathbb{A})$ are invertible.

A most interesting feature is that a similar structure (with a finite or infinite number of homomorphisms) was found in many sequence algebras, even if the operators are far from Toeplitz operators and the methods are different from the Finite Section Method. That it must be so was only recently proved by Steffen Roch [106] for separable C*-subalgebras of $\mathcal{F}$.

A further important observation derived from these earlier works is the notion of fractality applied to sequence algebras. Because the structure of the C*-algebra $\mathcal{A}/\mathcal{G}$ is given by the strong limit homomorphisms (that are in fact irreducible representations), all essential information is stored in each subsequence of the original sequence. Roch et al [103] attempted to describe rigorously the method of limit operators and to collect and unify known information about their applications in the study of Fredholm properties of band and band-dominated operators.

In [117], Roch and the author proved that all compact operators acting on $L^p(\mathbb{R})$ belong to the algebra generated by the operator of multiplication by the characteristic function of the positive half-axis and by the convolution operators with continuous generating function. This result, together with classical results, were used to prove that certain ideals of compact-like operator sequences (ideals of the type of $\mathcal{K}_0$ and $\mathcal{K}_1$ above), which are building blocks of the ideal $\mathcal{J}$, are included in the algebra generated by convolution and multiplication operators and the finite section projection sequence, in the case of Banach algebras.

In [36] the author, together with some students, showed in fact that the Banach algebra case also possessed a structure similar to the one referred above in this section, but considered a concrete example (see Figure 2). A key observation in that work was that it is possible to build the ideals $\mathcal{K}_j$ as the inverse of algebra automorphisms applied to the set of constant sequences $\mathcal{K} \subset \mathcal{F}$ containing the compact operators. The homomorphisms are built by applying the strong limit to the sequences resulting from the automorphisms.

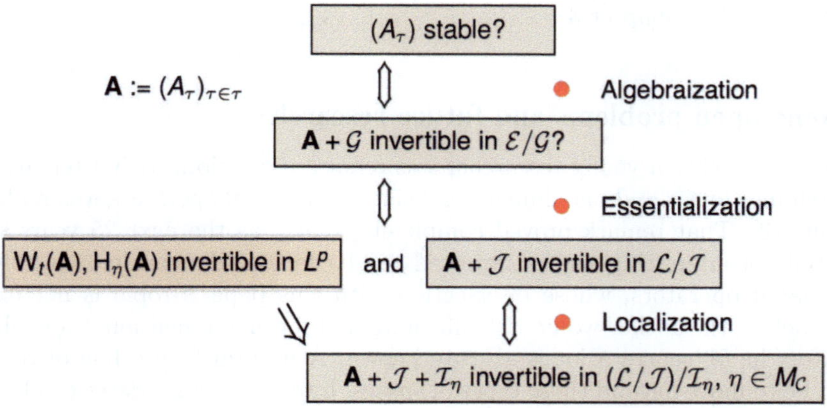

FIGURE 2. The algebraic approach to stability problems, as proposed in [36]

Here the usual role of the strong limit homomorphisms changes. While before the W homomorphisms served for the construction a sufficiently large ideal $\mathcal{J}$ such that the respective quotient $\mathcal{F}/\mathcal{J}$ has a large center, and the H homomorphisms provided the desired representatives for the identification of the local cosets, now the H homomorphisms are also used to enlarge the ideal $\mathcal{J}$, in such a way that the local algebras are now almost trivial, and invertibiity of the local cosets is implied by the invertibility of the essentialization homomorphisms...

There were many other attempts to understand and explore the structure of approximation algebras. Bernd Silbermann and the author [136, 137] generalized Coburn's theorem to approximating algebras where the operators being approximated are functions of a shift and the approximation method possess certain simple properties. It was possible in this abstract setting to recover many concrete results on applicability of concrete numerical methods. Roch [114] recently observed that (a minor modification of) the FSM-algebra $\mathcal{A}(\mathcal{T}(C))$ is also distinguished by a universal property: it is *-isomorphic to the universal C*-algebra generated by a power partial isometry, that is, an element v of a C*-algebra with the property that every power v^n is a partial isometry.

The theory for algebras generated by two idempotents, that originated independently with many authors (see [28, Remark 1.3]) was object of intense attention because, as was referred above, those structures appear when trying to identify local algebras related to singular integral operators. Important contributions included the generalization of Halmo's two projections theorem [61] to the case of two idempotents in Banach algebras, due to Roch and Silbermann [124], later completed by Gohberg and Krupnik [50, 53]. Generalization of the results for elements with massive spectra followed in [44] and the case of two idempotents and one flip in [45, 84] (there are several types of shifts, the original paper from Roch and Silbermann already considered one type). Algebras possessing more idempotents were also studied [16, 83]. Two gentle introductions to the subject can be found on [28] and [120, Chapter 3].

4. Some open problems and future research

In the early 1990s, a young researcher was refused a position, with a remark that research into linear and one-dimensional singular integral operators was no longer current [13]. That remark proved completely wrong, as the next 25 years saw a multitude of interesting and unexpected results for one-dimensional convolution and related operators, whose intersection with this paper's topic is not empty. Nevertheless, the author would risk affirming that for one-dimensional convolution operators in L^p spaces a lot is already known. A natural future line of research would be to move beyond those types of spaces. Interesting candidates can be variable exponent L^p type spaces, which are not rearrangement invariant. In the last few years advances in the study of algebras of operators in such spaces, commutator properties, etc (see, e.g., [77, 79]) opened the way for treating approximation methods in those spaces.

Another alternative are the super-Lebesgue spaces [40]. If we let $1 \leq p < q < \infty$, X be a suitable domain, one can define the super-Lebesgue space $L_{pq}(X)$ by

$$L_{pq}(X) := L^p(X) + L^q(X), \tag{15}$$

that is, the elements of $L_{pq}(X)$ are those functions[4] u that can be written as a sum $u = u_p + u_q$, with $u_p \in L^p(X)$ and $u_q \in L^q(X)$.

Super-Lebesgue spaces include the usual L^p spaces and have interesting properties. Torsten Ehrhardt and the author are working on a characterization of the Singular Integral Operator on those spaces, in such a way as to allow the further study of Numerical Methods for convolution operators in those types of spaces.

Numerical methods applications to real world problems need multi-dimensional domains with complex geometries. While tackling those complex geometries directly with Banach algebra techniques can be a difficult challenge, the study of multi-dimensional canonical problems, that "locally" can approximate those real world problem geometries will constitute a first step.

Not much is known yet about sequences of variable-coefficient Toeplitz matrices when the generating function is outside the Wiener algebra, or one considers l_p norms. As these matrices appear in many situations, this is also a promising area of future research.

Finally, the study of Banach algebras directly related to or just inspired by the applications to Numerical Analysis continues, with many recent papers on the subject, as was seen above. It would be interesting to be able to understand if the

[4] As is normal, functions which differ only on a set of measure 0 are identified.

different characterizations obtained so far can lead to a deeper understanding of these algebras's structure.

5. Conclusions

In this paper we tried to present an brief overview of the main lines of research connected with Banach algebra techniques in numerical analysis, which have been developed for more than 40 years. We tried to present, more than a dry account and enumeration of results, a lively account of the development of some of the key ideas behind those results, emphasising the cross-pollination of ideas between mathematicians across the globe.

The author's idea was to preserve some of the small details that are known by those who lived through what we described, but would be lost forever when only the published papers and results remained, of all those lifes dedicated to mathematics.

"Brief" here is the keyword. We set up with the objective of doing a 20 page paper with around 120 references, and have already overstepped those objectives. Nevertheless, much remains to be said, and the referred works are only a fraction of all that could be cited. We hope to have included all of the most relevant ones. If that is not the case, it is solely the author's fault.

Acknowledgement

The author would like to thank Albrecht Böttcher, Steffen Roch and Bernd Silbermann for many details and clarifications on the history of the subject during the 1980s, and for reading and commenting earlier versions of this document. This work has been partially supported by national funds through Fundação para a Ciêcia e a Tecnologia (FCT) with reference UID/CEC /50021/2013.

References

[1] G. Allan. Ideals of vector-valued functions. *Proc. London Math. Soc.*, 18(2):193–216, 1968.

[2] E. Basor. A brief history of the strong Szegő limit theorem. In *Mathematical methods in systems, optimization, and control*, volume 222 of *Oper. Theory Adv. Appl.*, pages 73–83. Birkhäuser/Springer Basel AG, Basel, 2012.

[3] E. Basor and T. Ehrhardt. Asymptotic formulas for determinants of a sum of finite Toeplitz and Hankel matrices. *Math. Nachr.*, 228:5–45, 2001.

[4] M. Bastos, P. Bravo, and Y.I. Karlovich. Convolution type operators with symbols generated by slowly oscillating and piecewise continuous matrix functions. volume 147 of *Oper. Theory Adv. Appl.*, pages 151–174. Birkhäuser, 2004.

[5] G. Baxter. A norm inequality for a "finite-section" Wiener–Hopf equation. *Illinois J. Math.*, 7:97–103, 1963.

[6] A. Böttcher. The finite section method for two-dimensional Wiener–Hopf integral operators in L^p with piecewise continuous symbols. *Math. Nachr.*, 116:61–73, 1984.

[7] A. Böttcher. The reduction method for Wiener–Hopf integral operators with piecewise-continuous symbol in L^p spaces. *Funktsional. Anal. i Prilozhen.*, 18(2):55–56, 1984.

[8] A. Böttcher. Pseudospectra and singular values of large convolution operators. *J. Integral Equations Applications*, 6(3):267–301, 1994.

[9] A. Böttcher. The Onsager formula, the Fisher–Hartwig conjecture, and their influence on research into Toeplitz operators. *J. Statist. Physics* (*Lars Onsager Festschrift*), 78:575–585, 1995.

[10] A. Böttcher. On the approximation numbers of large Toeplitz matrices. *Documenta Mathematica*, 2:1–29, 1997.

[11] A. Böttcher. C*-Algebras in numerical analysis. *Irish Math. Soc. Bulletin*, 45:57–133, 2000.

[12] A. Böttcher. Essay on Bernd Silbermann. In *Toeplitz matrices and singular integral equations* (*Pobershau,* 2001), volume 135 of *Oper. Theory Adv. Appl.*, pages 1–12. Birkhäuser, Basel, 2002.

[13] A. Böttcher. Linear and one-dimensional. In *Israel Gohberg and Friends*, pages 291–293. Birkhäuser, Basel, 2008.

[14] A. Böttcher, 2016. Private Communication.

[15] A. Böttcher, L. Fukshansky, S.R. Garcia, and H. Maharaj. Toeplitz determinants with perturbations in the corners. *J. Funct. Anal.*, 268(1):171–193, 2015.

[16] A. Böttcher, I. Gohberg, Y. Karlovich, N. Krupnik, S. Roch, B. Silbermann, and I. Spitkovsky. Banach algebras generated by N idempotents and applications. In *Singular integral operators and related topics*, volume 90 of *Oper. Theory Adv. Appl.*, pages 19–54. Birkhäuser, Basel, 1996.

[17] A. Böttcher and S.M. Grudsky. *Spectral Properties of Banded Toeplitz Matrices.* SIAM, Philadelphia, 2005.

[18] A. Böttcher, S.M. Grudsky, and B. Silbermann. Norms of inverses, spectra, and pseudospectra of large truncated Wiener–Hopf operators and Toeplitz matrices. *The New York Journal of Mathematics*, 3:1–31, 1997.

[19] A. Böttcher, N. Krupnik, and B. Silbermann. A general look at local principles with special emphasis on the norm computation aspect. *Integral Equations Operator Theory*, 11:455–479, 1988.

[20] A. Böttcher, S. Roch, and B. Silbermann. Local constructions and Banach algebras associated with Toeplitz operators on H^p. In *Seminar Analysis*, pages 23–30. Akad. Wiss. DDR, Berlin, 1986.

[21] A. Böttcher, S. Roch, B. Silbermann, and I. Spitkovsky. A Gohberg–Krupnik–Sarason symbol calculus for algebras of Toeplitz, Hankel, Cauchy, and Carleman operators. In *Topics in operator theory: Ernst D. Hellinger memorial volume*, volume 48 of *Oper. Theory Adv. Appl.*, pages 189–234. Birkhäuser, Basel, 1990.

[22] A. Böttcher and B. Silbermann. Über das Reduktionsverfahren für diskrete Wiener–Hopf-Gleichungen mit unstetigem Symbol. *Z. Analysis Anwendungen*, 1:1–5, 1982.

[23] A. Böttcher and B. Silbermann. The finite section method for Toeplitz operators on the quarter-plane with piecewise continuous symbols. *Math. Nachr.*, 110:279–291, 1983.

[24] A. Böttcher and B. Silbermann. Toeplitz matrices and determinants with Fisher–Hartwig symbols. *J. Funct. Anal.*, 63(2):178 – 214, 1985.

[25] A. Böttcher and B. Silbermann. *Introduction to Large Truncated Toeplitz Matrices.* Universitext. Springer-Verlag, New York, 1999.

[26] A. Böttcher and B. Silbermann. *Analysis of Toeplitz Operators.* Springer-Verlag, Berlin, Heidelberg, second edition, 2006.

[27] A. Böttcher, B. Silbermann, and H. Widom. A continuous analogue of the Fisher–Hartwig formula for piecewise continuous symbols. *J. Funct. Anal.*, 122(1):222–246, 1994.

[28] A. Böttcher and I. Spitkovsky. A gentle guide to the basics of two projections theory. *Lin. Alg. Appl.*, 432(6):1412–1459, 2010.

[29] L.A. Coburn. The C^*-algebra generated by an isometry. *Bull. Amer. Math. Soc.*, 73:722–726, 1967.

[30] W. Dahmen, S. Prössdorf, and R. Schneider. Wavelet approximation methods for pseudodifferential equations i: Stability and convergence. *Math. Zeit.*, 215(1):583–620, 1994.

[31] A. Devinatz and M. Shinbrot. General Wiener–Hopf operators. *Trans. Amer. Math. Soc.*, 145:467–494, 1969.

[32] V. Didenko and B. Silbermann. *Approximation of additive convolution-like operators: real C^*-algebra approach.* Birkhäuser, Basel, Boston, Berlin, 2008.

[33] V.D. Didenko, S. Roch, and B. Silbermann. Approximation methods for singular integral equations with conjugation on curves with corners. *SIAM J. Numer. Anal.*, 32(6):1910–1939, 1995.

[34] V.D. Didenko and B. Silbermann. On stability of approximation methods for the Muskhelishvili equation. *J. Comput. Appl. Math.*, 146(2):419–441, 2002.

[35] R.G. Douglas. Local Toeplitz operators. *Proc. London Math. Soc.* (3), 36(2):243–272, 1978.

[36] D. Dragičević, G. Preto, P. Santos, and M. Szamotulski. On a question by Markus Seidel. In *Operator Theory, Operator Algebras and Applications*, volume 242 of *Oper. Theory: Adv. Appl.*, pages 159–172. Birkhäuser, Basel, 2014.

[37] R. Duduchava. *Integral equations with fixed singularities.* B.G. Teubner Verlagsgesellschaft, Leipzig, 1979.

[38] T. Ehrhardt. A new algebraic approach to the Szegő–Widom limit theorem. *Acta Math. Hungar.*, 99(3):233–261, 2003.

[39] T. Ehrhardt, S. Roch, and B. Silbermann. A strong Szegő–Widom limit theorem for operators with almost periodic diagonal. *J. Funct. Anal.*, 260(1):30–75, 2011.

[40] T. Ehrhardt and P.A. Santos. On the singular integral operator in super-Lebesgue spaces. In preparation.

[41] T. Ehrhardt and B. Silbermann. Toeplitz determinants with one Fisher–Hartwig singularity. *J. Funct. Anal.*, 148(1):229 – 256, 1997.

[42] J. Elschner. On spline approximation for a class of non-compact integral equations. *Math. Nachr.*, 146:271–321, 1990.

[43] J. Favard. Sur les équations différentielles a coefficients presque périodiques. *Acta Math.*, 51:31–81, 1927.

[44] T. Finck, S. Roch, and B. Silbermann. Two projection theorems and symbol calculus for operators with massive local spectra. *Math. Nachr.*, 162:167–185, 1993.

[45] T. Finck, S. Roch, and B. Silbermann. Banach algebras generated by two idempotents and one flip. *Math. Nachr.*, 216:73–94, 2000.

[46] I. Gohberg and I. Feldman. *Convolution Equations and Projection Methods for their solution.* Amer. Math. Soc., Providence, RI., 1974. First published in Russian, Nauka, Moscow, 1971.

[47] I. Gohberg and N. Krupnik. The algebra generated by one-dimensional singular integral operators with piecewise continuous coefficients. *Funkcional. Anal. i Priložen*, 4(3):26–36, 1970 (in Russian).

[48] I. Gohberg and N. Krupnik. On singular integral operators on a composed curve. *Soobshch. Akad. Nauk. Gruz. SSR*, 64(3):21–24, 1971 (in Russian).

[49] I. Gohberg and N. Krupnik. On a local principle and algebras generated by Toeplitz matrices. *An. Şti. Univ. "Al. I. Cuza" Iaşi Secţ. I a Mat. (N.S.)*, 19:43–71, 1973. (in Russian; English translation in "Convolution equations and singular integral operators", *Oper. Theory Adv. Appl.*, vol. 206 (2010), 157–184).

[50] I. Gohberg and N. Krupnik. Extension theorems for invertibility symbols in Banach algebras. *Integral Equations Operator Theory*, 15:991–1010, 1992.

[51] I. Gohberg and N. Krupnik. *One-dimensional Linear Singular Integral Equations* (I). Birkhäuser, Basel, 1992.

[52] I. Gohberg and N. Krupnik. *One-dimensional Linear Singular Integral Equations* (II). Birkhäuser, Basel, 1992.

[53] I. Gohberg and N. Krupnik. Extension theorems for Fredholm and invertibility symbols. *Integral Equations Operator Theory*, 16:514–529, 1993.

[54] R. Hagen, S. Roch, and B. Silbermann. Stability of spline approximation methods for multidimensional pseudodifferential operators. *Integral Equations Operator Theory*, 19(1):25–64, 1994.

[55] R. Hagen, S. Roch, and B. Silbermann. *Spectral Theory of Approximation Methods for Convolution Equations.* Birkhäuser, Basel, 1995.

[56] R. Hagen, S. Roch, and B. Silbermann. *C^*-algebras and Numerical Analysis.* Marcel Dekker, Inc., New York, 2001.

[57] R. Hagen and B. Silbermann. A finite element collocation method for bisingular integral equations. *Applicable Anal.*, 19(2-3):117–135, 1985.

[58] R. Hagen and B. Silbermann. Local theory of the collocation method for the approximate solution of singular integral equations. II. In *Seminar Analysis (Berlin,* 1986/87), pages 41–56. Akad. Wiss. DDR, Berlin, 1987.

[59] R. Hagen and B. Silbermann. On the stability of the qualocation method. In *Seminar Analysis (Berlin,* 1987/1988), pages 43–52. Akademie-Verlag, Berlin, 1988.

[60] R. Hagen and B. Silbermann. A Banach algebra approach to the stability projection methods for singular integral equations. *Math. Nachr.*, 140(1):285–297, 1989.

[61] P. Halmos. Two subspaces. *Trans. Amer. Math. Soc.*, 144:381–389, 1969.

[62] P. Hartman. On completely continuous Hankel matrices. *Proc. Amer. Math. Soc.*, 9:862–866, 1958.

[63] R.E. Hartwig and M.E. Fisher. Asymptotic behavior of Toeplitz matrices and determinants. *Arch. Rational Mech. Anal.*, 32:190–225, 1969.

[64] P. Junghanns, R. Kaiser, and G. Mastroianni. Collocation for singular integral equations with fixed singularities of particular Mellin type. *Electron. Trans. Numer. Anal.*, 41:190–248, 2014.

[65] P. Junghanns, R. Kaiser, and D. Potts. Collocation-quadrature methods and fast summation for Cauchy singular integral equations with fixed singularities. *Linear Algebra Appl.*, 491:187–238, 2016.

[66] P. Junghanns, G. Mastroianni, and M. Seidel. On the stability of collocation methods for Cauchy singular integral equations in weighted L^p spaces. *Math. Nachr.*, 283(1):58–84, 2010.

[67] P. Junghanns and A. Rathsfeld. On polynomial collocation for Cauchy singular integral equations with fixed singularities. *Integral Equations Operator Theory*, 43(2):155–176, 2002.

[68] P. Junghanns and A. Rathsfeld. A polynomial collocation method for Cauchy singular integral equations over the interval. *Electron. Trans. Numer. Anal.*, 14:79–126, 2002. Orthogonal polynomials, approximation theory, and harmonic analysis (Inzel, 2000).

[69] P. Junghanns, S. Roch, and B. Silbermann. Collocation methods for systems of Cauchy singular integral equations on an interval. *Vychisl. Tekhnol.*, 6(1):88–124, 2001.

[70] P. Junghanns, S. Roch, and U. Weber. Finite sections for singular integral operators by weighted Chebyshev polynomials. *Integral Equations Operator Theory*, 21(3):319–333, 1995.

[71] P. Junghanns and B. Silbermann. Lokale Theorie des Kollokationsverfahrens zur näherungsweisen Lösung singulärer Integralgleichungen. Technical report, Institut für Mathematik, 1982. Preprint.

[72] P. Junghanns and B. Silbermann. Local theory of the collocation method for the approximate solution of singular integral equations, I. *Integral Equations Operator Theory*, 7(6):791–807, 1984.

[73] P. Junghanns and B. Silbermann. Numerical analysis for one-dimensional Cauchy singular integral equations. *Journal of Computational and Applied Mathematics*, 125(1–2):395–421, 2000. Numerical Analysis 2000. Vol. VI: Ordinary Differential Equations and Integral Equations.

[74] P. Junghanns and U. Weber. Local theory of projection methods for Cauchy singular equations on an interval. In *Boundary integral methods: numerical and mathematical aspects*, volume 1 of *Comput. Eng.*, pages 217–256. WIT Press/Comput. Mech. Publ., Boston, MA, 1999.

[75] A. Karlovich, H. Mascarenhas, and P.A. Santos. Finite section method for a Banach algebra of convolution type operators on $L^p(\mathbb{R})$ with symbols generated by PC and SO. *Integral Equations Operator Theory*, 67(4):559–600, 2010.

[76] A. Karlovich, H. Mascarenhas, and P.A. Santos. Erratum to: Finite section method for a Banach algebra of convolution type operators on $L^p(\mathbb{R})$ with symbols generated by PC and SO. *Integral Equations Operator Theory*, 69(3):447–449, 2011.

[77] A.Y. Karlovich. Fredholmness of singular integral operators with piecewise continuous coefficients on weighted Banach function spaces. *J. Integral Equations Applications*, 15(3):263–320, 09 2003.

[78] A.Y. Karlovich and P.A. Santos. On asymptotics of Toeplitz determinants with symbols of nonstandard smoothness. *J. Fourier Anal. Appl.*, 11(1):43–72, 2005.

[79] V. Kokilashvili, A. Meskhi, H. Rafeiro, and S. Samko. *Integral Operators in Non-Standard Function Spaces*, volume 248 of *Oper. Theory Adv. Appl.* Birkhäuser, Basel, 2016.

[80] A. Kozak. A certain projection method of solving operator equations in a Banach space. *Dokl. Akad. Nauk SSSR*, 211:1042–1045, 1973 (in Russian; English translation in *Soviet Math. Dokl.* 14:1159–1162, 1973).

[81] A. Kozak. A local principle in the theory of projection methods. *Dokl. Akad. Nauk SSSR*, 212:1287–1289, 1973 (in Russian; English translation in *Soviet Math. Dokl.* 14:1580–1583, 1974).

[82] N. Krupnik. *Banach algebras with symbol and singular integral operators.* Birkhäuser, Basel, 1987.

[83] N. Krupnik, S. Roch, and B. Silbermann. On C^*-algebras generated by idempotents. *J. Funct. Anal.*, 137(2):303–319, 1996.

[84] N. Krupnik and Y. Spigel. Invertibility symbols for a Banach algebra generated by two idempotents and a shift. *Integral Equations Operator Theory*, 17:567–578, 1993.

[85] H.J. Landau. On Szegő's eingenvalue distribution theorem and non-hermitian kernels. *Journal d'Analyse Mathématique*, 28(1):335–357, 1975.

[86] M. Lindner. *Infinite Matrices and their Finite Sections.* Birkhäuser, Basel, Boston, Berlin, 2006.

[87] M. Lindner and S. Roch. Finite sections of random Jacobi operators. *SIAM J. Numer. Anal.*, 50(1):287–306, 2012.

[88] H. Mascarenhas, P.A. Santos, and M. Seidel. Quasi-banded operators, convolutions with almost periodic or quasi-continuous data, and their approximations. *J. Math. Anal. Appl.*, 418(2):938–963, 2014.

[89] H. Mascarenhas, P.A. Santos, and M. Seidel. Approximation sequences to operators on banach spaces: a rich approach. *Journal of the London Mathematical Society*, 96(1):86–110, 2017.

[90] H. Mascarenhas and B. Silbermann. Convolution type operators on cones and their finite sections. *Math. Nachr.*, 278(3):290–311, 2005.

[91] H. Mascarenhas and B. Silbermann. Spectral approximation and index for convolution type operators on cones on $L^p(\mathbb{R}^2)$. *Integral Equations Operator Theory*, 65(3):415–448, 2009.

[92] H. Mascarenhas and B. Silbermann. Sequences of variable-coefficient Toeplitz matrices and their singular values. *Journal of Functional Analysis*, 270(4):1479 – 1500, 2016.

[93] S. Prössdorf. Ein Lokalisierungsprinzip in der Theorie der Spline-Approximationen und einige Anwendungen. *Math. Nachr.*, 119:239–255, 1984.

[94] S. Prössdorf. A localization principle in the theory of finite element methods. In *Problems and methods of mathematical physics* (*Karl-Marx-Stadt,* 1983), volume 63 of *Teubner-Texte Math.*, pages 169–177. Teubner, Leipzig, 1984.

[95] S. Prössdorf and A. Rathsfeld. A spline collocation method for singular integral equations with piecewise continuous coefficients. *Integral Equations Operator Theory,* 7(4):536–560, 1984.

[96] S. Prößdorf and A. Rathsfeld. On spline galerkin methods for singular integral equations with piecewise continuous coefficients. *Numer. Math.*, 48(1):99–118, 1986.

[97] S. Prössdorf and A. Rathsfeld. Stabilitätskriterien für Näherungsverfahren bei singulären Integralgleichungen in L^p. *Z. Anal. Anwendungen,* 6(6):539–558, 1987.

[98] S. Prössdorf and G. Schmidt. A finite element collocation method for singular integral equations. *Math. Nachr.*, 100:33–60, 1981.

[99] S. Prössdorf and R. Schneider. Spline approximation methods for multidimensional periodic pseudodifferential equations. *Integral Equations Operator Theory,* 15(4):626–672, 1992.

[100] S. Prössdorf and B. Silbermann. *Numerical Analysis for Integral and Related Operator Equations.* Birkhäuser, Basel, 1991.

[101] V. Rabinovich, S. Roch, and B. Silbermann. Fredholm theory and finite section method for band-dominated operators. *Integral Equations Operator Theory,* 30(4):452–495, 1998.

[102] V. Rabinovich, S. Roch, and B. Silbermann. Band-dominated operators with operator-valued coefficients, their Fredholm properties and finite sections. *Integral Equations Operator Theory,* 40(3):342–381, 2001.

[103] V. Rabinovich, S. Roch, and B. Silbermann. *Limit operators and their applications in operator theory,* volume 150 of *Oper. Theory Adv. Appl.* Birkhäuser Verlag, Basel, 2004.

[104] E. Reich. On non-hermitian Toeplitz matrices. *Mathematica Scandinavica,* 10:145–152, 1962.

[105] L. Reichel and L.N. Trefethen. Eigenvalues and pseudo-eigenvalues of Toeplitz matrices. *Linear Algebra and its Applications,* 162:153–185, 1992.

[106] S. Roch. Extension-restriction theorems for algebras of approximation sequences. to appear.

[107] S. Roch. Das Reduktionsverfahren für Produktsummen von Toeplitzoperatoren mit stückweise stetigen Symbolen. *Wiss. Z. Tech. Hochsch. Karl-Marx-Stadt,* 26(2):265–273, 1984.

[108] S. Roch. Das Reduktionsverfahren für Operatoren aus einer Toeplitzalgebra. *Wiss. Z. Tech. Hochsch. Karl-Marx-Stadt,* 27(1):121–126, 1985.

[109] S. Roch. Finite sections of operators generated by singular integrals with Carleman shift. Preprint 52. Technische Universität Karl-Marx-Stadt, 1987.

[110] S. Roch. Spline approximation methods cutting off singularities. *Z. Anal. Anwendungen,* 13(2):329–345, 1994.

[111] S. Roch. Spline approximation methods for Wiener–Hopf operators. In *Recent developments in operator theory and its applications* (*Winnipeg, MB,* 1994), pages 282–308. Birkhäuser, Basel, 1996.

[112] S. Roch. Spectral approximation of Wiener–Hopf operators with almost periodic generating function. *Numer. Funct. Anal. Optim.*, 21(1-2):241–253, 2000.

[113] S. Roch. *Finite Sections of Band-Dominated Operators*, volume 191/895 of *Memoirs of the American Mathematical Society.* AMS, 2008.

[114] S. Roch. The universal algebra generated by a power partial isometry. In *Large Truncated Toeplitz Matrices, Toeplitz Operators, and Related Topics*, volume 259 of *Oper. Theory Adv. Appl.*, pages 649–662. Birkhäuser, Basel, 2017.

[115] S. Roch and P.A. Santos. Two points, one limit: Homogenization techniques for two-point local algebras. *J. Math. Anal. Appl.*, 391(2):552–566, 2012.

[116] S. Roch and P.A. Santos. Finite section approximations in an algebra of convolution, multiplication and flip operators on $L^p(\mathbb{R})$. *Journal of Approximation Theory*, 186:64–97, 2014.

[117] S. Roch and P.A. Santos. A tour to compact type operators and sequences related to the finite sections projection. In *Operator theory, operator algebras and applications*, volume 242 of *Oper. Theory Adv. Appl.*, pages 311–323. Birkhäuser/Springer Basel AG, Basel, 2014.

[118] S. Roch, P.A. Santos, and B. Silbermann. Finite section method in some algebras of multiplication and convolution operators and a flip. *Z. Anal. Anwendungen*, 16(3):575–606, 1997.

[119] S. Roch, P.A. Santos, and B. Silbermann. A sequence algebra of finite sections, convolution and multiplication operators on $L^p(\mathbb{R})$. *Numer. Funct. Anal. Optim.*, 31(1):45–77, 2010.

[120] S. Roch, P.A. Santos, and B. Silbermann. *Non-commutative Gelfand Theories.* Universitext. Springer-Verlag, London, 2011.

[121] S. Roch, P.A. Santos, and B. Silbermann. Corrigendum: "A sequence algebra of finite sections, convolution, and multiplication operators on $L^p(\mathbb{R})$". *Numer. Funct. Anal. Optim.*, 34(1):113–116, 2013.

[122] S. Roch and B. Silbermann. Das Reduktionsverfahren für Potenzen von Toeplitzoperatoren mit unstetigem Symbol. *Wiss. Z. Tech. Hochsch. Karl-Marx-Stadt*, 24(3):289–294, 1982.

[123] S. Roch and B. Silbermann. Toeplitz-like operators, quasicommutator ideals, numerical analysis. I. *Math. Nachr.*, 120:141–173, 1985.

[124] S. Roch and B. Silbermann. Algebras generated by idempotents and the symbol calculus for singular integral operators. *Integral Equations Operator Theory*, 11:385–419, 1988.

[125] S. Roch and B. Silbermann. A symbol calculus for finite sections of singular integral operators with flip and piecewise continuous coefficients. *J. Funct. Anal.*, 78(2):365–389, 1988.

[126] S. Roch and B. Silbermann. Non-strongly converging approximation methods. *Demonstr. Math.*, 22(3):651–676, 1989.

[127] S. Roch and B. Silbermann. *Algebras of convolution operators and their image in the Calkin algebra. Report*, volume 157. R-MATH-05-90 des Karl-Weierstrass-Instituts für Mathematik, Berlin, 1990.

[128] S. Roch and B. Silbermann. Limiting sets of eigenvalues and singular values of Toeplitz matrices. *Asymptotic Analysis*, 8(3):293–309, 1994.

[129] S. Roch and B. Silbermann. Index calculus for approximation methods and singular value decomposition. *J. Math. Anal. Appl.*, 225:401–426, 1998.

[130] S. Roch and B. Silbermann. Szegő limit theorems for operators with almost periodic diagonals. *Oper. Mat.*, 1(1):1–29, 2007.

[131] A. Rogozhin and B. Silbermann. Approximation numbers for the finite sections of Toeplitz operators with piecewise continuous symbols. *J. Funct. Anal.*, 237(1):135–149, 2006.

[132] A. Rogozhin and B. Silbermann. Banach algebras of operator sequences: Approximation numbers. *Journal of Operator Theory*, 57(2):325–346, 2007.

[133] P.A. Santos. Spline approximation methods with uniform meshes in algebras of multiplication and convolution operators. *Math. Nachr.*, 232:95–127, 2001.

[134] P.A. Santos. Galerkin method with graded meshes for Wiener–Hopf operators with PC symbols in L^p spaces. In *Spectral Theory, Mathematical System Theory, Evolution Equations, Differential and Difference Equations*, volume 221 of *Oper. Theory: Adv. Appl.*, pages 587–605. Birkhäuser, Basel, 2012.

[135] P.A. Santos and B. Silbermann. Galerkin method for Wiener–Hopf operators with piecewise continuous symbol. *Integral Equations Operator Theory*, 38(1):66–80, 2000.

[136] P.A. Santos and B. Silbermann. An axiomatic approach to stability of approximation sequences. In *Singular integral operators, factorization and applications*, pages 369–381. Birkhäuser, Basel, 2003.

[137] P.A. Santos and B. Silbermann. An approximation theory for operators generated by shifts. *Numer. Funct. Anal. Optim.*, 27(3-4):451–484, 2006.

[138] D. Sarason. On products of Toeplitz operators. *Acta Sci. Math. (Szeged)*, 35:7–12, 1973.

[139] P. Schmidt and F. Spitzer. The Toeplitz matrices of an arbitrary Laurent polynomial. *Mathematica Scandinavica*, 8(1):15–38, 1960.

[140] M. Seidel. On an open question about the stability of the finite section method for a class of convolution type operators. *Integral Equations Operator Theory*, 75(2):249–255, 2013.

[141] M. Seidel and B. Silbermann. Banach algebras of operator sequences. *Oper. Matrices*, 6(3):385–432, 2012.

[142] B. Silbermann. Lokale Theorie des Reduktionsverfahrens für Toeplitzoperatoren. *Math. Nachr.*, 104:137–146, 1981.

[143] B. Silbermann. Lokale Theorie des Reduktionsverfahrens für Singuläre Integralgleichungen. *Z. Anal. Anwendungen*, 1(6):45–56, 1982.

[144] B. Silbermann. On the limiting set of singular values of toeplitz operators. *Lin. Alg. Appl.*, 182:35–43, 1993.

[145] B. Silbermann. Algebraic techniques in stability analysis for spline approximation methods. *Numer. Funct. Anal. Optim.*, 17(7-8):833–841, 1996.

[146] B. Silbermann and O. Zabroda. Asymptotic behavior of generalized convolutions: An algebraic approach. *J. Integral Equations Applications*, 18(2):169–196, 06 2006.

[147] I. Simonenko. A new general method of investigating linear operator equations of singular integral equation type. I. *Izv. Akad. Nauk SSSR Ser. Mat.*, 29:567–586, 1965 (in Russian).

[148] I. Simonenko. A new general method of investigating linear operator equations of singular integral equation type. II. *Izv. Akad. Nauk SSSR Ser. Mat.*, 29:757–782, 1965 (in Russian).

[149] I. Simonenko and C.N. Min′. *A local method in the theory of one-dimensional singular integral equations with piecewise continuous coefficients. Noethericity.* Rostov. Gos. Univ., Rostov-na-Donu, 1986 (in Russian).

[150] G. Szegő. Ein Grenzwertsatz über die Toeplitzschen Determinanten einer reellen positiven Funktion. *Math. Ann.*, 76(4):490–503, 1915.

[151] G. Szegő. On certain Hermitian forms associated with the Fourier series of a positive function. *Comm. Sém. Math. Univ. Lund* [*Medd. Lunds Univ. Mat. Sem.*], (Tome Supplémentaire):228–238, 1952.

[152] S. Treil. Invertibility of a Toeplitz operator does not imply its invertibility by the projection method. *Dokl. Akad. Nauk SSSR*, 292(3):563–567, 1987.

[153] I.È. Verbickiĭ and N.J. Krupnik. The applicability of the projection method to discrete Wiener–Hopf equations with a piecewise continuous symbol. *Mat. Issled.*, (Vyp. 45):17–28, 179, 1977 (in Russian).

[154] H. Widom. On the eigenvalues of certain Hermitian operators. *Trans. Amer. Math. Soc.*, 88:491–522, 1958.

[155] H. Widom. Asymptotic behavior of block Toeplitz matrices and determinants. *Advances in Math.*, 13:284–322, 1974.

[156] H. Widom. Asymptotic inversion of convolution operators. *Publications Mathématiques de l'Institut des Hautes Études Scientifiques*, 44(1):191–240, 1974.

[157] H. Widom. Asymptotic behavior of block Toeplitz matrices and determinants. II. *Advances in Math.*, 21:1–29, 1976.

[158] H. Widom. On the singular values of Toeplitz matrices. *Z. Anal. Anwendungen*, 8(3):221–229, 1989.

Pedro A. Santos
Departamento de Matemática / INESC-ID
Instituto Superior Técnico
Universidade Técnica de Lisboa
Av. Rovisco Pais
1049-001 Lisboa, Portugal
e-mail: `pedro.santos@tecnico.ulisboa.pt`

Operator Theory:
Advances and Applications, Vol. 267, 359–370

Linearizability of Multi-Control Systems of the Class C^1 by Additive Change of Controls

K.V. Sklyar, G.M. Sklyar and S.Yu. Ignatovich

Abstract. We give necessary and sufficient conditions for nonlinear control systems of the class C^1 with multi-dimensional control to be linearizable by means of changes of variables of the class C^2 and additive change of the control of class C^1.

Mathematics Subject Classification (2010). Primary 93B18; Secondary 93C10.
Keywords. Systems with multi-dimensional control, linearizability problem, C^1-smooth nonlinear control systems, Lie brackets technique.

1. Introduction and statement of the problem

We consider the linearizability problem for systems of the form

$$\dot{x} = f(x,u), \; x \in Q \subset \mathbb{R}^n, \; u \in \mathbb{R}^r, \tag{1.1}$$

where the vector function $f(x,u)$ is continuously differentiable, i.e., $f(x,u) \in C^1(Q \times \mathbb{R}^r)$. More specifically, we say that the system (1.1) is *locally linearizable in the domain* Q if there exists a change of variables

$$z = F(x) \in C^2(Q) \;\text{ such that }\; \det F_x(x) \neq 0, \; x \in Q, \tag{1.2}$$

which reduces the system (1.1) to a linear form

$$\dot{z} = Az + Bu, \quad z \in \mathbb{R}^n, \; u \in \mathbb{R}^r, \tag{1.3}$$

satisfying the controllability condition

$$\operatorname{rank}(B, AB, \ldots, A^{n-1}B) = n \;\text{ and }\; \operatorname{rank} B = r. \tag{1.4}$$

We say that (1.1) is *locally feedback linearizable in the domain* Q if there exists a change of variables (1.2) and a change of controls

$$\begin{aligned} &v = g(x,u) \in C^1(Q \times \mathbb{R}^r) \\ &\text{such that }\; g(x,\mathbb{R}^r) = \mathbb{R}^r, \; \det g_u(x,u) \neq 0, \; x \in Q, \; u \in \mathbb{R}^r, \end{aligned} \tag{1.5}$$

This work was financially supported by Polish National Science Centre grant no. 2017/25/B/ST1/01892.

which reduce (1.1) to a linear form

$$\dot{z} = Az + Bv, \quad z \in \mathbb{R}^n, \ v \in \mathbb{R}^r, \tag{1.6}$$

satisfying condition (1.4).

The linearizability problem was deeply and completely studied for the class of smooth systems, that is, for systems with $f(x, u) \in C^\infty(Q \times \mathbb{R}^r)$, during 70–80th [1, 2, 4, 6, 7, 13]. The main approaches were connected with differential-geometric tools, in particular, with Lie brackets technique.

Another ideas in the linearizability problem were proposed in [5] where a special subclass of linearizable systems (triangular systems) was described. Within this approach, smoothness requirements could be weaken: in fact, nonlinear systems of the class C^1 were treated.

In [11] the problem of linearizability in the class C^1 was considered and conditions of linearizability were formulated for systems with one-dimensional control. It turned out that in both cases – linearizability and feedback linearizability – well-known conditions for C^∞-smooth systems should be modified. For the latter case this adjustment was rather significant. In particular, it turned out that Lie brackets which were the basis of the whole analysis in C^∞-smooth case may not exist. Instead of Lie brackets, in [11] it was proposed to introduce new vector fields which were built recursively. In [10] and [9] these results were extended to the case of systems with multi-dimensional control.

Along with a change of controls of the general form, more specific cases are studied separately. In this paper we consider additive change of controls of the form

$$v = g(x) + u \ \text{ such that } \ g(x) \in C^1(Q), \ x \in Q. \tag{1.7}$$

Definition 1.1. We say that a control system of the form (1.1), where $f(x, u) \in C^1(Q \times \mathbb{R}^r)$, is *locally linearizable in the domain Q by additive change of controls* or, for brevity, *A-linearizable*, if there exists a change of variables of the form (1.2) and a change of controls of the form (1.7) which reduce the system (1.1) to a linear form (1.6), (1.4).

Such change of controls allows usage of results from linear control theory. For example, in [12] for A-linearizable systems the method of constructing a bang-bang control was proposed where the known methods for linear systems were applied.

Below we use the following common notation. By $[\cdot, \cdot]$ we denote the Lie bracket of vector fields, i.e., $[a(x), b(x)] = b_x(x)a(x) - a_x(x)b(x)$. Also, we use the notation $\mathrm{ad}_a^0 b(x) = b(x)$, $\mathrm{ad}_a^{k+1} b(x) = [a(x), \mathrm{ad}_a^k b(x)]$, $k \geq 0$. For a vector field $a(x)$ and a function $\varphi(x)$, a Lie derivative equals $L_a\varphi(x) = \varphi_x(x)a(x)$, and L_a^k denotes the kth power of the operator L_a, $k \geq 0$.

Conditions of A-linearizability for systems with one-dimensional control were obtained in [8], and also in [11] as a corollary of the general result. In this paper we consider multi-control systems. Instead of a direct generalization, we propose conditions of A-linearizability which are close to the conditions of linearizability [10]. In particular, for A-linearizability, a system should be affine, i.e., have the form (2.4). As in the general feedback linearizable case, Lie brackets $\operatorname{ad}_a^k b_i(x)$ may not exist. In fact, if $a(x)$ and $b_i(x)$ are of class $C^1(Q)$ then $\operatorname{ad}_a b_i(x) = (b_i)_x(x)a(x) - a_x(x)b_i(x)$ are necessarily of class $C(Q)$, however, *may not belong* to the class $C^1(Q)$. If this is the case, $\operatorname{ad}_a^2 b_i(x)$ cannot be found. This situation is shown below in Example 2.1. So, as in the general feedback linearizable case [9], vector fields $\chi_i^k(x)$ are introduced. However, unlike the general case, a recursive formula for $\chi_i^k(x)$ includes only $\chi_1^0(x), \ldots, \chi_r^0(x)$ (see (2.2)). Moreover, it turns out that, like in the linearizable case [10], condition (B4) of [11, Theorem 2] can be omitted.

2. A-linearizability conditions

Theorem 2.1. *A nonlinear system of the form* (1.1), *where* $f(x,u) \in C^1(Q\times\mathbb{R}^r)$, *is A-linearizable in the domain* Q *if and only if there exist integers* $n_1 \ge \cdots \ge n_r \ge 1$, $n_1 + \cdots + n_r = n$, *and a permutation* $\{m_1, \ldots, m_r\}$ *of the set* $\{1, \ldots, r\}$ *such that the following conditions hold:*

(A) $f(x,u) = a(x) + \sum\limits_{i=1}^{r} b_i(x)u_i$, *where* $a(x), b_1(x), \ldots, b_r(x) \in C^1(Q)$;

(B1) *there exist functions* $\mu_{ks}^p(x) \in C(Q)$ *such that the vector fields* $\chi_s^k(x)$, $s = 1, \ldots, r$, $k = 0, \ldots, n_s$, *which are defined recursively by*

$$\chi_s^0(x) = b_{m_s}(x), \tag{2.1}$$

$$\chi_s^{k+1}(x) = [a(x), \chi_s^k(x)] + \sum_{p=1}^{r} \mu_{ks}^p(x)\chi_p^0(x), \; k = 0, \ldots, n_s - 1, \tag{2.2}$$

exist and belongs to $C^1(Q)$;

(B2) $\operatorname{rank} M(x) = n$ *for* $x \in Q$, *where*

$$M(x) = \left(\chi_1^0(x), \ldots, \chi_1^{n_1-1}(x), \ldots, \chi_r^0(x), \ldots, \chi_r^{n_r-1}(x)\right),$$

and, additionally, for any $s = 1, \ldots, r$ *and any* $x \in Q$

$$\chi_s^{n_s}(x) \in \operatorname{Lin}\left\{\chi_p^i(x) : 1 \le p \le r, \; 0 \le i \le \min\{n_s, n_p - 1\}\right\}; \tag{2.3}$$

(B3) $[\chi_s^i(x), \chi_q^j(x)] = 0$, $x \in Q$, *for all* $s, q = 1, \ldots, r$, $i = 0, \ldots, n_s$, $j = 0, \ldots, n_q$.

In this case, the numbers $n_1, \ldots, n_r$ *are controllability indices of the resulting linear system.*

Proof. Necessity. Suppose the system (1.1) is A-linearizable in the domain Q. Namely, suppose that a change of variables (1.2) and a change of controls (1.7) reduce the system (1.1) to a linear form (1.6), (1.4). Then for new variables $z = F(x)$ we have

$$\dot z = F_x(x)f(x,u) = Az + Bv = AF(x) + Bg(x) + Bu.$$

Therefore,

$$F_x(x)f(x,u) = AF(x) + Bg(x) + Bu.$$

Differentiating this equality w.r.t. u we get $F_x(x)f_u(x,u) = B$. Hence,

$$f_u(x,u) = (F_x(x))^{-1}B,$$

what means that $f(x,u)$ is linear on u, i.e., it satisfies condition (A). So, the initial system has the form

$$\dot{x} = a(x) + \sum_{i=1}^{r} b_i(x)u_i, \tag{2.4}$$

where $a(x), b_1(x), \ldots, b_r(x) \in C^1(Q)$.

Now let us implement the change of controls (1.7) to the system (2.4), i.e., substitute $u = v - g(x)$. We get

$$\dot{x} = a(x) - \sum_{i=1}^{r} b_i(x)g_i(x) + \sum_{i=1}^{r} b_i(x)v_i,$$

and, by our supposition, this system is linearizable by a change of variables only, since the change of variables (1.2) reduces the system to a linear form (1.6), (1.4). So, we can apply known conditions of linearizability in the class C^1 [10, 11]. Namely, let us denote

$$\widetilde{a}(x) = a(x) - \sum_{i=1}^{r} b_i(x)g_i(x).$$

Taking into account [10, Theorem 2.1 and Remark 2.4], we get that all vector fields $\operatorname{ad}^k_{\widetilde{a}} b_s(x)$ exist, belong to the class $C^1(Q)$, and satisfy equalities

$$\left[\operatorname{ad}^k_{\widetilde{a}} b_s(x), \operatorname{ad}^j_{\widetilde{a}} b_q(x)\right] = 0.$$

Let us denote by $n_1 \geq \cdots \geq n_r \geq 1$ the controllability indices of the corresponding linear system. Then there exists a permutation $\{m_1, \ldots, m_r\}$ of the set $\{1, \ldots, r\}$ such that

$$\operatorname{rank}\{\operatorname{ad}^0_{\widetilde{a}} b_{m_1}(x), \ldots, \operatorname{ad}^{n_1-1}_{\widetilde{a}} b_{m_1}(x), \ldots, \operatorname{ad}^0_{\widetilde{a}} b_{m_r}(x), \ldots, \operatorname{ad}^{n_r-1}_{\widetilde{a}} b_{m_r}(x)\} = n$$

and

$$\operatorname{ad}^{n_s}_{\widetilde{a}} b_{m_s}(x) \in \operatorname{Lin}\left\{\operatorname{ad}^i_{\widetilde{a}} b_{m_p}(x) : 1 \leq p \leq r,\ 0 \leq i \leq \min\{n_s, n_p - 1\}\right\}.$$

Thus, let us denote

$$\chi^k_s(x) = \operatorname{ad}^k_{\widetilde{a}} b_{m_s}(x),\ s = 1, \ldots, r,\ k \geq 0, \tag{2.5}$$

then all these vector fields exist, belong to the class $C^1(Q)$, and satisfy conditions (B2) and (B3).

It remains to check that $\chi^k_s(x)$ defined by (2.5) satisfy the recursive formula (2.1), (2.2). Equality (2.1) follows directly from the definition of $\chi^k_s(x)$. For $k \geq 0$

using condition (B3) we get

$$\chi_s^{k+1}(x) = \mathrm{ad}_{\widetilde{a}}^{k+1} b_{m_s}(x) = [\widetilde{a}(x), \mathrm{ad}_{\widetilde{a}}^{k} b_{m_s}(x)] = [a(x) - \sum_{i=1}^{r} b_i(x) g_i(x), \chi_s^k(x)]$$
$$= [a(x), \chi_s^k(x)] - \sum_{p=1}^{r} [\chi_p^0(x) g_{m_p}(x), \chi_s^k(x)] \tag{2.6}$$
$$= [a(x), \chi_s^k(x)] - \sum_{p=1}^{r} \underbrace{[\chi_p^0(x), \chi_s^k(x)]}_{=0} g_{m_p}(x) + \sum_{p=1}^{r} \left((g_{m_p}(x))_x \chi_s^k(x)\right) \chi_p^0(x).$$

Let us denote $\mu_{ks}^p(x) = (g_{m_p}(x))_x \chi_s^k(x)$. Since $g_i(x) \in C^1(Q)$, we get $\mu_{ks}^p(x) \in C(Q)$; therefore, (2.6) leads to (2.2).

Sufficiency. Suppose conditions of the theorem are satisfied. For convenience, let us introduce the notation

$$\Omega = \{(p, i) : 1 \le p \le r,\ 0 \le i \le n_p - 1\},$$

then condition (B2) implies that the set of columns of the matrix $M(x)$

$$P(x) = \{\chi_p^i(x) : (p, i) \in \Omega\}$$

is a basis of $\mathbb{R}^n$ for any $x \in Q$. Let us expand $\chi_s^{n_s}(x)$ w.r.t. the basis $P(x)$ and take into account condition (2.3). We get

$$\chi_s^{n_s}(x) = \sum_{(p,i)\in\Omega,\ i\le n_s} \nu_{p,i}^s(x) \chi_p^i(x), \quad x \in Q,$$

where $\nu_{p,i}^s(x)$ are some functions. Since all functions $\chi_p^i(x)$ and $\chi_s^{n_s}(x)$ belong to the class $C^1(Q)$ and $P(x)$ is a basis, we get $\nu_{p,i}^s(x) \in C^1(Q)$. Now we prove that $\nu_{p,i}^s(x) = \mathrm{const}$. Let us consider any $(m, k) \in \Omega$. Using (B3), we get

$$[\chi_m^k(x), \chi_s^{n_s}(x)] = \sum_{(p,i)\in\Omega,\ i\le n_s} [\chi_m^k(x), \nu_{p,i}^s(x) \chi_p^i(x)]$$
$$= \sum_{(p,i)\in\Omega,\ i\le n_s} \left(\underbrace{[\chi_m^k(x), \chi_p^i(x)]}_{=0} \nu_{p,i}^s(x) + \left((\nu_{p,i}^s(x))_x \chi_m^k(x)\right) \chi_p^i(x) \right) = 0.$$

Since $P(x)$ is a basis, we get $(\nu_{p,i}^s(x))_x \chi_m^k(x) = 0$ for any $(p, i) \in \Omega$, $i \le n_s$. Since this equality holds for any $(m, k) \in \Omega$ and $P(x)$ is a basis, we get $(\nu_{p,i}^s(x))_x = 0$. Hence, functions $\nu_{p,i}^s(x)$ do not depend on x, i.e., $\nu_{p,i}^s(x) = \nu_{p,i}^s$, and

$$\chi_s^{n_s}(x) = \sum_{(p,i)\in\Omega,\ i\le n_s} \nu_{p,i}^s \chi_p^i(x), \quad x \in Q,\ s = 1, \ldots, r. \tag{2.7}$$

Now, for any fixed $q = 1, \ldots, r$, we consider the system of partial differential equations with coefficients in $C^1(Q)$ of the form

$$\begin{aligned} &(\varphi(x))_x \chi_s^j(x) = 0, \quad (s, j) \in \Omega,\ (s, j) \ne (q, n_q - 1), \\ &(\varphi(x))_x \chi_q^{n_q-1}(x) = (-1)^{n_q-1}. \end{aligned} \tag{2.8}$$

In a matrix form, the system can be written as $(\varphi(x))_x M(x) = (-1)^{n_q-1} e_p$, where e_p is the unit row vector with 1 on the pth place, $p = n_1 + \cdots + n_q$. Then (2.8) is equivalent to the system

$$(\varphi(x))_x = (-1)^{n_q-1} e_p M^{-1}(x),$$

where the right-hand side belongs to the class $C^1(Q)$. It can be shown analogously to [10] that the conditions of the Frobenius Theorem [3, Chapter VI] are satisfied and there exists a solution $\varphi_q(x)$ of this system such that $\varphi_q(x) \in C^2(Q)$. Our goal is to prove that for $k = 0, \ldots, n_q - 1$

$$L_a^k \varphi_q(x) \in C^2(Q). \tag{2.9}$$

Moreover, we also prove that

$$\begin{aligned} &(L_a^k \varphi_q(x))_x X_s^j(x) = (-1)^k y_{q,s}^{k+j} = \text{const},\\ &s = 1, \ldots, r,\ j = 0, \ldots, n_s, \end{aligned} \tag{2.10}$$

for $k = 0, \ldots, n_q - 1$, where, in particular, for $s = 1, \ldots, r$

$$y_{q,s}^m = 0 \ \text{ for } \ 0 \le m \le n_q - 2,\ s \ne q, \tag{2.11}$$
$$y_{q,q}^m = 0 \ \text{ for } \ 0 \le m \le n_q - 2, \tag{2.12}$$
$$y_{q,q}^{n_q-1} = (-1)^{n_q-1}, \tag{2.13}$$
$$y_{q,s}^m = 0 \ \text{ for } \ 0 \le m \le n_s - 1,\ s \ne q. \tag{2.14}$$

For $k = 0$, (2.9) was justified above and (2.10) follows immediately from (2.8) and (2.7), as well as (2.12)–(2.14). Let us prove that, additionally,

$$y_{q,s}^{n_s} = 0 \ \text{ if } \ n_s \le n_q - 2,\ s \ne q. \tag{2.15}$$

In fact, in this case $n_q - 1 \ge n_s + 1$, therefore, (2.7), (2.8), and (2.14) imply

$$y_{q,s}^{n_s} = (\varphi_q(x))_x X_s^{n_s}(x) = \sum_{(p,i)\in\Omega,\ i\le n_s} \nu_{p,i}^s (\varphi_q(x))_x X_p^i(x) = 0.$$

Thus, (2.14) and (2.15) imply

$$y_{q,s}^m = 0 \ \text{ for } \ 0 \le m \le \min\{n_s, n_q - 2\},\ s \ne q.$$

Now suppose that for some d such that $0 \le d \le n_q - 2$, relations (2.9), (2.10) and

$$y_{q,s}^m = 0 \ \text{ for } \ 0 \le m \le \min\{n_s + k, n_q - 2\},\ s \ne q, \tag{2.16}$$

hold for $k = 0, \ldots, d$. Let us prove that they are true also for $k = d + 1$.

First, consider any $(s,j)\in\Omega$. Notice that (2.16) for $k=d$ gives $y^d_{q,p}=0$ for any $p=1,\dots,r$. Using (2.2) and the induction supposition we get

$$
\begin{aligned}
(L_a^{d+1}\varphi_q(x))_x\chi_s^j(x) &= \big((L_a^d\varphi_q(x))_x a(x)\big)_x \chi_s^j(x)\\
&= \underbrace{\big((L_a^d\varphi_q(x))_x\chi_s^j(x)\big)_x}_{=0} a(x) - (L_a^d\varphi_q(x))_x[a(x),\chi_s^j(x)]\\
&= \underbrace{-(L_a^d\varphi_q(x))_x\chi_s^{j+1}(x)}_{=(-1)^{d+1}y_{q,s}^{d+j+1}} + \sum_{p=1}^{r}\mu_{js}^p(x)\underbrace{(L_a^d\varphi_q(x))_x\chi_p^0(x)}_{=(-1)^d y_{q,p}^d=0}\\
&= (-1)^{d+1}y_{q,s}^{d+j+1}.
\end{aligned}
$$

Thus, for any $(s,j)\in\Omega$ we have

$$(L_a^{d+1}\varphi_q(x))_x\chi_s^j(x) = (-1)^{d+1}y_{q,s}^{d+j+1}. \tag{2.17}$$

Using the matrix form we get $(L_a^{d+1}\varphi_q(x))_x = yM^{-1}(x)$, where y is a constant row vector with components $(-1)^{d+1}y_{q,s}^{d+j+1}$. Since $M^{-1}(x)\in C^1(Q)$, we get

$$L_a^{d+1}\varphi_q(x)\in C^2(Q). \tag{2.18}$$

Now let us consider any $s=1,\dots,r$ and $j=n_s$. Applying (2.7) and (2.17), we get

$$(L_a^{d+1}\varphi_q(x))_x\chi_s^{n_s}(x) = \sum_{(p,i)\in\Omega,\ i\le n_s}\nu_{p,i}^s\underbrace{(L_a^{d+1}\varphi_q(x))_x\chi_p^i(x)}_{=(-1)^{d+1}y_{q,p}^{d+i+1}} = (-1)^{d+1}y_{q,s}^{d+n_s+1} \tag{2.19}$$

where

$$y_{q,s}^{d+n_s+1} = \sum_{(p,i)\in\Omega,\ i\le n_s}\nu_{p,i}^s y_{q,p}^{d+i+1} = \text{const}\,. \tag{2.20}$$

Thus, (2.17)–(2.20) imply that (2.9) and (2.10) hold for $k=d+1$.

Now let us notice that if $n_s+d\ge n_q-2$ then (2.16) for $k=d$ implies (2.16) for $k=d+1$ and, moreover, in this case (2.11) holds.

Suppose d is such that $n_s+d+1\le n_q-2$. Let us consider $y_{q,p}^{d+i+1}$ from (2.20): we have $d+i+1\le d+n_s+1\le n_q-2$ since $i\le n_s$, and $d+i+1\le d+n_p$ since $(p,i)\in\Omega$. Hence, $d+i+1\le\min\{n_p+d,n_q-2\}$, therefore, each term in the right-hand side of (2.20) vanishes due to the induction supposition. Hence, (2.16) holds for $k=d+1$.

The induction arguments complete the proof of (2.11) and (2.9), (2.10) for $k=0,\dots,n_q-1$.

Let us denote $\sigma_1=0$ and $\sigma_q=n_1+\cdots+n_{q-1}$ for $q=2,\dots,r$. We consider the change of variables $z=F(x)$ where

$$F_{\sigma_q+k}(x) = L_a^{k-1}\varphi_q(x),\ q=1,\dots,r,\ k=1,\dots,n_q.$$

As was shown above, $F(x)\in C^2(Q)$. One can prove analogously to [10] that the map $z=F(x)$ is locally invertible, i.e., $\det F_x(x)\ne 0$ for $x\in Q$.

In the new variables, for $k = 1, \dots, n_q$ we have

$$\dot z_{\sigma_q+k} = (F_{\sigma_q+k}(x))_x \dot x = \left(L_a^{k-1}\varphi_q(x)\right)_x \left(a(x) + \sum_{i=1}^r b_{m_i}(x)u_{m_i}\right)$$

$$= L_a^k\varphi_q(x) + \sum_{i=1}^r \left(L_a^{k-1}\varphi_q(x)\right)_x \chi_i^0(x)u_{m_i} = L_a^k\varphi_q(x) + \sum_{i=1}^r (-1)^{k-1} y_{q,i}^{k-1} u_{m_i}.$$

Taking into account (2.11)–(2.14) we get $y_{q,i}^{k-1} = 0$ for $k = 1, \dots, n_q - 1$, $i = 1, \dots, r$. For $k = n_q$, we get $y_{q,i}^{n_q-1} = 0$ if $n_i = n_q$, $i \neq q$, and $y_{q,q}^{n_q-1} = (-1)^{n_q-1}$. Also, (2.14) implies that $y_{q,i}^{n_q-1} = 0$ if $n_i \geq n_q$, $i \neq q$. Hence,

$$\dot z_{\sigma_q+k} = z_{\sigma_q+k+1}, \; k = 1, \dots, n_q - 1,$$
$$\dot z_{\sigma_q+n_q} = L_a^{n_q}\varphi_q(x) + u_{m_q} + \sum_{n_i \leq n_q-1} (-1)^{n_q-1} y_{q,i}^{n_q-1} u_{m_i}.$$

Now we define a change of controls. Let us introduce the constant matrix $S = \{S_{qi}\}_{q,i=1}^r$ with elements $S_{qi} = (-1)^{n_q-1} y_{q,i}^{n_q-1}$. Since $n_1 \geq \dots \geq n_r$, this matrix is upper triangular with unit diagonal elements. Denote by $h(x)$ the vector with elements $h_q(x) = L_a^{n_q}\varphi_q(x) \in C^1(Q)$, then

$$\dot z_{\sigma_q+n_q} = h_q(x) + \sum_{i=1}^r S_{qi} u_{m_i}.$$

Let us define

$$g_{m_i}(x) = (S^{-1}h(x))_i \in C^1(Q), \quad i = 1, \dots, r,$$

and introduce the additive change of controls of the form $v = g(x) + u$. Then

$$\dot z_{\sigma_q+n_q} = h_q(x) + \sum_{i=1}^r S_{qi}u_{m_i} = \sum_{i=1}^r S_{qi}v_{m_i} = v_{m_q} + \sum_{n_i \leq n_q-1} (-1)^{n_q-1} y_{q,i}^{n_q-1} v_{m_i}.$$

Thus, the change of variables $z = F(x)$ and the additive change of controls $v = g(x) + u$ reduce the initial system to the linear form

$$\dot z_{\sigma_q+k} = z_{\sigma_q+k+1}, \; k = 1, \dots, n_q - 1,$$
$$\dot z_{\sigma_q+n_q} = v_{m_q} + \sum_{n_i \leq n_q-1} (-1)^{n_q-1} y_{q,i}^{n_q-1} v_{m_i}, \; q = 1, \dots, r.$$

Finally, we note that the numbers $n_1, \dots, n_r$ are controllability indices of this linear system. □

Example 2.1. Consider the system of the class C^1

$$\begin{aligned} \dot x_1 &= x_2 + x_2^2|x_2|, \\ \dot x_2 &= x_4|x_4| + \tfrac{1}{1+3x_2|x_2|}u_1, \\ \dot x_3 &= \tfrac{x_4}{1-3|x_3|x_3}, \\ \dot x_4 &= u_1 + u_2, \end{aligned} \qquad (2.21)$$

in the domain $Q = \{x \in \mathbb{R}^4 : x_2 > -\frac{1}{\sqrt{3}},\ x_3 < \frac{1}{\sqrt{3}}\}$. For brevity, denote $\alpha(x) = x + x^2|x| \in C^2$, $\beta(x) = x - x^2|x| \in C^2$, $\gamma(x) = x|x| \in C^1$, then the system can be rewritten as

$$\dot{x}_1 = \alpha(x_2),\quad \dot{x}_2 = \gamma(x_4) + \frac{1}{\alpha'(x_2)}u_1,$$
$$\dot{x}_3 = \frac{x_4}{\beta'(x_3)},\quad \dot{x}_4 = u_1 + u_2.$$

We have

$$a(x) = \begin{pmatrix} \alpha(x_2) \\ \gamma(x_4) \\ \frac{x_4}{\beta'(x_3)} \\ 0 \end{pmatrix},\quad \chi_1^0(x) = b_1(x) = \begin{pmatrix} 0 \\ \frac{1}{\alpha'(x_2)} \\ 0 \\ 1 \end{pmatrix},\quad \chi_2^0(x) = b_2(x) = e_4.$$

Then

$$[a(x), \chi_1^0(x)] = \begin{pmatrix} -1 \\ \gamma(x_4)\left(\frac{1}{\alpha'(x_2)}\right)' - \gamma'(x_4) \\ -\frac{1}{\beta'(x_3)} \\ 0 \end{pmatrix}, [a(x), \chi_2^0(x)] = \begin{pmatrix} 0 \\ -\gamma'(x_4) \\ -\frac{1}{\beta'(x_3)} \\ 0 \end{pmatrix}.$$

Obviously, $[a(x), \chi_1^0(x)]$ and $[a(x), \chi_2^0(x)]$ do not belong to the class $C^1(Q)$. Since $\chi_i^0(x) = b_i(x)$, we conclude that $\mathrm{ad}_a b_i(x) = [a(x), b_i(x)]$ do not belong to $C^1(Q)$, therefore, $\mathrm{ad}_a^2 b_i(x)$ do not exist.

However, let us introduce $\mu_{01}^1(x) = -\mu_{01}^2(x) = -\alpha'(x_2)\Big(\gamma(x_4)\left(\frac{1}{\alpha'(x_2)}\right)' - \gamma'(x_4)\Big) \in C(Q)$ and $\mu_{02}^1(x) = -\mu_{02}^2(x) = \alpha'(x_2)\gamma'(x_4) \in C(Q)$, then

$$\chi_1^1(x) = [a(x), \chi_1^0(x)] + \mu_{01}^1(x)\chi_1^0(x) + \mu_{01}^2(x)\chi_2^0(x) = -e_1 - \frac{1}{\beta'(x_3)}e_3,$$
$$\chi_2^1(x) = [a(x), \chi_2^0(x)] + \mu_{02}^1(x)\chi_1^0(x) + \mu_{02}^2(x)\chi_2^0(x) = -\frac{1}{\beta'(x_3)}e_3,$$

so, $\chi_1^1(x), \chi_2^1(x) \in C^1(Q)$. One easily finds $\chi_1^2(x) = [a(x), \chi_1^1(x)] = 0$ and $\chi_2^2(x) = [a(x), \chi_2^1(x)] = 0$. Obviously, vector fields $\chi_1^0(x)$, $\chi_2^0(x)$, $\chi_1^1(x)$, $\chi_2^1(x)$ are linearly independent and all their Lie brackets vanish. Therefore, all conditions of Theorem 2.1 are satisfied (with $n_1 = n_2 = 2$), hence, the system is A-linearizable.

Let us find linearizing changes of variables and controls. We have two systems of first-order partial differential equations

$$(\varphi_1(x))_x\chi_1^0(x) = (\varphi_1(x))_x\chi_2^0(x) = (\varphi_1(x))_x\chi_2^1(x) = 0,\ (\varphi_1(x))_x\chi_1^1(x) = -1,$$
$$(\varphi_2(x))_x\chi_1^0(x) = (\varphi_2(x))_x\chi_1^1(x) = (\varphi_2(x))_x\chi_2^0(x) = 0,\ (\varphi_2(x))_x\chi_2^1(x) = -1,$$

and we may choose

$$\varphi_1(x) = x_1,\ \varphi_2(x) = \beta(x_3) - x_1 = x_3 - x_3^2|x_3| - x_1$$

(the solutions are unique up to a constant). Let us define $z = F(x)$ as

$$z_1 = F_1(x) = \varphi_1(x) = x_1,\qquad z_2 = F_2(x) = L_a\varphi_1(x) = \alpha(x_2),$$
$$z_3 = F_3(x) = \varphi_2(x) = \beta(x_3) - x_1,\quad z_4 = F_4(x) = L_a\varphi_2(x) = -\alpha(x_2) + x_4.$$

Then $\det F_x(x) = \alpha'(x_2)\beta'(x_3) = (1+3|x_2|x_2)(1-3|x_3|x_3) \neq 0$ for $x \in Q$. In the new variables, we have

$$\dot z_1 = z_2, \quad \dot z_2 = \underbrace{\alpha'(x_2)\gamma(x_4) + u_1}_{=v_1},$$
$$\dot z_3 = z_4, \quad \dot z_4 = \underbrace{-\alpha'(x_2)\gamma(x_4) + u_2}_{=v_2}.$$

So, we choose $v = g(x) + u$, where

$$g_1(x) = \alpha'(x_2)\gamma(x_4) = (1+3x_2|x_2|)x_4|x_4|, \quad g_2(x) = -g_1(x),$$

then $g(x) \in C^1(Q)$. Thus, the change of variables $z = F(x)$ and the additive change of controls $v = g(x) + u$ reduce the system (2.21) to the linear form

$$\dot z_1 = z_2, \quad \dot z_2 = v_1, \quad \dot z_3 = z_4, \quad \dot z_4 = v_2.$$

Remark 2.1. The true source of the results like Theorem 2.1 is the class of triangular systems first proposed in the remarkable paper [5]. Namely, a control system is called triangular if it is of the form

$$\begin{aligned} \dot x_i &= f_i(x_1,\ldots,x_{i+1}), \ i = 1,\ldots,n-1,\\ \dot x_n &= f_n(x_1,\ldots,x_n,u), \end{aligned}$$

where $x \in \mathbb{R}^n$, $u \in \mathbb{R}$. Such systems arise in a plenty of applications; for example, many studies were initiated by control problems for satellites. It was proved in [5] that a triangular system is (globally) feedback linearizable under rather weak requirements on smoothness of the functions f_i; namely, $f_i \in C^{n+1-i}(\mathbb{R}^{i+1})$, $i = 1,\ldots,n$, i.e., in fact, the right-hand side of the system is of class $C^1(\mathbb{R}^{n+1})$. It follows from a constructive proof [5] that the change of a control becomes additive, i.e., takes the form $v = g(x) + u$, iff $\frac{\partial f_1}{\partial x_2}\cdots\frac{\partial f_n}{\partial u} = c_1(x_1)$, where $c_1 \in C^n(\mathbb{R})$, $c_1(x_1) \neq 0$.

Example 2.2. In [5] a generalization for multi-control systems was mentioned where x_i and u are m-dimensional vectors rather than numbers. As a very simple example, we consider a partial case of a triangular system with two controls

$$\begin{aligned} \dot x_1 &= f_1(x_1,x_2) + x_3,\\ \dot x_2 &= f_2(x_1,x_2) + x_4,\\ \dot x_3 &= f_3(x_1,x_2,x_3,x_4) + u_1,\\ \dot x_4 &= f_4(x_1,x_2,x_3,x_4) + u_2, \end{aligned} \tag{2.22}$$

where $f_1, f_2 \in C^2(\mathbb{R}^2)$, $f_3, f_4 \in C^1(\mathbb{R}^4)$. A special form of the system allows us to see immediately that it is A-linearizable by the change of variables $y_1 = x_1$, $y_2 = x_2$, $x_3 = f_1(x_1,x_2) + x_3$, $x_4 = f_2(x_1,x_2) + x_4$, and by the change of controls $v_1 = (f_1)_{x_1}(f_1+x_3) + (f_1)_{x_2}(f_2+x_4) + f_3 + u_1$, $v_2 = (f_2)_{x_1}(f_1+x_3) + (f_2)_{x_2}(f_2+x_4) + f_4 + u_2$.

Concerning conditions of Theorem 2.1, we have $\chi_1^0 = b_1 = e_3$, $\chi_2^0 = b_2 = e_4$, and

$$[a(x), \chi_1^0(x)] = \begin{pmatrix} -1 \\ 0 \\ -(f_3)_{x_3} \\ -(f_4)_{x_3} \end{pmatrix}, \quad [a(x), \chi_2^0(x)] = \begin{pmatrix} 0 \\ -1 \\ -(f_3)_{x_4} \\ -(f_4)_{x_4} \end{pmatrix}.$$

Obviously, $[a(x), \chi_i^0(x)]$ are continuous, however, they may be non-differentiable. We choose $\chi_1^1(x) = [a(x), \chi_1^0(x)] + (f_3)_{x_3}\chi_1^0(x) + (f_4)_{x_3}\chi_2^0(x) = -e_1$, $\chi_2^1(x) = [a(x), \chi_2^0(x)] + (f_3)_{x_4}\chi_1^0(x) + (f_4)_{x_4}\chi_2^0(x) = -e_2$, which satisfy conditions of Theorem 2.1.

Acknowledgement

We are grateful to the anonymous referee for his/her careful review of our manuscript and helpful comments.

References

[1] R.W. Brockett, *Feedback invariance for nonlinear systems.* Proc. 7 World Cngress IFAC, Helsinki, 1978, 1115–1120.

[2] W. Dayawansa, W.M. Boothby, and D.L. Elliot, *Global state and feedback equivalence of nonlinear systems.* Systems and Control Lett. **6** (1985), 229–234.

[3] P. Hartman, *Ordinary Differential Equations.* John Wiley and Sons, New York, London, Sydney, 1964.

[4] B. Jakubczyk and W. Respondek, *On linearization of control systems.* Bull. Acad. Sci. Polonaise Ser. Sci. Math. **28** (1980), 517–522.

[5] V.I. Korobov, *Controllability, stability of some nonlinear systems.* Differential Equations **9** (1973), 466–469 (1975).

[6] A. Krener, *On the equivalence of control systems and the linearization of non-linear systems.* SIAM J. Control **11** (1973), 670–676.

[7] W. Respondek, *Global aspects of linearization, equivalence to polynomial forms and decomposition of nonlinear control systems.* In: Algebraic and Geometric Methods in Nonlinear Control Theory, D. Reidel P.C., 1986, 257–284.

[8] E.V. Sklyar, *On the class of non-linear control systems mapping onto linear systems.* Mat. Fiz. Anal. Geom. **8** (2001), 205–214.

[9] K.V. Sklyar and S.Yu. Ignatovich, *Linearizability of systems of the class C^1 with multi-dimensional control.* Systems and Control Lett. **94** (2016), 92–96.

[10] K.V. Sklyar, S.Yu. Ignatovich, and V.O. Skoryk, *Conditions of linearizability for multi-control systems of the class C^1.* Commun. Math. Anal. **17** (2014), 359–365.

[11] G.M. Sklyar, K.V. Sklyar, and S.Yu. Ignatovich, *On the extension of the Korobov's class of linearizable triangular systems by nonlinear control systems of the class C^1.* Systems and Control Lett. **54** (2005), 1097–1108.

[12] K.V. Sklyar, G.M. Sklyar, and Yu.I. Karlovich, *On bang-bang controls for some nonlinear systems.* Commun. Math. Anal. **14** (2013), 163–178.

[13] R. Su, *On the linear equivalents of nonlinear systems*. Systems and Control Lett. **2** (1982), 48–52.

K.V. Sklyar and G.M. Sklyar
University of Szczecin
Wielkopolska 15
Szczecin 70-451, Poland
e-mail: `sklar@univ.szczecin.pl`
`sklar@univ.szczecin.pl`

S.Yu. Ignatovich
Department of Applied Mathematics
V.N. Karazin Kharkiv National University
Svobody sqr. 4,
Kharkiv 61022, Ukraine
e-mail: `ignatovich@ukr.net`

Operator Theory:
Advances and Applications, Vol. 267, 371–376

A Distance Formula Related to a Family of Projections Orthogonal to Their Symmetries

Ilya M. Spitkovsky

Abstract. Let u be a Hermitian involution, and e an orthogonal projection, acting on the same Hilbert space $\mathcal{H}$. We establish the exact formula, in terms of $\|eue\|$, for the distance from e to the set of all orthogonal projections q from the algebra generated by e, u, and such that $quq = 0$.

Mathematics Subject Classification (2010). Primary 47A05; Secondary 47A30.

Keywords. Orthogonal projection, involution, C^*-algebra, W^*-algebra.

1. Introduction

Let $\mathcal{H}$ be a Hilbert space and let $\mathcal{B}(\mathcal{H})$ stand for the C^*-algebra of all bounded linear operators acting on $\mathcal{H}$. Given a Hermitian involution $u \in \mathcal{B}(\mathcal{H})$, denote by $\mathcal{Q}_u$ the set of all orthogonal projections $q \in \mathcal{B}(\mathcal{H})$ for which $quq = 0$.

Theorem 1.2 of [6] can be stated as follows:

Theorem 1. *Let $e \in \mathcal{B}(\mathcal{H})$ be an orthogonal projection such that*

$$\|eue\| < \xi,$$

where $\xi(\approx 0.455)$ is the positive root of

$$x^2 \left(2 + \frac{4}{1+\sqrt{1-4x^2}}\right) = 1.$$

Then there exists $q \in \mathcal{Q}_u$ for which

$$\|e - q\| \le \frac{1}{2} \|eue\| + 4 \|eue\|^2 . \tag{1}$$

Further, q is in the C^*-subalgebra of $\mathcal{B}(\mathcal{H})$ generated by e, ueu^*.

The author was supported in part by Faculty Research funding from the Division of Science and Mathematics, New York University Abu Dhabi.

Note that the distance between any two orthogonal projections does not exceed one. So, estimate (1) is useful only when $\|eue\|$ is smaller than the positive root of $8x^2 + x - 2$, that is, approximately 0.441 ($< \xi$).

We will provide an explicit formula for the distance from e to the intersection of $\mathcal{Q}_u$ with the W^*-algebra $\mathcal{W}(e,u)$ generated by e, u, as well as for the element on which this distance is attained. No a priori restriction on $\|eue\|$ is needed, and the respective q indeed lies in the C^*-algebra $\mathcal{C}(e, ueu^*)$ generated by e, ueu^* whenever $\|eue\| < 1$.

Theorem 2. *Let $e, u \in \mathcal{B}(\mathcal{H})$ be, respectively, an orthogonal projection and a Hermitian involution. Denote by $\mathcal{H}_\pm$ the eigenspace of u corresponding to its eigenvalue ± 1. Then the distance d from e to $\mathcal{Q}_u \cap \mathcal{W}(e,u)$ is one if the range of e has a non-trivial intersection with $\mathcal{H}_+$ or $\mathcal{H}_-$, and is given by the formula*

$$d = \sqrt{\frac{1}{2}\left(1 - \sqrt{1 - \|eue\|^2}\right)} \tag{2}$$

otherwise.

For small values of $\|eue\|$, it is instructive to compare (1) with the Taylor expansion of (2):

$$d = \tfrac{1}{2}\|eue\| + \tfrac{1}{16}\|eue\|^3 + \cdots$$

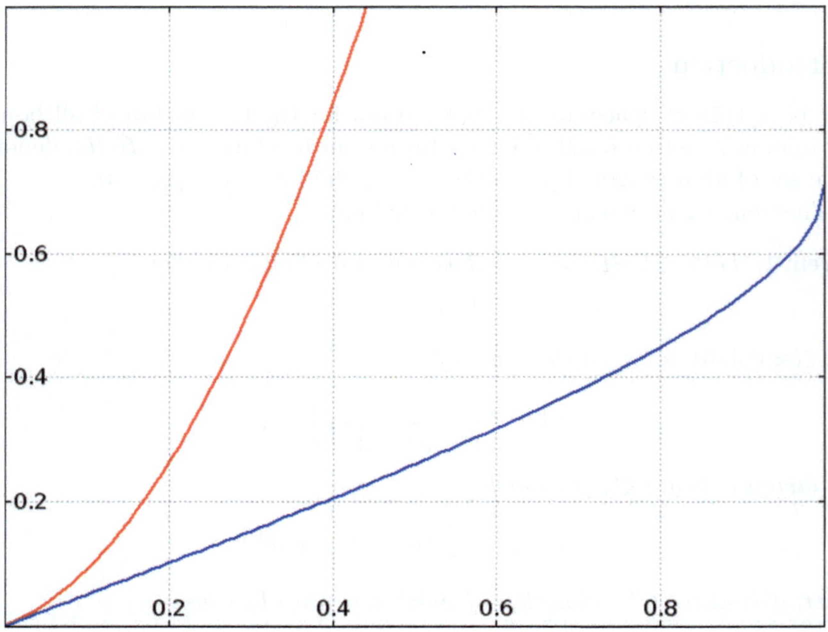

FIGURE 1. Estimate (1) versus formula (2) as functions of $\|eue\|$

2. Proof of the main result

Using the canonical representation [3] (see also [4] or a more recent survey [1]) of the pair $e, (u+I)/2$ of orthogonal projections, we can find an orthogonal decomposition of $\mathcal{H}$ into six summands,

$$\mathcal{H} = \mathcal{M}_{00} \oplus \mathcal{M}_{01} \oplus \mathcal{M}_{10} \oplus \mathcal{M}_{11} \oplus (\mathcal{M} \oplus \mathcal{M}), \tag{3}$$

with respect to which

$$\begin{aligned} u &= I \oplus I \oplus (-I) \oplus (-I) \oplus \operatorname{diag}[I, -I], \\ e &= I \oplus 0 \oplus 0 \oplus I \oplus \begin{bmatrix} H & \sqrt{H(I-H)} \\ \sqrt{H(I-H)} & I-H \end{bmatrix}. \end{aligned} \tag{4}$$

(Here and in what follows we use the notation $\operatorname{diag}[X_1, \dots, X_k]$ for block diagonal matrices with $X_1, \dots X_k$ as their diagonal blocks.) Note that in (3) the subspaces $\mathcal{M}_{00}$ and $\mathcal{M}_{11}$ (resp, $\mathcal{M}_{01}$ and $\mathcal{M}_{10}$) are the intersections of the range (resp, the kernel) of e with $\mathcal{H}_+$ and $\mathcal{H}_-$. The (Hermitian) operator H is the compression of e onto $\mathcal{M} := \mathcal{H}_+ \ominus (\mathcal{M}_{00} \oplus \mathcal{M}_{01})$. By construction, H has its spectrum Δ lying in $[0,1]$ and $0, 1$ are not its eigenvalues.

Elements of $\mathcal{W}(e,u)$ with respect to the same decomposition (3) look as

$$q = a_{00}I \oplus a_{01}I \oplus a_{10}I \oplus a_{11}I \oplus \Phi(H), \tag{5}$$

where $\Phi = \begin{bmatrix} \phi_{00} & \phi_{01} \\ \phi_{10} & \phi_{11} \end{bmatrix}$, $a_{ij} \in \mathbb{C}$, and the functions ϕ_{ij} are Borel-measurable and essentially bounded on Δ, in the sense of the spectral measure of H ([2], see also [1, 4]). Consequently, $q \in \mathcal{W}(e,u)$ is an orthogonal projection if and only if $a_{ij} \in \{0,1\}$, the functions ϕ_{00}, ϕ_{11} are real-valued, while ϕ_{01}, ϕ_{10} are complex conjugate, and

$$\phi_{00} - \phi_{00}^2 = \phi_{11} - \phi_{11}^2 = |\phi_{01}|^2, \quad (\phi_{00} + \phi_{11} - 1)\phi_{01} = 0. \tag{6}$$

On the other hand, direct computations immediately reveal that condition $quq = 0$ is equivalent to

$$\phi_{00}^2 = \phi_{11}^2 = \phi_{01}\phi_{10}, \quad (\phi_{00} - \phi_{11})\phi_{01} = (\phi_{00} - \phi_{11})\phi_{10} = 0. \tag{7}$$

Solving the system of equations (6)–(7) yields

$$\phi_{00} = \phi_{01} = \frac{1}{2}\chi, \; \phi_{01} = \frac{1}{2}\chi\omega, \; \phi_{10} = \frac{1}{2}\chi\overline{\omega}$$

with χ being a characteristic function of some subset of Δ and unimodular ω.

So, elements of $\mathcal{Q}_u \cap \mathcal{W}(e,u)$ have the form

$$q = 0 \oplus 0 \oplus 0 \oplus 0 \oplus \frac{1}{2}\begin{bmatrix} \chi & \chi\omega \\ \chi\overline{\omega} & \chi \end{bmatrix}(H). \tag{8}$$

The rest of the reasoning depends on whether or not the subspaces $\mathcal{M}_{00}$, $\mathcal{M}_{11}$ are actually present in the decomposition (3).

Case 1. At least one of the subspaces $\mathcal{M}_{00}$, $\mathcal{M}_{11}$ is different from zero, that is, the range of e contains some eigenvectors of u.

Since for any q of the form (8) the restriction of $e-q$ on $\mathcal{M}_{00}\oplus\mathcal{M}_{11}$ is the identity, we then have $\|e-q\|=1$. Consequently, $d=1$. Note that in this case also $\|eue\|=1$.

Case 2. $\mathcal{M}_{00}=\mathcal{M}_{11}=\{0\}$. Since both e given by (4) and q given by (8) have zero restrictions onto $\mathcal{M}_{01}\oplus\mathcal{M}_{10}$, we may without loss of generality suppose that in place of (3) simply $\mathcal{H}=\mathcal{M}\oplus\mathcal{M}$, and respectively

$$e=\begin{bmatrix} H & \sqrt{H(I-H)} \\ \sqrt{H(I-H)} & I-H\end{bmatrix},\quad q=\frac{1}{2}\begin{bmatrix}\chi & \chi\omega \\ \chi\overline{\omega} & \chi\end{bmatrix}(H). \tag{9}$$

So, $e-q=\Phi_{\chi,\omega}(H)$, where

$$\Phi_{\chi,\omega}(t)=\begin{bmatrix} t-\frac{1}{2}\chi(t) & \sqrt{t(1-t)}-\frac{1}{2}\chi(t)\omega(t) \\ \sqrt{t(1-t)}-\frac{1}{2}\chi(t)\overline{\omega(t)} & 1-t-\frac{1}{2}\chi(t)\end{bmatrix}.$$

Consequently,

$$\|e-q\|=\operatorname{ess\,sup}_{t\in\Delta}\lambda_{\chi,\omega}(t),$$

where $\lambda_{\chi,\omega}(t)$ is the positive eigenvalue of $\Phi_{\chi,\omega}(t)$, and *ess* is understood in the sense of the spectral measure of H.

If $\chi(t)=0$ for some $t\in\Delta$, then the respective $\lambda_{\chi,\omega}(t)$ equals one, guaranteeing $\|e-q\|=1$. We should concentrate therefore on elements q with $\chi(t)\equiv 1$. Then we have

$$\Phi_{1,\omega}(t)=\begin{bmatrix} t-\frac{1}{2} & \sqrt{t(1-t)}-\frac{1}{2}\omega(t) \\ \sqrt{t(1-t)}-\frac{1}{2}\overline{\omega(t)} & \frac{1}{2}-t\end{bmatrix},$$

and

$$\lambda_{1,\omega}(t)=\sqrt{\frac{1}{2}-\sqrt{t(1-t)}\operatorname{Re}\omega(t)}.$$

Since ω is unimodular, to minimize $\lambda_{1,\omega}(t)$ for any given t we should take $\omega(t)=1$. The respective element q is simply

$$q_0=\frac{1}{2}\begin{bmatrix} I & I \\ I & I\end{bmatrix}, \tag{10}$$

$\lambda_{1,1}(t)=\sqrt{\frac{1}{2}-\sqrt{t(1-t)}}$, and

$$\|e-q_0\|=\sqrt{\frac{1}{2}-\min_{t\in\Delta}\sqrt{t(1-t)}}=\sqrt{\frac{1}{2}\left(1-\sqrt{1-\max_{t\in\Delta}|2t-1|^2}\right)}.$$

In order to justify (2), it remains only to observe that

$$\max_{t\in\Delta}|2t-1|=\|eue\|. \tag{11}$$

But this is indeed the case, since $eue=\Phi(H)$ with the matrix

$$\Phi(t)=(2t-1)\begin{bmatrix} t & \sqrt{t(1-1)} \\ \sqrt{t(1-1)} & 1-t\end{bmatrix},$$

the eigenvalues of which are zero and $2t-1$.

3. Additional remarks

1. As in the case of W^*-algebras, the C^*-algebra $\mathcal{C}(e,u)$ generated by e and u is the same as the C^*-algebra generated by the orthogonal projections $e, (u+I)/2$. Invoking the description from [5] we observe that elements of $\mathcal{C}(e,u)$ are those of the form (5) for which the functions ϕ_{ij} are continuous on Δ and such that $\phi_{01}(j) = \phi_{10}(j) = 0$, $a_{ij} = \phi_{ii}(1-j)$ if $j \in \Delta$ $(i,j = 0,1)$. From (10) we therefore conclude that the element $q_0 \in \mathcal{W}(e,u)$ on which the distance from e to $\mathcal{Q}_u$ is attained does not lie in $\mathcal{C}(e,u)$ if the spectrum of H contains 0 or 1.

On the other hand, due to (11) condition $\|eue\| < 1$ guarantees that $0, 1 \notin \Delta$, and thus the invertibility of the operator $H(I-H)$. Moreover, $\mathcal{M}_{00} = \mathcal{M}_{11} = \{0\}$, as was observed in Section 2. So, without loss of generality e is given by the first formula in (9), while $u = \operatorname{diag}[I, -I]$. From here:

$$z := \frac{1}{2}(e - ueu^*) = \begin{bmatrix} 0 & \sqrt{H(I-H)} \\ \sqrt{H(I-H)} & 0 \end{bmatrix} \in \mathcal{C}(e, ueu^*),$$

$z^2 = \operatorname{diag}[H(I-H), H(I-H)]$ is positive definite and also lies in $\mathcal{C}(e,ueu^*)$, and therefore so does $(z^2)^{-1/2} = \operatorname{diag}[((H(I-H))^{-1/2}, (H(I-H))^{-1/2}]$. Along with z and $(z^2)^{-1/2}$, the algebra $\mathcal{C}(e,ueu^*)$ contains their product

$$w := z(z^2)^{-1/2} = \begin{bmatrix} 0 & I \\ I & 0 \end{bmatrix}.$$

Since $w^2 = I$, the algebra $\mathcal{C}(e,ueu^*)$ is unital, and we conclude from (10) that $q_0 = \frac{1}{2}(I+w) \in \mathcal{C}(e,ueu^*)$.

2. The distances from e to the sets $\mathcal{Q}_u$ and $\mathcal{Q}_u \cap \mathcal{W}(e,u)$ may not coincide. To illustrate, consider $\mathcal{H} = \mathbb{C}^3$, $u = \operatorname{diag}[1,1,-1]$ and $e = \operatorname{diag}[1,0,0]$. Then $\mathcal{Q}_u$ consists of zero and all matrices of the form

$$q_{x,y} = \frac{1}{2}\begin{bmatrix} |x|^2 & x\overline{y} & x \\ \overline{x}y & |y|^2 & y \\ \overline{x} & \overline{y} & 1 \end{bmatrix},$$

with the parameters $x, y \in \mathbb{C}$ satisfying $|x|^2 + |y|^2 = 1$. An easy computation shows that

$$\|e - q_{x,y}\| = \sqrt{(1+|y|^2)/2}.$$

So, the distance from e to $\mathcal{Q}_u$ equals $1/\sqrt{2}$ and is attained on all the matrices $q_{\omega,0}$ with $|\omega| = 1$, that is, having the form

$$\frac{1}{2}\begin{bmatrix} 1 & 0 & \omega \\ 0 & 0 & 0 \\ \overline{\omega} & 0 & 1 \end{bmatrix}.$$

On the other hand, the algebra generated by e and u consists simply of all 3-by-3 diagonal matrices. The only diagonal matrix lying in $\mathcal{Q}_u$ is 0, and $d = 1$ in full agreement with Theorem 2.

Note that in this example $\|eue\| = 1$, and so the right-hand side of (2) coincides with the distance d_0 from e to $\mathcal{Q}_u$. It would be interesting to figure out if in fact d_0 always equals the right-hand side of (2), thus implying that $d = d_0$ when the intersections of the range of e with $\mathcal{H}_\pm$ are trivial.

Acknowledgement

The author is thankful to the anonymous referees for their valuable comments.

References

[1] A. Böttcher and I.M. Spitkovsky, *A gentle guide to the basics of two projections theory.* Linear Algebra Appl. **432** (2010), no. 6, 1412–1459.

[2] R. Giles and H. Kummer, *A matrix representation of a pair of projections in a Hilbert space.* Canad. Math. Bull. **14** (1971), no. 1, 35–44.

[3] P.L. Halmos, *Two subspaces.* Trans. Amer. Math. Soc. **144** (1969), 381–389.

[4] I.M. Spitkovsky, *Once more on algebras generated by two projections.* Linear Algebra Appl. **208/209** (1994), 377–395.

[5] N. Vasilevsky and I. Spitkovsky, *On the algebra generated by two projections.* Doklady Akad. Nauk Ukrain. SSR, Ser. A **8** (1981), 10–13 (in Russian).

[6] S. Walters, *Projection operators nearly orthogonal to their symmetries.* J. Math. Anal. Appl. **446** (2017), no. 2, 1356–1361.

Ilya M. Spitkovsky
Division of Science
New York University Abu Dhabi (NYUAD)
Saadiyat Island
P.O. Box 129188
Abu Dhabi, UAE

MIX
Papier aus verantwortungsvollen Quellen
Paper from responsible sources
FSC® C105338

If you have any concerns about our products,
you can contact us on
ProductSafety@springernature.com

In case Publisher is established outside the EU,
the EU authorized representative is:
Springer Nature Customer Service Center GmbH
Europaplatz 3, 69115 Heidelberg, Germany

Printed by Libri Plureos GmbH
in Hamburg, Germany